KU-539-752

DYNAMICS OF MULTIPHASE MEDIA

#22113944

DYNAMICS OF MULTIPHASE MEDIA

Revised and Augmented Edition

Volume 2

R. I. Nigmatulin

Moscow State University
Moscow, USSR

English Edition Editor

J. C. Friedly

University of Rochester
Rochester, New York

Translator

M. A. Piterman

Rochester Institute of Technology
Rochester, New York

HEMISPHERE PUBLISHING CORPORATION

A member of the Taylor & Francis Group

New York   Washington   Philadelphia   London

UNIVERSITY OF STRATHCLYDE
24 DEC 1991
UNIVERSITY LIBRARY

Originally published by Nauka Moscow under the title "Dinamika mnogofaznykh sred."

DYNAMICS OF MULTIPHASE MEDIA: Volume 2

Copyright © 1991 by Hemisphere Publishing Corporation. All rights reserved. Printed in the United States of America. Except as permitted under the United States Copyright Act of 1976, no part of this publication may be reproduced or distributed in any form or by any means, or stored in a database or retrieval system, without the prior written permission of the publisher.

1 2 3 4 5 6 7 8 9 0 E B E B 9 8 7 6 5 4 3 2 1 0

Cover design by Debra Eubanks Riffe.
A CIP catalog record for this book is available from the British Library.

Library of Congress Cataloging-in-Publication Data

Nigmatulin, Robert Iskanderovich.
[Dinamika mnogofaznykh sred. English]
Dynamics of multiphase media / R. I. Nigmatulin ; translated by Mark A. Piterman.—Rev. and augm. ed.
v. <2 > ; cm.
Translation of: Dinamika mnogofaznykh sred.
Includes bibiographical references and index.

1. Multiphase flow. 2. Fluid dynamics. I.Title.
TA357.N52513 1990
620.1'064—dc20 90-43638
ISBN 1-56032-207-1 (set) CIP
ISBN 0-89116-328-X (Vol. 2)

D
620·1064
NIG
m.f.f.

CONTENTS

FOREWORD TO ENGLISH TRANSLATION

Multiphase media are the rule not the exception in modern industrial practice. Understanding the dynamics of such systems, including traveling and stationary wave phenomena, oscillations and transients, offers an even greater challenge to researchers. In spite of their difficulty, important industrial problems demand attention. This monograph pulls together, reviews and analyzes the major research contributions in this field made by Russian researchers, as well as those in the West. This monograph is a valuable source of current knowledge on phenomena, theory and analysis of multiphase systems. It will be of interest to scientists, researchers, and advanced students working in such diverse fields as nuclear power, ceramics, petroleum production, chemical processing, and space technology, among others.

R. I. Nigmatulin has been a prolific researcher in a number of rather diverse fields. He has worked on multiphase flow, combustion and detonation, solid phase transitions, and boiling heat transfer. The underlying theme in all of this work has been the multiphase nature of the media being studied. The common approach to the problems has been the study of how multiphase mixtures respond to the passage of transient waves. The propagation of shock waves and their attendant effects on exchange between phases and phase transitions has been especially emphasized. Such studies are much more common in the Russian literature than in the West. As a result, too many of Nigmatulin's contributions, together with those of other Russian investigators, have remained undiscovered by English-speaking researchers.

In the two-volume monograph ''Dynamics of Multiphase Media,'' R. I. Nigmatulin has attempted to consolidate knowledge gained during his years of research. The approach has been both comprehensive, covering rather generally all transport processes of importance in multiphase mixtures, and narrow, emphasizing the behavior of these systems when subjected to the passage of transient waves. Clearly, the coverage has been selected carefully to reflect the author's expertise. However, the comprehensive nature of the monograph reflects how broad that expertise is and how his diverse research interests mesh together. It is indeed unusual to see a comprehensive treatment of energy, mass and momentum exchange in multiphase mixtures along with such diverse applications as shock hardening of metals, boiling heat transfer, critical flow of two-phase mixtures, combustion of fuel droplets, tertiary oil recovery and detonation of powders. Nigmatulin has emphasized theory, but gives ample coverage to experimental and numerical results illustrating the phenomena.

In this era of free scientific interchange among nations, it is perhaps surprising that a new translation can open up a new way of thinking that has previously been largely overlooked. There have been some notable previous examples in fields with some relation to this monograph. One might include the Scripta Technica, Inc., translation of V. G. Levich's *Physicochemical Hydrodynamics,* published by Prentice-Hall in 1962. Nathaniel Thon's translation of D. A. Frank-Kamenetskii's *Diffusion and Heat Exchange in Chemical Kinetics,* first published by Princeton University Press in 1955, might be another. This translation by Mark Piterman reveals a large amount of the work the Russian scientists have long been known for. It is a timely contribution, though, in that Nigmatulin's original monographs included research up through the mid-1980's in subjects in which there is currently intense interest around the world. In addition, Nigmatulin has contributed several completely new sections of even more recent work just for the English translation.

Mark Piterman has attempted to reproduce faithfully not only Nigmatulin's ideas, but also his style. Where possible the original phraseology has been retained, skillfully adapting the differing sentence structures common in the two languages. In working between the Russian and English texts it has been a pleasure to debate the nuances of Russian words whose English counterpart may not be nearly so precise. Indeed, Ned Rorem's view "the art of translation lies less in knowing the other language than in knowing your own"[1] is true. Mark Piterman has treated both languages as his own.

John C. Friedly

[1]Ned Rorem (1967), "Random Notes from a Diary. Music From Inside Out." *New York diary,* G. Braziller, NY.

PREFACE

This book presents nonstationary flows including waves, vibration and filtration flows, and also steady flow of various heterogeneous or multiphase mixtures generally observed in natural conditions and often encountered in various fields of human activities.

The effects of the phase multiplicity complicate to a considerable extent the study of such flows, especially in the process of wave propagation resulting from shock and vibrational loading. It is these particular waves that are given detailed consideration in Chapters 3 to 6. The specific features of wave propagation in mixtures of gas with drops or particles (gas suspensions), in liquids containing gas- or vapor-bubbles (bubbly liquids), and in condensed media (liquids, minerals, metals, etc.), which undergo phase transitions effected by these waves, are also discussed in detail. Knowledge of the principal laws and regularities of the indicated processes is of great importance from the standpoint of establishing the scientific foundation for the analysis of energy-generating installations safety and explosion-consequences in various media, as well as for the development of new ways of using the explosions not only for military but also for technological purposes.

High pressures and velocities of a substance arising in the vicinity of detonation of condensed explosives enable us to use explosion for welding and hardening of metals, and also for creating new materials having unique properties (diamond, borazon, etc.). The theory of intensive wave motions characterized by strong shock waves that cause physical-and-chemical transformations in liquid and solid bodies is presented in Chapter 3.

Development of motors of new type and system (internal combustion engines, gas turbine engines, air-jet engines, and rocket engines), improvement of their performance, development of novel explosives and high-energy fuels, analysis of safety of some industries and works force us to further thorough investigation of the heterogeneous combustion of suspensions of atomized liquid or solid fuels, detonation, explosion and other gas-dynamic phenomena in gas suspensions. The outcome of such studies is especially important for the analysis of fire- and explosion-safety of engineering devices, in which the combustible, hazardous gas-particle suspensions are produced. It is the gas-particle suspensions where the large-scale detonation can be observed, with pressures across the detonation wave ranging from 1 to 10 MPa for a gas mixture, and from 1,000 to 10,000 MPa in the liquid or solid explosive. In this context, Chapter 5 presents the processes of combustion and detonation in gas suspensions.

A number of long-range technological processes both on the Earth and in space are associated with utilization of vibrations on multi-phase liquids. The vibrational effects may account for a significant intensification of the heat and mass exchange. These effects may become especially significant if the resonance modes are employed. The fundamentals of the theory of non-linear oscillations of gas-liquid media are treated in Sections 4.6 and 6.12.

A new aerodynamics concerned with the investigation of two-phase flow past bodies is being intensively developed in recent years. Particular interest is attached to a study of flow over bodies by a gas with high mass concentration of drops or particles, when the dispersed phase significantly affects the gas-parameter distribution around the body, and the drops or particles themselves are involved in an intensive bombardment of the body. The first results of such investigation are presented in Sections 4.7 and 4.8.

A considerable part of this book (Chapter 7) is devoted to one-dimensional steady gas-liquid flows. This chapter essentially presents from a unified standpoint such disciplines as: hydraulics and thermophysics of disperse-film flows with heat-and-mass-exchange and force interactions of the wall liquid film with the gas-drop flow, the theory of heat-exchange crises, the theory of stationary critical maximum-discharge outflows, and the theory of hydrocarbon raw material heating in tube furnaces. Considerable attention is attached to these types of flows because of the fact that they are associated with many problems inherent to power engineering, reactor engineering, chemical technology, the petroleum industry, etc.

Chapter 8 is concerned with some class of filtration of a multiphase multicomponent mixture of several mutually insoluble liquids (e.g., petroleum, water, micellar solutions, etc.) in a porous medium accompanied by formation of kinematic waves in application to the analysis of one of the promising methods of oil yield enhancement, namely the micellar-polymeric flooding of the oil-bearing bed.

This book includes a few sections whose development is at the very initial stage and, therefore, a vigorous discussion is anticipated; with new facts recognized, the described theories may undergo a significant modification. Such are the sections presenting the theoretical scheme of interaction between incident and reflected dispersed particles in the process of gas-suspension flow past bodies (Section 4.8), and some theoretical results on the anomalistic strong pressure oscillations in shock waves in liquids with bursting and breaking-up bubbles (Sections 6.8 and 6.10).

The material selected for this book is intended to illustrate the ongoing integration of various branches of mechanics and physics (acoustic, shock-wave physics, gas dynamics, physics of explosion and high-speed shock, hydraulics, thermophysics, theory of filtration), which many researchers go in for though they often relate themselves (traditionally, rather than in principle) to different branches of science and engineering.

The heterogeneous mixtures, their motion, the results of various effects exerted upon them, and the generated waves are extremely diversified. This is a consequence of varied phase combinations, their structures, the variety of interphase and intraphase interactions and processes (viscosity and interphase friction, heat conduction and interphase heat exchange, phase transitions and chemical reactions, fragmentation and coagulation of drops and bubbles, difference in phase compressibility, strength, capillary forces, etc.), and, also, the variety of various forms of effects upon mixtures. For example, in gas suspensions spread waves are often formed whose structure and attenuation are mainly defined by forces of interphase friction and by fragmentation of drops or particles. In liquids with gas or vapor, bubbles—because of the bubble radial pulsations—in addition to spread waves with oscillatory structure, are also generic; this structure strongly depends on the heat and mass exchange processes, as well as on the bubble fragmentation. Furthermore, the phase transitions initiated in condensed media by strong shock waves may lead to multiple-front waves occurring due to non-monotonic variation of the medium compressibility in the process of phase transitions. Unusual wave flows with kinematic waves are also generated in the process of multiphase-liquid filtration.

The study of motion of heterogeneous mixtures with an account of the original structure of the mixture and the phase physical properties is associated with the development and employment of new parameters, and solutions of more complicated equations compared to those which are ordinarily used in mechanics of a single-phase (homogeneous) media. In this case, the detailed description of the intraphase and interphase interactions in heterogeneous media is sometimes extremely intricate, and obtaining sound and comprehensible results, the reasonable schematizations of which may lead to solvable equations, is especially essential.

Chapter 1 deals with the derivation of principal equations of mechanics, as well as methods of description of both intraphase and interphase processes, with consideration given to disperse mixtures (gas suspensions, bubbly liquids) and condensed elastic-plastic media which undergo a polymorphic phase transition of the type graphite → diamond, -iron → -iron, etc. These equations—depending on the medium and processes under consideration—are generalized and specified, e.g., for investigation of the gas-suspension combustion, disperse-film flow of the gas-liquid mixture in a pipe, and flow of mixtures of several mutually soluble liquids in a porous medium. The mathematical modelling and equations of heterogeneous mixtures are described in the previous book by the author (R. I. Nigmatulin, 1978).

In presenting the subject of mechanics of heterogeneous media, the author aimed at those issues that are common, and are encountered in the traditional branches of mechanics of continuous single-phase media; at the same time, the emphasis is placed on subjects which are specific for the multiphase mixtures of various kinds.

The author tried to make the principal concepts contained in each particular section be conceivable without thorough study of previous sections or chapters, in order to make the book more useful for a broader spectrum of readers rather than for those who can afford a detailed study of the whole material.

Although the major emphasis is on theoretical approaches, the author also gives consideration to experimental investigation of the studied processes and, in particular, to methods of determination and measurement of parameters used in equations; the experimental and theoretical data are subjected to a detailed analysis and discussion.

The book is based on research the author and his coworkers carried out in the Laboratory of Mechanics of Multiphase Media, Institute of Mechanics, Moscow University, and also on lectures delivered in the Mechanical-Mathematical Department, Moscow University.

Sections 7.1 to 7.7, and 7.10 were written by B. I. Nigmatulin.

A special debt is owed by the author to his teachers and colleagues in the Mechanical-Mathematical Department and the Institute of Mechanics of Moscow University.

The author is thankful to his colleagues, graduate students and participants of seminars conducted in the Laboratory of Mechanics of Multiphase Media, Institute of Mechanics, Moscow University, and in the laboratories of the same name of the Bashkir Branch of the Academy of Sciences, U.S.S.R., and at the Institute of Development of the Northern Regions of U.S.S.R., Siberian Branch of Academy of Sciences of U.S.S.R., for fruitful cooperation.

The author is grateful to the following individuals who have generously helped with the preparation of a number of sections of this book: P. B. Vainshtein (Ch. 5), I. Kh. Enikeev (Section 4.8), A. I. Ivandaev (Sections 4.2 and 4.5), K. M. Fedorov (Sections 8.2 and 8.3), N. S. Khabeev (Sections 6.5 and 6.10), V. Sh. Shagapov (Section 6.2), and N. A. Gumerov (Section 2.6).

Thanks are also due to V. P. Aleshin for his generous, lasting help.

SYMBOLS

a	radius of a particle, drop or bubble (m)
C_i	speed of sound in the i-th phase (m/s)
$C_* = \sqrt{p_0/\rho_l^\circ}$ (m/s)	
$C(\omega)$ and $C(k)$	phase velocities of sound (m/s)
C_{ij}, C_W	coefficients of hydrodynamic interaction (friction) between i-th and j-th components (species) of flow, and also, between flow and the tube wall (see Eqs. (7.3.22) and (7.3.3))
$c_i \equiv c_{pi}$	specific heat at constant pressure ($m^2/(s^2 \cdot K)$)
c_{Vi}	specific heat at constant volume ($m^2/(s^2 \cdot K)$)
$c_{g(k)}$	specific heat at constant pressure of the k-th gas component (Section 7.9)
$c_{i(k)}$	mass concentration of the k-th component (species) in the i-th phase (i = 1, 2, 3, 4, p, w, . . .) (Ch. 8)
D	shock wave velocity (m/s)
F_{ij}, F_W	respectively, friction force between the i-th and j-th species, and friction force between flow and the tube wall per unit length (kg/s^2)
g	acceleration of external body forces, and, of gravity in particular (m/s^2)
Im	imaginary part of complex number
i	imaginary unit

*For a more complete list of symbols see Vol. 1.

i_i	enthalpy (m^2/s^2)
J_{ij}	intensity of phase transition or mass transfer in the unit volume of mixture from the i-th phase into the j-th phase ($kg/(m^3 \cdot s)$)
$J_{ij(k)}$	intensity of phase transition, $i \rightarrow j$ for the k-th component ($kg/(m^3 \cdot s)$ or $kg/(m \cdot s)$)
J°_{ij}	component J_{ij}, corresponding to mass transfer from i-th to j-th phase ($kg/(m^3 \cdot s)$ or $kg/(m \cdot s)$)
j and j_{ij}	intensity of phase transition for a dispersed particle, drop, bubble (kg/s)
k	permeability of porous medium (m^2)
$k_* = k + ik_{**}$	complex wave number (m^{-1})
$k_{i(k)})$	mass concentration of the k-th component in the i-th phase (Section 7.9)
L	characteristic linear macroscopic dimension, wavelength (m)
l and l_{ij}	heat of evaporization and that of the phase transition $i \rightarrow j$ (m^2/s^2), respectively
m	porosity
m and m_i	mass flow of mixture and that of the i-th component ($i = 1, 2, 3$) in pipe (kg/s)
$m^{\circ}_i = \rho_i v_i$ or $\rho_i w_i$	mass flow of i-th phase per unit area of cross section ($kg/(m^2 \cdot s)$)
m°	mass flow of mixture per unit area of pipe ($kg/(m^2 \cdot s)$)
n	number of droplets, particles or bubbles per unit volume of mixture (m^{-3})
p	pressure ($kg/(m \rightarrow s^2)$)
$Q_{i(\Sigma j)}$	heat flux from the i-th phase ($i = 1, 2, 3$) to the interface of drops ($j = 2$), or a film ($j = 3$) per unit length of a channel ($kg \cdot m/s^3$)
$Q_W(Q_{Wj})$	heat transfer from the channel wall (W) to the flow (to the j-th component) ($j = 1, 2, 3$) per unit length of the channel length ($kg \cdot m/s^3$)
$q_W = Q_W/(\pi D)$	heat transfer from the channel wall to the flow per unit heating surface (kg/s^3)
q_i, $q_{\Sigma i}$, q_{ji}	heat fluxes to i-th phase, from Σ-phase to i-th phase, and from j-th phase to i-th phase per unit dispersed particle, drop, bubble ($kg \cdot m^2/s^3$)
$\mathcal{R} = 8.31 \cdot 10^3$ J/(kmol·K)	universal gas constant ($m^2/(s^2 \cdot K)$)
R_i and $R_{i(k)}$	gas constants of the i-th phase and of k-th component of i-th phase ($i = g, 1, 2$) ($m^2/(s^2 \cdot K)$)

Re	real part of complex number
S_i	cross-section of a channel (Ch. 7) per i-th component (species) ($i = 1, 2, 3, c, f$) (m^2)
$S_i = m\alpha_i$	saturation of the porous medium with i-th liquid (Ch. 8)
T	absolute temperature (K)
$v(v_i)$	velocity (of the i-th phase) (m/s)
v_{ij}	velocity of mass undergoing the phase transition $i \rightarrow j$ (m/s)
$W_i = \alpha_i v_i$	reduced velocity (m/s)
w, w_{ia}, w_a	radial velocities of the medium, that of the i-th phase on the interface, and that of the interface around spherical droplet or bubble, respectively (m/s)
$\mathbf{w}_{12} = \mathbf{v}_1 - \mathbf{v}_2$	velocity of relative macroscopic phase flow (m/s)
x_i	fraction of mass flow of mixture (in a tube) related to various components of phases: gas ($x_1 \equiv x_g$), drop ($x_2 \equiv x_d$) and film ($x_3 \equiv x_f$)
α_i	volumetric concentration of the i-th phase
γ, γ_g, and γ_i	respectively, adiabatic index, adiabatic index of gas phase and that of the$i$-th phase
δ	wall film thickness (m)
$\epsilon \equiv \tilde{\overset{\circ}{\rho}}_g \equiv \overset{\circ}{\rho}_g / \overset{\circ}{\rho}_l$	vapor-density to liquid-density ratio
$\varkappa$	polytropic exponent
δ_i	coefficient of thermal conductivity (kg·m/(s^3·K))
μ_i	coefficient of dynamic viscosity (kg/(m·s))
$\nu_i^{(v)}$	coefficient of kinematic viscosity (m^2/s)
$\nu_i^{(T)}$	coefficient of thermal diffusivity (m^2/s)
$\nu_i^{(k)}$	diffusivity (m^2/s)
ρ	density of medium or mixture (kg/m^3)
ρ_i	reduced density of i-th phase (mass of i-th phase in unit volume of mixture (kg/m^3))
$\tilde{\rho}_2 = \rho_2/\rho_2$ ($\tilde{\rho}_g = \rho_g/\rho_l$)	relative mass concentration in the second (gas) phase
$\overset{\circ}{\rho}_i$	true density of the i-th phase equal to mass of i-th phase in unit volume of i-th phase (kg/m^3)
$\tilde{\overset{\circ}{\rho}}_2 \equiv \overset{\circ}{\rho}_2/\overset{\circ}{\rho}_1$ ($\tilde{\overset{\circ}{\rho}}_g \equiv \epsilon \equiv \overset{\circ}{\rho}_g/\overset{\circ}{\rho}_l$)	ratio of true phase densities (relative density of gas phase)
ϵ_i	intensity of phase transition to i-th phase per unit area of interface (kg/(m^2·s))
Σ	surface tension (kg/s^2)
τ, τ_{13}, τ_W	shear stress, shear stress between the flow

	core and film, and the pipe-wall shear stress, respectively
τ	dimensionless time
$\varphi^{(1)}$, $\varphi^{(2)}$, $\varphi^{(3)}$, $\varphi_*^{(1)}$, $\varphi_*^{(2)}$	coefficients that take into account the bubble-non-zero volume in Rayleigh-Lamb equation (see Section 1.3)
χ_m	coefficient of the associated (apparent) mass factor ($\chi_m = 1/2$ for sphere)
$\omega_* = \omega + i\omega_{**}$	complex frequency (s^{-1})

DIMENSIONLESS CRITERIA (NUMBERS)

$\mathrm{Ja} = (c_l \Delta T/l)(\rho_l^\circ/\rho_g^\circ)$	Jacob Number
$\mathrm{Nu} = 2a/\delta^{(T)}$	Nusselt Number where $\delta^{(T)}$ is thickness of the temperature boundary layer ($\delta^{(T)} = 4\pi\alpha\delta\Delta T/q$)
$\mathrm{Pe} = 2av/\nu^{(T)}$	Peclet Number
$\mathrm{Pr} = \nu^{(v)}/\nu^{(T)}$	Prandtl Number
$\mathrm{Re} = 2av/\nu^{(v)}$	Reynolds Number
$\mathrm{We} = 2a\rho v^2/\Sigma$	Weber Number

SUBSCRIPTS

a	parameters on the drop wall or bubble wall
b (boundary)	parameters of the gas-drop nucleus of the dispersed-film flow
d (dispersed)	parameters of dispersed phase
e (equilibrium)	equilibrium parameters behind the wave
f (film)	film parameters in dispersed-film flow
g (gas)	gas parameters
i	phase number ($i = 0, 1, 2, \ldots, N$); in a disperse and film mixture $i = 1$ relates to the carrying (continuous) phase, $i = 2$ to the dispersed phase, $i = 3$ to film; in a porous medium (Ch. 8) $i = 0$ relates to solid phase. See also $i = g, l, p, w$
j	the same as subscript *i*
(*k*)	component (species) number
l (liquid)	parameter of condensed (liquid or solid) phase
0	parameters of an initial or reference state
0	parameters of solid phase in saturated porous medium (Ch. 8)
p (petroleum)	parameters of hydrocarbon fluids
S (Saturated)	parameters of saturated phase
W (Wall)	parameters on the pipe wall

w (water)	aquatic liquid parameters
Σ	parameters on interface (Σ-phase)

SUPERSCRIPTS

$'$	microparameters
$\bar{\ }$	value of a corresponding parameter normalized by its value in original state ($\bar{\rho} = \rho/\rho_0$, $\bar{T}_i = T_i/T_{i0}$, $\bar{p} = p/p_0$)
$\tilde{\ }$	ratio of corresponding parameter value in liquid and gas phases ($\tilde{\rho}_g = \overset{\circ}{\rho}_g/\overset{\circ}{\rho}_l$, $\tilde{\nu}_g^{(T)} = \nu_g^{(T)}/\nu_l^{(T)}$, $\tilde{c} = \overset{\circ}{\rho}_g c_g/\overset{\circ}{\rho}_l \overset{\circ}{c}_l$, . . .)
$*$	dimensionless parameters associated with physical properties of phase (δ_i^*, μ_i^*, Σ^*, l^*, B^*, β^*, . . .)

CHAPTER

SIX

WAVE DYNAMICS OF BUBBLY LIQUIDS

This chapter presents results of the investigation of nonstationary motions of liquids containing gas or vapor bubbles, including flows with compression shock waves and rarefaction waves, flows affected by vibrations, and effluxions out of high-pressure vessels.

§6.1 A SHOCK TUBE ARRANGEMENT FOR STUDY OF WAVES IN BUBBLY LIQUIDS. SPECIFIC FEATURES OF SHOCK WAVES.

The processes of the propagation of waves in bubbly liquids are experimentally investigated in vertical shock tubes. Their generic layout is shown in Fig. 6.1.1. The shock tube consists of a high-pressure chamber (HPCh) and an operating part, or a low-pressure chamber (LPCh), which are separated by a diaphragm. The gas is pumped into the HPCh, where a high pressure is built up, and the LPCh is filled with liquid to a level slightly lower than the diaphragm; bubbles of a predetermined radius are dispersed in the liquid so that a mixture resembling a monodisperse system is produced. The bubble radii a varied in separate experiments within a range 0.2–2 mm, and their volume concentration α_2, measured by the rise of the column of liquid, ranged from 0.01–0.1. Following the diaphragm rupture affected by gas contained in the HPCh, a shock wave which propagates through the bubbly mixture downward is produced in the LPCh. The disturbance duration is defined by the size of HPCh, and its intensity by pressure in the

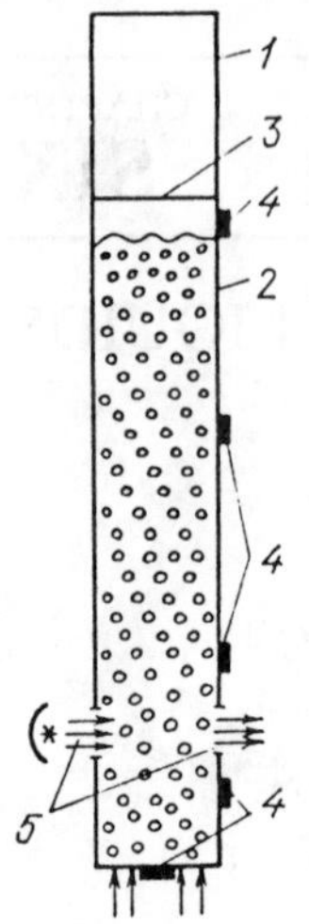

Figure 6.1.1 A schematic drawing of a shock tube designed for investigation of bubbly liquids: *1*) high-pressure chamber (HPCh), *2*) low-pressure chamber (LPCh), *3*) diaphragm, *4*) sensors, *5*) windows for taking photographs.

HPCh. The pressure evolution in the mixture is recorded by quick-response pressure indicators installed in several locations throughout the liquid column in the LPCh wall; the bubble behavior was studied using the method of rapid filming through a window in the wall of the tube.

The characteristic pressure oscillograms for mixtures during the shock-wave propagation are schematically shown in Fig. 6.1.2, and are discussed below. The primary property of such shock waves in bubbly liquids (with

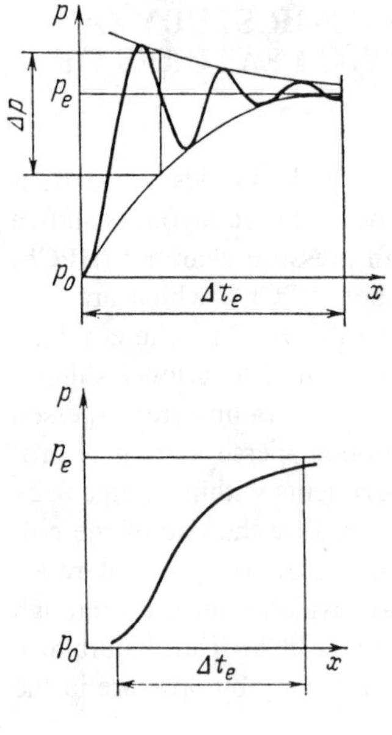

Figure 6.1.2 The characteristic pressure oscillograms for shock-wave propagation in bubbly liquids.

either gas or vapor) is that at some conditions they are characterized by an oscillatory structure associated with the bubble-volume pulsations due to both liquid inertia and gas elasticity (G. Batchelor (1967), L. Wijngaarden (1972), V. E. Nakoryakov et al. (1983), R. Nigmatulin (1982)).

§6.2 LINEAR THEORY OF PROPAGATION OF WEAK DISTURBANCES IN LIQUIDS WITH GAS BUBBLES

We shall now analyze (as in §§4.1 and 4.2) the characteristics of differential equations and the linear approximation for describing the propagation of weak disturbances in a homogeneous (when external body forces are insignificant, and all parameters of the mixture in an undisturbed state are independent of the x-coordinate), monodisperse mixture of a a low-compressibility liquid with gas bubbles. The single-velocity approach with a polytropic gas and effective viscosity will be used in order to take into account all possible dissipative effects.

Characteristics. Let us find the characteristics and associated conditions for the system of equations (1.5.4) governing the behavior of bubbly liquids under the simplifications indicated above. The medium momentum equation, mass-conservation equation, and the equation for the acoustic compressibility of the carrying liquid, given Eq. (1.3.6), may be represented as

$$\rho\frac{\partial v}{\partial t}+\rho v\frac{\partial v}{\partial x}+\frac{\partial p}{\partial x}=0,\qquad \rho\approx\overset{\circ}{\rho}_1\alpha_1$$

$$\alpha_1\frac{\partial\overset{\circ}{\rho}_1}{\partial t}+\alpha_1 v\frac{\partial\overset{\circ}{\rho}_1}{\partial x}+\overset{\circ}{\rho}_1\frac{\partial v}{\partial x}=\frac{3\alpha_2 w_{1a}}{a}\overset{\circ}{\rho}_1,\qquad \overset{\circ}{\rho}_1=\overset{\circ}{\rho}_{10}+\frac{p-p_0}{C_1^2}\tag{6.2.1}$$

The remaining equations of the system (1.5.4) define the variation of α_2, a, and w, and close the represented equations. If the derivatives of $\overset{0}{\rho}_1$ are expressed in terms of p, then two differential equations are obtained which, in contrast to the single-phase media, give consideration to compressibility not only due to that of the carrying liquid but also due to volumetric deformation of bubbles, which is defined by the radial velocity w_{1a}. These two equations contain derivatives of only two functions: v and p. Therefore, some characteristics of the complete system of equations may be determined based only on Eqs. (6.2.1). The equations for determining the characteristics are written in the form

$$\rho\frac{\partial v}{\partial t}+\rho v\frac{\partial v}{\partial x}+\frac{\partial p}{\partial x}=0$$

$$\overset{\circ}{\rho}_1 \frac{\partial v}{\partial x} + \frac{\alpha_1 v}{C_1^2}\frac{\partial p}{\partial x} + \frac{\alpha_1}{C_1^2}\frac{\partial p}{\partial t} = \frac{3\alpha_2 w_{1a}}{a}\overset{\circ}{\rho}_1$$

$$dt\frac{\partial v}{\partial t} + dx\frac{\partial v}{\partial x} = dv \qquad (6.2.2)$$

$$dx\frac{\partial p}{\partial x} + dt\frac{\partial p}{\partial t} = dp$$

If the characteristic determinant of the coefficients of partial derivatives of v and p is zero, we arrive at expressions for two characteristic directions; if the determinant in which (as distinct from the characteristic determinant) one of the columns is replaced by a column of free terms is zero, the conditions on characteristics are obtained. After simple mathematical transformations, we obtain

$$\frac{dx}{dt} = v \pm C_f, \quad \frac{dp}{dt} \pm \rho C_f\frac{dv}{dt} = \frac{3\alpha_2 w_{1a}}{a}\rho C_f^2 \quad \left(C_f = \frac{C_1}{\alpha_1}\right) \qquad (6.2.3)$$

Thus, expressions (6.2.1) are hyperbolic equations. The remaining equations (1.5.4) involve in the capacity of characteristics the streamlines ($dx/dt = v$), along which equations (as the conditions on characteristics) defining the substantial derivatives α_2, a, and w_{1a} are used.

Linear equations for weak disturbances. In the coordinate system associated with an undisturbed medium whose parameters have a subscript 0 ($v_0 = 0$), the linearized equations for mass, momentum, and number of bubbles are written in the form

$$\rho_0\frac{\partial v}{\partial t} = -\frac{\partial p}{\partial x}, \quad \frac{\partial \rho}{\partial t} = -\rho_0\frac{\partial v}{\partial x}, \quad \frac{\partial n}{\partial t} = -n_0\frac{\partial v}{\partial x} \qquad (6.2.4)$$

Upon simple transformations, we obtain

$$\frac{\partial^2 \rho}{\partial t^2} = \frac{\partial^2 p}{\partial x^2}, \quad dn = \frac{n_0}{\rho_0}d\rho \qquad (6.2.5)$$

Based on the mixture density definition $\rho = \rho_1(1 - \alpha_2)$, bubble volumetric concentration $\alpha_2 = 4/3\pi a^3 n$, and the equation for the liquid linear compressibility (the last equation in (6.2.1)), we have

$$d\rho = -\overset{\circ}{\rho}_{10}d\alpha_2 + \alpha_{10}C_1^{-2}dp$$

$$d\alpha_2 = 4\pi a_0^2 n_0 da + {}^4/_3\pi a_0^3 dn = \alpha_{20}(3da/a_0 + d\rho/\rho_0) \qquad (6.2.6)$$

whence, we obtain

$$d\rho = \alpha_{10}^2 C_1^{-2}dp - 3\rho_0\alpha_{20}a_0^{-1}da \qquad (6.2.7)$$

Let us now write the linear equation of the radial motion with consideration given to both the finiteness of the bubble volumetric concentration and phase transitions

$$\left(1-\varphi_0^{(1)}\right) a_0 \overset{\circ}{\rho}_{10} \frac{\partial w_{1a}}{\partial t} + \frac{4\mu}{a_0} w_{1a} = p_2 - p + \frac{2\Sigma}{a_0^2}(a-a_0)$$

$$\left(w_{1a} - \frac{\partial a}{\partial t}\right) \overset{\circ}{\rho}_{10} = \left(w_{2a} - \frac{\partial a}{\partial t}\right) \overset{\circ}{\rho}_{20} = \xi_{21} \qquad (6.2.8)$$

Weak sinusoidal disturbances in a liquid with bubbles of nonsoluble gas. First, we will examine the case when both phase transitions and surface tension may be ignored; the gas polytropy condition ($p_2/p_{20} = (a_0/a)^{3\varkappa}$) and effective viscosity may be used in order to take into account the dissipation

$$\xi_{21} = 0, \quad 2\Sigma/a_0 \ll p_0, \quad dp_2 = -3\varkappa p_0 a_0^{-1} da, \quad \mu = \mu_{\text{eff}} \qquad (6.2.9)$$

Then, from Eqs. (6.2.6)–(6.2.9), two independent linear homogeneous equations in variables a and p may be readily obtained

$$\frac{\alpha_{10}^2}{C_1^2}\frac{\partial^2 p}{\partial t^2} - \frac{\partial^2 p}{\partial x^2} - \frac{3\rho_0\alpha_{20}}{a_0}\frac{\partial^2 a}{\partial t^2} = 0$$

$$\left(1-\varphi_0^{(1)}\right) a_0 \overset{\circ}{\rho}_{10} \frac{\partial^2 a}{\partial t^2} + \frac{4\mu_{\text{eff}}}{a_0}\frac{\partial a}{\partial t} + \frac{3\varkappa p_0}{a_0}(a-a_0) + (p-p_0) = 0 \qquad (6.2.10)$$

It is evident that, in the absence of bubbles ($\alpha_{20} = 0$), the pressure equation assumes the form of a linear wave equation from acoustics. Let us find the solution of the obtained system in the form of sinusoidal waves of the type (4.1.11), which are defined by the wave number k_* and complex frequency ω_*. Then, from the condition of existence of the nonzero solutions of the indicated type (or nonzero amplitudes $A^{(p)}$ and $A^{(a)}$), we obtain a characteristic equation interrelating k_* and ω_*

$$\frac{k_*^2}{\omega_*^2} = \frac{\alpha_{10}^2}{C_1^2} + \frac{1}{C_0^2\left[1-(\omega_*^2/\omega_r^2) + i(\omega_*/\varkappa\omega_\mu)\right]}$$

$$\left(\omega_r = \omega_a\sqrt{\varkappa}, \quad C_0 = C_\alpha\sqrt{\varkappa}, \quad \omega_a = \sqrt{\frac{3p_0}{\left(1-\varphi_0^{(1)}\right)\overset{\circ}{\rho}_{10} a_0^2}}\right. \qquad (6.2.11)$$

$$\left. C_\alpha = \sqrt{\frac{p_0}{\overset{\circ}{\rho}_1 \alpha_{10}\alpha_{20}}}, \quad \omega_\mu = \frac{3p_0}{4\mu_{\text{eff}}}\right)$$

We shall confine ourselves to ω-waves ($\omega > 0$, $\omega_{**} = 0$; see Eq. (4.1.19)), which correspond to the forced oscillations initiated by the outside generator. It can be readily proven that the imaginary part of k_*^2 is negative. Therefore, only the $k \geqslant 0$, $k_{**} < 0$, or $k \leqslant 0$, $k_{**} \geqslant 0$ cases are possible. This means that the ω-wave amplitudes in a bubbly liquid do not grow in the direction of their phase velocity, because two wave numbers $k_*^{(1)}$ and $k_*^{(2)}$ are available, which, because of $k_*^{(1)} = -k_*^{(2)}$, yield two symmetrical ω-waves propagating in opposite directions.

In the dissipation-free case ($\mu_{\text{eff}} = 0$), the following values of the linear damping decrement and phase velocity are obtained (these are shown schematically by dashed lines in Fig. (6.2.1))

$$k_{**} = 0, \quad C(\omega) = -\frac{\omega}{k} = \pm\sqrt{\frac{C_f^2 C_0^2 (1 - \omega^2/\omega_r^2)}{C_f^2 + C_0^2 (1 - \omega^2/\omega_r^2)}}$$

$$(\omega < \omega_r \quad \text{or} \quad \omega > \omega_C)$$

$$k_{**} = \mp \frac{\alpha_{10}\omega}{C_1}\sqrt{\frac{\omega_C^2 - \omega^2}{\omega^2 - \omega_r^2}}, \quad C(\omega) = \infty \quad (\omega_r < \omega < \omega_C) \tag{6.1.12}$$

$$\left(C_f \doteq \frac{C_1}{\alpha_{10}}, \quad \omega_C^2 = \omega_r^2\left(1 + \frac{C_f^2}{C_0^2}\right), \quad \frac{1}{C_e^2} = \frac{1}{C_f^2} + \frac{1}{C_0^2}\right)$$

The velocity C_f corresponds to the phase velocity $C(\omega)$ at $\omega \to \infty$, and is referred to as the *frozen speed of sound,* and C_e corresponds to $C(\omega)$ at $\omega = 0$, and is referred to as the *equilibrium speed of sound,* the magnitude C_f coinciding with the speed C_1 of sound in pure liquid. The values of both C_f and C_e are dissipation-independent. In the absence of dissipation, the phase velocity—with ω approaching the resonance frequency ω_r of the bubble natural oscillations—diminishes to zero, complying with the degenerated ($L = 0$) stationary ($C = 0$) ω-wave. In the frequency range $\omega_r \leqslant \omega \leqslant \omega_C$ (which sometimes is referred to as the range of "nontransparency" because of the large values of the damping decrement k_{**}), both phase velocity C and wavelength L are formally equal to infinity. The phase velocities beyond the range of nontransparency, except for $\omega = \omega_C + 0$, are finite, and nonzeros, as is the linear decrement $k_{**} = 0$ that corresponds to time-independent progressive waves with constant amplitudes along the axis x. Thus, in the absence of dissipation, the dispersion curves are characterized by three regions: low-frequency ($\omega < \omega_r$) region, a "band of nontransparency" ($\omega_r \leqslant \omega \leqslant \omega_C$), and a high-frequency ($\omega > \omega_C$) region.

In the ω-waves under consideration, the bubbles undergo steady, forced, radial oscillations with a predetermined frequency; therefore, the effective viscosity of the mixture must, evidently, be found using formula (1.6.50). It should be kept in mind that in water-like liquids of low viscosity with the bubble size $a \gtrsim 0.1$ mm, the "viscosity" $\mu^{(T)}$ is much higher than the liquid viscosity μ_1 because of thermal dissipation. In this case, $\mu_{\text{eff}} = \mu^{(T)} + \mu_1$ depends on frequency ($\mu^{(T)} \sim \omega^{1/2}$) and the properties of gas. In a number of works, the thermal dissipation in the process of wave propagation in bubbly liquids was not taken into consideration at all.

The account of dissipation, which is a characteristic feature of indicated mixtures, significantly modifies the functions $C(\omega)$ and $k_{**}(\omega)$, which describe, respectively, the phase speed of sound and damping decrement (see Fig. 6.2.1), and enhances k_{**} to a considerable extent. Dispersion curves become smooth and continuous over all three frequency bands indicated above, and the mentioned range of "nontransparency," realized due to strong attenuation of oscillations, expands in the direction of high frequencies. In this case, there is no degeneration that might have taken place because of wavelength approaching infinity. The asymptotic for both phase velocity and damping decrement in the low-frequency range $\omega \ll \omega_r$, at $C_1 \gg C_0$ is

$$C(\omega) = -\frac{\omega}{k} = C_0\left(1 - \frac{\omega^2}{2\omega_r^2} + \frac{\omega^2}{2\varkappa^2\omega_\mu^2}\right), \qquad k_{**}(\omega) = \frac{\omega^2}{2\varkappa C_0\omega_\mu} \qquad (6.2.13)$$

The experimental data on $C(\omega)$ and $k_{**}(\omega)$ for water with air bubbles are obtained by F. Fox, S. Curley, and G. Larson (1955), and for boiling water with steam bubbles the data are reported by E. V. Stekol'shchikov and A. S. Fedorov (1974). These data are characterized by a substantial scatter, which is accounted for by both mixture polydispersity and nonstability of gas-concentration. As was shown by V. M. Shaganov (1977), the relations $C(\omega)$ and $k_{**}(\omega)$—with the finite set of the bubble sizes in a polydisperse mixture—are of a significantly irregular appearance. V. K. Kedrinskii (1968) proved that, for a polydisperse mixture with a smooth bubble-size distribution function, the relation $C(\omega)$ (but not $k_{**}(\omega)$) assumes the same appearance as the solid line in Fig. 6.2.1, even in the absence of dissipation.

Double-wave equation, Boussinesq and Klein-Gordon equations. In the absence of dissipation ($\mu_{\text{eff}} = 0$), the linear equations (6.2.10) may be reduced to a single pressure equation. Indeed, having expressed the term $\partial^2 a/\partial t^2$ from the second equation followed by substitution into the first one, and then expressing da in terms of $d\rho$ and dp from (6.2.7), we finally obtain

$$\frac{1}{C_f^2}\frac{\partial^2 p}{\partial t^2} - \frac{\partial^2 p}{\partial x^2} + \omega_r^2\left[\frac{p - p_0}{C_e^2} - (\rho - \rho_0)\right] = 0$$

The double differentiation of this equation with respect to t, with an account of (6.2.6), leads to the double-wave equation which realizes the disperse relations discussed above (V. E. Nakoryakov, and I. R. Shreiber, 1983).

$$\frac{\partial^2 p}{\partial t^2} - C_e^2\frac{\partial^2 p}{\partial x^2} + \frac{C_e^2}{\omega_r^2 C_f^2}\frac{\partial^2}{\partial t^2}\left(\frac{\partial^2 p}{\partial t^2} - C_f^2\frac{\partial^2 p}{\partial x^2}\right) = 0 \qquad (6.2.14)$$

In conformity with this equation, the low-frequency part of the signal ($\omega \ll \omega_r$) propagates with small velocities, whose values resemble the equilibrium speed of sound $C_e \approx C_0$, and the high-frequency part ($\omega \gg \omega_C$) propagates with high velocities (in the form of a high-speed preindicator associated with the elastic compressibility of the carrying liquid), which are

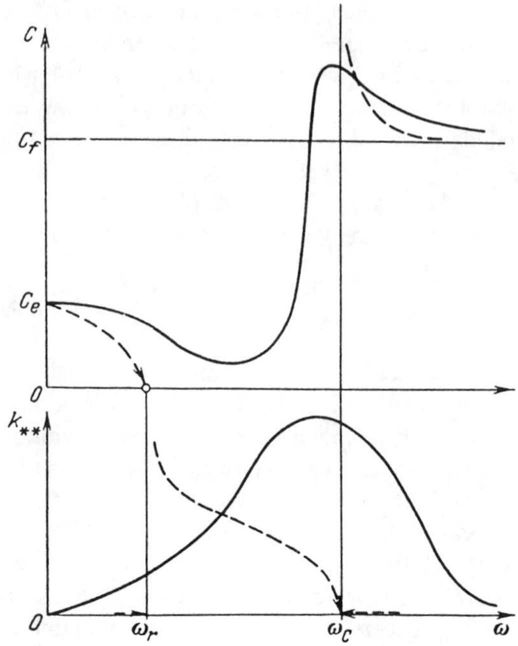

Figure 6.2.1 The relation of phase velocity $C(\omega)$ of sinusoidal forced oscillations (in the form of ω-waves), and the relation of the linear decrement $k_{**}(\omega)$ of their attenuation or growth with length in a liquid with gas bubbles. The dashed lines correspond to the solution (6.2.8) in the absence of dissipation ($\mu_{\text{eff}} = 0$), the solid line relates to the same solution in the presence of dissipation ($\mu_{\text{eff}} > 0$).

almost equal the speed C_1 of sound in pure liquid. For the indicated low-frequency and high-frequency parts of the spectrum, the double-wave equation (6.2.14) simplifies. Indeed, if $\varepsilon = \omega^2/\omega_r \ll 1$ and $C_0 \ll C_1$, then from (6.2.12) we have

$$\beta k_*^2 = \left(\frac{\omega_*}{\omega_r}\right)^2 + \left(\frac{\omega_*}{\omega_r}\right)^4 + O(\varepsilon^3) = \left(\frac{\omega_*}{\omega_r}\right)^2 + O(\varepsilon^2) \qquad \left(\beta = \frac{C_0^2}{\omega_r^2} = \frac{a_0^2}{3\varkappa\alpha_{10}\alpha_{20}}\right)$$

whence, in the adopted range of accuracy, it may be assumed

$$\beta k_*^2 - \frac{\omega_*^2}{\omega_r^2} = \beta^2 k_*^4 + O(\varepsilon^3) \tag{6.2.15}$$

Once consideration is given to the fact that the values k_*^2 and ω_*^2 correspond to operators $-\partial^2/\partial x^2$ and $-\partial^2/\partial t^2$, the obtained disperse relation becomes consistent with the dissipation-free linear Boussinesq equation, which is discussed in more general form in §6.6

$$\frac{\partial^2 p}{\partial t^2} - C_0^2 \frac{\partial^2 p}{\partial x^2} = \beta C_0^2 \frac{\partial^4 p}{\partial x^4} \tag{6.2.16}$$

This equation describes the evolution of the low-frequency part of the signal when compressibility of the carrying liquid does not manifest itself.

For the high-frequency branch, when $\omega^2/\omega_r^2 \gg 1$, we have from (6.2.11)

$$k_*^2 = \frac{\omega_*^2}{C_f^2} - \frac{\omega_r^2}{C_0^2} \tag{6.2.17}$$

This disperse relation is in compliance with the linear telegraph equation which is sometimes referred to as the Klein-Gordon equation

$$\frac{\partial^2 p}{\partial t^2} - C_f^2 \frac{\partial^2 p}{\partial x^2} = -\beta C_f^2 (p - p_0) \tag{6.2.18}$$

This equation describes waves propagating with the velocity C_f, which coincides with the speed of sound in pure liquid; it has been used by N. V. Malykh and I. A. Ogorodnikov (1977), and V. E. Nakoryakov (1983) for describing the wave behavior in bubbly liquids.

Weak sinusoidal disturbances in a liquid with vapor bubbles. The presence of both phase transitions and surface tension may lead to fundamentally new wave-propagation effects. Analysis of these effects involves a more detailed account of the interphase heat and mass exchange compared to the scheme of polytropic gas and effective-viscosity liquid, which was used above in Eq. (6.2.9). We shall assume spherically symmetrical arrangement of the trial bubble (see §1.6), which describes the interphase heat and mass exchange affecting variation of both bubble radius a and pressure p_2 within bubbles with an account of the temperature distribution variation inside and around the trial bubble.

For further description of the liquid with vapor-bubbles, it is worthwhile to introduce a linear parameter a_L, associated with the surface tension, phase thermal energy, and phase transition heat, and also the dimensionless parameter $\bar{\Sigma}_L$ and the inherent velocity C_L (which is sometimes referred to as the Landau speed of sound[1])

$$a_L = \frac{2\Sigma_0}{p_0(1+2\Sigma^*)} \frac{\bar{c}_1}{\gamma_2 l^*} \frac{\overset{\circ}{\rho}_1}{\overset{\circ}{\rho}_{20}} \quad \left(\Sigma^* = \frac{\Sigma_0}{a_0 p_0},\quad \bar{c}_1 = \frac{c_1}{\gamma_2 R_2},\quad l^* = \frac{l_0}{\gamma_2 R_2 T_0}\right)$$

$$\bar{\Sigma}_L = \frac{\alpha_1 a_L}{\alpha_2 a_0}, \quad C_L = \frac{\rho_{20}^0}{\overset{\circ}{\rho}_1} \frac{l_0}{\sqrt{c_1 T_0}} \tag{6.2.19}$$

Here $\bar{c}_1$ and l^* are, respectively, dimensionless heat capacity of liquid and the phase transition heat already employed in Chapters 2 and 4, with the only difference that the subscript 1 here relates to liquid, and subscipt 2 to gas (vapor). The subscript 0, which refers to parameters in an undisturbed state, will be omitted.

[1]As will be shown below, the velocity C_L coincides with the equilibrium speed of sound in a vapor-liquid medium only if $\Sigma_L \ll 1$, which is equivalent to the condition $\Sigma^* \ll \rho_2/\rho_1$.

The following relations are involved

$$a_L = \frac{{}^2/_3\Sigma}{\overset{\circ}{\rho}_1 C_L^2}, \quad \bar{\Sigma}_L = \frac{2\bar{c}_1}{3\gamma_2 l^{*2}(1+2\Sigma^*)} \frac{\Sigma^*}{\tilde{\rho}_2} = \frac{2\Sigma^*}{3} \frac{\alpha_1 C_\alpha^2}{C_L^2} \quad \left(\tilde{\rho}_2 = \frac{\overset{\circ}{\rho}_2 \alpha_2}{\overset{\circ}{\rho}_1 \alpha_1}\right)$$

It is evident that the parameter $\bar{\Sigma}_L$ characterizes the ratio of two small quantities: dimensionless coefficient of surface tension Σ^* to the relative mass concentration $\tilde{\rho}_2$ of vapor. In this case, $\bar{\Sigma}_L \gg \Sigma^*$. Although the vapor density ρ_2^0 is governed by pressure $p_2 = p + 2\Sigma/a = p_S(T)$ within the bubble (the pressure for a particular liquid being dependent on the bubble size a at the fixed liquid pressure p), generally $\Sigma^* \ll 1$, and the effect of a on ρ_2^0, C_L, and a_L is very weak. Therefore, a_L and C_L, as well as $\bar{c}_1$ and l^*, may be regarded as dependent only on p. For water at $p = 0.1$ MPa, we have $C_L = 1.1$ m/s, $a_L = 35$ μcm ($\bar{c}_1 = 7.25$, $l^* = 10.41$), and at $p = 1.0$ MPa, we have $C_L = 8.3$ m/s, $a_L = 0.47$ μm ($\bar{c}_1 = 8.67$, $l^* = 8.72$).

The linearization of equations (1.6.3)–(1.6.10) leads to the same equations (2.7.9) which were obtained for a single bubble, but in which instead of the dimensionless functions

$$\theta_1(\bar{t}_1, \eta), \quad \theta_2(\bar{t}_2, \eta), \quad P_1(\bar{t}_p), \quad P_2(\bar{t}_p), \quad A(\bar{t}_p)$$

the following functions should be kept in mind

$$\theta_1(t, x, \eta), \quad \theta_2(t, x, \eta), \quad P_1(t, x), \quad P_2(t, x), \quad A(t, x)$$

The latter define, respectively, the disturbances T_1', T_2', p_2, p_1, and a, and instead of the isothermal boundary condition at infinity ($\eta \to \infty$: $\theta_1 = 0$), the adiabatic condition (1.6.10) on the cell boundary should be used ($\eta = \alpha_2^{-1/3}$: $\partial\theta_1/\partial\eta = 0$). As a result, the corrected equations (2.7.9), together with Eq. (6.2.4), constitute a closed linear system of equations in partial derivatives with respect to three variables: t, x, and η. The corresponding dispersion equation is similar to (6.2.11), with the only distinction that the first addend in brackets in the denominator is not equal to unity but is a function of ω_w. This function uses the nonequilibrium of heat exchange and surface tension which, as it will be shown below, may (in spite of the smallness of the relevant parameter Σ^*) have an essential bearing on the acoustics of vapor-liquid bubbly media. The indicated dispersion equation is written in the form

$$\frac{k_*^2}{\omega_*^2} = \frac{\alpha_1^2}{C_1^2} + \frac{\alpha_1^2}{C_L^2\{\Pi(\omega_*) - \bar{\Sigma}_L[1 + (\omega_*/\omega_{\Sigma a})^2 - i\omega_*/\omega_{\Sigma\mu}]\}}$$

$$\Pi(\omega_*) = \{\Pi_1(z_1) + \tilde{\rho}_2 l^{*2}\bar{c}_1^{-1}[1 + (\gamma_2' - 1)\Pi_2(z_2)]\}^{-1}$$

$$\Pi_1(z_1) = \frac{3\alpha_2\{Bz_1 + [(B+1)z_1^2 - 1]\,\mathrm{th}\,Bz_1\}}{\alpha_1 z_1^2[(B+1)z_1 - \mathrm{th}\,Bz_1]} \qquad (6.2.20)$$

$$\Pi_2(z_2) = \frac{3(z_2 \operatorname{cth} z_2 - 1)}{z_2^2}, \quad \gamma_2' = \frac{\gamma_2}{\varkappa_{gS}} = 1 + (\gamma_2 - 1)\left(1 - \frac{1}{(\gamma_2 - 1)\, l^*}\right)^2$$

$$\left(B = \frac{1}{\alpha_2^{1/3}} - 1, \quad z_j = \left(\omega_* t_j^{(\lambda)}\right)^{1/2} \frac{1+i}{\sqrt{2}}, \quad t_j^{(\lambda)} = \frac{a^2}{\nu_j^{(T)}} \quad (j = 1, 2)\right.$$

$$\left.\omega_{\Sigma a} = \frac{1}{a}\sqrt{\frac{2\Sigma}{a\rho_1^{\circ}(1 - \varphi^{(1)})}}, \quad \omega_{\Sigma\mu} = \frac{\Sigma}{2\mu_1 a}\right)$$

Here, $\Pi(\omega_*)$ is a function reflecting the effects of interphase heat and mass exchange, $\omega_{\Sigma a}$, similar to ω_a, (see (6.2.11)) defines the natural oscillation frequency of the isothermal ($\varkappa = 1$) vapor bubble in a liquid with an infinite heat conductivity ($z = 0$) which is in agreement (as will be shown below) with $\Pi = 1$; $\omega_{\Sigma a}$ is a parameter analogous to ω_μ in (6.2.11) (for a vapor bubble), which characterizes the dissipation effects due to liquid viscosity.

The following asymptotic forms apply for the transcendental functions $\Pi_1(z_1)$ and $\Pi_2(z_2)$

$$|z_1| \ll 1:\ \Pi_1 \approx 1; \quad |z_1| \gg 1:\ \Pi_1 \approx 3\alpha_2/(\alpha_1 z_1) \approx 0$$
$$|z_2| \ll 1:\ \Pi_2 \approx 1; \quad |z_2| \gg 1:\ \Pi_2 \approx 3/z_2 \approx 0 \qquad (6.2.21)$$

It should be understood, in this case, that generally $|z_2| \ll |z_1|$, and the terms containing Π_2 in the expression for Π are small (because of ρ_2^0/ρ_1^0 smallness). The latter is indicative of the insignificant effect of the vapor heat conduction (within bubbles) upon the propagation of weak disturbances.

Two limiting asymptotes may be written for $\Pi(\omega_*)$ which correspond to almost equilibrium (low frequencies ω_*), and almost "frozen" (high frequencies ω_*) heat and mass exchange. The first, a "near-equilibrium" asymptote, fits frequencies of such a low level at which the temperature inside a cell of radius $r_b \equiv R = a/\alpha_2^{1/3}$ have sufficient time to equalize

$$\Pi(\omega_*) = 1 + i\overline{R}' t_1^{(\lambda)} \omega_* \qquad \left(\left(|\omega_*|\, t_{R1}^{(\lambda)}\right)^{1/2} \ll 1\right.$$

$$\left. t_{R1}^{(\lambda)} = \frac{R^2}{\nu_1^{(T)}} = \frac{t_1^{(\lambda)}}{\alpha_2^{2/3}}, \quad \overline{R}' = \frac{\left(1 - \alpha_2^{2/3}\right)\left(5 + 6\alpha_2^{1/3} + 3\alpha_2^{2/3} + \alpha_2\right)}{15\alpha_2^{1/3}(1 - \alpha_2)}\right) \qquad (6.2.22)$$

We recall that $\tilde{\rho}_2 \ll 1$ and $\tilde{\rho}_2 l^{*2}/\bar{c}_1 \ll 1$. As a consequence, the thermal processes in vapor (inside the bubbles) are insignificant. For a liquid with vapor bubbles, i.e., at boiling temperature, the following evaluations take place

$$t_1^{(\lambda)} \gg \omega_{\Sigma a}^{-1} \gg \omega_{\Sigma\mu}^{-1}, \quad C_1^2 \gg C_L^2 \qquad (6.2.23)$$

Therefore, at the low frequencies indicated in (6.2.22), the radial inertia, viscosity, and compressibility of liquid are nonessential, i.e.

$$(|\omega_*|/\omega_{\Sigma a})^2, \quad |\omega_*|/\omega_{\Sigma\mu} \ll \Pi, \quad C_1^2 \gg C_L^2 \Pi \qquad (6.2.24)$$

The second, an intermediate asymptote, corresponds to frequencies (low but comparable to $(t_{R1}^{(\lambda)})^{-1}$, when conditions (6.2.24) are satisfied, i.e., when, during the oscillation period, the temperature disturbances from the bubble in liquid extend over distances comparable to the bubble radii or to distances between bubbles, which are of order of the cell radius

$$\Pi = \frac{\alpha_1}{\alpha_2}\left\{\frac{\tilde{\overset{\circ}{\rho}}_2 l^{*2}}{\bar{c}_1}\left[1 + (\gamma_2' - 1)\,\Pi_2^{\cdot}(z_2)\right] + \frac{3}{z_1}\left(1 + \frac{1}{z_1}\right)\right\}^{-1}$$

$$\left(B\,|z_1| \sim \left(|\omega_*|\,t_1^{(\lambda)}\right)^{1/2}\alpha_2^{-1/3} = \omega t_{R1}^{(\lambda)} > 3\right) \qquad (6.2.25)$$

At even higher frequencies

$$|z_1| \gg \frac{3\bar{c}_1}{l^{*2}}\frac{\overset{\circ}{\rho}_1}{\overset{\circ}{\rho}_2} \quad \left(\text{usually } \frac{3\bar{c}_1}{l^{*2}}\frac{\overset{\circ}{\rho}_1}{\overset{\circ}{\rho}_2} \gg 1\right) \qquad (6.2.26)$$

(when, generally, $|z_1| \gg 1$ and $|z_2| \gg 1$), in conformity with (6.2.21), we have

$$\Pi_1,\ \Pi_2 \to 0 \text{ and } \Pi = \tilde{\rho}_2 l^{*2}/\bar{c}_1 \qquad (6.2.27)$$

The latter fits very thin (compared to the bubble radii) boundary layers in liquids and vapor when the interphase heat and mass exchange are practically negligible ("frozen"). In this case, we obtain

$$\left(\Pi - \bar{\Sigma}_L\right) C_L^2 = C_\alpha^2 \alpha_1^2 \gamma_2'', \quad \gamma_2'' = \gamma_2\left[1 + 2\Sigma^*\left(1 - 1/(3\gamma_2)\right)\right] \approx \gamma_2 \qquad (6.2.28)$$

With the indicated frequencies, the dispersion relation (6.2.20) coincides with the relation (6.2.11) for a gas-liquid mixture discussed above. Thus, the phase transitions qualitatively affect the propagation of sound only at sufficiently low frequencies ($|\omega_*|t_1^{(\lambda)} \lesssim 1$), when Eq. (6.2.24) is relevant, i.e., when all other dispersion effects are insignificant, and the dispersion relation $k^*(\omega_*)$ is governed by the function $\Pi(\omega_*)$, and by parameter $\bar{\Sigma}_L$

$$\frac{k_*^2}{\omega_*^2} = \frac{\alpha_1^2}{C_L^2\left[\Pi(\omega_*) - \bar{\Sigma}_L\right]} \qquad (6.2.29)$$

We shall first review the k-waves ($k^* = k > 0$; see §4.1), whose evolution enables us to judge the stability of the equilibrium state of liquid with gas bubbles. The investigation of Eq. (6.2.20), made by V. Sh. Shagapov, proves that, from the standpoint of the identification of unstable states, a low-frequency approximation (6.2.29) is enough, when for $\Pi(\omega_*)$, the asymptotes (6.2.22) or (6.2.25) hold good. In support of this statement, the fact that at higher frequencies, the vapor-liquid medium resembles the acoustic properties of a gas-liquid system which is stable, may be used. Consequently, for k-waves we obtain the dispersion equation for ω_* in the form

$$\Pi(\omega_*) - \left(\alpha_1^2 C_L^{-2} k^{-2}\right)\omega_*^2 - \bar{\Sigma}_L = 0 \quad (\omega_* = \omega + i\omega_{**}) \qquad (6.2.30)$$

Investigation shows that the roots of this equation are imaginary ($\omega_* = i \cdot \omega_{**}$). If

$$\bar{\Sigma}_L < 1 \tag{6.2.31}$$

then, $\omega_{**}(k) > 0$, which is indicative of the stability of the equilibrium state under consideration. Otherwise, if

$$\bar{\Sigma}_L > 1 \tag{6.2.32}$$

then, $\omega_{**}(k) < 0$, and the disturbance amplitude grows with time. Thus, the equilibrium vapor-liquid media in a condition $\bar{\Sigma}_L > 1$ may, as distinct from "cold" gas-liquid media, be unstable due to vaporization or condensation, which leads to either the growth or disappearance of bubbles, the susceptibility to state of instability increasing with the growth of $\bar{\Sigma}_L$, i.e., with the vapor-bubble concentration and size reduction.

Analysis of the relation $\omega_{**}(k)$ in compliance with Eq. (6.2.30) shows that, with variation of k from 0 to ∞, the disturbance-growth indicator ω_{**} increases from 0 to a maximum value which is approached asymptotically; this maximum value (denoted by ω_∞) defines the maximum possible rate of the disturbance growth in a given state. The values of ω_∞ in two limiting cases with respect to the parameter $\bar{\Sigma}_L$ are outlined below. At $\bar{\Sigma}_L \approx 1$, the relation (6.2.22) may be used for $\Pi(\omega_*)$, and at $\bar{\Sigma}_L \gg 1$, the formula (6.2.25), where the terms containing Π_2 may be neglected, may hold good. Then, we obtain

$$\begin{gathered}
(\omega_\infty t_1^{(\lambda)})^{-1} = \overline{R}'/(\bar{\Sigma}_L - 1) \qquad (0 < \bar{\Sigma}_L - 1 \ll 1) \\
(\omega_\infty t_1^{(\lambda)})^{-1} = \frac{1}{2}\left[1 + \frac{2a}{a'_L} - \sqrt{1 + \frac{4a}{a'_L}}\right] \qquad (\bar{\Sigma}_L \gg 1) \\
\left(a'_L = \frac{a_L}{1 - 2\Sigma^*/(3\gamma_2)}, \quad \Sigma^{*\prime} = \frac{\Sigma^*}{1 + 2\Sigma^*}\right)
\end{gathered} \tag{6.2.33}$$

For the water-steam media at $p = 0.1$ MPa, $\alpha_2 = 0.33$ for $a = 1$ mm ($\bar{\Sigma}_L = 1.17$), the first formula yields $\omega_\infty^{-1} \approx 200$ s. At the same pressure but with $\alpha_2 < 10^{-3}$ for $a = 1$ mm and 0.1 mm so that the state of the mixture is far in the region of instability ($\bar{\Sigma}_L \gg 1$), the second formula yields $\omega_\infty^{-1} = 50$ and 0.024, respectively. It is obvious that for coarse-grain disperse equilibrium mixtures ($a \gtrsim 1$ mm), the inherent time of the disturbance growth is great even if the mixture is unstable because of low steam-bubble volumetric concentrations; in this case, the equilibrium is being disturbed in a slow mode.

We shall now move to the analysis of the propagation of forced oscillation in the form of ω-waves ($\omega_* = \omega < 0$; see §4.1). Figure 6.2.2 schematically depicts the relations $C(\omega)$ and $k_{**}(\omega)$ in conformity with Eq. (6.2.20).

For stable vapor-liquid media ($\bar{\Sigma}_L < 1$) in the near-equilibrium range of frequencies, according to Eqs. (6.2.29) and (6.2.22), we have the following expressions for phase velocity and attenuation coefficient

$$C(\omega) = \pm C_e\left(1 + O\left(\omega^2 t_{R1}^{(\lambda)2}\right)\right), \qquad C_e = C_L \alpha_1^{-1}\sqrt{1 - \bar{\Sigma}_L}$$
$$k_{**}(\omega) = \pm \frac{\bar{R}'\omega^2 t_1^{(\lambda)}}{2C_L\sqrt{1-\bar{\Sigma}_L}} \qquad \left(\omega t_{R1}^{(\lambda)} \sim \bar{R}'\omega t_1^{(\lambda)} \ll 1\right) \tag{6.2.34}$$

It is seen that the equilibrium speed of sound C_e in a vapor-liquid bubbly ($\alpha_1 \approx 1$) mixture equals the Landau speed of sound only if its state is located extremely deep in the stability zone ($\bar{\Sigma}_L \ll 1$).

The respective expressions for unstable mixtures ($\bar{\Sigma}_L > 1$) are written in the form

$$C(\omega) = \pm \frac{2C_L\left(\bar{\Sigma}_L - 1\right)^{3/2}}{\alpha_1 \bar{R}'\left(\omega t_1^{(\lambda)}\right)}, \qquad k_{**}(\omega) = \mp \frac{\alpha_1 \omega}{C_L\left(\bar{\Sigma}_L - 1\right)^{1/2}} \tag{6.2.35}$$

i.e., at $\omega \to 0$, the phase velocity $C(\omega) \to C_e \to \infty$. In other words, the equilibrium speed of sound for unstable vapor-liquid mixtures tend to infinity.

Figure 6.2.2 also shows the relations $C(\omega)$ and $k_{**}(\omega)$ related to both equilibrium ($\lambda_1, \lambda_2 \to \infty$) and frozen ($\lambda_1 = \lambda_2 = 0$) (with respect to heat conduction) dissipation-free ($\mu_1 = 0$) approximations or asymptotes represented, respectively, by dashed and dot-dashed lines. There is a natural frequency for vapor bubbles at the equilibrium heat conduction (similar to ω_r in Eq. (6.2.11) for gas bubbles), which will is denoted by $\omega_{\Sigma r}$

$$\omega_{\Sigma r} = \omega_{\Sigma a}\sqrt{\varkappa'}, \qquad \varkappa' = (1 - \bar{\Sigma}_L)/\bar{\Sigma}_L \tag{6.2.36}$$

In this approximation ($\lambda_1, \lambda_2 \to \infty$, $\mu_1 = 0$), the oscillations of a gas bubble and sound dispersion are defined by both radial inertia of liquid and the resultant elasticity of surface tension (2Σ) and vapor ($\varkappa^1$), which depends, in particular, on the function $T_S(p_2)$ and mass concentration $\tilde{\rho}_2$ of vapor.

In the frozen heat-conduction approximation ($\lambda_1 = \lambda_2 = 0$, $\mu_1 = 0$), the phase speed of sound at $\omega = 0$ ("partial-equilibrium" speed of sound) and natural frequency of the bubble oscillations, which are denoted, respectively, by C_0 and ω_0, are calculated similar to the gas-liquid mixture with adiabatic bubbles (see Eq. (6.2.11))

$$C_0 = C_a\sqrt{\gamma_2}, \qquad \omega_0 = \omega_a\sqrt{\gamma_2} \tag{6.2.37}$$

Both the frozen speed of sound C_f and the frequency ω_C associated with the liquid compressibility are calculated analogous to the gas-liquid medium (see Eq. (6.2.12), where ω_0 should be taken as ω_r).

It must be kept in mind that recognition of the stability region ($\bar{\Sigma}_L < 1$) of the vapor-liquid bubbly media is accounted for by the adiabatic nature of the cell boundaries (see the discussion of (1.6.11)). If, instead of the adi-

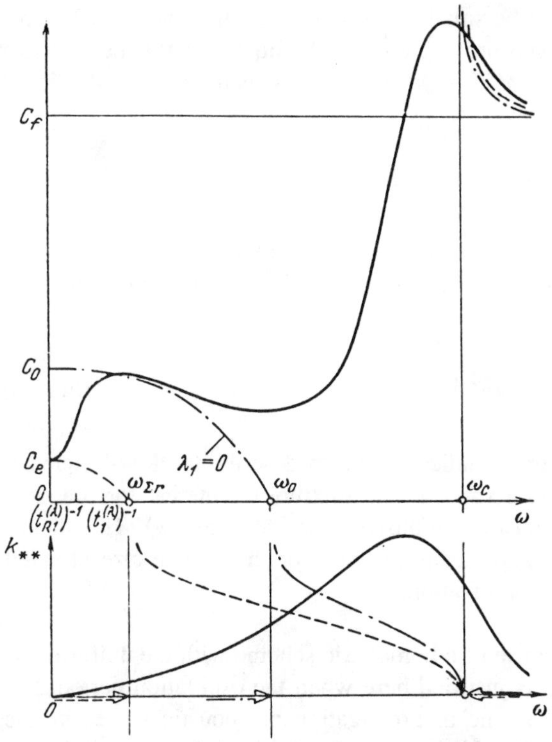

Figure 6.2.2 Schematic of relationships between the speed C and frequency ω of forced oscillations (ω-waves) generated by an exterior source in a stable ($\bar{\Sigma}_L < 1$), equilibrium (in the original state) vapor-liquid bubbly medium, and between the attenuation coefficient and the above-mentioned frequency ω. Solid lines correspond to the most characteristic media in states far from critical. Dashed lines, defined by values C_e (see Eq. (6.2.34)) and $\omega_{\Sigma r}$ (see Eq. (6.2.36)), correspond to a single-temperature (homothermic) dissipation-free scheme with uniform and equal phase temperatures in a cell ($\mu_1 = 0$, $\lambda \to \infty$, and $\lambda_2 \to \infty$). The dot-dashed lines, defined by C_0 and ω_0 (see Eq. (6.2.37)), correspond to a dissipation-free scheme with no heat conduction ($\mu_1 = 0$, $\lambda_1 = \lambda_2 = 0$). The values C_f and ω_C, which depend on the liquid compressibility and characterize the dispersion relations at high frequencies, are defined by formulas (6.2.12).

abatic mode, the isothermality of liquid is assumed at a remote distance from the bubble in accordance with (1.6.10a), the dispersion analysis results in the disturbance instability in the form of k-waves at and $\bar{\Sigma}_L$; for ω-waves at $\omega \to 0$, we always obtain $C(\omega) \to \infty$.

Also, it should be kept in mind that at low near-equilibrium frequencies, when the phase thermal equilibrium is realized, restraints more rigorous than $\Delta p/p \ll 1$ must be imposed upon the pressure amplitudes in order to ensure the validity of the linear dispersion analysis implemented here. Indeed, the bubble mass variation must be small: $\Delta m/m_2 \ll 1$, where $m_2 = 4/3\pi\alpha^3\rho_2^0$. The heat of phase transition is provided mainly by liquid, and in the presence of thermal equilibrium $\Delta T_i = \Delta T_1 = \Delta T_2 = \Delta T_S(p)$. Then, denoting the liquid mass in a cell by m_1, we have

$$l\Delta m = c_1 \Delta T_S m_1, \qquad m_1 \approx m \approx (^4/_3\pi a^3/\alpha_2)\,\overset{\circ}{\rho}_1$$

Now, using the condition of smallness of the bubble-mass variation, the Clausius-Clapeyron equation ($\Delta T_S \approx T_0 \Delta p/(\rho_2^0 l)$), and equation of state ($p_0 \sim c_2 T_0 \rho_{20}^0$), we arrive at

$$\frac{\Delta p}{p_0} \ll \frac{\overset{\circ}{\rho}_2 \alpha_2}{\overset{\circ}{\rho}_1} \frac{l^2}{c_1 c_2 T^2} \sim \frac{\overset{\circ}{\rho}_2 \alpha_2}{\overset{\circ}{\rho}_1} \approx \tilde{\rho}_2 \ll 1 \tag{6.2.38}$$

This extremely strong restraint indicates that to describe the low-frequency disturbances ($\omega t_{R1}^{(\lambda)} \ll 1$) in a vapor-liquid mixture of interest for practical use, the nonlinear theory must be employed (see §6.10 below). This theory gives consideration to the significant variation of the bubble size affected by either condensation or vaporization.

Elastic preindicator. The use of a homobaric scheme with the uniform gas pressure inside the bubble is justified here when the oscillation period $2\pi/\omega$ is much longer than the time of propagation of acoustic waves in gas within the bubble (a/C_g). Employment of the Rayleigh-Lamb equation in which the radial inertia of liquid is produced by the entire associated mass, which is characteristic for an incompressible liquid, is justified when the oscillation period $2\pi/\omega$ is much longer than the time of propagation of acoustic waves in liquid at distances of the order of the cell of radius R per unit bubble

$$\omega \ll \omega_g \sim C_g/a, \quad \omega \ll \omega_l \sim C_1 \alpha_2^{1/3}/a \tag{6.2.39}$$

For bubbly mixtures with parameters $\alpha_2 \sim 10^{-2}$, $a \sim 1$ mm, when the speeds of sound in phases $C_g \approx 400$ m/s, $C_1 \approx 1500$ m/s, we have $\omega_g \sim 10^5 \text{s}^{-1}$, $\omega_l \sim 10^6 \text{s}^{-1}$, and at $p \sim 0.1$ MPa, $\rho_1^0 \sim 10^3$ kg/m^3, we have $\omega_r \sim 10^4 \text{s}^{-1}$, $\omega_C \sim 10^5 \text{s}^{-1}$. Thus, at indicated conditions, the entire high-frequency branch of the dispersion curve ($\omega > \omega_C$) does not satisfy the requirement (6.2.39). Therefore, it is unlikely that the system of adopted equations, including the obtained double-wave equation and telegraph equation, will properly describe the evolution of the high-frequency and high-velocity elastic preindicator which propagates with the speed of sound in pure liquid. As far as nonsupershort impulses are concerned (whose length is many times larger than both bubble size and distances between them), this imperfection of the model associated mainly with using the Rayleigh-Lamb equation, is generally not very important, because the fraction of energy of such impulses accounting for the high-frequency part of the spectrum ($\omega > \omega_C$), is small, and therefore the preindicator amplitude is also negligible. Moreover, in the adopted model, in which the radial motion is described (because of liquid incompressibility) by the Rayleigh-Lamb equation, the elastic preindicator amplitude does not attenuate (at $\omega \to \infty$ we have $C \to C_f$, $k_{**} \to 0$). Actually, this amplitude rapidly attenuates at distances of order of distances $2R$ between bubbles because of the acoustic unloading of the high-frequency compression waves on bubbles; there is no time for bubble size to undergo any change during the indicated unloading process.

Let us evaluate the attenuation of the elastic preindicator described in the form of a shock propagating along the characteristic across the undisturbed medium (see (6.2.3))

$$x = C_f t, \quad \frac{dp_f}{dt} + \rho C_f \frac{dv_f}{dt} = \frac{3\alpha_1 \alpha_2 w}{a} \rho C_f^2 \tag{6.2.40}$$

Let both the longitudinal velocity v and pressure p undergo the shock discontinuity in conformity with the momentum conservation law

$$\rho_0 C_f v_f = p_f - p_0 \qquad (\rho_f \approx \rho_0,\ \alpha_{1f} \approx \alpha_{10},\ \alpha_{2f} \approx \alpha_{20},\ a_f \approx a_0) \tag{6.2.41}$$

Then, substituting this equation into the condition on the characteristic, we obtain

$$\frac{dp_f}{dx} = \frac{3\alpha_{10}\alpha_{20} w_f}{2a_0} \overset{\circ}{\rho}_1 C_f \tag{6.2.42}$$

Note that in the obtained equaiton for the pressure variation across the preindicator front, the Rayleigh-Lamb equation was not used. In accordance with the frozen scheme, the radial velocity w remains zero ($w_f = 0$) because of the finite radial associated mass of liquid during shock-like change of pressure from p_0 to p_f. Then, in compliance with (6.2.42), the elastic preindicator will not decay.

To determine w_f using the liquid compressibility, we shall use the acoustic-approximation solution of a problem concerned with the spherical liquid unloading from pressure p_f to the pressure p_0 within a bubble. In accordance with this solution, there is a spherical domain $r < R_C$ within a cell; this domain experiences disturbances induced by unloading, and the pressure distribution inside the cell is

$$\begin{aligned} p' &= p_f + (p_0 - p_f)\, a_0/r \qquad (a_0 < r < R_C = a_0 + C_1 t) \\ p' &= p_f \qquad (r > R_C) \end{aligned} \tag{6.2.43}$$

The radial velocity distribution is found by means of momentum equations. In particular, for $r = a$, we have

$$w'_f = w_{f0} - \frac{1}{\overset{\circ}{\rho}_1}\int_0^t \frac{\partial p'}{\partial r}\,dt = \frac{p - p_f}{\overset{\circ}{\rho}_1 C_1}\frac{R_C(t)}{a} \tag{6.2.44}$$

where the initial value w_{f0} was found from the relation for an instantaneous unloading, which is analogous to (6.2.41)

$$\overset{\circ}{\rho}_1 C_1 w_{f0} = p_0 - p_f \tag{6.2.45}$$

As a characteristic value of radial velocity w_f, defining attenuation of the elastic preindicator in Eq. (6.2.42), we take the average value which it assumes during the period of time it takes for the preindicator to cover the average distance $L \approx 2R = 2a_0/\alpha_{20}^{1/3}$ between bubbles

$$w_f = \frac{1}{2}\frac{p_0 - p_f}{\overset{\circ}{\rho}_1 C_1}\left(1 + \frac{2}{\alpha_{20}^{1/3}}\right) \approx \frac{p_0 - p_f}{\overset{\circ}{\rho}_1 C_1 \alpha_{20}^{1/3}} \tag{6.2.46}$$

Then, upon integrating Eq. (6.2.42), we obtain

$$\Delta p_f = \Delta p_0 \exp\left(-\frac{x}{x_f}\right), \quad x_f = \frac{2a_0}{3\alpha_{20}^{2/3}} \tag{6.2.47}$$

where x_f is the distance over which the amplitude of the front shock $\Delta p_f = p_f - p_0$ reduces by a factor of e.

Equation (6.2.47) for the elastic preindicator decay may be also obtained form some other considerations proposed by V. Sh. Shagapov. The average liquid pressure within a cell is

$$p = \frac{1}{{}^4/_3\pi\,(R^3 - a^3)}\int_a^R 4\pi r^2 p'\,dr \qquad \left(R = \frac{a}{\alpha_2^{1/3}} \approx \frac{L}{2}\right) \tag{6.2.48}$$

When the preindicator-wave passes through the bubble, there occurs a zone behind (upstream) the wave, which experiences unloading on the bubble. Therefore, the average pressure across the front $p_f(x + L)$ when it approaches the neighboring bubble, is lower than $p_f(x)$. Let us assume a spherical distribution of pressure p' in a cell, which is governed by Eq. (6.2.43). Then, having integrated this distribution in accordance with (6.2.48), ignoring $\alpha_2^{2/3}$ and α_2 (comparing them to unity), and letting

$$p_f(x + R) = p_f(x) + R\frac{dp_f}{dx} \tag{6.2.49}$$

we obtain Eq. (6.2.47).

Experimental data on attenuation of the elastic preindicator in bubbly liquids are reported by V. K. Kedrinskii (1968, 1980), and by V. V. Kuznetsov and B. G. Pokusaev (1978). Some results of the latter for a mixture of water with air bubbles are represented in Figs. 6.2.3 and 6.2.4. Authors point out that the speed of elastic preindicator practically coincides with the speed of sound $C_l \approx 1500$ m/s in pure liquid. The experimental value of

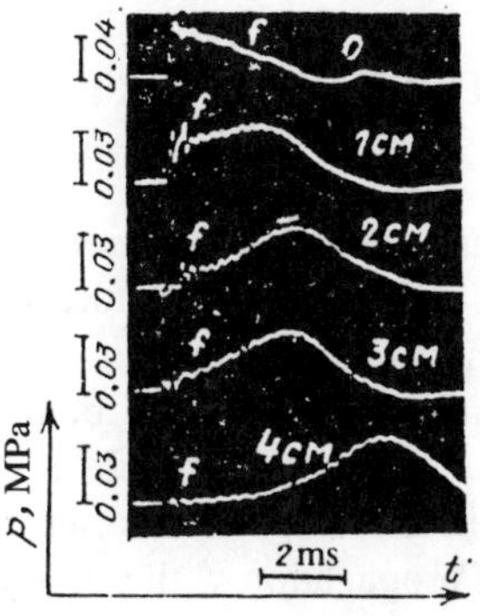

Figure 6.2.3 Pressure oscillograms obtained (V. V. Kuznetsov and B. G. Pokusaev, 1978) in the shock-tube experiments which identify the elastic preindicator f at various distances x = 0, 1, 2, 3, and 4 cm (correspond to labels on curves) from the entrance of a triangular shock impulse (Δp_0 = 0.051 MPa) into a mixture of an aqueous-glycerine solution ρ_1^0 = 1080 kg/m³) with air-bubbles (p_0 = 0.1 MPa, T_0 = 293 K, a_0 = 1.5 mm, $\alpha_{20} \approx 0.006$).

x_f may be readily found from the represented curves. For a medium with a_0 = 1.5 mm, α_{20} = 0.006, $x_f \approx 2$ cm is obtained, whereas $x_f \approx 3$ cm is found from formula (6.2.47); for a_0 = 1.5 mm, α_{20} = 0.002, the experiment yields x_f = 7 cm, and calculation using (6.2.47) gives x_f = 6.7 cm. It is evident that the obtained theoretical evaluation is in satisfactory agreement with the experimental data. Some quantitive disagreement may be partly attributed to errors in measurements of extremely small gas concentrations ($\alpha_2 = (2\text{–}6) \cdot 10^{-3}$) near the upper free surface of liquid where the elastic preindicator manifests itself. As to the main part of the wave, its amplitude does not noticeably alter in a region where the elastic preindicator is encountered to any extent. We shall further place the most emphasis upon the disturbances whose characteristic duration is much longer than $2\pi/\omega_C$. In this case, the high-frequency branch of the dispersion curve, associated with compressibility of the carrying liquid, will be nonessential.

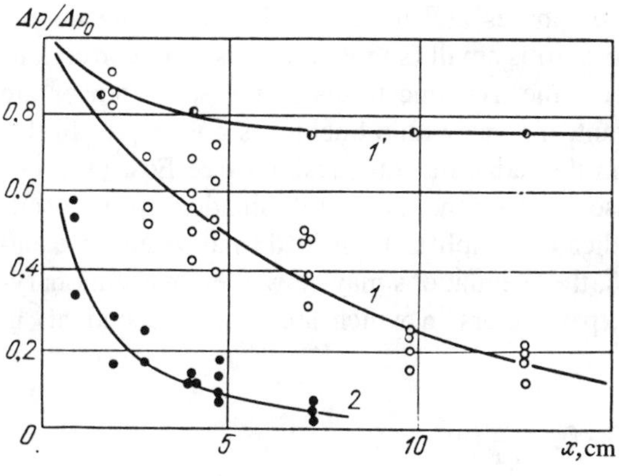

Figure 6.2.4 Variation of pressure disturbances across an elastic preindicator (points *1* and *2*), and in the main signal (points *1'*) with a distance x at the same conditions as in Figure 6.2.2. Points *1* and *1'* are for α_{20} = 0.002, points *2* are for α_{20} = 0.006.

§6.3 THE EFFECTIVE-VISCOSITY AND POLYTROPIC-GAS APPROACH FOR DESCRIBING THE STATIONARY SHOCK WAVES IN LIQUID WITH GAS BUBBLES

We shall examine the conditions of existence and the structure of a stationary wave which may be established either during flow of a bubbly mixture in a tube (the wave in this case is said to be stationary), or as a result of a "piston" action initiated by the high-pressure chamber, the "piston" being thrust with a constant velocity, or constant pressure, into a immobile homogeneous mixture when, with long HPCh and LPCh, the wave, after a transient regime, establishes a stationary regime of propagation with a fixed velocity D_0 without any change of its structure. Then, as was described in §4.4, the process is stationary in the wave related coordinate system, and the undisturbed medium enters the wave with velocity $v_0 = -D_0$. As before, the direction of axis x coincides with the wave propagation direction relative to the undisturbed medium (see (4.4.1)).

We shall confine ourselves to the phase-transition absence case (i.e., when the bubble mass does not alter). The capillary effects and external body forces are

$$j = 0 \quad (\overset{\circ}{\rho}_2 a^3 = \mathrm{const}), \qquad \Sigma = 0, \qquad g_1 = g_2 = 0 \tag{6.3.1}$$

The following three sections (§§6.3–6.5) present the analysis of the stationary shock waves under these conditions in the framework of various schemes available for describing bubbly liquids.

Equations for a stationary one-dimensional flow. Analysis of a stationary wave structure in a simplified formulation is of special interest for further studies; in this case, one assumes that: the carrying (continuous) liquid is incompressible, there is no fragmentation and no coalescence of bubbles, there is no relative motion of phases and no capillary effects, the bubble volumetric concentration (holdup) is small (so that the constraint corrections $\varphi^{(1)}$ and $\varphi^{(2)}$ may be ignored), the dynamic terms in the stress tensor are small, and, the most important, the gas within bubbles is polytropic. In this case, the effective viscosity in the radial-motion equations (see Eqs. (1.6.48), (1.6.50), and (1.6.51)) is used to take the thermal dissipation during oscillations into account. The indicated simplifications lead to a system of equations (1.5.16). Using (6.3.1), these equations may be written in a stationary-wave-related frame system (parameters in which are time-independent) in the following form

$$\frac{d}{dx}(\alpha_1 v) = 0, \quad \frac{d}{dx}(\overset{\circ}{\rho}_2 \alpha_2 v) = 0, \quad \frac{d}{dx}(nv) = 0$$

$$v\frac{da}{dx} = w_{1a}, \quad av\frac{dw_{1a}}{dx} = \frac{p_2 - p_1}{\overset{\circ}{\rho}_1} - \frac{4\mu_{\text{эф}} w_{1a}}{a\overset{\circ}{\rho}_1} - \frac{3w_{1a}^2}{2}$$

$$\rho v \frac{dv}{dx} + \frac{dp}{dx} = 0, \quad \frac{\overset{\circ}{\rho}_2}{\overset{\circ}{\rho}_{20}} = \left(\frac{a_0}{a}\right)^3, \quad \frac{p_2}{p_{20}} = \left(\frac{a_0}{a}\right)^{3\varkappa} \tag{6.3.2}$$

$$\rho = \overset{\circ}{\rho}_1\alpha_1 + \overset{\circ}{\rho}_2\alpha_2 \approx \overset{\circ}{\rho}_1\alpha_1, \quad \alpha_2 = {}^4/_3\pi a^3 n$$

The equilibrium parameters before the wave ($x \to +\infty$), (they must be predetermined), are denoted by a subscript 0, and those behind the wave ($x \to -\infty$) (where equilibrium is also established) by a subscript e. Therefore, the boundary conditions are

$$\begin{gathered} x \to +\infty: \; v = v_0 = -D_0 < 0, \quad p = p_2 = p_0, \quad w_0 = 0 \\ \alpha_2 = \alpha_{20}, \quad \overset{\circ}{\rho}_2 = \overset{\circ}{\rho}_{20}, \quad a = a_0 \\ x \to -\infty: \; p = p_2 = p_e, \quad w_e = 0 \end{gathered} \tag{6.3.3}$$

The system of equations (6.2.2) has the following four first integrals

$$\begin{gathered} \alpha_1 v = \alpha_{10} v_0, \quad \alpha_2 \overset{\circ}{\rho}_2 v = \alpha_{20} \overset{\circ}{\rho}_{20} v_0, \quad nv = n_0 v_0 \\ \overset{\circ}{\rho}_1 \alpha_{10} v_0 v + p = \overset{\circ}{\rho}_1 \alpha_{10} v_0^2 + p_0. \end{gathered} \tag{6.3.4}$$

These integrals enable us to determine the equilibrium parameters behind the wave based on parameters before the wave

$$\alpha_{1e} v_e = \alpha_{10} v_0, \quad \alpha_{2e} v_e \left(\frac{p_e}{p_0}\right)^{1/\varkappa} = \alpha_{20} v_0, \quad v_0 - v_e = \frac{p_e - p_0}{\overset{\circ}{\rho}_1 \alpha_{10} v_0} \tag{6.3.5}$$

Let us introduce dimensionless variables denoted by bars

$$\begin{gathered} \bar{p}_i = \frac{p_i}{p_0}, \quad \bar{a} = \frac{a}{a_0}, \quad \bar{v} = \frac{v}{C_\alpha}, \quad \bar{w} = \frac{w}{C_*}, \quad \bar{x} = \frac{x}{L_\alpha} \\ \left(C_* = \sqrt{\frac{p_0}{\overset{\circ}{\rho}_1}}, \quad C_\alpha = \frac{C_*}{\sqrt{\alpha_{10}\alpha_{20}}}, \quad L_\alpha = \frac{a_0}{\sqrt{\alpha_{10}\alpha_{20}}}\right) \end{gathered} \tag{6.3.6}$$

Then, we obtain the following equation for dimensionless pressure $\bar{p}_e$ behind the wave, which may be referred to as the shock adiabat of the medium under consideration in coordinates

$$\bar{p}_e - 1 = \bar{D}_0^2 \left[1 - \bar{p}_e^{-1/\varkappa}\right] \qquad (\bar{D}_0 = -\bar{v}_0 = D_0/C_\alpha) \tag{6.3.7}$$

where $\bar{D}_0$ is the dimensionless wave velocity relative to the undisturbed medium.

The minimum value $D_0 = C_0$ of the velocity, which corresponds to the stationary compression wave ($\bar{p}_e > 1$), partially defines the equilibrium, polytropic for gas, speed of sound in the mixture

$$\bar{C}_0 = \sqrt{\varkappa}, \quad C_0 = C_\alpha \sqrt{\varkappa} = \sqrt{\frac{\varkappa p_{10}}{\overset{\circ}{\rho}_1 \alpha_{10} \alpha_{20}}} \tag{6.3.8}$$

For an isothermal gas ($\varkappa = 1$) and strong waves ($\bar{p}_e \gg 1$, the pressure behind the wave is independent of the gas polytrope), we have

$$\bar{p}_e = \bar{v}_0^2 = \bar{D}_0^2 \tag{6.3.9}$$

For weak waves ($\Delta\bar{p}_e = \bar{p}_e - 1 \ll 1$), the pressure behind the wave may be expressed in terms of $\bar{D}_0$ by means of expansion of right-hand side Eq. (6.3.7) in the Taylor series

$$\bar{p}_e^{-1/\varkappa} = 1 - \frac{1}{\varkappa}\Delta\bar{p}_e + \frac{\varkappa+1}{2\varkappa^2}(\Delta\bar{p}_e)^2 + O\left((\Delta\bar{p}_e)^2\right)$$

Then, (with an accuracy of O ($\Delta\bar{p}_e$)), we obtain

$$\Delta\bar{p}_e = \frac{2\varkappa}{\varkappa+1}\frac{D_0^2 - C_0^2}{D_0^2}, \quad \frac{D_0}{C_0} = 1 + \frac{\varkappa+1}{4\varkappa}\Delta\bar{p}_e \tag{6.3.10}$$

From the mass conservation equation for both liquid and gas phases, and also from integrals (6.3.4), we obtain

$$\begin{gathered}\frac{d\bar{v}}{d\bar{x}} = \frac{\bar{v}}{\alpha_1}\frac{d\alpha_2}{d\bar{x}} = \frac{3\alpha_2\bar{w}}{\bar{a}}, \quad \frac{\alpha_2}{\alpha_{20}} = \frac{\overset{\circ}{\rho}_{20}v_0}{\overset{\circ}{\rho}_2 v} = \frac{\bar{v}_0\bar{a}^3}{\bar{v}} \\ \bar{v} = \bar{v}_0\left(\alpha_{10} + \alpha_{20}\bar{a}^3\right), \quad \bar{p} = 1 + \bar{v}_0^2\left(1 - \bar{a}^3\right)\end{gathered} \tag{6.3.11}$$

The following evaluations take place in this case

$$\bar{v} - \bar{v}_0 = \frac{\alpha_{10}\bar{v}_0 - \alpha_1\bar{v}}{\alpha_1} = \frac{v_0}{\alpha_1}(\alpha_2 - \alpha_{20}) \ll \bar{v}_0, \quad \bar{p} < 1 + \bar{D}_0^2 \tag{6.3.12}$$

Upon some elementary transformations, and using (6.3.11), the system of equations (6.3.2) may be represented in a form which expresses the parameter gradients in terms of $\bar{w}$ and $\bar{p}_{21} = \bar{p}_2 - \bar{p}$. The fact that the values $\bar{w}$ and $\bar{p}_{21}$ are not zeros is said to be a measure of the interphase nonequilibrium and the parameter nonhomogeneity along axis x

$$\begin{gathered}\frac{d\bar{w}}{d\bar{x}} = \frac{1}{\bar{a}\bar{v}}(\bar{p}_2 - \bar{p}) - \frac{4\mu^*}{\bar{a}^2\bar{v}}\bar{w} - \frac{3}{2\bar{a}\bar{v}}\bar{w}^2 \qquad \left(\mu^* = \frac{\mu_{\text{эф}}}{\overset{\circ}{\rho}_1 a_0 C_*}\right) \\ \frac{d\bar{p}}{d\bar{x}} = -\frac{3\bar{v}_0^2\bar{a}}{\bar{v}}\bar{w}, \quad \frac{d\bar{p}_2}{d\bar{x}} = -\frac{3\varkappa\bar{p}_2}{\bar{a}\bar{v}}\bar{w} \\ \bar{a} = \bar{p}_2^{-\varkappa/3}, \quad \bar{v} = \bar{v}_0 - \alpha_{20}(\bar{p} - 1)/\bar{v}_0 \approx \bar{v}_0\end{gathered} \tag{6.3.13}$$

Investigation of the solution behavior before (upstream) and behind (downstream) the shock wave. Points o and e correspond to the equilibrium states before ($x \to +\infty$) and behind ($x \to -\infty$) the wave; these are the singular points of the applicable system of differential equations in the space of physical parameters $\bar{w}$, $\bar{p}$, and $\bar{p}_2$. We shall investigate the behavior of the solution in the neighborhood of these points using the linearization approach.

To describe the zone of the shock-wave front where deviations of variables $\bar{\upsilon}$, $\bar{a}$, $\bar{p}$, $\bar{p}_2$, and $\bar{w}$ from their values in the initial state $o(\upsilon = \bar{\upsilon}_0, \bar{a} = \bar{p} = \bar{p}_2 = 1, \bar{w} = 0)$ are sufficiently small, the variable coefficients of $\bar{p}_2 - \bar{p}$ and $\bar{w}$ may be assumed as constant quantities defined by the values of variables in state o. Then, we obtain a system of linear equations

$$\frac{d\bar{w}}{d\bar{x}} = -\frac{4\mu^*}{\bar{v}_0}\bar{w} + \frac{1}{\bar{v}_0}\bar{p}_{21} \qquad (\bar{p}_{21} = \bar{p}_2 - \bar{p})$$
$$\frac{d\bar{p}_{21}}{d\bar{x}} = 3\left(\bar{v}_0 - \frac{\varkappa}{\bar{v}_0}\right)\bar{w}, \quad \frac{d\bar{p}}{d\bar{x}} = -3\bar{v}_0\bar{w} \tag{6.3.14}$$

The fundamental solutions of this linear system are

$$\bar{w} = A^{(w)} \exp k_0(\bar{x} - \bar{x}_b)$$
$$\bar{p}_1 = 1 + A_1^{(p)} \exp k_0(\bar{x} - \bar{x}_b), \quad \bar{p}_2 = 1 + A_2^{(p)} \exp k_0(x - x_b) \tag{6.3.15}$$

where k_0 is in general an imaginary number. Only the real part of these expressions is understood as a solution of the physical problem. Since the parameter deviation from an equilibrium state before the wave at $x \to +\infty$ must approach zero: $\bar{p}_{21}$, $\bar{w} \to 0$, the real part k_0 should be negative

$$\mathrm{Re}\,\{k_0\} < 0 \tag{6.3.16}$$

If the fundamental solutions are substituted into (6.3.14), we obtain the characteristic equation for k_0, which specifies the condition for existence of nontrivial solutions of the type (6.3.15) with nonzero constants $A^{(w)}$, $A^{(P)}$, and $A_2^{(P)}$

$$K_0^2 - 4\mu^* K_0 - 3\varkappa(\bar{D}_0^2 C_0^{-2} - 1) = 0 \qquad (K_0 = k_0\bar{D}_0 = -k_0\bar{v}_0) \tag{6.3.17}$$

In this case, the constants themselves are expressed by linear relations

$$A^{(w)} = -\frac{\bar{v}_0 k_0}{3\varkappa} A_2^{(p)}, \quad A_1^{(p)} = -\frac{\alpha_{10}\alpha_{20}\bar{v}_0^2}{\varkappa} A_2^{(p)} \tag{6.3.18}$$

The condition (6.3.16) leads to $D_0 > C_0$. Thus, in the medium under discussion, only stationary compression waves ($p_e > p_0$) are feasible, and the stationary rarefaction waves are impossible. Only one root of the characteristic equation

$$\bar{D}_0 k_0 = 2\mu^* - \sqrt{4\mu^{*2} + 3\varkappa(D_0^2 C_0^{-2} - 1)} \qquad (D_0 > C_0 > 1) \tag{6.3.19}$$

may be used as k_0; this expression is a real number which is in conformity with the monotonic profile, or the absence of oscillations before the wave (see Fig. 6.3.1) where dashed lines $f^{(1)}o^{(1)}$ and $f^{(2)}o^{(2)}$ are in agreement with

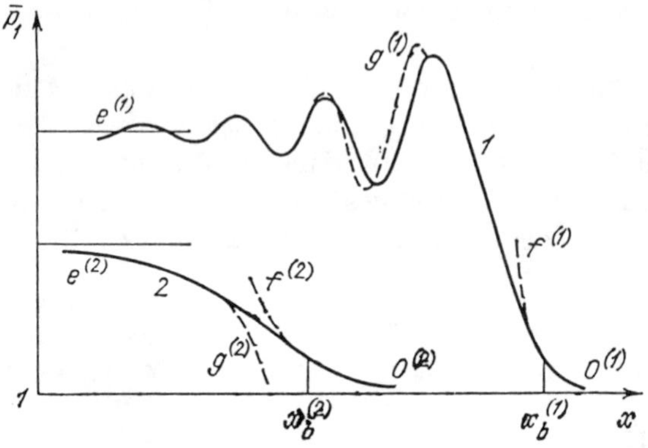

Figure 6.3.1 Characteristic shock-wave structures, and the asymptote of solutions of equations before and behind a wave; *1*) shock wave with an oscillating structure; *2*) shock wave with a monotonic structure.

relations (6.3.15), which determine the asymptotes at $x \to +\infty$ for the wave $o^{(1)}e^{(1)}$ and wave $o^{(2)}e^{(2)}$.

For weak waves ($\Delta\bar{p}_e = \bar{p}_e - 1 \ll 1$), when the shock adiabat is realized in the form (6.3.10), the root k_0 may be expressed in terms of the wave intensity

$$\bar{D}_0 k_0 = 2\mu^* \left[1 - \sqrt{1 + \frac{3(\varkappa+1)}{8}\frac{\Delta\bar{p}_e}{\mu^{*2}}}\right]$$
$$\left(\bar{D}_0 = \bar{C}_0\left(1 + \frac{\varkappa+1}{4\varkappa}\Delta\bar{p}_e\right),\ \bar{C}_0 = \sqrt{\varkappa}\right) \tag{6.3.20}$$

The asymptotes of the wave-tail parameter variation during the approach to an ultimate state of equilibrium, i.e., for a zone where deviations of variables $\bar{v}$, $\bar{a}$, $\bar{p}$, $\bar{p}_2$, and $\bar{w}$ from their values in the final state ($\bar{v} = \bar{v}_e$, $\bar{a} = \bar{a}_e$, $\bar{p} = \bar{p}_2 = \bar{p}_e$, and $\bar{w} = 0$) are sufficiently small, are investigated in a similar manner. For this particular zone, we obtain from (6.3.13) a system of equations linearized with respect to the ultimate state in the form

$$\frac{d\bar{w}}{d\bar{x}} = -\frac{4\mu^*}{\bar{a}_e^2\bar{v}_e}\bar{w} + \frac{1}{\bar{a}_e\bar{v}_e}\bar{p}_{21} \qquad (\bar{p}_{21} = \bar{p}_2 - \bar{p})$$
$$\frac{d\bar{p}_{21}}{d\bar{x}} = \frac{3}{\bar{a}_e}\left(\frac{\alpha_{1e}\alpha_{2e}}{\alpha_{10}\alpha_{20}}\bar{v}_e - \frac{\varkappa\bar{p}_e}{\bar{v}_e}\right)\bar{w} \tag{6.3.21}$$

The fundamental solutions of this linear system are

$$\bar{p}_{21} = A^{(p)}\exp k_e(\bar{x} - \bar{x}_b), \quad \bar{w} = A^{(w)}\exp k_e(\bar{x} - \bar{x}_b)$$
$$(A^{(p)},\ A^{(w)},\ \bar{x}_b = \text{const}) \tag{6.3.22}$$

Since deviations from the state e downstream the wave at $x \to -\infty$ must approach zero ($\bar{p}_{21}$, $\bar{w} \to 0$), the real part k_e should be positive

$$\mathrm{Re}\,\{k_e\} > 0 \tag{6.3.23}$$

Similar to (6.3.17), we obtain for k_e the characteristic equation in the form

$$K_e^2 + \frac{4\mu^*}{\bar{a}_e} K_e - 3\left(\frac{\alpha_{1e}\alpha_{2e}}{\alpha_{10}\alpha_{20}} \bar{v}_e^2 - \varkappa \bar{p}_e\right) = 0 \qquad (K_e = k_e \bar{v}_e \bar{a}_e, \quad \bar{v}_e < 0) \tag{6.3.24}$$

Both roots of the following equation may be used as k_e

$$\bar{a}_e \bar{v}_e k_e = -\frac{2\mu^*}{\bar{a}_e} \pm \sqrt{\frac{4\mu^{*2}}{\bar{a}_e^2} + 3\left(\frac{\alpha_{1e}\alpha_{2e}}{\alpha_{10}\alpha_{20}} \bar{v}_e^2 - \varkappa \bar{p}_e\right)} \tag{6.3.25}$$

For weak shock waves ($\Delta\bar{p}_e = \bar{p}_e - 1 \ll 1$), when the shock-wave adiabat is written in the form (6.3.10), and $\bar{a}_e \approx 1$, $\bar{v}_e \approx \bar{v}_0 = -\bar{D}_0 \approx -\sqrt{\varkappa}$, we obtain, analogous to (6.3.20)

$$\bar{D}_0 k_e = 2\mu^*[1 \pm \sqrt{1 - {}^3/_8(\varkappa + 1)\Delta\bar{p}_e/\mu^{*2}}] \tag{6.3.26}$$

If the values k_e are complex conjugate numbers, then the wave tail may have an oscillatory structure. When k_e are real numbers, and both are positive, then the oscillations in the wave are impossible. The dashed lines $g^{(1)}e^{(1)}$ and $g^{(2)}e^{(2)}$ in Fig. 6.3.1 fit the relations (6.3.22) representing the parameter-distribution asymptote at $x \to -\infty$ for two waves, viz., for the wave $o^{(1)}e^{(1)}$ having an oscillating structure, and the wave $o^{(2)}e^{(2)}$ characterized by a monotonic structure.

The realization of a stationary shock wave with an oscillating structure is stipulated by

$$\Delta\bar{p}_e > \frac{8}{3(\varkappa+1)}\mu^{*2} = \frac{8\mu_{\mathrm{eff}}}{3(\varkappa+1)\rho_1^{\circ} a_0 p_0} \tag{6.3.27}$$

In a number of cases, with sufficiently large bubbles, the impact of dissipation on the oscillation frequency is small ($\mu^{*2} \ll \Delta\bar{p}_e$); then, both length L_e of oscillatory waves and oscillation frequency ω_e of bubbles are defined by simple formulas

$$\frac{L_e}{L_\alpha} = \frac{2\pi}{\mathrm{Im}\,\{k_e\}} = \frac{2\pi \bar{D}_0}{\sqrt{{}^3/_2(\varkappa+1)\Delta\bar{p}_e}}, \qquad \frac{\omega_e}{2\pi} = \frac{D_0}{L_e} = \sqrt{\frac{3(\varkappa+1)}{2}\Delta\bar{p}_e}\,\frac{C_*}{a_0} \tag{6.3.28}$$

At small wave intensity, the oscillations in stationary waves are not effected, i.e., the wave has a monotonic structure.

Determination of the effective viscosity of a bubbly medium in a shock wave. Recalling Eq. (6.3.28), we can find from formula (1.6.50) the expression for a part ($\mu_e^{(T)}$) of the effective viscosity μ_{eff}, which is associated with heat conductivity at radial oscillations at frequency ω_e behind the shock wave, through the mixture characteristics and wave intensity

$$\frac{\mu_e^{(T)}}{\mu_r^{(T)}} = \sqrt{\frac{\omega_e}{\omega_r}} = \left[\frac{\varkappa + 1}{2\varkappa}(\bar{p}_e - 1)\right]^{1/4} \tag{6.3.29}$$

In the monotonic wave-structure case (when the monotonic bubble contraction is affected), the method of predetermining either μ_{eff} or $\mu^{(T)}$ is not obvious (see Notes at the close of §1.6).

Finding a solution for shock-wave structure. Assuming a predetermined μ_{eff}, we shall outline the procedure for finding a solution of a system of equations (6.3.2) or (6.3.13) with the boundary conditions (6.3.3). This solution passes through two singular points which relate to the initial and final states before and behind the shock wave. It describes the structure of this wave.

The integral curves of the system of equations (6.3.2) or (6.3.13) allow a displacement along axis x. Therefore, we shall fix a certain value of the gas dimensionless pressure within bubbles $\bar{p}_2 > 1$, or $A_2^{(P)} = \bar{p}_2(\bar{x}_b) - 1$ at $\bar{x} = \bar{x}_b$ (see Fig. 6.3.1, where $\bar{x}_b^{(1)} = \bar{x}_b$ for wave $o^{(1)}e^{(1)}$ and $\bar{x}_b^{(2)} = \bar{x}_b$ for wave $o^{(2)}e^{(2)}$); the quantity $\bar{p}_2(\bar{x}_b)$ must be sufficiently close to unity in this case for the linear solution (6.3.15) in the domain $\bar{x} > \bar{x}_b$ to be satisfied. Using formulas (6.3.18) and final relations for $\bar{v}$ and $\bar{a}$ in (6.3.13) the values of remaining functions sought at $\bar{x} = \bar{x}_b$ can be readily determined. The values $\bar{p}_2$, $\bar{p}_1$, $\bar{v}$, and $\bar{w}$ at the point $\bar{x} = \bar{x}_b$ yield the boundary conditions for a numerical solution of a nonlinear Cauchy problem for a system of equations (6.3.13) in the domain $\bar{x} < \bar{x}_b$. As a result of this solution, the structure is determined for the entire shock wave, in which the medium transits from the equilibrium state o into an equilibrium state e.

The impact of effective viscosity. As may be learned from (6.3.18), the value k_0 decreases, i.e., the steepness of the wave front reduces. In this case, as may be seen from (6.3.24), the value k_e decreases in the absence of oscillations (when k_e is a real number), and the steepness of the back part of the wave also reduces. When oscillations are present (when k_e is a complex number), the real part of k_e increases with a μ_{eff} increase, which is indicative of a more intensive attenuation of oscillations. It should be kept in mind that reduction of k_0 or reduction of the wave front steepness leads, in turn, to weaker oscillations because of the reduction of an amplitude of the first oscillation. Some examples illustrating the parameter distribution in stationary waves based on the solution of a more general system of equations, which give consistent consideration to both thermal and dynamic effects, are given below.

§6.4 STATIONARY WAVES IN A LIQUID WITH GAS BUBBLES. THE TWO-TEMPERATURE AND TWO-VELOCITY SCHEMES.

We shall now proceed to a more consistent investigation of a stationary shock wave in a bubbly liquid by taking account of: thermal irreversibility in the gas phase, translatory motion of bubbles relative to liquid, and liquid compressibility. We will employ the two-temperature and two-velocity model.

Equations of a stationary, one-dimensional motion. The constituent equations of the two-velocity and two-temperature models of bubbly medium (1.5.4) in a coordinate system associated with a wave (whose parameters are time-independent), in the absence of phase transitions and external body forces (see Eqs. (6.3.1) and (6.3.2)), are written in the form

$$\frac{d(\rho_1 v_1)}{dx} = 0, \quad \frac{d(\rho_2 v_2)}{dx} = 0, \quad v_2 \frac{da}{dx} = w_{1a}$$

$$\left(1-\varphi^{(1)}\right) a v_2 \frac{dw_{1a}}{dx} = \frac{p_2 - p_1 - 2\Sigma/a}{\overset{\circ}{\rho}_1} - \frac{4\mu_1 w_{1a}}{a\overset{\circ}{\rho}_1} - \left(1-\varphi^{(2)}\right)\frac{3w_{1a}^2}{2}$$

$$+\left(1-\varphi^{(3)}\right)\frac{w_{12}^2}{4}$$

$$\frac{\overset{\circ}{\rho}_2}{\overset{\circ}{\rho}_{20}} = \left(\frac{a_0}{a}\right)^3, \quad \alpha_2 = \frac{\rho_2}{\overset{\circ}{\rho}_2}, \quad \alpha_1 = 1-\alpha_2, \quad n = \frac{3\alpha_2}{4\pi a^3} \quad \left(\frac{d(nv_2)}{dx} = 0\right)$$

$$\rho_1 v_1 \frac{dv_1}{dx} = \frac{d\sigma_{1*}}{dx}, \quad v_2 \frac{dv_2}{dx} = \frac{3}{\rho_1}\frac{d\sigma_{1*}}{dx} + \frac{2K_\mu(\mathrm{Re}_{12})\mu_1}{\overset{\circ}{\rho}_1 a^2} w_{12} + \frac{3\alpha_1^2}{a} w_{1a} w_{12} \tag{6.4.1}$$

$$v_2 \frac{dp_2}{dx} = \frac{3(\gamma_2 - 1)}{2a^2}\lambda_2 \mathrm{Nu}_2 (T_0 - T_2) - \frac{3\gamma_2 p_2 w_{1a}}{a}$$

$$p_1 - p_0 = C_1^2\left(\overset{\circ}{\rho}_1 - \overset{\circ}{\rho}_{10}\right), \quad T_2 = p_2/\left(R_2\overset{\circ}{\rho}_2\right)$$

$$\sigma_{1*} = -p_1 - \alpha_2\left\{(p_2 - p_1 - 2\Sigma/a) + \overset{\circ}{\rho}_1\left(w_{1a}^2 + {}^1/_2\, w_{12}^2\right)\right\}$$

where $K_\mu(\mathrm{Re}_{12})$ and $\mathbf{Nu}_2$ must be calculated in accordance with formulas presented in §§1.5 and 1.6. As a consequence, the represented nonlinear system of equations is closed. It is used for analysis (similar to §6.3) of a stationary wave in which the medium transits from the initial equilibrium state *o* to another equilibrium state *e*. The indicated states *o* and *e* may generally be realized asymptotically at $x \to \mp\infty$. The appropriate boundary conditions analogous to (6.3.3) are

$$x \to +\infty: \; v_{10} = v_{20} = v_0 = -D_0, \quad w_{1a} = 0, \quad T_{20} = T_0 \qquad (6.4.2)$$
$$p_{20} = p_{10} + 2\Sigma/a_0$$
$$x \to -\infty: \; v_{1e} = v_{2e} = v_e, \quad w_{1a} = 0, \quad T_{2e} \approx T_0$$
$$p_{2e} = p_{1e} + 2\Sigma/a_e$$

From the equation of state of liquid, using the mass-conservation equations for phases and the bubble, we successively have

$$\frac{d\overset{\circ}{\rho}_1}{dx} = \frac{1}{C_1^2}\frac{dp_1}{dx}, \quad \alpha_1 v_1 \frac{d\overset{\circ}{\rho}_1}{dx} = \overset{\circ}{\rho}_1 v_1 \frac{d\alpha_2}{dx} - \overset{\circ}{\rho}_1 \alpha_1 \frac{dv_1}{dx}$$
$$\overset{\circ}{\rho}_2 v_2 \frac{d\alpha_2}{dx} = -\overset{\circ}{\rho}_2 \alpha_2 \frac{dv_2}{dx} - \alpha_2 v_2 \frac{d\overset{\circ}{\rho}_2}{dx}, \quad a^3 \frac{d\overset{\circ}{\rho}_2}{dx} = -\frac{3\overset{\circ}{\rho}_2 a^2 w_{1a}}{v_2}$$

As a result, the differential equation is obtained, which is used instead of the acoustic equation of compressibility of the carrying (continuous) liquid

$$\frac{\overset{\circ}{\rho}_1}{v_1}\frac{dv_1}{dx} + \frac{\overset{\circ}{\rho}_1 \alpha_2}{\alpha_1 v_2}\frac{dv_2}{dx} + \frac{1}{C_1^2}\frac{dp_1}{dx} = \frac{3\overset{\circ}{\rho}_1 \alpha_2 w_{1a}}{a\alpha_1 v_2} \qquad (6.4.3)$$

The system of differential equations (6.4.1) has three first integrals

$$\rho_1 v_1 = \rho_{10} v_{10} = m_1 = \text{const}, \quad \rho_2 v_2 = \rho_{20} v_{20} = m_2 = \text{const} \qquad (6.4.4)$$
$$m_1 v_1 + m_2 v_2 - \sigma_{1*} = (m_1 + m_2) v_0 + p_{10} \qquad (\sigma_{1*}(o) = -p_{10})$$

Here, the first equation is the integral of the first phase mass, the second is the integral of the second phase mass, and the third is the mixture momentum integral. All of these integrals, as can be seen from (1.3.69), are valid as well in the case when there are some stationary shocks in the wave structure in a coordinate system related to the stationary wave.

Below, we will use the dimensionless variables (6.3.6) denoted by a superscore (bar)

$$\overline{T}_i = T_i/T_0, \quad \overline{\overset{\circ}{\rho}}_i = \overset{\circ}{\rho}_i/\overset{\circ}{\rho}_{i0} \qquad (6.4.5)$$

and fixed (denoted by a subscript 0) dimensionless (denoted by either a bar or asterisk) parameters characterizing physical properties of phases, and expressed in terms of inherent velocities C^* and C_α, defined in (6.3.6)

$$\overline{C}_{10} = \frac{C_1}{C_\alpha}, \quad \lambda_2^* = \frac{\lambda_2}{\overset{\circ}{\rho}_{20} c_2 a_0 C_*}, \quad \mu_1^* = \frac{\mu_1}{a_0 \overset{\circ}{\rho}_{10} C_*}, \quad \Sigma_0^* = \frac{\Sigma}{a_0 p_0} \qquad (6.4.6)$$

In the interest of simplification of calculations, and recalling the comments to (1.5.12), we assume

$$-\sigma_{1*} = p_1 \tag{6.4.7}$$

From (6.4.1), and given Eqs. (6.4.3) and (6.4.7), we obtain the following system of six ordinary differential equations with respect to $\bar{p}_1$, $\bar{p}_2$, $\bar{v}_1$, $\bar{v}_2$, $\bar{a}$, and $\bar{w}$, represented in the form of derivatives in dimensionless parameters

$$\frac{d\bar{p}_1}{d\bar{x}} = \frac{\Delta_{p1}}{\Delta}, \quad \frac{d\bar{v}_1}{d\bar{x}} = \frac{\Delta_{v1}}{\Delta}, \quad \frac{d\bar{v}_2}{d\bar{x}} = \frac{\Delta_{v2}}{\Delta}$$

$$\bar{v}_2\frac{d\bar{p}_2}{d\bar{x}} = \bar{q} - b_p, \quad \bar{v}_2\frac{d\bar{a}}{d\bar{x}} = \bar{w}, \quad \bar{v}_2\frac{d\bar{w}}{d\bar{x}} = b_w \tag{6.4.8}$$

$$\left(\Delta = \frac{\bar{v}_2}{\bar{v}_0\bar{v}_1}\left(\frac{\alpha_1\bar{v}_1^2}{\bar{C}_{10}^2} - 1 - \frac{3\alpha_2\bar{v}_1^2}{\alpha_1\bar{v}_2^2}\right), \quad \Delta_{p1} = -\frac{\alpha_2}{\alpha_{20}\alpha_{10}\alpha_1}\left(\frac{2\bar{f}\bar{v}_1}{\bar{v}_2} - \frac{3\bar{w}}{\bar{a}}\right)\right.$$

$$\left.\Delta_{v1} = -\frac{\alpha_{20}\Delta_{p1}}{\bar{v}_0}, \quad \Delta_{v2} = \frac{2}{\alpha_{10}\bar{v}_0}\left[\left(\frac{\alpha_1\bar{v}_1^2}{\bar{C}_{10}^2} - 1\right)\bar{f} - \frac{9\alpha_2\bar{v}_1\bar{w}}{2\alpha_1\bar{a}\bar{v}_2}\right]\right)$$

where the right-hand sides associated with interphase interactions are determined by the following dimensionless quantities

$$\bar{f} = \left(\frac{\bar{\mu}_{10}K_\mu}{\bar{a}^2} + \frac{3\alpha_1^2}{2\bar{a}}\right)(\bar{v}_1 - \bar{v}_2)$$

$$\bar{q} = \frac{3\gamma_2\lambda_2^*\,\mathrm{Nu}_2}{2\bar{a}^2}(\bar{p}_{20} - \bar{p}_2\bar{a}^3), \quad b_p = \frac{3\gamma_2\bar{p}_2\bar{w}}{\bar{a}} \tag{6.4.9}$$

$$b_{\bar{w}} = \frac{1}{(1-\varphi^{(1)})\,\bar{a}}\left[\left(p_2 - p_1 - \frac{2\bar{\Sigma}_0}{\bar{a}}\right) - (1-\varphi^{(2)})\frac{3\bar{w}^2}{2} - \frac{4\mu_1^*\bar{w}}{\bar{a}}\right.$$

$$\left. + \frac{1-\varphi^{(3)}}{\alpha_{10}\alpha_{20}}\frac{(\bar{v}_1 - \bar{v}_2)^2}{4}\right]$$

Here, it is no use to give consideration to the small difference between the carrying liquid density ρ_1^0 and its original value ρ_{10}^0, i.e., the difference between $\bar{\rho}_1^0$ and 1, since $\bar{f}$ and $\bar{q}$ involve the empirical relations for K_μ and $\mathbf{Nu}_2$. The obtained equations include the dimensionless values which are determined by the final relations

$$\bar{\rho}_1^0 = 1 + \alpha_{10}\alpha_{20}\frac{\bar{p}_1 - 1}{\bar{C}_1^2}, \quad \bar{\rho}_2^0 = \frac{1}{\bar{a}^3}, \quad \alpha_2 = \frac{\alpha_{20}v_{20}}{\bar{\rho}_2^0 v_2}, \quad \alpha_1 = 1 - \alpha_2 \tag{6.4.10}$$

Furthermore, the integrals of the system of equations (6.4.8) should also be kept in mind: the continuous-liquid mass integral and the mixture momentum integral (the latter is written using the gas momentum)

$$\bar{\overset{\circ}{\rho}}_1 \alpha_1 v_1 = \alpha_{10} v_{10} \tag{6.4.11}$$

$$\bar{v}_1 + \tilde{\rho}_{20}\bar{v}_2 + \frac{\alpha_{20}\bar{p}_1}{\bar{v}_{10}} = \bar{v}_{10} + \tilde{\rho}_{20}\bar{v}_{20} + \frac{\alpha_{20}}{\bar{v}_{10}} \qquad \left(\tilde{\rho}_{20} = \frac{m_2}{m_1} = \frac{\alpha_{20}\overset{\circ}{\rho}_{20}}{\alpha_{10}\overset{\circ}{\rho}_{10}} \ll 1\right)$$

The values of dimensionless variables in both initial (o) and final (e) equilibrium states in conformity with (6.4.2) are

$$\bar{v}_{10} = \bar{v}_{20} = \bar{v}_0, \quad \bar{w}_0 = 0, \quad \bar{T}_{10} = \bar{T}_{20} = \bar{p}_{10} = \bar{a}_0 = \bar{\overset{\circ}{\rho}}_{10} = \bar{\overset{\circ}{\rho}}_{20} = 1$$

$$\bar{p}_{20} = 1 + 2\Sigma_0^* \tag{6.4.12}$$

$$\bar{v}_{1e} = \bar{v}_{2e} = \bar{v}_e, \quad \bar{w}_e = 0, \quad \bar{T}_{1e} = \bar{T}_{2e} \approx 1, \quad \bar{p}_{2e} = \bar{p}_{1e} + 2\Sigma_0^*/\bar{a}_e$$

Thus, the required dimensionless solution, which describes the plane stationary waves in a bubbly liquid, is defined by: the wave velocity $\bar{D}_0 = -\bar{v}_0$ relative to the quiescent medium, initial volumetric concentration α_{20} of bubbles, and parameters γ_2, $\tilde{\rho}_{20}$, $\bar{C}_{10}$, λ_2^*, μ_1^*, and Σ_0^*, which characterize the phase properties and bubble size.

An equilibrium shock adiabat of the mixture and the conditions of existence of stationary compression waves. The values of parameters behind the wave in state e are determined from the final relations (6.4.10)–(6.4.12) using the predetermined parameters before the wave α_{20} and $\bar{v}_0$.

In the case of slight capillary effects ($\Sigma_0^* \ll 1$), which always takes place in a case of relatively large bubbles ($a_0 \gtrsim 1$ mm) and relatively high pressure ($p_0 \gtrsim 0.01$ MPa), these relations lead to an expression for the wave velocity $\bar{D}_0 = -\bar{v}_0$ in terms of intensity $\bar{p}_e$

$$\bar{D}_0^2 = \bar{p}_e \frac{1 + \alpha_{10}\alpha_{20}(\bar{p}_e - 1)\bar{C}_{10}^{-2}}{(1 + \tilde{\rho}_{20})(1 + \alpha_{10}(\bar{p}_e - \alpha_{20})\bar{C}_{10}^{-2})} \tag{6.4.13}$$

The minimum value of velocity D_0, at which the stationary compression wave may be realized ($\bar{p}_e > 1$) leads to a condition

$$\bar{D}_0 > \bar{C}_e = [(1 + \alpha_{10}/\bar{C}_{10}^2)(1 + \tilde{\rho}_{20})]^{-1/2}$$

where $\bar{C}_e$ corresponds to the equilibrium speed of sound (see (6.2.12) at $\varkappa = 1$, $\tilde{\rho}_{20} \ll 1$). If

$$\bar{C}_{10}^{-2} \ll 1, \quad \bar{p}_e\bar{C}_{10}^{-2} \ll 1 \qquad \left(\alpha_{20} \gg \alpha_{C0} = \frac{p_0}{\overset{\circ}{\rho}_{10} C_1^2}, \ \bar{p}_e \ll \frac{\alpha_{20}}{\alpha_{C0}}\right) \tag{6.4.14}$$

then, compressibility of the carrying liquid may be ignored (the gas volume α_2 is sufficient that the entire compression of the mixture in the wave under consideration be realized due to the gas compression), and then, for the bubbly mixture (given the smallness of the gas mass concentration ($\tilde{\rho}_{20} \ll 1$), we have

$$p_e = \alpha_{20}\alpha_{10}\overset{\circ}{\rho}_{10}D_0^2, \quad D_0 > C_e = C_\alpha = \left(\frac{p_0}{\overset{\circ}{\rho}_{10}\alpha_{10}\alpha_{20}}\right)^{1/2}$$

$$(\bar{p}_e = \bar{D}_0^2, \quad \bar{D}_0 > 1) \tag{6.4.15}$$

Thus, the compression wave moves relative to the quiescent medium before the wave with a greater velocity compared to the equilibrium speed C_e of sound, which equals the phase velocity of propagation of weak harmonic perturbations $C(\omega)$, whose frequency $\omega \to 0$ (see (6.2.12)). The obtained expression for C_e in a bubbly liquid coincides with the formula (4.2.20) for gas with drops if one takes into consideration the fact that the effective adiabatic index of a mixture liquid-bubbles $\gamma \approx 1$. This resemblance is accounted for by the fact that equilibrium parameters behind the stationary wave are independent of the mixture structure.

From the first two equations (6.4.8), it follows that the integral curve in plane $\bar{v}_1\bar{p}_1$ satisfies the equation

$$\frac{d\bar{p}_1}{d\bar{v}_1} = \frac{\Delta_{p1}}{\Delta_{v1}} = -\frac{\bar{v}_0}{\alpha_{20}} \quad \text{or} \quad \alpha_{20}(\bar{p}_1 - 1) = \bar{v}_0(\bar{v}_0 - \bar{v}_1) \tag{6.4.16}$$

i.e., it is a straight line interconnecting the initial point o and the final point e. The algebraic equation $\Delta(v_1, v_2, \alpha_1) = 0$ determines the points where the gradients of parameters v_1, v_2, and p_1 along the x-coordinate become equal infinity and change their sign, except for the singular point where, in addition, $\Delta_{v1} = \Delta_{v2} = \Delta_{p1} = 0$. The solution of equation $\Delta = 0$ is written in the form

$$\bar{v}_1 = \bar{C}_1(1 + O(\alpha_2)) \tag{6.4.17}$$

that yields a sonic, or characteristic curve, shown by a dashed line mn in Fig. 6.4.1. Intersection of this line with a line of initial states ($\bar{p}_1 = 1$) determines the speed C_f

$$\bar{C}_f = \bar{C}_{10}(1 + O(\alpha_2)) \approx \bar{C}_{10}, \quad \text{or} \quad C_f = \bar{C}_f C_\alpha \approx C_{10}$$

which specifies two different regimes of shock waves.

The first regime is realized when

$$D_0 = |v_0| < C_f \tag{6.4.18}$$

and the integral curve in the plane v_1p_1 does not intersect the sonic line. Therefore, all x-derivatives of the required functions are finite quantities. As is proved by the investigation, there exists, similar to the relaxing mixture of gas and particles (see §4.4), a unique continuous solution of the system (6.4.8), which interconnects the initial o and the final e equilibrium states (see oe in Fig. 6.4.1a and b).

For the second regime when

$$D_0 = |v_0| > C_f \tag{6.4.19}$$

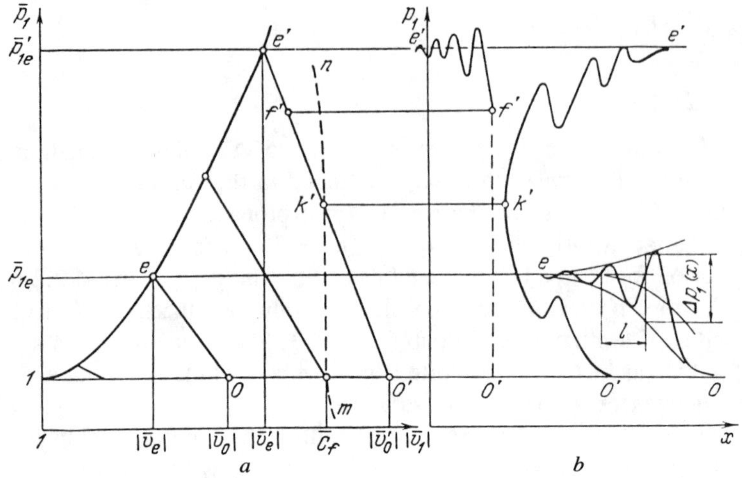

Figure 6.4.1 Integral curves for a structure of a stationary shock wave in a bubbly liquid at various wave velocities.

the integral curve $o'e'$ in the plane v_1p_1 must intersect the sonic line at point k'. If point k' is not a singular point (when $\Delta_{v1} = \Delta_{v2} = \Delta_{p1} = 0$), the pertinent continuous solution has virtually no physical meaning since an "overturned wave" in a plane xp_1 would correspond to this solution (see $o'k'e'$ in Fig. 6.4.1*b*). Indeed, at the point k', the derivative dp/dx varies from $+\infty$ to $-\infty$, and the solution "overturns", or becomes ambiguous (equivocal), i.e., each x conforms to two values p_1. The physically feasible solution (see $o'f'e'$ in Fig. 6.4.1*b*) may be realized by passing over the sonic line by means of a shock of type $o'f'$ (using adequate equations for this shock), after which the process is described by a continuous integral curve $f'e'$ corresponding to the relaxation zone.

The proof of either the existence or the absence of a continuous solution for the wave structure when $D_0 > C_f$ (when the integral curve intersects the sonic line at a singular point where $\Delta_{v1} = \Delta_{v2} = \Delta_{p1} = \Delta = 0$) has to do with the investigation of a system of six independent differential equations. We omit the discussion of this issue, since the case $D_0 > C_f$ at discernible volumetric bubble concentrations $\alpha_2 \geqslant 10^{-2}$ may be realized only in extremely strong shock waves when the bubble fragmentation, phase transitions, and other physiochemical processes must necessarily be given consideration; i.e., the model of the process must be made something more involved. Thus, in water with air bubbles at pressure $p_0 = 0.1$ MPa, and $\alpha_{20} = 0.02$, the regime $D_0 > C_f$ is realized at pressure $p_e > 370$ MPa. In the case of incompressible continuous liquid ($C_f \to \infty$), the stationary shock wave does not exist, and only a stationary wave with a continuous structure may be realized.

Thus, in analogy with relaxing gas and a mixture of gas with drops or particles, the equilibrium speed of sound C_e obtained from the condition of existence of a stationary compression wave coincides with the phase velocity of propagation of weak harmonic perturbations $C(\omega)$ (with the frequency $\omega \to 0$), and the speed of sound C_f obtained from the condition of existence of a stationary wave with a shock coincides with phase velocity of harmonic perturbations $C(\omega)$ with the frequency $\omega \to \infty$, i.e., it resembles the frozen speed of sound.

The investigated stationary solutions with or without a shock are the limiting solutions which are approached by the nonstationary perturbations with a shock if the conditions before (o) and behind (e) the wave are stationary. For instance, during the motion of a piston with a fixed velocity V_0 into a quiescent medium, a shock is generated near the piston at the initial time; both initial amplitude and propagation velocity are independent of the bubble presence and are defined only by the liquid properties. In particular, the shock propagation velocity is equal the speed of sound C_1 in pure liquid. Further, the front-shock diffraction on bubbles and the shock-unloading due to bubble compressibility begin to display themselves. The shock intensity (the shock being the disturbance propagation front) will now be reduced. In this case, the major disturbance must lag behind the shock. With the piston velocity V_0 maintained constant, a stationary wave configuration will asymptotically be established at $t \to \infty$. If $V_0 = |v_0 - v_e| > V^{(f)}$, the front shock is characterized by a limiting nonzero amplitude, which implies a stationary regime $D_0 > C_f$; if $V_0 = |v_0 - v_e| < V^{(f)}$, the shock intensity diminishes to zero, which is consistent with a stationary regime $C_e < D_0 < C_f$. Similar regimes will take place at an instantaneous rise of pressure from p_0 to p_e when it is maintained constant at some location. If $p_e < p^{(f)}$, the limiting wave will be of a continuous structure.

For bubbly mixtures which have the following properties

$$\alpha_C \ll \alpha_{20} \ll 1 \quad (\alpha_C = \bar{C}_{10}^{-2}), \quad \tilde{\rho}_{20} \ll 1, \quad \Sigma_0^* \ll 1 \qquad (6.4.20)$$

(all these properties were already discussed and they led to a shock adiabat in the form (6.4.150)), both the medium pressure and velocity variation in a stationary shock-free wave ($D_0 = C_f$) of maximum intensity are

$$p^{(f)} - p_0 = \overset{\circ}{\rho}_{10}\alpha_{20}^{1/2}C_1^2, \quad V^{(f)} \approx \alpha_{20}^{1/2}C_1 \qquad (6.4.21)$$

It should be kept in mind that, with the conditions (6.4.20) satisfied, an evaluation for fairly weak shock waves ($\bar{p}_e < \bar{C}_{10}^2$) for the upper limit of pressure within the wave analogous to (6.3.12) exists

$$\bar{p} = \bar{p}_e + 1 - (\bar{a}/\bar{a}_e)$$
$$\bar{p} \leqslant \bar{p}_{max} < \bar{p}_e + 1, \quad \text{or} \quad p \leqslant p_{max} < p_e + p_0 \qquad (6.4.22)$$

We shall emphasize that formula (6.4.15) defining the pressure behind the wave in terms of its velocity, the condition (6.4.21) for realization of front shock in the compression wave, and evaluation (6.4.22) for maximum pressure in the wave are valid only for stationary waves. In cases when the stationary configuration of the wave over the experimental segment does not have enough time to be established, these formulas may not hold even if the assumptions (6.4.20) used in derivation of indicated relations are satisfied.

We will further examine a very interesting case of moderate stationary shock-free waves when the wave velocity $D_0 = -v_0$ is higher than the equilibrium velocity C_e but lower than the frozen speed of sound C_f $(1 < \bar{D}_0 < \bar{C}_f)$.

Analysis of a continuous structure of a shock wave. Similar to the investigation of a simpler system of equations (6.3.13), we will now consider the asymptotic behavior of the solution of a system (6.4.8) in the neighborhood of its singular points o and e related, respectively, to the equilibrium states before $(x \to +\infty)$ and behind $(x \to -\infty)$ the shock wave. In the interest of simplifying the analysis, both indicated equilibrium states are denoted by a subscript b, signifying $b = o$ and $b = e$.

The fundamental solutions of the pertinent two linear systems of equations (near state o and state e) are determined, analogous to (6.3.15) and (6.3.22), by characteristic numbers k_b $(b = 0, e)$, which must satisfy the conditions (6.3.16) and (6.3.23). Given (6.4.20), i.e., ignoring the liquid compressibility $(\bar{v}_0^2 = \bar{D}_0^2 \ll \bar{C}_1^2)$, capillary effects $(\Sigma_0^* \ll 1)$, and volumetric and mass concentration of gas $(\alpha_{20} \ll 1,\ \tilde{\rho}_{20} \ll 1)$, the characteristic equations for k_0 and k_e are reduced to

$$K_b^4 + E_b^{(3)} K_b^3 + E_b^{(2)} K_b^2 + E_b^{(1)} K_b + E_b^{(0)} = 0$$

$$K_b = \bar{v}_b \bar{a}_b k_b \qquad (b = 0,\ e)$$

$$E_b^{(3)} = \beta_{2b} \bar{a}_b + 4(1 + \chi)\, \mu_1^* / \bar{a}_b$$

$$E_b^{(2)} = 3\left(\gamma_2 - \bar{p}_b - (\alpha_{2b}/\alpha_{20})\, \bar{v}_b^2\right) + 4(1 + \chi)\, \mu_1^* \beta_{2b} + 16 \mu_1^{*2} \chi \bar{a}_b^{-2} \qquad (6.4.23)$$

$$E_b^{(1)} = 3\beta_{2b} \bar{a}_b \left(\bar{p}_b - \frac{\alpha_{2b}}{\alpha_{20}} + \frac{16\chi\mu_1^{*2}}{3\bar{a}_b^2}\right) + \frac{12\chi\mu_1^*}{\bar{a}_b} \left(\gamma_2 \bar{p}_b - \frac{\alpha_{2b}}{\alpha_{20}} \bar{v}_b^2\right)$$

$$E_b^{(0)} = 12\chi\mu_1^* \beta_{2b} \left(\bar{p}_b - \frac{\alpha_{2b}}{\alpha_{20}} v_b^2\right)$$

$$\mu_1^* = \frac{\mu_1}{\overset{\circ}{\rho}_1 a_0} \sqrt{\frac{\overset{\circ}{\rho}_1}{p_0}}, \quad \beta_{2b} = \frac{3\gamma_2 \lambda_2^* \mathrm{Nu}_{2b}}{2} = \frac{3\gamma_2 \lambda_2 \mathrm{Nu}_{2b}}{2\overset{\circ}{\rho}_{2b} c_2 a_b} \sqrt{\frac{\overset{\circ}{\rho}_1}{p_0}}, \quad \chi = \frac{3K_\mu}{4\pi}$$

Here, it is taken into consideration that effective values of Nusselt numbers $\mathbf{Nu}_{20}$ and $\mathbf{Nu}_{2e}$ may be different, respectively, on the wave front (in the vicinity of state o), and in the wave tail (in the vicinity of state e), and, consequently, the corresponding dimensionless coefficients β_{20} and β_{2e} of heat exchange may also be different. Note that the liquid viscosity manifests itself through the dimensionless coefficient μ_1^*, the two-velocity effects through μ_1^*, and the gas heat conductivity through the dimensionless heat exchange coefficient β_2.

For low-viscosity liquids with fairly large bubbles ($a_0 \sim 1$ mm) at relatively high pressure ($p_0 \gtrsim 1$ MPa), the following evaluations hold good

$$\mu_1^* \ll 1, \quad \lambda_2^* \ll 1, \quad \beta_2 \ll 1 \tag{6.4.24}$$

For smaller gas-containing bubbles with high heat conductivity, and at small pressure the cases $\beta_2 >> 1$ are possible. In very viscous liquids like glycerine, $\mu_1^* >> 1$ may hold.

First, let us analyze the characteristic root for the initial state, i.e., $K_b = K_0$ ($b = 0$). In the limiting case when $\mu_1^* = 0$ and $\beta_{20} = 0$, and expressing K_0 as $K_0(0,0)$, we obtain

$$K_0^4(0,0) + 3(\gamma_2 - \bar{D}_0^2)K_0^2(0,0) = 0$$
$$K_0^{(1,2)}(0,0) = 0, \quad K_0^{(3,4)}(0,0) = \pm\sqrt{3(\bar{D}_0^2 - \gamma_2)}$$

Note that, because of the first condition (6.3.17), only the third root is relevant.

Given the case of small μ_1^* and β_{20}, we shall look for the four characteristic roots in the form of expansion with respect to small parameters

$$K_0 = K_0(0,0) + B\beta_{20} + M\mu_1^* + \ldots$$

Differentiating the characteristic equations with respect to β_{20}, we obtain an equation

$$B\left(4K_0^3 + 3E_0^{(3)}K_0^2 + 2E_0^{(2)}K_0 + E_0^{(1)}\right)$$
$$+ K_0^3\frac{\partial E_0^{(3)}}{\partial\beta_{20}} + K_0^2\frac{\partial E^{(2)}}{\partial\beta_{20}} + K_0\frac{\partial E^{(1)}}{\partial\beta_{20}} + \frac{\partial E^{(0)}}{\partial\beta_{20}} = 0$$

In order to find B, the expressions for the remaining coefficients of $\beta_{20} = \mu_1^* = 0$ should be substituted into this equation, viz.

$$K_0^{(q)} = K_0^{(q)}(0,0) \qquad (q = 1, 2, 3, 4)$$
$$E_0^{(3)} = 0, \quad E_0^{(2)} = 3(\gamma_2 - \bar{D}_0^2), \quad E_0^{(1)} = 0, \quad E_0 = 0$$
$$\frac{\partial E_0^{(3)}}{\partial\beta_{20}} = 1, \quad \frac{\partial E_0^{(2)}}{\partial\beta_{20}} = 0, \quad \frac{\partial E_0^{(1)}}{\partial\beta_{20}} = 3(1 - \bar{D}_0^2), \quad \frac{\partial E_0^{(0)}}{\partial\beta_{20}} = 0$$

In the same manner, differentiating the characteristic equation with respect to μ_1, M is found. As a result, the roots of this equation may be represented in the form

$$K_0^{(1)} = K_0^{(2)} = \frac{\bar{D}_0^2 - 1}{\gamma_2 - \bar{D}_0^2}\beta_{20}$$

$$K_0^{(3,4)} = \pm\sqrt{3(\bar{D}_0^2 - \gamma_2)} + \frac{\gamma_2 - 1}{2(\bar{D}_0^2 - \gamma_2)}\beta_{20} - \mu_1^*$$

Because of condition (6.3.17) stipulating the sign of k_0, it is imperative that $K_0 > 0$. Hence, the stationary waves are possible only in the compression-wave case ($\bar{p}_e = \bar{D}_0^2 > 1$), and, depending on the wave intensity ($\bar{D}_0^2 < \gamma_2$ or $\bar{D}_0^2 > \gamma_2$), only either the first or third root fits. It is evident that the linear expansion with respect to β_{20} is irrelevant in the case of waves with intensity $\bar{p}_e = \bar{D}_0^2 \approx \gamma_2$. For such waves at $\mu_1 = 0$, the characteristic equation is reduced to the form

$$K_0^{(1)} = 0, \quad K_0^3 + \beta_{20}K_0^2 - 3(\gamma_2 - 1)\beta_{20} = 0$$

One of the roots of the latter (we shall refer to it as the second root $K_0^{(2)}$) equals

$$K_0^{(2)} = \sqrt[3]{3(\gamma_2 - 1)\beta_{20}} + O(\beta_{20}^{2/3})$$

and the remaining ($K_0^{(3)}$ and $K_0^{(4)}$) satisfy the equation

$$K_0^2 + (\beta_{20} + \Gamma\beta_{20}^{1/3})K_0 + \Gamma^2\beta_{20}^{2/3} + \Gamma\beta_{20}^{4/3} = O(\beta_{20}^{5/3}) \quad (\Gamma = \sqrt[3]{3(\gamma_2 - 1)})$$

It can be readily proven that the roots $K_0^{(3)}$ and $K_0^{(4)}$ are negative and, consequently, are not suited.

Thus, to develop a solution on the shock wave front, the following values of parameter k_0 at $\beta_{20} \ll 1$ and $\mu_1^* \ll 1$ are suitable

$$1 \leqslant \bar{D}_0^2 = \bar{p}_e < \gamma_2, \quad k_0 = -\frac{\bar{p}_e - 1}{(\gamma_2 - \bar{p}_e)\bar{D}_0}\beta_{20} + \Delta_1$$

$$\bar{D}_0^2 = \bar{p}_e > \gamma_2, \quad k_0 = -\sqrt{\frac{3(\bar{p}_e - \gamma_2)}{\bar{p}_e}}$$

$$-\frac{\gamma_2 - 1}{2(\bar{p}_e - \gamma_2)\bar{D}_0}\beta_{20} + \frac{\mu_1^*}{\bar{D}_0} + \Delta_2 \tag{6.4.25}$$

$$\bar{D}_0^2 = \bar{p}_e \approx \gamma_2, \quad k_0 = -\frac{(3(\gamma_2 - 1)\beta_{20})^{1/3}}{\sqrt{\gamma_2}} + O(\beta_{20}^{2/3})$$

$$\Delta_1, \Delta_2 = O\left(\left(\frac{\beta_{20}}{\bar{p}_e - \gamma_2}\right)^2, \mu_1^{*2}, \beta_{20}\mu_1^*\right)$$

In a similar manner, the roots of the characteristic equation for cases $\beta_{20} \gg 1$ (bubble behavior resembles the isothermal state) and $\mu_1^* \ll 1$ are found in the form of expansion with respect to β_{20}^{-1} and μ_1^*. In this case also, only one root is suitable

$$k_0 = -\sqrt{\frac{3(\bar{p}_e - 1)}{\bar{p}_e} + \frac{3(\gamma_2 - 1)}{2\bar{D}_0\beta_{20}} + \frac{\mu_1^*}{\bar{D}_0}} + \Delta'$$
$$\Delta' = O\left(\beta_{20}^{-2}, \mu_1^{*2}, \mu_1^*/\beta_{20}\right) \qquad (6.4.26)$$

For any compression wave there exists only one root k_0, which corresponds to perturbation attenuation before the wave; i.e., this root is real and negative, indicating the absence of oscillations near the initial state o before the wave. With the gas adiabatic behavior ($\beta_{20} = 0$), a suitable root k_0 exists only at $\bar{D}_0^2 = p_e > \gamma_2$, and in the presence of heat exchange ($\beta_{20} > 0$), it exists at all $\bar{D}_0^2 = \bar{p}_e > 1$ complying with compression waves. Stationary rarefaction waves are not possible. The dependence of the indicated root K_0 on both wave intensity $\bar{p}_e = \bar{D}_0^2$ and heat-exchange coefficient β_{20} is illustrated in Fig. 6.4.2. With both β_{20} and $\bar{p}_e$ increases, the value of k_0 increases, which leads to the steaper wave-front. With sufficiently small β_{20}, when gas resembles adiabatic behavior, the most intensive growth of both k_0 and front steepness with increase of $\bar{p}_e$ takes place at $\bar{p}_e \approx \gamma_2$. Considering that disposition to an oscillatory structure is stronger the steeper the wave front, the value γ_2 for qualitative evaluations may be viewed as an approximate boundary dividing stationary waves with relatively gentle-slope front, behind which the oscillations are not realized ($\bar{p}_e < \gamma_2$) from stationary waves with a steeper front, behind which the oscillations are well-defined ($p_e > \gamma_2$).

The bubbles together with the liquid move across the wave front almost uniformly ($x - x_e \approx v_0 t$), and in accordance with (6.3.15), the exponential compression of bubbles is realized

$$a - a_0 = A^{(a)} \exp k_0(\bar{x} - \bar{x}_b) = A^{(a)} \exp(-\varepsilon t)$$
$$\varepsilon = \frac{k_0 D_0}{L_\alpha} = -\frac{k_0 \bar{v}_0 C_*}{a_0} \qquad (6.4.27)$$

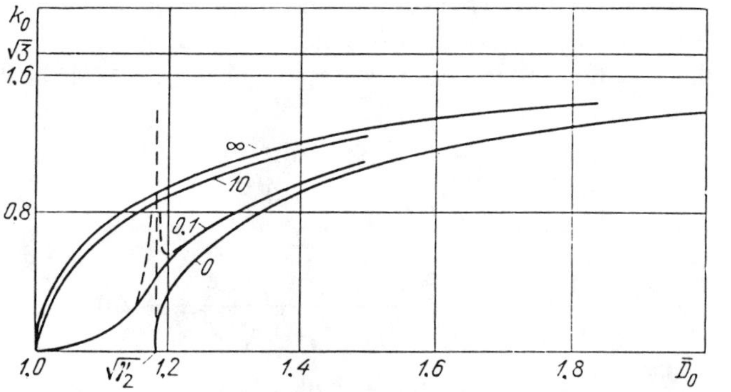

Figure 6.4.2 Dependence of the characteristic root k_0 (defining the medium behavior before the wave) on both wave velocity $\bar{D}_0 = \sqrt{\bar{p}_e}$ and parameter $\beta_{20}^{(T)}$ of heat exchange within a bubble (labels on curves indicate magnitudes of this parameter). The gas adiabatic index $\gamma_2 = 1.4$; ($\beta_{20}^{(T)} = 0; 0.1; 10; \infty$).

In this case (see the discussion of (1.6.40)), the heat exchange is realized with a fixed Nusselt number $\mathbf{Nu}_2$, which may be determined from formulas (1.6.43) and (1.6.41). In the case of sufficiently large numbers $\mathbf{Pe}_2$, when there is a thin (compared to a_0) thermal boundary layer in gas, we obtain (using the first formula (1.6.43)) the expression defining heat exchange in terms of parameter k_0, which characterizes both the intensity of the bubble compression across the wave front and the steepness of this front

$$\mathrm{Nu}_{20} = 2\sqrt{k_0\bar{v}_0/\lambda_2^*}, \quad \beta_{20} = 3\gamma_2\sqrt{\lambda_2^* k_0 \bar{v}_0} \tag{6.4.28}$$

The relationship between β_{20} and k_0 and the characteristic equation for k_0 enable us to uniquely determine k_0, $\mathbf{Nu}_{20}$, and β_{20} as functions of the wave intensity and the medium properties. We will illustrate it using the most interesting case when $\beta_{20} \ll 1$ (the bubble compression is nearly adiabatic) and when the solutions in the form of (6.4.27) hold good.

In the weak wave case ($\bar{p}_e < \gamma_2 - \delta$, $\delta \approx 0.05$, but $\mathbf{Nu}_{20} \gg 1$, $\beta_{20} \ll 1$). By substituting (6.4.28) into the first formula (6.4.25), we obtain

$$\sqrt{K_0} = 3\gamma_2\sqrt{\lambda_2^*}\,\frac{\bar{p}_e - 1}{\gamma_2 - \bar{p}_e}, \quad \mathrm{Nu}_{20} = 6\gamma_2\,\frac{\bar{p}_e - 1}{\gamma_2 - \bar{p}_e}$$

$$\beta_{20} = 9\gamma_2\lambda_2^*\,\frac{\bar{p}_e - 1}{\gamma_2 - \bar{p}_e} \tag{6.4.29}$$

We shall place emphasis on the fact that these formulas are valid only at sufficiently high $\mathbf{Pe}_2 > 100$ when $\mathbf{Nu}_{20} > 10$, making formula (6.4.28) acceptable. At very small wave intensities $\bar{p}_e$, this condition is violated. Then, in conformity with (1.6.43), the Nusselt number $\mathbf{Nu}_{20} \approx 10$ should be taken for evaluations instead of (6.4.28).

In a similar manner, the characteristic equation (6.4.23) may be investigated for the number k_e ($b = e$), which defines the behavior of the medium in the wave tail. Using the described small-parameter method, an expression for the roots of this equation in the limiting case of a weak ($\beta_{20} \ll 1$) heat exchange can be obtained

$$k_e^{(1)} = \frac{4\chi\mu_1^*}{\bar{a}_e^2\sqrt{\bar{p}_e}} + \Delta^{(1)}, \quad k_e^{(2)} = \frac{(\bar{p}_e - 1)\,\beta_{2e}}{(\gamma_2\bar{p}_e - 1)\sqrt{\bar{p}_e}} + \Delta^{(2)}$$

$$k_e^{(3,4)} = \pm i\,\frac{\sqrt{3(\gamma_2\bar{p}_e - 1)}}{a_e\sqrt{\bar{p}_e}} + \frac{(\gamma_2 - 1)\,\beta_{2e}}{2(\gamma_2\bar{p}_e - 1)\sqrt{\bar{p}_e}} + \frac{2\mu_1^*}{\bar{a}_e^2\sqrt{\bar{p}_e}} + \Delta^{(3)} \tag{6.4.30}$$

$$\left(\pm i = \sqrt{-1},\ \Delta^{(1)},\ \Delta^{(2)},\ \Delta^{(3)} = O\left(\beta_{2e}^2,\ \mu_1^{*2},\ \beta_{2e}\mu_1^*\right)\right)$$

and in the limit case of a strong heat exchange

$$k_e^{(1)} = 4\chi\mu_1^*/\left(\bar{a}_e^2\sqrt{\bar{p}_e}\right) + \Delta^{(1)}, \quad k_e^{(2)} = \beta_{2e}/\sqrt{\bar{p}_e} + O(1)$$

$$k_e^{(3,4)} = \pm i \frac{\sqrt{3(\bar{p}_e - 1)}}{\bar{a}_e \sqrt{\bar{p}_e}} + \frac{3(\gamma_2 - 1)}{2\bar{a}_e^2 \sqrt{\bar{p}_e}\,\beta_{2e}} - \frac{2\mu_1^*}{\bar{a}_e^2 \sqrt{\bar{p}_e}} + \Delta^{(3)} \qquad (6.4.31)$$

$$\Delta^{(1)},\ \Delta^{(3)} = O(\beta_{2e}^{-2},\ \mu_1^{*2},\ \beta_{2e}\mu_1^*)$$

All four roots k_0 have a positive real part, i.e., they satisfy the condition for the perturbation decay behind the wave (at $x \to -\infty$). Thus, the solution near the ultimate equilibrium state e is generally a superposition of four fundamental solutions describing a monotonic exponential approximation to the state e (roots $k_e^{(1)}$ and $k_e^{(2)}$), and the exponentially decaying harmonic oscillations (complex conjugate roots $k_e^{(3)}$ and $k_e^{(4)}$). Figure 6.4.3 depicts the dependence of the real parts of roots k_e on the value β_{2e}. The attention should be focused on the nonmonotonic nature of dependence of real parts of complex roots $k_e^{(3)}$ and $k_e^{(4)}$ on β_{2e}; these real values define the damping decrement of pulsations.

Note that, at $\beta_2 = 0$, the solutions under consideration are in agreement with solutions based on the effective-viscosity scheme (see (6.3.19) and (6.3.26) if $\varkappa = \gamma_2$ is assumed. The same agreement takes place also at $\beta_2 \to \infty$, if $\varkappa = 1$ in the effective-viscosity scheme.

The results of the analysis (R. I. Nigmatulin and V. Sh. Shagapov, 1974) based on the above-discussed two-temperature scheme for two statonary shock waves in an aqueous 1:1 solution of glycerine containing air bubbles (see the thermophysical parameters of phases in Appendix) are outlined below.

The friction coefficient K_μ between phases has been predetermined in accordance with (1.5.8), and the heat exchange coefficient was defined by a fixed Nusselt number $\mathbf{Nu}_2 = 30$, which is consistent with the evaluation (1.6.18).

Figure 6.4.4 depicts an example of a shock wave with a pulsating structure: Fig. 6.4.5 depicts an example of a shock wave with a monotonic structure. The wave-intensity reduction p_e, the decrease of both bubble size a_0 and the volumetric concentration α_{20} of bubbles, the increase of viscosity μ_1, and the reduction of the heat exchange parameter $\mathbf{Nu}_{20}$, or β_{20}, over the wave initial segment enhance the tendency towards the monotonic structure.

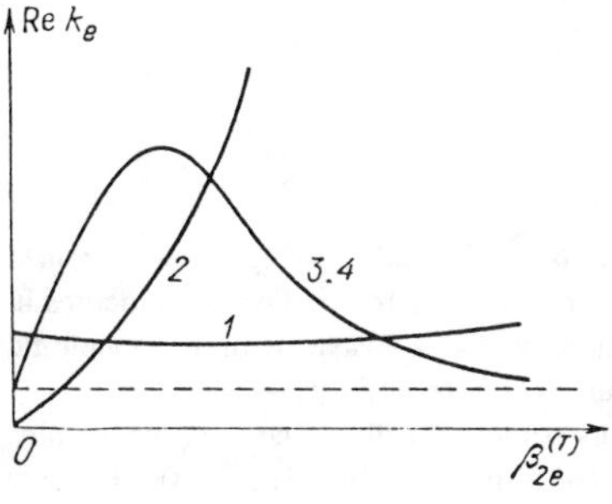

Figure 6.4.3 Dependence of real parts of roots $k_e^{(1)}$, $k_e^{(2)}$ and $k_e^{(3,4)}$, defining the intensity of the parameters approach to the ultimate state behind the wave, on the value $\beta_{2e}^{(T)}$, which characterizes the gas heat exchange within a bubble. Labels on curves correspond to the root number.

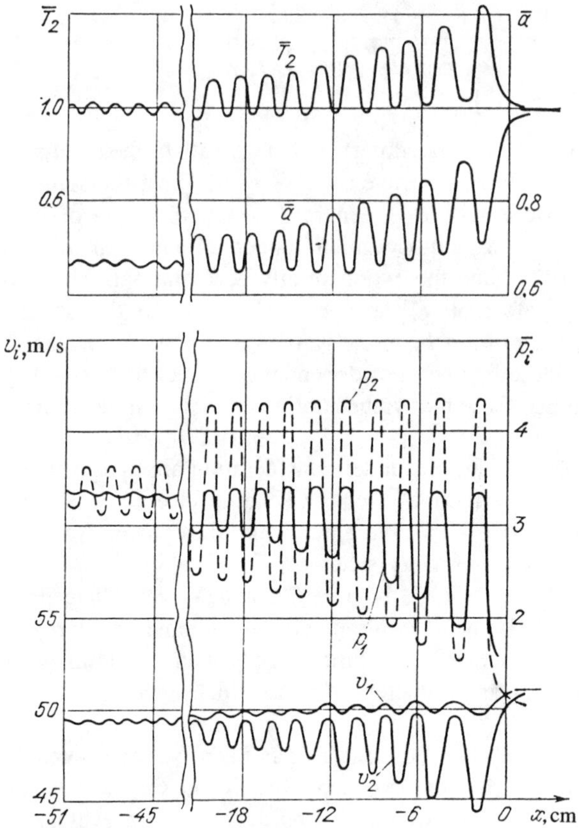

Figure 6.4.4 Variation of gas temperature, bubble radius, pressure, and phase velocities in a stationary shock wave with an oscillatory structure (D_0 = 51 m/s, $\bar{p}_e$ = 3.3) in an 1:1 aqueous solution of glycerine containing air bubbles (a_0 = 1.5 mm, α_{20} = 0.042, p_0 = 0.036 MPa).

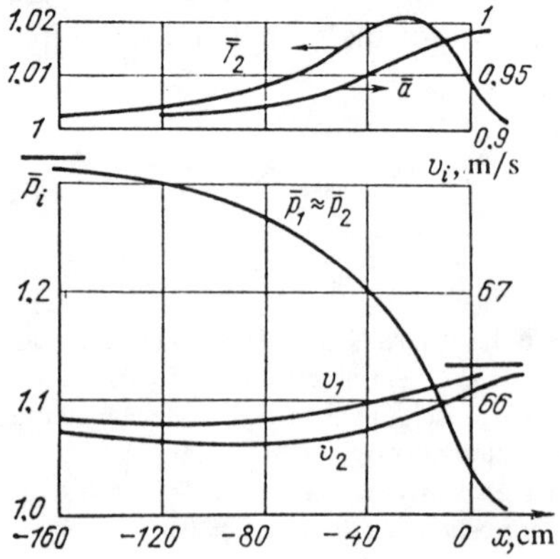

Figure 6.4.5 Variation of gas temperature, bubble radius, pressure, and phase velocities in a stationary shock wave with a monotonic structure (D_0 = 66 m/s, $\bar{p}_e$ = 1.32) in an 1:1 aqueous solution of glycerine with air bubbles (a_0 = 1.4 mm, α_{20} = 0.025, p_0 = 0.09 MPa).

The results of these analyses were compared with data obtained in experiments on vertical shock tubes by L. Noordzij (1971, 1973), V. E. Nakoryakov (1983), and other authors (1983) (see §6.1). In these experiments, the pressure variation $p_1(t)$ was measured at a fixed section x_0 of the tube. If a wave is stationary, then all parameters, including pressure in liquid, are functions of only one argument

$$p_1 = p_1(\xi), \qquad \xi = x - D_0 t$$

the experimental oscillogram $p_1(t)$ may be easily transformed into $p_1(x)$. Accordingly, we shall analyze the calculated curves $p_1(x)$ in more detail. The nature of these curves is depicted in Fig. 6.4.1*b* by the line *oe*. In order to compare different variants, it is convenient to characterize the oscillatory relationship $p_1(x)$ by three functions: $p_m(x)$ is the averaged pressure in liquid, $\Delta p_1(x)$ is the pulsation amplitude, and $L_e(x)$ is the period, or the pulsating wave length. In this case, the oscillation decay has an exponential pattern $\Delta p_1 \approx A^{p_1} \exp(-x/d_e)$, where d_e is the inherent length of the pulsation decay. In the case of a monotonic structure ($\Delta p_1 = 0$), L_e and d_e are of no use, and $p_1(x) = p_m(x)$. From Eqs. (6.4.30) and (6.4.31), expressions may be obtained for limiting values $L_e = 2\pi/\mathrm{Im}\{k\}$ and $d_e = 1/\mathrm{Re}\{k\}$ behind the wave (at $x \to -\infty$). In particular, for the case $\beta_{2e} \ll 1$, we have

$$L_e = \frac{2\pi \bar{v}_e a_e}{\sqrt{3(\gamma_2 \bar{p}_e - 1)}}, \qquad d_e = 2\left[\frac{(\gamma_2 - 1)\beta_{2e}}{2(\gamma_2 \bar{p}_e - 1)} + \frac{2\mu_1^*}{\bar{a}_e^2}\right]^{-1} \bar{v}_e a_e \tag{6.4.32}$$

The values of parameters β_{2e} and μ_1 which appear in these formulas are determined by expressions (6.4.23); note that, $\bar{v}_e \approx v_0 = \sqrt{\bar{p}_e}$, $a_e \approx a_0$. Figures 6.4.6 and 6.4.7 represent the variation of the values $L(x)$ and $\Delta p_1(x)$ for waves under consideration at various parameters K_μ and $\mathbf{Nu}_2$, which define friction and heat exchange between phases. In this case, to find out the impact of two-velocity effects for the value K_μ, two magnitudes were used: formula (1.5.8) and $K_\mu \to \infty$, which is consistent with a single-velocity scheme with equal phase velocities ($v_1 = v_2$).

To assess the impact of the temperature effects upon the parameter $\mathbf{Nu}_2$, the following values were used: $\mathbf{Nu}_2 = 0, 30, 300, 3000$, and ∞. Note that $\mathbf{Nu}_2 \to \infty$ corresponds to the isothermic gas behavior ($T_1 = T_2 = T_0$), and $\mathbf{Nu}_2 = 0$ to the adiabatic regime.

Effects of the interphase heat exchange. The results of analysis indicate the significant effects of interphase heat exchange and respective parameter $\mathbf{Nu}_2$ (or β_2) on the shock wave structure. This can be seen from a comparison of curves related to various $\mathbf{Nu}_2$ in Figs. 6.4.5 and 6.4.7. For a weak wave (see Figs. 6.4.5 and 6.4.7 for a wave with $\bar{p}_e = 1.3 < \gamma_2 = 1.4$) at sufficiently great values $\mathbf{Nu}_2 \gtrsim 1000$, an oscillatory structure is obtained, and at smaller values $\mathbf{Nu}_2 < 100$, a monotonic structure is obtained when the

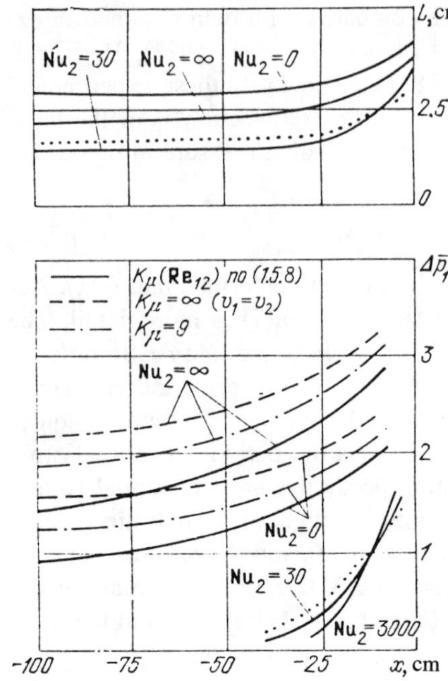

Figure 6.4.6 Variation of the oscillatory wave lengths L and their amplitudes Δp_1 in a stationary shock wave with an intensity $\bar{p}_e = 3.3$ (for the rest of the parameters, see Fig. 6.4.4) at various values of coefficients of interphase friction K_μ and heat exchange $\mathbf{Nu}_2$. The dotted lines without labels correspond to the analysis based on a nonstationary spherically-symmetrical temperature distribution within a bubble (see §6.5).

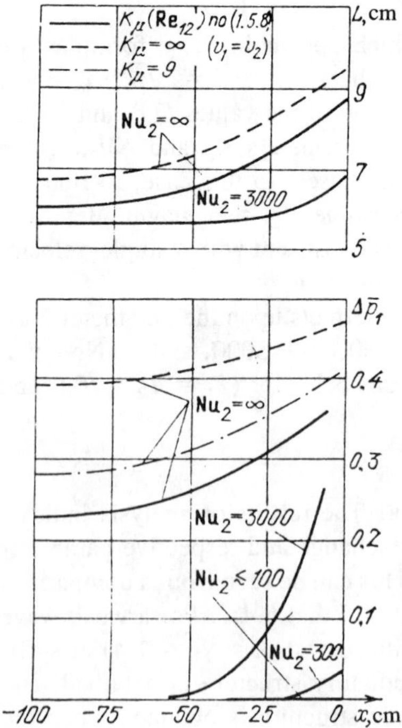

Figure 6.4.7 Variation of the oscillatory wave lengths L and their amplitude Δp_1 in a stationary shock wave with an intensity $\bar{p}_e = 1.32$ (for the rest of the parameters, see Fig. 6.4.5) at various values of coefficients of interphase friction K_μ and heat exchange $\mathbf{Nu}$ (notations are the same as in Fig. 6.4.6).

phase pressures, velocities, and temperatures are almost the same. This is accounted for by a fact that, at indicated small $\bar{p}_e < \gamma_2$ and $\mathbf{Nu}_2$ (or β_{20}), small values of k_0 are obtained which provide the slow variation of parameters over the initial section of the wave (near the state o, $x \to +\infty$), and the phase pressures p_1 and p_2 have enough time to equalize between each other. It is sufficient to increase the Nusselt number $\mathbf{Nu}_2$ in calculations to 100 and more merely over the initial section of the wave (where a has enough time to reduce only by 3–5%), as the predicted wave structure becomes of a pulsating nature.

The solution of a heat-conduction problem for a pulsating bubble in a liquid, and the qualitative analysis of the circulating-motion effects within a bubble showed that it is the Nusselt number values $\mathbf{Nu}_2 \sim 10$–100 ($\beta_2 \sim 10$–1) that apply at $p \sim 0.1$ MPa for bubbles with $a \sim 1$ mm both in the process of their accelerating compression across the wave front and at decaying pulsations when the heat exchange is qualitatively described by formulas (1.6.18) or (1.6.16), and governed quantitively by formulas (1.6.45) and (1.6.46). The gas behavior, in this case, is rather more adiabatic than isothermic.

Thus, the sufficiently weak stationary wave ($\bar{p}_e < \gamma_2$) (e.g., a wave depicted in Fig. 6.4.5) should not have pulsations. In a nonstationary regime in a wave of the same intensity, bubble pulsations gradually decaying with the wave transition to a stationary mode (see §6.7 below) may be realized.

The analysis of stationary waves in bubbly mixtures reported by L. Noordzij and L. Wijngaarden (1974) does not consider thermal dissipation, and therefore, concludes that stationary waves of low intensity $\bar{p}_e > 1 + 2\alpha_{20}$ must have an oscillatory structure. Waves of such intensity (in particular, a wave whose predicted stationary structure is depicted in Fig. 6.4.5), which were observed in experiments conducted by L. Noordzij (1971, 1973), were indeed of an oscillatory structure, but this is accounted for only by their nonstatonary nature. To observe stationary waves in those experiments was impossible because of the insufficient length of the shock tube. The pressure-sensing devices in these experiments have been installed at a distance of 1–1.5 meters from the liquid surface where the shock wave generated. At the same time, the thicknesses of the oscillatory relaxation zones of studied shock waves also range within 1–1.5 meters. Therefore, waves over these distances do not obey the stationary analysis, since they do not have sufficient time to turn into a stationary regime when a wave moves with no change of its structure. The same comment applies to experimental data reported by other authors for mixtures with the bubble siz $a \sim 1$ mm. An adequate analysis may only be accomplished based on the nonstationary theory, which is given consideration in §§6.6 and 6.7 below. The higher the wave intensity, the more distinctively the oscillations manifest themselves. This fact is illustrated in Fig. 6.4.8, where structures of waves of various intensities are plotted at fixed reference parameters of the mixture,

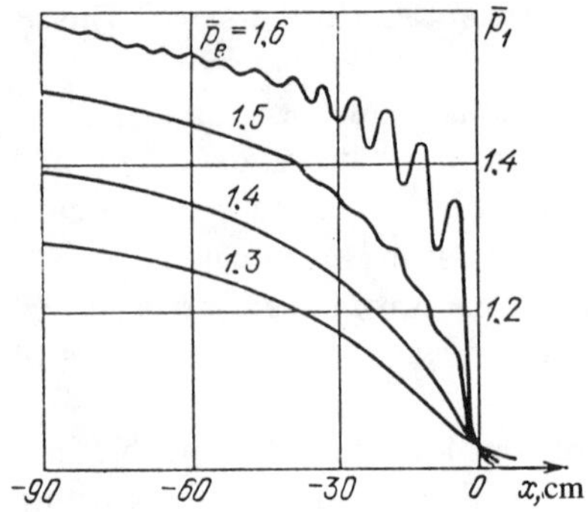

Figure 6.4.8 The effect of intensity $\bar{p}_e$ on the stationary wave structure in water with air bubbles ($a_0 = 1.4$ mm, $\alpha_{20} = 0.025$, $p_0 = 0.09$ MPa).

which are the same as in Fig. 6.4.5. The nonmonotonic dependence of the structure, in particular, on parameters such as wavelength L and the inherent length d of pulsation decay on the parameter $\mathbf{Nu}_2$ defining the heat exchange in the wave oscillation zone should be pointed out. This can be seen from Figs. 6.4.6 and 6.4.7; moreover, it follows from the nonmonotonic relationship between the pulsation damping decrement in the linear solution (see Fig. 1.6.2). In two limiting cases of adiabatic ($\mathbf{Nu}_2 = 0$) and isothermal ($\mathbf{Nu}_2 \to \infty$) gas behavior within a bubble, and also for any polytropic gas behavior with a fixed polytropic index $\varkappa$, dissipation of kinetic energy is due only to the liquid viscosity. At finite values of $\mathbf{Nu}_2$, dissipation also stems from the irreversibility of interphase heat exchange when the kinetic energy of liquid transforms into the gas thermal energy which irreversibly dissipates in liquid. At some value $\mathbf{Nu}_2$, this dissipation reaches its maximum. As a consequence, the shock wave structure and its characteristics (such as L and d) do not fall between the respective values for adiabatic and isothermic regimes. The interphase heat exchange and variation of the related parameter $\mathbf{Nu}_2$ exert a strong impact upon the display of oscillations and their amplitude and have a relatively weak effect on the oscillatory wavelength L. The latter was measured in experiments conducted by L. Noordzij (1971, 1973), as was mentioned above, for nonstationary waves. In particular, for waves whose predicted stationary structures are depicted in Figs. 6.4.4 and 6.4.5, the measurements yield, respectively, $L = 2.6$ cm and $L = 6.4$ cm, which is in fairly good agreement with the results of the analysis presented based on the stationary theory (see Figs. 6.4.6 and 6.4.7), and which is indicative of only a slight alteration of the oscillatory wave lengths in the process of their transition to a stationary regime.

Effects of the phase relative motion. The relative longitudinal motion (slip) of phases (two-velocity effects) is of significance only when the thermal dissipation is not taken into account, i.e., the gas polytropic behavior, e.g., isothermic or adiabatic, is anticipated. This is evident from comparison of dashed ($K_\mu \to \infty$) and solid ($K_\mu = K_\mu(\mathbf{Re}_{12})$) curves depicted in Figs. 6.4.6 and 6.4.7 calculated using formula (1.5.8) for the cases $\mathbf{Nu}_2 = 0$ and $\mathbf{Nu}_2 \to \infty$. With consideration given to an actual thermal nonequilibrium ($10 < \mathbf{Nu}_2 < 1000$), the role of two-velocity effects on the background of much stronger thermal dissipation becomes imperceptible, and variation of the parameter K_μ from 0 to ∞ has practically no effect upon the wave structure and, in particular, upon the oscillation amplitude Δp_1 and the oscillatory wave lengths L. Note that the relative translational phase motion in a wave with oscillatory structure is mainly defined by the inertia effects, since the interphase forces affected by the associated mass are much greater than forces due to viscosity ($f_m \gg f_\mu$). The effects of friction force f_μ lead to the phase velocity flattening and, also, to an additional thermal dissipation unobtrusive on the background of thermal dissipation; it gradually exhibits itself after a great number of pulsations. Therefore, variation of f_μ stemming from variation of K_μ exerts almost no effect on the wave behavior.

It should be understood that, in accordance with (6.4.32), with the wave intensity p_e increasing, the oscillatory wave length L reduces and may become equal by an order of magnitude smaller than the distance between bubbles. This may result in violation of one of the most important assumptions (see Introduction, assumption 2); i.e., significant improvement of the model for analysis of stronger waves may be required by giving special consideration to the bubble nonsphericity and fragmentation as their extreme manifestation.

Effects of longitudinal flow over a bubble on heat exchange. The liquid flow over a bubble with a velocity $w_{12} = \mathrm{v}_1 - \mathrm{v}_2$, accounting for a circulatory motion within a bubble, distorts the spherical symmetry of the pressure distribution in it, and, because of this circulation, intensifies the interphase heat exchange. This effect is characterized by the Peclet number $\mathbf{Pe}_2^{(\mathrm{V})} = 2aw_{12}/\nu_2^{(T)}$, which, for the wave depicted in Fig. 6.4.4, does not exceed 100, and for the wave in Fig. 6.4.5, is not higher than 10. The evaluations based on the formulas outlined in §2.3 indicate that the heat exchange parameter $\mathbf{Nu}_2$ affected by the flow-over increases by a value of order $(\mathbf{Pe}_2^{(\mathrm{V})})^{1/2}$. Such an increase of $\mathbf{Nu}_2$ cannot significantly affect the analysis, since in the variants discussed above the inherent values $\mathbf{Nu}_2$ were approximately equal 30. The fact that the bubble displacement relative to liquid per unit period of pulsations (which can be readily determined based on the predicted profiles of v_1 and v_2 across the wave (see Fig. 6.4.4)) is considerably smaller compared to the bubble radius is additional evidence that translational motion of a bubble relative to liquid is fairly low and the distortion of spherical symmetry of the temperature distribution in a bubble is also small. Moreover, in insufficiently purified liquids, surface-active agents may also be present which, when accumulated on the bubble surface, hinder the development of circulation within them. Nevertheless, it should be kept in mind that in stronger waves, the interphase heat exchange may be more intensive than is predicted by a spherically-symmetrical bubble scheme, because of the increase of interface due to the bubble nonsphericity, the development of an interior circulation, and their fragmentation.

Effect of polydispersity. In all the experiments discussed above the bubble radius was determined to an accuracy of +(10–20)%, and, in addition, there was always a set of fractions of various size in a mixture. In this regard, it is of considerable interest to identify the effect of polydispersity on both wave propagation and its structure. The pertinent research was undertaken by V. Sh. Shagapov (1976) in the framework of a model with a limited number of fractions (1.8.6). In particular, there were predicted structures of stationary shock waves with the same parameters as in Figs. 6.4.4 and 6.4.5, but in mixtures with three bubble sizes when the gas phase in a reference equilibrium state was divided into three equal by mass and volume fractions with noticeably different bubble radii

$$\alpha_{20} = \alpha_{30} = \alpha_{40} = {}^1/_3\alpha_{g0},$$
$$a_{20} = 1\,\text{mm}, \quad a_{30} = 1.5\,\text{mm}, \quad a_{40} = 2\,\text{mm}, \tag{6.4.33}$$

where α_{g0} is the volumetric concentration of all bubbles (of gas). In this case, both the total gas-content α_{g0} and the mass-average (with respect to all fractions) bubble radius a_0 were maintained the same as in already described variants of analysis intended for a monodisperse mixture.

The analyses recognized more unpredictible gas-pressure oscillations in various fractions. The "fine" bubbles ($i = 2$) are responding better to pressure in liquid, and those of a "coarse" size "sway" in a stronger manner than bubbles of medium ($i = 3$) size, or the bubbles in an appropriate monodisperse mixture. Once the pressure-diagrams in mixture or liquid ($p \approx p_1$) are compared across the wave (it is pressure p that is to be measured in experiments), it becomes clear that, from the standpoint of comparison with the experiment, polydispersity of type (6.4.33) (when various bubbles differ in size, approximately, by a factor of 2–3) has little effect on these diagrams. This implies that waves in such polydisperse mixtures may be described in the same manner as those of a monodisperse medium model.

§6.5 USING THE NONSTATIONARY TEMPERATURE DISTRIBUTION WITHIN BUBBLES FOR ANALYSIS OF STATIONARY SHOCK WAVES

For a more consistent account of the nonstationary heat-exchange effects within a deformable gas bubble in a shock wave, and for verification of the two-temperature model, we shall now examine the heat-exchange model for a bubbly mixture, which makes use of the spherically-symmetrical distribution of both temperature T_2' and density $\rho_2^{0\prime}$ of gas within bubbles (see §1.6). In conformity with a stationary wave, both T_2' and $\rho_2^{0\prime}$ depend on the longitudinal coordinate x, which defines the bubble-center position. The stationarity condition requires that at a fixed point (x, r), all parameters, including the microparameters T_2' and $\rho_2^{0\prime}$, are time-independent. However, for each bubble the process is said to be nonstationary.

Analogous to §2.5, the distribution of both T_2' and $\rho_2^{0\prime}$ within a trial bubble is described by means of the Lagrangian coordinate ξ which defines the distance from the gas microparticle to the bubble center in an undisturbed equilibrium state o when both temperature T_2' and density $\rho_2^{0\prime}$ were uniform and were equal, respectively, to T_0 and ρ_{20}^0. The heat-conduction and the mass-conservation equations (1.6.6) for spherically-symmetrical processes within a trial bubble are reduced to Eq. (2.5.1), the only difference being that instead of

$$T_2(t, \xi), \quad \rho_2^{\circ}(t, \xi), \quad r(t, \xi), \quad p_2(t), \quad (\partial/\partial t)_\xi \tag{6.5.1}$$

the following must, respectively, be kept in mind

$$T_2'(x, \xi), \quad \rho_2^{\circ\prime}(x, \xi), \quad r(x, \xi), \quad p_2(x), \quad v_2(\partial/\partial x) \tag{6.5.1a}$$

Here, $r(x, \xi)$ determines the Eulerian radial coordinate, or the position of the gas microparticle (in a bubble with a center having the coordinate x), which in its undisturbed state o was located at distance ξ from the center of the trial bubble. In the undisturbed state o, we have $r = \xi$.

Later on, we will use—together with dimensionless variables (6.3.6) and (6.4.5)—additional dimensionless variables

$$\bar{\xi} = \frac{\xi}{a_0}, \quad \bar{r} = \frac{r}{a_0}, \quad \bar{T}_2' = \frac{T_2}{T_0}, \quad \bar{\rho}_2^{\circ\prime} = \frac{\rho_2^{\circ\prime}}{\rho_{20}^{\circ}} \tag{6.5.2}$$

Let us now write the complete system of equations in partial derivatives, which governs the one-dimensional stationary motion of a mixture of liquid with gas bubbles (with no phase transitions), when the two-velocity effects ($v_1 = v_2 = v$), liquid compressibility and viscosity (ρ_1^0 = constant, $\mu_1 = 0$), surface tension ($\Sigma = 0$), and mass concentration of gas in comparison with mass concentration of liquid ($\rho_2 \ll \rho_1$) may all be ignored. As a result, instead of systems of equations (6.3.2) and (6.4.1), we obtain a more complicated system in which the equations for r and T_2' are equations in partial derivatives

$$\bar{v}\frac{d\bar{a}}{d\bar{x}} = \bar{w}, \quad \bar{a}\bar{v}\frac{d\bar{w}}{d\bar{x}} = \bar{p}_2 - \bar{p}_1 - \frac{3\bar{w}^2}{2}$$

$$\bar{v}\frac{d\bar{p}_2}{d\bar{x}} = \bar{q} - b_p \qquad \left(\bar{q} = 3\gamma_2\bar{\lambda}_2^*\bar{a}\left(\frac{\partial \bar{T}_2'}{\partial \bar{\xi}}\right)_{\bar{\xi}=1}, \quad b_p = \frac{3\gamma_2\bar{p}_2\bar{w}}{\bar{a}}\right)$$

$$\bar{v}\frac{\partial \overline{T}_2'}{\partial \bar{x}} = \frac{\lambda_2^*}{\bar{\xi}^2}\frac{\partial}{\partial \bar{\xi}}\left(\bar{\overset{\circ}{\rho}}{}_2' \frac{\bar{r}^4}{\bar{\xi}^2}\frac{\partial T_2'}{\partial \bar{\xi}}\right) + \frac{\gamma_2 - 1}{\gamma_2}\frac{\bar{v}}{\bar{\overset{\circ}{\rho}}{}_2'}\frac{d\bar{p}_2}{d\bar{x}}$$

$$\frac{\partial \bar{r}}{\partial \bar{\xi}} = \frac{\bar{\xi}^2}{\bar{\overset{\circ}{\rho}}{}_2' \bar{r}^2}, \quad \bar{\overset{\circ}{\rho}}{}_2' = \frac{\bar{p}_2}{\bar{T}_2'} \qquad (0 \leqslant \bar{\xi} \leqslant 1) \tag{6.5.3}$$

$$\bar{p} = 1 + \bar{v}_0^2(1 - \bar{a}^3), \quad \alpha_2 = \frac{\alpha_{20}\bar{a}^3}{\alpha_{10} + \alpha_{20}\bar{a}^3}, \qquad \bar{v} = \frac{\alpha_{10}\bar{v}_0}{1 - \alpha_2}$$

$$\left(\lambda_2^* = \frac{\lambda_2}{\overset{\circ}{\rho}_{20} c_2 a_0}\sqrt{\frac{\overset{\circ}{\rho}_1}{p_0}}\right)$$

The boundary conditions in the center and on the bubble wall are

$$\xi = 0\colon\ \partial T_2'/\partial \bar{\xi} = 0, \quad r = 0; \quad \bar{\xi} = 1\colon\ T_2' = 1 \tag{6.5.4}$$

The latter condition implies the temperature constancy on the interface between gas and liquid (see discussion of (1.6.15)).

As far as stationary shock-wave structure is concerned, we have the equilibrium conditions (6.4.12) upstream and downstream the wave, which should be supplemented by

$$\bar{x} \to \infty\colon\ \overline{T}_2' = 1, \quad \bar{r} = \bar{\xi}; \qquad \bar{x} \to -\infty\colon\ \overline{T}_2' = 1, \quad \bar{r} = \bar{a}_e\bar{\xi} \tag{6.5.5}$$

The solution of this problem in the represented dimensionless form is defined by the gas adiabatic index γ_2, dimensionless heat conductivity λ_2^*, initial volumetric concentration of gas (void fraction) α_{20}, and the wave intensity $\bar{p}_e = \bar{D}_0^2$ (see (6.4.15)).

To investigate the asymptotic behavior of a solution of the obtained system of equations in the neighborhood of the initial state o, this system is linearized relative to parameter values at point o, and the solution is sought in the form of a vanishing exponent of type (6.3.15) at $\bar{x} \to \infty$. In this case, instead of a function of one variable $\bar{T}_2(\bar{x})$, here the function of two variables $T_2'(\bar{x}, \bar{\xi})$ is sought. The exponential asymptotic for both this function and a function $\bar{r}(\bar{x}, \bar{\xi})$ is

$$\overline{T}_2' = 1 + A_2^{(T)}(\bar{\xi})\exp k_0(\bar{x} - \bar{x}_b), \quad \bar{r} = \bar{\xi} + A^{(r)}(\xi)\exp k_0(\bar{x} - \bar{x}_b) \tag{6.5.6}$$

Upon linearization, we obtain an algebraic linear equations in terms of $A^{(a)}$, $A^{(w)}$, $A^{(v)}$, $A_1^{(p)}$, and $A_2^{(p)}$

$$\bar{v}_0 k_0 A^{(a)} = A^{(w)}$$

$$\bar{v}_0 k_0 A^{(w)} = A_2^{(p)} - A_1^{(p)}, \qquad A_1^{(p)} = -\bar{v}_0 A^{(v)}/\alpha_{20}, \quad A^{(v)} = 3\alpha_{20}\bar{v}_0 A^{(a)} \tag{6.5.7}$$

$$\bar{v}_0 k_0 A_2^{(p)} = 3\gamma_2\lambda_2^*\left(\frac{dA_2^{(T)}}{d\bar{\xi}}\right)_{\bar{\xi}=1} - 3\gamma_2 A^{(w)}$$

and, also, a linear differential equation for $A_2^{(T)}(\xi)$ with the following boundary conditions on the wall and in the center of a trial bubble

$$\bar{v}_0 k_0 A_2^{(T)}(\xi) = \frac{\lambda_2^*}{\xi^2}\frac{d}{d\xi}\left(\xi^2 \frac{dA_2^{(T)}}{d\xi}\right) + \frac{\gamma_2 - 1}{\gamma_2}\bar{v}_0 k_0 A_2^{(p)}$$
$$A_2^{(T)}(1) = 0, \quad \left(\frac{dA_2^{(T)}}{d\xi}\right)_{\xi=0} = 0 \tag{6.5.8}$$

The solution of this equation may be written in the form

$$A_2^{(T)} = \frac{\gamma_2 - 1}{\gamma_2}\left[1 - \frac{\mathrm{sh}\,(H^{1/2}\xi)}{\xi\,\mathrm{sh}\,H^{1/2}}\right] A_2^{(p)} \quad \left(H = \frac{k_0 \bar{v}_0}{\lambda_2^*}\right) \tag{6.5.9}$$

The derivative appearing in Eqs. (6.5.7) equals

$$\left(\frac{dA_2^{(T)}}{d\xi}\right)_{\xi=1} = \frac{\gamma_2 - 1}{\gamma_2}\left(1 - \sqrt{H}\,\mathrm{cth}\,\sqrt{H}\right) A_2^{(p)} \tag{6.5.10}$$

As a result, the condition for the existence of nonzero solutions of a linear homogeneous system of algebraic equations (6.5.7) for $A^{(a)}, \ldots, A_2^{(p)}$ is reduced to a transcendental characteristic equation in terms of $H = k_0\bar{v}_0/\lambda_2^* = K_0/\lambda_2^*$

$$\varphi(H) \equiv H + \frac{3\gamma_2 H}{\lambda_2^* H^2 - 3\bar{p}_e} + 3(\gamma_2 - 1)\left[H^{1/2}\,\mathrm{cth}\,H^{1/2} - 1\right] = 0 \tag{6.5.11}$$

For a solution of type (6.5.8), only those roots H that have positive real part ($\mathrm{Re}\,\{H\} > 0$, $\mathrm{Re}\,\{k_0\} < 0$) are suitable. Having introduced the complex plane $H = \mathrm{Re}\,\{H\} + i\,\mathrm{Im}\,\{H\}$, we shall prove both the existence and uniqueness of the root of Eq. (6.5.11) in the right half-plane in the compression-wave case ($\bar{p}_e > 1$), and also the fact that this root is a real and, consequently, positive number.

The relation $\varphi(H)$ is a meromorphic function, for which, in conformity with the argument-principle (see M. A. Lavrent'ev and B. V. Shabat, 1973), the difference between the number of zeroes (roots) N and number of poles P in a domain bounded by a closed curve C is determined by the increment of its argument when rounding the domain along its boundary C counterclockwise

$$N - P = \Delta_C\{\arg\varphi(H)\}/(2\pi) \equiv \Delta_C/(2\pi)$$

As the above-indicated curve C, we shall use a contour consisting of two semicircles and two segments on the imaginary axis, shown in Fig. 6.5.1a. We will now calculate the indicated increment of the argument, which equals

$$\Delta_C = \Delta_{NML} + \Delta_{LS} + \Delta_{SQW} + \Delta_{WN}$$

On the semicircle NML at $R \to \infty$, we have

$$\varphi(H)=H(1+O(R^{-1/2})),\quad \Delta_{NML}=\pi(1+O(R^{-1/2})) \tag{6.5.12}$$

On the semicircle SQW at $\varepsilon \to 0$, we have

$$\varphi(H)=\frac{\bar{p}_e-1}{\bar{p}_e}\gamma H(1+O(\varepsilon)),\quad \Delta_{SQW}=-\pi(1+O(\varepsilon)) \tag{6.5.13}$$

It can readily be shown that the operator φ and the conjugation denoted by an asterisk are commutative, i.e.

$$\varphi^*(H)=\varphi(H^*)$$
$$(H^*=\operatorname{Re}\{H\}-i\operatorname{Im}\{H\},\quad \varphi^*=\operatorname{Re}\{\varphi\}-i\operatorname{Im}\{\varphi\}) \tag{6.5.14}$$

Hence, $\Delta_{LS} = \Delta_{WN}$, and using the evaluations obtained above, we have

$$\Delta_C=2\Delta_{LS}\qquad (R\to\infty,\quad \varepsilon\to 0) \tag{6.5.15}$$

Introducing parameter y for the segment LS: $H = iy^2/2$, it may be proven that at $y \neq 0$, the following holds

$$\frac{\operatorname{Re}\{\varphi\}}{3(\gamma_2-1)}=-2+y\frac{\operatorname{sh} y+\sin y}{\operatorname{ch} y-\cos y}>0 \tag{6.5.16}$$

Thus, $L'S' = \varphi(LS)$ and $W'N' = \varphi(WN)$ are located in the right half-plane (see Fig. 6.5.1). And, in conformity with (6.5.13) and (6.5.16), at $\bar{p}_e > 1$, point $S' = \varphi(S)$ lies in the upper right (first) quadrant (see Fig. 6.5.1*b*), and at $\bar{p}_e < 1$, point S' lies in the lower right (fourth) quadrant (see Fig. 6.5.1*c*). Then, in the limiting cases $R \to \infty$ and $\varepsilon \to 0$, we have

$$\bar{p}_e>1:\ \Delta_{LS}=0;\quad \bar{p}_e<1:\ \Delta_{LS}=-\pi$$

As a result, recalling that $\varphi(H)$ in the right half-plane (inside the contour C) has one pole ($H = \sqrt{3\bar{p}_e/\lambda_2^*}$), i.e., $P = 1$, we obtain

$$N=\frac{\Delta_C}{2\pi}+1=\frac{\Delta_{LS}}{\pi}+1$$

For rarefaction waves ($\bar{p}_e < 1$), we have $N = 0$, and the required root does not exist, which indicates that stationary rarefaction waves are not possible. For rarefaction waves ($\bar{p}_e > 1$), we have $N = 1$, and the sought root does exist and is unique; this root is real since (given Eq. (6.5.12)) if H is a root, then, the complex conjugate number H^* is also a root. Because it is unique, $H = H^*$, i.e., this root is a real positive number.

For limiting polytropic regimes of gas behavior (adiabatic ($\lambda_2^* = 0$, $\varkappa = \gamma_2$), and isothermal ($\lambda_2^* \to \infty$, $\varkappa = 1$)) the characteristic equation transforms to a quadratic equation (6.3.18) at $\mu_1^* = 0$.

Figure 6.5.2 represents the dependence of the characteristic root $K_0 = k_0\bar{v}_0 = \lambda_2^* H$, defining the wave-front steepness, on λ_2^* and $\bar{p}_e$ for $\gamma_2 = 1.4$; this relationship is obtained through a numerical approach.

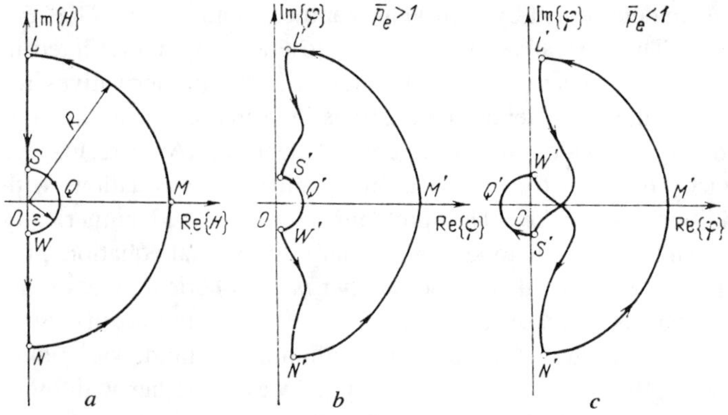

Figure 6.5.1 Schematics of contours designed to determine the increment of argument in Eq. (6.5.11).

Having determined k_0, it becomes possible to start from the singular point (point o), i.e., to prescribe the boundary conditions at $\bar{x} = \bar{x}_b$, and then to integrate the system of equations in the domain $\bar{x} < \bar{x}_b$ using numerical methods. With a numerical approach, the interior of the trial bubble $\bar{\xi} = [0,1]$ is divided into m spherical layers $\bar{\xi}_1, \bar{\xi}_2, \ldots, \bar{\xi}_m$, and all functions

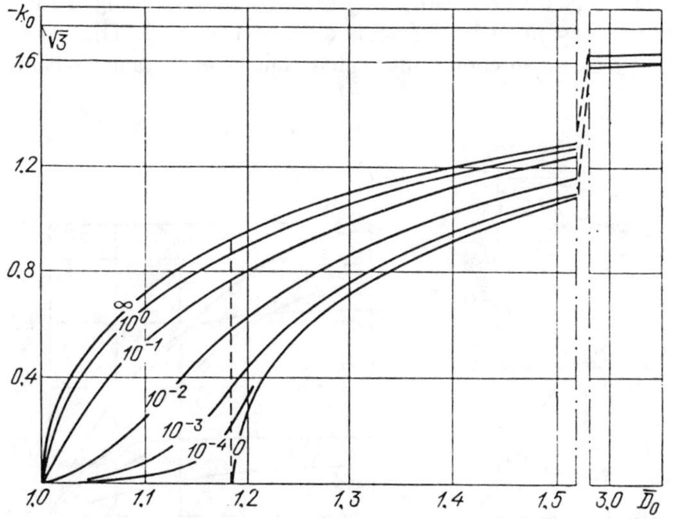

Figure 6.5.2 Dependence of the root of a characteristic equation (6.5.13), defining the steepness of a stationary shock-wave front, on the dimensionless heat conductivity λ_2^* of gas and wave velocity $\bar{D}_0 = \sqrt{\bar{p}_e}$ in a bubbly gas-liquid mixture with the gas adiabatic index $\gamma_2 = 1.4$. Labels on curves indicate values λ_2^*.

depending on ξ are determined by their values at these points, e.g., $T_2'(x, \xi_j)$, $j = 1, 2, \ldots, m$. The derivatives with respect to ξ are now represented in the form of finite differences. Then, the equation in partial derivatives for T_2' divides into m ordinary differential equations in terms of $\bar{x}$, and the continuity equation for $\bar{r}$ divides into m algebraic equations. As a result, we obtain the Cauchy problem for $m + 3$ ordinary differential equations with boundary conditions at $\bar{x} = \bar{x}_b$. This problem may be solved numerically using one of the integration methods for ordinary differential equations; for instance, the Runge-Kutta method. The number m of spherical layers must be sufficient so that its increase has virtually no effect on the results.

The dashed lines in Fig. 6.4.6 show the oscillation amplitude variations, as well as their lengths along the oscillatory shock wave together with other (already discussed) curves obtained according to the two-temperature scheme. Figure 6.5.3 depicts the wave structure whose prediction based on the two-temperature scheme is shown in Fig. 6.4.5. Figure 6.5.3 also illustrates both variation of the effective Nusselt number in the wave and the mass-average gas temperature in bubbles

$$\mathrm{Nu}_2 = \frac{2a}{T_0 - T_2}\left(\frac{\partial T_2'}{\partial r}\right)_{r=a}, \qquad T_2(x) = \frac{3}{a^3}\int_0^{a_0} T_2'(x, \xi)\, \xi^2 d\xi$$

In the case of a shock wave with an oscillatory structure (see Fig. 6.5.3), the instantaneous Nusselt number also fluctuates, assuming at some points in time even negative values which are accounted for by occurrence of "thermal pits" in a bubble (see §2.5). However, the $\mathbf{Nu}_2$ values averaged for a time-period are well described by formulas discussed in §1.6. The wave-pressure variation predicted in conformity with a simpler temperature scheme

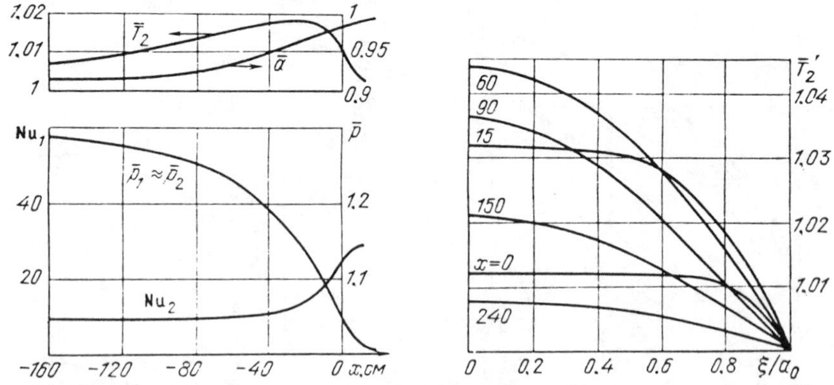

Figure 6.5.3 Change of the pressure phase, $\mathbf{Nu}_2$, and the temperature inside of the bubbles in a shock wave in a solution of glycerine and water (1:1) with air bubbles having the same parameters as in Fig. 6.4.5 (a_0 = 1.4 mm, α_{20} = 0.025, p_0 = 0.09 MPa, $\bar{p}_e$ = 1.32).

is in satisfactory agreement with the shown curves plotted in accordance with a more accurate model. In particular, the pressure fluctuation amplitude, frequency, their damping decrement, and the wave thickness are in agreement.

As has been indicated before, the front steepness across the initial section in the neighborhood of the initial state o affects the wave structure to a significant extent. Therefore, it proved to be reasonable to employ the characteristic equation (6.5.11) instead of (6.4.23) for determining both k_0 and $\mathbf{Nu}_{20}$ in the framework of a two-temperature scheme. Even though the Nusselt number $\mathbf{Nu}_2$ in the remaining part of the wave is different from $\mathbf{Nu}_{20}$ (see Fig. 6.5.3), failure to take this fact into consideration, as is proved by analysis, does not result in significant errors, and, therefore, allows the analysis to be substantially simplified.

§6.6 THE BOUSSINESQ AND BURGERS-CORTEVEG-DE VRIES EQUATIONS FOR INVESTIGATING WEAK NONLINEAR DISTURBANCES IN BUBBLY LIQUIDS

The theory of nonlinear (as distinct from §§6.3–6.5) waves is essential for bubbly liquids from the viewpoint of a better understanding of wave-evolution laws in the process of wave propagation. The nonstationarity of these waves may be linked to both the nonstationarity of boundary conditions at the location of their initiation responsible for occurrence of rarefaction waves and relatively great distances over which waves change their structure even at stationary boundary conditions tending to a stationary structure investigated in §§6.3–6.5. In this section, in conformity with the investigation of sufficiently weak one-dimensional waves in a quiescent mixture of incompressible liquid with "polytropic" bubbles initially at equilibrium, a theoretical approach of nonlinear wave dynamics is presented, which is widely used for analysis of both stationary and nonstationary plane one-dimensional waves in various media (gravitational waves on water surface, waves in viscous compressible gas, waves in plasma in magnetic field, electromagnetic waves in conducting media and dielectrics, etc.). This approach is based on reducing the analysis of the process to a solution of Boussinesq and Burgers-Corteveg-de Vries (BCdV) equations, which at present are investigated in detail.

The characteristic value of the pressure-related disturbance is determined by a number ε, whose square is assumed to be small

$$\varepsilon \sim \frac{p - p_0}{p_0}, \quad \varepsilon^2 \ll 1 \tag{6.6.1}$$

For stationary shock waves, the quantity $\varepsilon = \bar{p}_e - 1$ may be taken as ε.

The volumetric concentration of bubbles is also considered to be small, so that $\alpha_2 < \varepsilon$.

As before, the subscript 0 will signify parameters (υ_0, ρ_0, p_0, a_0, n_0, α_{20}) in an undisturbed state. And, x, t implies a system of coordinates associated with a homogeneous undisturbed medium. Therefore, $\upsilon_0 = 0$, and the velocity perturbation $\Delta\upsilon = \upsilon$.

Nonholonomic equation of state of bubbly liquid. Coefficients of dispersion and dissipation. This section employs the same model of a bubbly liquid but for nonstationary flows only) and the same simplifications that are used in §.6.3. The equations consistent with this model are written in the form (1.5.6). In addition to major simplifications, stemming from the absence of relative phase motion, liquid incompressibility, and the gas polytropic properties, we shall ignore, as in §6.3, the capillarity effects and those of external body forces (i.e., small Σ^* and $\bar{g}$; for more detail see Eqs. (6.7.18) and (6.7.17) below. Giving consideration to phase transitions (unless they result in the bubble disappearance), capillary effects, and bubble constraints through use of coefficients $\varphi^{(1)}$ and $\varphi^{(2)}$ does not present any problem for a variety of cases, including the method being discussed below.

Let us express the values appearing in the Rayleigh-Lamb equation in terms of mixture density and its derivatives. From equations constituting the system (1.6.16), we have

$$\rho = \overset{\circ}{\rho}_1\left(1 - \frac{4}{3}\pi a^3 n\right), \quad \frac{n}{n_0} = \frac{\rho}{\rho_0} \quad \left(\frac{1}{\rho}\frac{d\rho}{dt} = \frac{1}{n}\frac{dn}{dt}\right) \tag{6.6.2}$$

when, using the polytropic equation for gas, we readily obtain

$$\left(\frac{a}{a_0}\right)^3 = \frac{\overset{\circ}{\rho}_1 - \rho}{\overset{\circ}{\rho}_1\alpha_{20}}\frac{\rho_0}{\rho}, \quad p_2 = p_0\left(\frac{\overset{\circ}{\rho}_1\alpha_{20}}{\rho_0}\right)^{\varkappa}\left(\frac{\rho}{\overset{\circ}{\rho}_1 - \rho}\right)^{\varkappa} \tag{6.6.3}$$

Having differentiated the first formula (6.6.2) with respect to t, and expressing dn/dt in terms of $d\rho/dt$, we arrive at

$$\frac{1}{\rho}\frac{d\rho}{dt} = -4\pi a^2 n\frac{da}{dt} \tag{6.6.4}$$

Differentiating once more with respect to t, expressing da/dt, n, dn/dt in terms of ρ and $d\rho/dt$ by means of (6.6.2) and (6.6.4), we obtain

$$\frac{d^2a}{dt^2} = \frac{dw_{1a}}{dt} = -\frac{a}{3\alpha_2\rho}\frac{d^2\rho}{dt^2} - (1 - 3\alpha_2)\frac{2a}{9\alpha_2^2\rho^2}\left(\frac{d\rho}{dt}\right)^2 \tag{6.6.5}$$

Substituting the last equations into the Rayleigh-Lamb equation (1.6.47) using formula (1.6.50) and (1.6.51) for μ_{eff}, we obtain

$$\frac{p-p_0}{p_0}=\left[\left(\frac{\mathring{\rho}_1\alpha_{20}}{\mathring{\rho}_1-\rho}\right)^{\varkappa}\left(\frac{\rho}{\rho_0}\right)^{\varkappa}-1\right]+\frac{a^2\mathring{\rho}_1}{3\alpha_2\rho p_0}\frac{d^2\rho}{dt^2}+\frac{4\mu_{\text{эф}}}{3\alpha_2\rho p_0}\frac{d\rho}{dt}$$
$$+\left(\frac{1}{6}-2\alpha_2\right)\frac{a^2\mathring{\rho}_1}{3\alpha_2^2\rho^2 p_0}\left(\frac{d\rho}{dt}\right)^2 \qquad (6.6.6)$$

The obtained equation for pressure is

$$p=p(\rho,\ \dot{\rho},\ \ddot{\rho}) \qquad (\dot{\rho}=d\rho/dt,\quad \ddot{\rho}=d^2\rho/dt^2) \qquad (6.6.7)$$

This equation may be viewed as a nonholonomic equation of state of a medium whose pressure depends not only on the medium density but also on the first (because of the liquid radial motion and viscosity) and the second (because of the radial inertia of liquid) time derivatives of density. The latter has been indicated by B. S. Kogarko (1961). The relationship between p and ρ reflects the local compression and tension deformation inertia. This is characteristic for bubbly media. By virtue of such inertia, the medium elementary volume may, under its own momentum, continue to undergo compression ($\rho > 0$) for a period of time immediately after the compression is removed, or, more specifically, compression may go on even though $p < p_2$.

If the gas-polytropy simplification is dropped, then, instead of the second equation (6.6.3) and the equation of type (6.6.7), we have

$$\frac{p_2}{p_0}=\frac{T_2}{T_0}\frac{\rho}{\rho_0}\frac{\mathring{\rho}_1\alpha_{20}}{\mathring{\rho}_1-\rho},\qquad p=p\,(\rho,\ \dot{\rho},\ \ddot{\rho},\ T_2)$$

and, in a similar manner, for T_2 the differential heat-influx equation which results in additional nonholonomicity because of the increase of order of equation should be used. In this case, the nonholonomic equations of state of bubbly medium with consideration given to local deformation inertia and nonequilibrium interphase heat exchange, as follows from Eqs. (6.6.6) and (1.6.29), may represented in the form

$$m\ddot{\rho}=p-p_2-k^{(1)}\dot{\rho}-k^{(2)}(\dot{\rho})^2$$
$$\dot{p}_2=N^{(1)}\left(1-\frac{p_2}{p_0}\left(\frac{a}{a_0}\right)^3\right)+N^{(2)}p_2\dot{\rho} \qquad (6.6.8)$$

where coefficients m, $k^{(1)}$, $k^{(2)}$, $N^{(1)}$, $N^{(2)}$, and a/a_0 depend on the variable density ρ and the initial parameters of mixture (ρ_1^0, α_{20}, a_0, μ_1, λ_2, T_0, $\mathbf{Nu}_2$), and, in the case $\alpha_2 \ll 1$, are expressed by the following formulas

$$m=\frac{a_0^3}{3\mathring{\rho}_1\alpha_{20}^{2/3}\alpha_2^{1/3}},\qquad \left(\frac{a}{a_0}\right)^3=\frac{\alpha_2}{\alpha_{20}}\qquad \left(\alpha_2(\rho)=1-\frac{\rho}{\mathring{\rho}_1}\right)$$

$$k^{(1)}=\frac{4\mu_{\text{эф}}}{3\mathring{\rho}_1\alpha_2},\qquad k^{(2)}=\frac{2a_0^2}{3\mathring{\rho}_1\alpha_{20}^{2/3}}\,\frac{1/12-\alpha_2}{\alpha_2^{4/3}} \qquad (6.6.8a)$$

$$N^{(1)} = \frac{3(\gamma_2 - 1)\lambda_2 T_0 \mathrm{Nu}_2}{2a_0^2}\left(\frac{\alpha_{20}}{\alpha_2}\right)^{2/3}, \quad N^{(2)} = \frac{\gamma_2}{\alpha_2}$$

In the case of the gas polytropic behavior, the differential equation for p_2 is reduced to an algebraic expression

$$p_2 = p_0(\alpha_{20}/\alpha_2)^{\varkappa}$$

From Eqs. (6.6.6) and (6.6.8), it is obvious that, for sufficiently slow processes ($m\ddot{\rho}$, $k^{(1)}\dot{\rho}$, $k^{(2)}(\dot{\rho})^2 \ll p$), the quasi-static approximation is obtained in the form $p = p_2$, which, together with an approximation of gas polytropicity, leads to a barotropic equation of state in the form $p = p(\rho)$.

From experimental results, and an analysis of equations and above-indicated solutions based on them (§§6.2–6.5), it can be seen that sufficiently weak waves propagate with a velocity slightly higher than the equilibrium speed of sound C_e, and the values of measured lengths of oscillatory waves are fairly well described by formulas (6.3.28) or (6.4.32) obtained for stationary waves

$$L = \frac{2\pi}{\sqrt{3}}\frac{a_0}{\sqrt{\bar{p}_e - 1}}\frac{1}{\sqrt{\alpha_{10}\alpha_{20}}} \approx \frac{4a_0}{\sqrt{\varepsilon\alpha_{20}}}$$

Therefore, the values L_* and τ_* may be adopted as the characteristic length and time, respectively, over which the medium characteristics are varying

$$L_* = \frac{a_0}{\sqrt{\alpha_{20}\varepsilon}}, \quad \tau_* = \frac{L_*}{C_\alpha} = a_0\sqrt{\frac{\mathring{\rho}_1}{p_0}}$$
$$\left(\varepsilon = \Delta p_0/p_0, \quad C_\alpha = \sqrt{p_0/\mathring{\rho}_1\alpha_{20}}\right) \tag{6.6.9}$$

Here, Δp_0 is the inherent disturbance of pressure.

Now, upon introducing the dimensionless values

$$x_* = x/L_*, \quad t_* = t/\tau_*, \quad v_* = v/\alpha_{20}C_\alpha$$
$$p_* = (p - p_0)/p_0, \quad \rho_* = (\rho - \rho_0)/(\alpha_{20}\rho_0) = (\alpha_{20} - \alpha_2)/\alpha_{20} \tag{6.6.10}$$

from (1.5.16), we obtain the equation for mass, momentum, and the state of a bubbly mixture in a dimensionless form

$$\frac{\partial \rho_*}{\partial t_*} + \alpha_{20}v_*\frac{\partial \rho_*}{\partial x_*} + (1 + \alpha_{20}\rho_*)\frac{\partial v_*}{\partial x_*} = 0$$

$$\frac{\partial v_*}{\partial t_*} + \alpha_{20}v_*\frac{\partial v_*}{\partial x_*} + \frac{1}{1 + \alpha_{20}\rho_*}\frac{\partial p_*}{\partial x_*} = 0$$

$$p_* = p_{2*} + 2\left(\frac{a^2}{a_0^2}\frac{\alpha_{20}}{\alpha_2}\frac{1}{\rho}\right)\beta_*\frac{d^2\rho_*}{dt_*^2} + 2\left(\frac{\alpha_{20}}{\alpha_2}\frac{1}{\rho}\right)\mu_*\sqrt{\varkappa}\frac{d\rho_*}{dt_*} \tag{6.6.11}$$

$$+2\left(\frac{1}{6}-2\alpha_2\right)\left(\frac{a^2}{a_0^2}\frac{\alpha_{20}}{\alpha_2}\frac{1}{\bar{\rho}^2}\right)\alpha_{20}\beta_*\left(\frac{d\rho_*}{dt_*}\right)^2 \qquad \left(\frac{d}{dt_*}\equiv\frac{\partial}{\partial t_*}+\bar{v}\frac{\partial}{\partial x_*}\right)$$

$$p_{2*}=\left(\frac{1+\alpha_{20}\rho_*}{1-\rho_*}\right)^{\varkappa}-1, \quad \bar{\rho}=1+\alpha_{20}\rho_*, \quad \bar{v}=\alpha_{20}v_*$$

Here, the dimensionless coefficients of dispersion β_* and dissipation μ_* are introduced

$$\beta_*=\frac{a_0^2}{6L_*\alpha_{20}}, \quad \mu_*=\frac{2\mu_{\text{eff}}C_\alpha}{3L_*p_0\sqrt{\varkappa}} \tag{6.6.12}$$

From (6.6.9) and expressions for p_* and p_{2*} in (6.6.11), it follows

$$p_*\sim\rho_*\sim\varepsilon \tag{6.6.13}$$

The values p_*, ρ_*, and v_* vary over distances $\Delta x\sim L_*(\Delta x_*\sim 1)$, and during a time interval $\Delta t\sim\tau_*(\Delta t_*\sim 1)$. Therefore, the following evaluations hold good for derivatives

$$\frac{\partial p_*}{\partial x_*}\sim\varepsilon, \quad \frac{\partial\rho_*}{\partial t_*}\sim\frac{\partial\rho_*}{\partial x_*}\sim\varepsilon, \quad \frac{\partial v_*}{\partial t_*}\sim\frac{\partial v_*}{\partial x_*}\sim v_* \tag{6.6.14}$$

Then, from the momentum equation (the second equation (6.6.11)), the evaluation follows for the velocity perturbation

$$v_*\sim\varepsilon \tag{6.6.15}$$

Thus, in a bubbly mixture with a small volumetric concentration α_2 of gas at small pressure perturbations $p_*\sim\varepsilon$, the dimensionless perturbations of density ρ_* and velocity v_* are of the same order of magnitude as ε.

Further, for the terms appearing in the dimensionless mass-conservation equation for the medium (the first equation (6.6.11)), in accordance with (6.6.14), we have the following evaluations

$$\alpha_{20}v_*\frac{\partial\rho_*}{\partial x_*}=O(\alpha_{20}\varepsilon^2), \quad (1+\alpha_{20}\rho_*)\frac{\partial v_*}{\partial x_*}=\frac{\partial v_*}{\partial x_*}+O(\alpha_{20}\varepsilon^2) \tag{6.6.16}$$

The values defining the variation of the medium density in time, and appearing in the equation of state (the third equation of (6.6.11)), are of order

$$\frac{d\rho_*}{dt_*}\sim\varepsilon, \quad \frac{d^2\rho_*}{dt_*^2}\sim\varepsilon \tag{6.6.17}$$

The dispersion coefficient β_*, appearing in equation of state, is a small value of order of ε. Indeed, given (6.6.9), we have

$$\beta_*\sim a_0^2/(L_*\alpha_{20})\sim\varepsilon \tag{6.6.18}$$

The dissipation coefficient μ_*, according to (6.6.12), (1.6.50), and (1.6.51) at small radial pulsations of bubbles with frequencies near their natural oscillations, is determined by a formula

$$\mu_* \sim \left[\frac{2}{3} \frac{\mu_1/\rho_1}{a_0 \sqrt{\varkappa p_0/\rho_1}} + \frac{(3\varkappa)^{1/4} (\gamma_2 - 1)}{2\sqrt{2\varkappa}} \left(\frac{\nu_2^{(T)}}{a_0 \sqrt{p_0/\rho^\circ}} \right)^{1/2} \right] \varepsilon^{1/2} \qquad (6.6.19)$$

As the evaluations show, for all important cases affected by radial inertia of liquid, we may assume

$$\mu_* \leqslant \varepsilon \qquad (6.6.20)$$

We will now represent the relation $p_{2*}(\rho_*)$ in a form of two terms of the Taylor series with respect to exponents of ρ_*

$$p_{2*} = \overline{C}_0^2 \left(1 + \frac{\varkappa + 1}{2} \rho_*\right) \rho_* + O(\varepsilon^3) \qquad (\overline{C}_0^2 = \varkappa) \qquad (6.6.21)$$

Acoustic equations for a perfect linear medium of low compressibility. Simple waves. The equation of state for a medium may be represented accurate to terms of order ε in the form of linear relation $p_* = \overline{C}_0^2\rho_*$; then, the system (6.6.11) simplifies and reduces to the acoustic equations of a perfect linear compressible medium

$$\frac{\partial \rho_*}{\partial t_*} + \frac{\partial v_*}{\partial x_*} = \alpha_{20} O(\varepsilon^2) \approx 0, \quad \frac{\partial v_*}{\partial t_*} + \overline{C}_0^2 \frac{\partial \rho_*}{\partial x_*} = O(\varepsilon^2) \approx 0$$
$$p_* = \overline{C}_0^2 \rho_* + O(\varepsilon^2) \qquad (6.6.22)$$

These equations, upon differentiating with respect to t_* and x_*, transforms into linear wave equations

$$\frac{\partial^2 u_*}{\partial t_*^2} - \overline{C}_0^2 \frac{\partial^2 u_*}{\partial x_*^2} = O(\varepsilon^2) \approx 0 \qquad (u_* = (\rho_*, v_*, p_*)) \qquad (6.6.23)$$

The general soluton of these equations, if quantities $O(\varepsilon^2)$ are dropped, is

$$\rho_* = P_\rho (x_* - \overline{C}_0 t_*) + Q_\rho (x_* + \overline{C}_0 t_*)$$
$$v_* = P_v (x_* - \overline{C}_0 t_*) + Q_v (x_* + \overline{C}_0 t_*) \qquad (6.6.24)$$

For unidirectional waves propagating through an undisturbed medium (e.g., propagating along the positive axis x), it should be assumed that $Q_\rho = Q_v = 0$. Then

$$\rho_* = P_\rho (\xi), \quad v_* = P_v (\xi), \quad \xi = x_* - \overline{C}_0 t_* \qquad (6.6.25)$$

which is consistent with the fixed-velocity wave propagation when there occurs no change of the wave shape defined by functions p_ρ and P_v. Substituting these solutions into (6.6.22), we obtain the relationship between the density perturbation and velocity

$$-\overline{C}_0 \frac{dP_\rho}{d\xi} + \frac{dP_v}{d\xi} = 0 \quad \text{or} \quad P_v = \overline{C}_0 P_\rho + \text{const}$$

Understanding that at $\xi > 0$ we have an undisturbed state ($\rho_* = P_\rho = 0$, $v_* = P_v = 0$), for a linear solution we obtain

$$v_* = \overline{C}_0 \rho_* = p_* / \overline{C}_0 \tag{6.6.26}$$

Waves for which there exist a single-valued relation between v and ρ, or between v and p, are referred to as simple waves.

Thus, the solution of a system of equations (1.5.16) or (6.6.11) for a bubbly mixture of an incompressible liquid with a polytropic gas to an accuracy of values of order ε for a wave propagating through an undisturbed medium along axis x renders a simple wave moving with a fixed velocity

$$\begin{gathered} v_* = P_v(\xi) + \delta_v, \quad \rho_* = v_* / \overline{C}_0 + \delta_\rho, \quad p_* = \overline{C}_0 v_* + \delta_p \\ \delta_v \sim \delta_\rho \sim \delta_p = O(\varepsilon^2), \quad \xi = x_* - \overline{C}_0 t_* \end{gathered} \tag{6.6.27}$$

Differentiating v_* along the characteristic ξ = constant, we obtain an equation for a simple wave along axis x, which is employed subsequently

$$\frac{\partial v_*}{\partial t} + \overline{C}_0 \frac{\partial v_*}{\partial x_*} = O(\varepsilon^2) \tag{6.6.28}$$

The Boussinesq approximation for weakly nonlinear waves. From Eqs. (6.6.11), and given (6.6.22), there follow expressions for quantities defining the variation of density

$$\begin{aligned} \frac{d\rho_*}{dt_*} &= \frac{\partial \rho_*}{\partial t_*} + \alpha_{20} v_* \frac{\partial \rho_*}{\partial x_*} = -\frac{\partial v_*}{\partial x_*} + O(\alpha_{20}\varepsilon^2) \\ \frac{d^2\rho_*}{dt_*^2} &= -\frac{\partial}{\partial t_*}\left(\frac{\partial v_*}{\partial x_*}\right) - \alpha_{20} v_* \frac{\partial}{\partial x_*}\left(\frac{\partial v_*}{\partial x_*}\right) + \alpha_{20} O(\varepsilon^2) \\ &= -\frac{\partial}{\partial x_*}\left(\frac{\partial v_*}{\partial t_*} + \alpha_{20} v_* \frac{\partial v_*}{\partial x_*}\right) + \alpha_{20} O(\varepsilon^2) = \frac{\partial^2 p_*}{\partial x_*^2} + O(\alpha_{20}\varepsilon^2) = \overline{C}_0^2 \frac{\partial^2 \rho_*}{\partial x_*^2} + O(\varepsilon^2) \end{aligned} \tag{6.6.29}$$

Using (6.6.21) and other evaluations for values appearing in a system of equations (6.6.11), we shall write it in a quadratic (with respect to ε) approximation; i.e., we retain the terms of order ε and ε^2, and drop the terms of order ε^3 and $\alpha_{20}\varepsilon^2$, thereby obtaining the so-called Boussinesq approximation when applied to a bubbly medium

$$\begin{gathered} \frac{\partial \rho_*}{\partial t_*} + \frac{\partial v_*}{\partial x_*} = O(\alpha_{20}\varepsilon^2) \approx 0, \quad \frac{\partial v_*}{\partial t_*} + \frac{\partial p_*}{\partial x_*} = O(\varepsilon^3) \approx 0 \\ p_* - \overline{C}_0^2 \rho_* = \frac{\varkappa + 1}{2} \overline{C}_0^2 \rho_*^2 + 2\overline{C}_0^2 \beta_* \frac{\partial^2 \rho_*}{\partial x_*^2} - 2\overline{C}_0 \mu_* \frac{\partial v_*}{\partial x_*} + O(\varepsilon^3) \end{gathered} \tag{6.6.30}$$

The second equation may be rewritten in the form

$$\frac{\partial v_*}{\partial t_*} + \overline{C}^2 \frac{\partial \rho_*}{\partial x_*} = -2\overline{C}_0^2 \beta_* \frac{\partial^3 \rho_*}{\partial x_*^3} + 2\overline{C}_0 \mu_* \frac{\partial^2 v_*}{\partial x_*^2} + O(\varepsilon^3)$$

$$\overline{C}^2 = \overline{C}_0^2 [1 + (\varkappa + 1)\rho_*] + O(\varepsilon^2) = \overline{C}_0^2 \left[1 + \frac{\varkappa + 1}{\overline{C}_0^2} p_*\right] + O(\varepsilon^2) \tag{6.6.30a}$$

Terms on the left-hand side of equations (6.6.30) and (6.6.30a) are of an order of ε. The relationship between speed of sound $\bar{C}$ and ρ_* or p_* takes into account nonlinear compressibility (physical nonlinearity), which contributes to the second equation of order of ε^2. When expressing $\bar{C}$ in terms of p_*, the last equation in (6.6.22) is taken into consideration. Terms on the left-hand side of equation of state (the third equation in (6.6.30)) are of order of ε^2. They give consideration to the nonlinearity of the bubble incompressibility (the first term), the radial inertia of liquid near bubbles (the second term), and the dissipation due to viscosity and other effects (the third term).

Having differentiated the first and second equations (6.6.30) with respect to t_* and x_*, respectively, we obtain a linear equation

$$\frac{\partial^2 \rho_*}{\partial t_*^2} = \frac{\partial^2 p_*}{\partial x_*^2} + O(\varepsilon^3, \alpha_{20}\varepsilon^2) \tag{6.6.31}$$

Let us now differentiate the equation of state (6.6.30) two times with respect to t_*

$$\frac{\partial^2 p_*}{\partial t_*^2} - \overline{C}_0^2 \frac{\partial^2 \rho_*}{\partial t_*^2} = \frac{\varkappa + 1}{2} \overline{C}_0^2 \frac{\partial^2 (\rho_*)^2}{\partial t_*^2} + 2\overline{C}_0^2 \beta_* \frac{\partial^4 \rho_*}{\partial t_*^2 \partial x_*^2}$$

$$- 2\overline{C}_0 \mu_* \frac{\partial^3 v_*}{\partial t_*^2 \partial x_*} + O(\varepsilon^3)$$

We shall now transform this equation into an equation in terms of p_*. In accordance with (6.6.22), (6.6.23), and (6.6.29), (6.6.31), we arrive at

$$\frac{\partial^2 (\rho_*)^2}{\partial t_*^2} = \frac{\partial^2}{\partial t_*^2} \left[\frac{p_*}{\overline{C}_0^2} + O(\varepsilon^2)\right]^2 = \frac{1}{\overline{C}_0^4} \frac{\partial^2}{\partial t_*^2} [p_*^2 + O(\varepsilon^3)]$$

$$= \frac{2}{\overline{C}_0^4} \left[\left(\frac{\partial p_*}{\partial t_*}\right)^2 + p_* \frac{\partial^2 p_*}{\partial t_*^2} + O(\varepsilon^3)\right] = \frac{2}{\overline{C}_0^4} \left[\left(\frac{\partial p_*}{\partial t_*}\right)^2 + \overline{C}_0^2 p_* \frac{\partial^2 p_*}{\partial x_*^2} + O(\varepsilon^3)\right]$$

$$\frac{\partial^3 v_*}{\partial t_* \partial x_*} = \frac{\partial^2}{\partial t_*^2} \left(-\frac{\partial \rho_*}{\partial t_*} + O(\alpha_{20}\varepsilon)\right) = -\frac{\partial^3 p_*}{\partial t_* \partial x_*^2} + O(\alpha_{20}\varepsilon^2)$$

As a result, we obtain a generalization of a well-known nonlinear Boussinesq equation of wave dynamics in application to bubbly liquids (cf. (6.2.16))

$$\frac{\partial^2 p_*}{\partial t_*^2} - \overline{C}_0^2 \frac{\partial^2 p_*}{\partial x_*^2} = \frac{\varkappa + 1}{\varkappa}\left[\left(\frac{\partial p_*}{\partial t_*}\right)^2 + \varkappa p_* \frac{\partial^2 p_*}{\partial x_*^2}\right] + 2\overset{\circ}{\beta}_* \frac{\partial^4 p_*}{\partial x_*^4} + 2\overset{\circ}{\mu}_* \frac{\partial^3 p_*}{\partial x_*^2 \partial t_*}$$

$$\left(\overline{C}_0^2 = \varkappa,\ \overset{\circ}{\beta}_* = \beta_* \overline{C}_0 = {}^1/_6 a_0^2 \sqrt{\varkappa}/(L_*^2 \alpha_{20})\right. \tag{6.6.32}$$

$$\left.\overset{\circ}{\mu}_* = \mu_* \overline{C}_0 = {}^2/_3 \mu_{э\phi} C_\alpha/(L_* p_0)\right)$$

We will now analyze the stationary solutions of this equation of type of a moving wave, setting $p_* = p_*(\xi)$, $\xi = x_* - \bar{D}_0 t_*$ (where difference between $\bar{D}_0$ and $\bar{C}_0$ is small, and has an order of ε), when before the wave ($\xi \to \infty$, $p_* = 0$), and behind the wave ($\xi \to -\infty$) homogeneous equilibrium states ($dp_*/d\xi = d^2p_*/d\xi^2 = d^3p_*/d\xi^3 = 0$) are available. Then, after simple transformations and double integration, the Boussinesq equation assumes the following form

$$p_*\left[(\overline{D}_0^2 - \overline{C}_0^2) - \frac{\varkappa + 1}{2} p_*\right] = 2\overset{\circ}{\beta}_* \frac{d^2 p_*}{d\xi^2} - 2\overset{\circ}{\mu}_* \overline{C}_0 \frac{dp_*}{d\xi} \tag{6.6.33}$$

At $\mu^0 > 0$, this equation has a solution of the stationary shock compression-wave type, pressure $\bar{p}_e = 1 + p_{*e}$ behind which is determined by formula (6.3.10) (which is consistent with weak waves when $\bar{p}_e - 1 \ll 1$). The indicated solution is discussed below in connection with investigation of the Burger-Corteveg-de Vries equation (see (6.6.71)).

In the absence of dissipation ($\mu_* = 0$), upn integration, we have

$$2\overset{\circ}{\beta}_* \left(\frac{dp_*}{d\xi}\right)^2 = p_*^2\left[(\overline{D}_0^2 - \overline{C}_0^2) - \frac{\varkappa + 1}{3} p_*\right] \tag{6.6.34}$$

This equation has a solution of the soliton stationary wave type, discussed below in connection with the Corteveg-de Vries equation.

Nonlinear dispersion-free and dissipation-free simple waves. The terms on the right-hand side of Eq. (6.6.30a) associated with dispersion (β_*) and dissipation (μ_*) are of a higher order of smallness than the terms on the left-hand side of this equation. We will, therefore, indicate first the general solution of the nonlinear system for $\beta_* = \mu_* = 0$, namely

$$\frac{\partial \rho_*}{\partial t_*} + \frac{\partial v_*}{\partial x_*} = 0, \quad \frac{\partial v_*}{\partial t_*} + \overline{C}^2 \frac{\partial \rho_*}{\partial x_*} = 0 \tag{6.6.35}$$

Thus, for a bubbly liquid with a low volumetric concentration of bubbles, the nonlinearity of equations is attributed only to the variable nature of speed of sound (physical nonlinearity due to nonlinear compressibility of the medium). The convective nonlinearity accounted for by the smallness of the velocity perturbations ($\bar{v} = \alpha_2 v_* \sim \alpha_{20}\varepsilon$) is, in this case, insignificant. The system of equations (6.6.35) has two families of characteristics

1) $$\frac{dx_*}{dt_*} = \bar{C}, \qquad \frac{d\rho_*}{dv_*} = -\frac{1}{\bar{C}}$$

2) $$\frac{dx_*}{dt_*} = -\bar{C}, \qquad \frac{d\rho_*}{dv_*} = \frac{1}{\bar{C}}$$

For the unidirectional case of wave propagation (e.g., along the positive direction of axis x) when at $x \to \infty$, we have an undisturbed case ($\upsilon_* = \rho_* = 0$), all characteristics of the second family (running in the plane xt in the opposite direction of axis x) are passing through the point $\upsilon_* = \rho_* = 0$ in the plane $\rho_*\upsilon_*$, and, therefore, merge into a single so-called Riemann line

$$\rho_* = \int_0^{v_*} \frac{dv_*}{\bar{C}} = \frac{1}{\bar{C}_0} \int_0^{v_*} \left[1 - \frac{\varkappa + 1}{2} \frac{v_*}{\bar{C}_0}\right] dv_* + O(\varepsilon^3)$$

Here, the expression for $\bar{C}$ is written in terms of υ_* using both (6.6.30a) and (6.6.27). Upn simple transformations in the framework of adopted accuracy, we obtain the Riemann solution, which provides an unambiguous relation $\rho_*(\upsilon_*)$

$$\rho_* = \frac{v_*}{\bar{C}_0}\left[1 - \frac{\varkappa + 1}{4} \frac{v_*}{\bar{C}_0}\right] + O(\varepsilon^3) \tag{6.6.36}$$

Each characteristic of the first family in the plane $\rho\upsilon$ degenerates into a single point on characteristics of the second family, or Riemann lines, i.e., on each of them the values ρ_*, υ_*, $\bar{C}$ remain constant. The characteristics of the first family in the plane xt are straight lines.

Thus, the solution of a system (6.6.30) for a wave propagating along axis x through an undisturbed, physically nonlinear but dispersion-free and dissipation-free medium realizes an unambiguous relation $\rho(\upsilon)$ defined by Eq. (6.6.36), i.e., it corresponds to a simple wave.

Quasi-simple waves with the presence of both dispersion and dissipation. The solution of a system of equations (6.6.30) will be sought in the form of a so-called quasi-simple unidirectional wave representing a superposition of both simple wave $\rho_*^{(1)}$ and an additional disturbance $\rho_*^{(2)}$ caused by both dispersion and dissipation

$$\rho_* = \rho_*^{(1)}(v_*) + \rho^{(2)}(x_*, t_*) \tag{6.6.37}$$

Here, $\rho_*^{(1)}(\upsilon_*)$ is a Riemann solution for a simple wave (6.6.36)

$$\rho_*^{(1)} = \frac{v_*}{\bar{C}_0}\left[1 - \frac{\varkappa + 1}{4} \frac{v_*}{\bar{C}_0}\right] + O(\varepsilon^3) \tag{6.6.38}$$

which follows from the differential equation

$$\frac{d\rho^{(1)}}{dv_*} = \frac{1}{\overline{C}} \tag{6.6.39}$$

Given that both dispersion and dissipation are determined by terms of order of ε^2, it is natural to assume that disturbance $\rho_*^{(2)}$, specifying the distinction between the sought wave and simple wave, is of the same order of smallness

$$\rho_*^{(2)} = O(\varepsilon^2), \quad \rho_*^{(1)} = O(\varepsilon) \tag{6.6.40}$$

and the major part of disturbance is determined by the solution of a simple wave.

Since both representation of the value ρ_* (see (6.6.37)) in the form of two addends and identification of the principal part $\rho_*^{(1)}$ may be accomplished by an infinite number of methods, let us assume that such representation of (6.6.37) is possible when an additional disturbance $\rho_*^{(2)}(x, t)$ (having a higher order of smallness than $\rho_*^{(1)}$ propagates with a velocity C_0 without any change of its profile

$$\rho_*^{(2)} = \rho_*^{(2)}(\xi), \qquad \xi = x_* - \overline{C}_0 t_*$$

i.e., $\rho_*^{(2)}$ satisfies the linear equation

$$\frac{\partial \rho_*^{(2)}}{\partial t_*} + \overline{C}_0 \frac{\partial \rho_*^{(2)}}{\partial x_*} = 0 \tag{6.6.41}$$

Thus, the problem of determining both $\rho_*(x_*, t_*)$ and $v_*(x_*, t_*)$ satisfying the system of equations (6.6.30), we reduce to the determination of $v_*(x_*, t_*)$ and $\rho_*^{(2)}(x_*, t_*)$ where $\rho_*^{(2)}$ satisfies conditions (6.6.40) and (6.6.41); this is expected to be achieved by separating the principal part of the disturbance in the form of a simple wave. It will be shown below that such representation of a solution of a system (6.6.30) is feasible.

Let us substitute (6.6.37) into equations (6.6.30) using (6.6.39). Then, ignoring values $\sim \varepsilon^3$, we obtain

$$\begin{aligned} \frac{\partial \rho_*^{(2)}}{\partial t_*} &= -\frac{1}{\overline{C}} \left(\frac{\partial v_*}{\partial t_*} + \overline{C} \frac{\partial v_*}{\partial x_*} \right) \\ \overline{C}_0 \frac{\partial \rho_*^{(2)}}{\partial x_*} &= -\frac{1}{\overline{C}} \left(\frac{\partial v_*}{\partial t_*} + \overline{C} \frac{\partial v_*}{\partial x_*} \right) - 2\beta_* \frac{\partial^3 v_*}{\partial x_*^3} + 2\mu_* \frac{\partial^2 v_*}{\partial x_*^2} \end{aligned} \tag{6.6.42}$$

Summing these two equations, and taking into consideration the condition (6.6.41), we obtain, in the framework of adopted accuracy, the equation for velocity

$$\frac{\partial v_*}{\partial t_*} + \overline{C} \frac{\partial v_*}{\partial x_*} - \mu_* \overline{C}_0 \frac{\partial^2 v_*}{\partial x_*^2} + \beta_* \overline{C}_0 \frac{\partial^3 v_*}{\partial x_*^3} = 0 \tag{6.6.43}$$

Subtracting the second equation (6.6.42) from the first, we obtain an equation for $\rho_*^{(2)}$ using the velocity v_* distribution by means of linear differential operator

$$\frac{\partial \rho_*^{(2)}}{\partial t_*} - C_0 \frac{\partial \rho_*^{(2)}}{\partial x_*} = 2\mathscr{L}[v_*], \qquad \mathscr{L}[v_*] = \beta_* \frac{\partial^3 v_*}{\partial x_*^3} - \mu_* \frac{\partial^2 v_*}{\partial x_*^2} \tag{6.6.44}$$

or, solving for derivatives using (6.6.41), we have

$$\frac{\partial \rho_*^{(2)}}{\partial t_*} = \mathscr{L}[v_*], \quad \frac{\partial \rho_*^{(2)}}{\partial x_*} = -\frac{1}{C_0}\mathscr{L}[v_*] \tag{6.6.45}$$

It is evident that, because of the smallness of both and β_*, the operator ℓ raises the order of smallness by one

$$\mathscr{L}[\varphi] = \varepsilon O(\varphi) \tag{6.6.46}$$

The condition for the existence of a function $\rho_*^{(2)}$, which satisfies two differential equations (6.6.45) simultaneously (or, equations (6.6.41) and (6.6.44) that are the same), is that it is necessary and sufficient that the following equality holds

$$\frac{\partial^2 \rho_*^{(2)}}{\partial x_* \partial t_*} = \frac{\partial^2 \rho_*^{(2)}}{\partial t_* \partial x_*} \tag{6.6.47}$$

In conformity with (6.6.45), it is satisfied only when

$$\mathscr{L}\left[\frac{\partial v_*}{\partial t_*} + C_0 \frac{\partial v_*}{\partial x_*}\right] = 0 \tag{6.6.48}$$

This condition within the adopted accuracy is satisfied based on (6.6.28) and (6.6.46).

Integrating (6.6.45), and given the absence of disturbance at $x_* \to \infty$, we obtain

$$\rho_*^{(2)} = \frac{1}{C_0}\left(\mu_* \frac{\partial v_*}{\partial x_*} - \beta_* \frac{\partial^2 v_*}{\partial x_*^2}\right) \tag{6.6.49}$$

which, together with (6.6.37) and (6.6.38), defines the distribution of density ρ disturbances in terms of velocity v disturbances.

From Eq. (6.6.30) of state, by substituting (6.6.37) into this equation, and using $\rho_*^{(2)} \sim \varepsilon^2$, we obtain

$$p_* = \overline{C}_0^2\left(\rho_*^{(1)} + \rho_*^{(2)}\right) + \frac{\varkappa + 1}{2}\overline{C}_0^2\left(\rho_*^{(1)}\right)^2$$

$$+ 2\overline{C}_0^2\beta_* \frac{\partial^2 \rho_*^{(1)}}{\partial x_*^2} - 2\overline{C}_0\mu_* \frac{\partial v_*}{\partial x_*} + O(\varepsilon^3)$$

Expressing $\rho_*^{(1)}$ and $\rho_*^{(2)}$ in terms of v_*, and using the expressions for $\bar{C}$ in (6.6.31), we obtain formulas interrelating p_* and v_* in terms of each other in a quasi-simple wave

$$p_* = \overline{C}_0 \left[v_* + \frac{\varkappa + 1}{4\overline{C}_0} v_*^2 - \mu_* \frac{\partial v_*}{\partial x_*} + \beta_* \frac{\partial^2 v_*}{\partial x_*^2} \right] + O(\varepsilon^3)$$
$$v_* = \frac{1}{\overline{C}_0} \left[p_* - \frac{\varkappa + 1}{4\overline{C}_0^2} p_*^2 + \mu_* \frac{\partial p_*}{\partial x_*} - \beta_* \frac{\partial^2 p_*}{\partial x_*^2} \right] + O(\varepsilon^3) \tag{6.6.50}$$

Having differentiated with respect to x_* and t_*, we have

$$\frac{\partial v_*}{\partial x_*} = \left\{ \frac{1}{\overline{C}} \frac{\partial p_*}{\partial x_*} - \mathscr{L}[p_*] \right\} + O(\varepsilon^3) = \frac{1}{\overline{C}_0} \frac{\partial p_*}{\partial x_*} + O(\varepsilon^2)$$
$$\frac{\partial v_*}{\partial t_*} = \left\{ \frac{1}{\overline{C}} \frac{\partial p_*}{\partial t_*} + \mathscr{L}[p_*] \right\} + O(\varepsilon^3) = \frac{1}{\overline{C}_0} \frac{\partial p_*}{\partial t_*} + O(\varepsilon^2)$$

Here, it is taken into consideration that in accordance with (6.6.27) and (6.6.28)

$$\mathscr{L}[v_*] = \frac{1}{C_0} \mathscr{L}[p_*] + O(\varepsilon^2), \quad \frac{\partial p_*}{\partial t_*} = -C_0 \frac{\partial p_*}{\partial x_*} + O(\varepsilon^2)$$

Substituting these expressions into (6.6.43), we obtain that, within the adopted accuracy, the differential equation for p_* is of the same form that Eq. (6.6.43) for v_*

$$\frac{\partial p_*}{\partial t_*} + \overline{C} \frac{\partial p_*}{\partial x_*} + \mathscr{L}[p_*] = 0 \tag{6.6.51}$$

Using another set of variables, we obtain

$$t_*, \quad \xi = x_* - \overline{C}_0 t_*$$
$$\tilde{v} = \frac{\varkappa + 1}{2} v_* = \frac{(\varkappa + 1)\, v}{2\sqrt{\alpha_{20} p_0 / \rho_1^\circ}}, \quad \tilde{p} = \frac{\varkappa + 1}{2\sqrt{\varkappa}} \approx \frac{p - p_0}{p_0} \tag{6.6.52}$$

Here, consideration is given to the fact that, for the gas polytropic exponent $1 \le \varkappa \le 5/3$, it may be assumed

$$\tfrac{1}{2}(\varkappa + 1)/\sqrt{\varkappa} \approx 1 \tag{6.6.53}$$

As a consequence, we obtain that Eqs. (6.6.43) and (6.6.51) assume the form of the Burgers-Corteveg-de Vries equation governing propagation of the reduced disturbances of both velocity $\tilde{v}$ and pressure $\tilde{p}$ in a system of coordinates ξ, t_*, which is moving along axis x with an equilibrium speed of sound C_0 relative to an undisturbed medium

$$\frac{\partial u}{\partial t_*} + u\frac{\partial u}{\partial \xi} - \mu_* \frac{\partial^2 u}{\partial \xi^2} + \beta_* \frac{\partial^3 u}{\partial \xi^3} = 0 \qquad (u = \tilde{v},\ \tilde{p}) \tag{6.6.54}$$

To solve the obtained equations, the boundary conditions at $\xi = 0$ and the initial data defining the initial disturbance of either velocity or pressure at $t_* = 0$ must be prescribed. Let us consider the initial pressure disturbance in the form

$$t_* = 0:\ \Delta\bar{p} = \Delta\bar{p}_0 \cdot \varphi(\xi) \tag{6.6.55}$$

where, $\Delta\bar{p}_0 = \Delta p_0/p_0$ is the pressure disturbance amplitude ($\Delta p_0 \sim \varepsilon$), $\varphi(\xi)$ characterizes the configuration of the original disturbance which is assumed to be fairly long. The latter implies that the inherent length L_0 of the original disturbance and, accordingly, its characteristic duration satisfy conditions that follow from Eq. (6.6.9)

$$L_0 \gtrsim L_*, \quad \Delta t_0 = L_0/C_\alpha \gtrsim \tau_*$$

Then, there may be adopted as linear L_* and time τ_* scales, relative to which are determined the independent variables ξ and t_* as well as coefficients β_* and μ_*, the following quantities

$$L_* = L_0, \quad \tau_* = \tau_0$$

This permits the use of the smallness of coefficients μ_* and β_* to be retained. Thus, L_0 (or τ_0), characterizing the original disturbance, appears in relations for determining μ_* and β_* also with a fixed configuration of disturbance $\varphi(\xi)$; however, with its various amplitudes Δp_0 and lengths L_0, its evolution (given the fact of nonlinearity, and presence of dissipation and dispersion) is determined by three dimensionless parameters: μ_*, β_*, and Δp_0. It turns out that evolution of the impulse form is defined by only two independent dimensionless parameters constituted of these three parameters indicated above. To prove it, let us make use of other dimensionless variables

$$\tilde{t} = \tilde{p}_0 t_* = \frac{\Delta p_0}{p_0}\frac{C_\alpha}{L_*}\,t, \quad \xi = \frac{x - C_0 t}{L_*} \qquad (L_* = L_0)$$

$$P = \frac{\tilde{p}}{\tilde{p}_0} = \frac{\Delta p}{\Delta p_0} = O(1) \quad \left(\tilde{p}_0 = \frac{\varkappa + 1}{2\sqrt{\varkappa}}\,\Delta\bar{p}_0 \approx \frac{\Delta p_0}{p_0}\right) \tag{6.6.56}$$

Here, P is the pressure disturbance related to the original disturbance Δp_0 amplitude. Then, from (6.6.54), we obtain

$$\frac{\partial P}{\partial \tilde{t}} + P\frac{\partial P}{\partial \xi} - \frac{1}{\mathrm{Re}}\frac{\partial^2 P}{\partial \xi^2} + \frac{1}{\sigma^2}\frac{\partial^3 P}{\partial \xi^3} = 0$$

$$\tilde{t} = 0:\ P(0, \xi) = \varphi(\xi) \tag{6.6.57}$$

$$\text{Re} = \frac{\tilde{p}_0}{\mu_*} = \frac{3(\varkappa+1)}{4}\frac{L_* p_0}{\mu_{\text{эф}} C_\alpha}\frac{\Delta p_0}{p_0} \geqslant O(1)$$

$$\sigma = \sqrt{\frac{\tilde{p}_0}{\beta_*}} = \frac{L_*}{a_0}\sqrt{\frac{6\alpha_{20}\Delta p_0}{p_0}} \geqslant O(1)$$

Thus, with the predetermined original configuration $\varphi(\xi)$ of disturbance, evolution of this configuration is defined by two parameters **Re** and σ in the B-C-dV equation; these parameters characterize, respectively, dissipation and dispersion, and are determined by physical properties and state of medium ($\varkappa$, μ_{eff}, α_{20}, $\overset{0}{\rho}_1$, and p_0), and, also, by characteristics of the initial impulse, in particular, by its length (L_0) and amplitude (Δp_0).

In the dispersion-free case ($\beta_* = 0$, or $\sigma^{-2} = 0$), the B-C-dV equation is reduced to the Burgers equation

$$\frac{\partial u}{\partial t_*} + u\frac{\partial u}{\partial \xi} - \mu_* \frac{\partial^2 u}{\partial \xi^2} = 0 \tag{6.6.58}$$

and, in the dissipation-free case ($\mu_* = 0$, or $\mathbf{Re}^{-1} = 0$), the B-C-dV equation is reduced to the Corteveg-de Vries equation

$$\frac{\partial u}{\partial t_*} + u\frac{\partial u}{\partial \xi} + \beta_* \frac{\partial^3 u}{\partial \xi^3} = 0 \tag{6.6.59}$$

The B-C-dV equation may be written in the divergent form

$$\frac{\partial u}{\partial t_*} = \frac{\partial S}{\partial \xi}, \quad S = -\frac{u^2}{2} + \mu_* \frac{\partial u}{\partial \xi} - \beta_* \frac{\partial^2 u}{\partial \xi^2} \tag{6.6.60}$$

whence, in case the principal moment has a limit

$$\int_{-\infty}^{\infty} u(0, \xi)\, d\xi = M, \qquad |M| < \infty \tag{6.6.61}$$

the "impulse" integral follows if the consideration is givien to the fact that $S(t_*, -\infty) = S(t_*, +\infty) = 0$, i.e.

$$\int_{-\infty}^{\infty} u(t_*, \xi)\, d\xi = M = \text{const} \tag{6.6.62}$$

Because of numerous practical applications in physics and mechanics (gravitational waves on a liquid surface, electromagnetic waves in plasma, in dielectrics, etc.), both the Burgers equation and the Corteveg-de Vries equation are investigated in detail. We shall now briefly discuss the properties of solutions of these equations (V. I. Karpman, 1973).

Properties of solutions of Burgers equation. The general solution of the Burgers equation may be written in quadratures of initial conditions. Asymptotes of this solution for compression waves ($M > 0$), and rarefaction waves ($M < 0$) at $t_* \to \infty$ are shown in Fig. 6.6.1. In this case, the effect of initial conditions is exhibited only through the principal moment M, and the configuration of the initial impulse at $t = 0$ is insignificant. With $\mu_* \to 0$, the asymptotic profile is a triangle (curves 2 in Fig. 6.6.1) with a shock in the front (for compression waves: $M > 0$), or in the back (for rarefaction waves: $M < 0$) part of the profile. The dimension of the shock equals $\sqrt{2M/t_*}$, and the wave front moves in conformity with the law $\sqrt{t_*}$, so that the total area of the profile according to (6.6.62) does not vary. Viscosity contributes to the shock spreading in such a manner that its thickness δ varies proportional to $\mu_* t_*^{1/2}$ (curves *1* in Fig. 6.6.1).

The Burgers equation has a stationary solution related to a uniform motion of the compression wave ($M \to +\infty$) through an undisturbed medium with velocity w_0 (in a system of coordinates $\xi\tau$, which moves with a speed C_0 relative to an undisturbed medium) with no variation of the wave configuration

$$u = u(\xi - w_0 t_*) \qquad (6.6.63)$$

$$u(\xi) = u_e \left[1 + \exp \frac{u_e \xi}{2\mu_*} \right]^{-1}, \quad w_0 = \frac{u_e}{2} = \text{const}$$

We shall emphasize that this wave moves with a velocity $\bar{D}_0 = \bar{C}_0 + 1/2 u_e$ relative to the undisturbed medium. Thickness of a transient zone of the indicated wave is $\delta \sim \mu_* L_0 / u_e$.

In the case under consideration, the nonstationary disturbance which satisfies conditions at $t = 0$

$$\begin{aligned} &u(\xi) \to 0 \quad \text{at} \quad \xi \to \infty \\ &u(\xi) \to u_e > 0 \quad \text{at} \quad \xi \to -\infty \qquad (M \to \infty) \end{aligned} \qquad (6.6.64)$$

assumes, at $t_* \to \infty$, the form determined by the stationary solution (6.6.63).

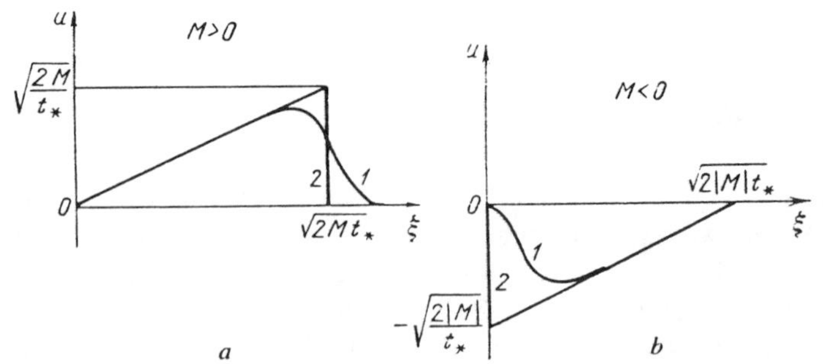

Figure 6.6.1 Asymptotic configuration of the solution of Burgers equation at $t \to \infty$ and finite M for wave compression (*a*) and rarefaction (*b*). Curves *1* relate to finite μ_*, and curves *2* to an asymptote at $\mu_* \to 0$.

Properties of solutions of the Corteveg-de Vries equation. The Corteveg-de Vries equation has an analytical solution adequate to a stationary solitary wave (soliton) propagating with a fixed velocity with no variation of its profile (Fig. 6.6.2*a*)

$$u = u(\xi - w_0 t_*)$$
$$u(\xi) = \tilde{p}_0 \left[\operatorname{ch} \xi \sqrt{\frac{\tilde{p}_0}{12\beta_*}} \right]^{-2}, \quad w_0 = \frac{\tilde{p}_0}{3} = \frac{\varkappa + 1}{6\sqrt{\varkappa}} \Delta\bar{p}_0 \qquad (\Delta\bar{p}_0 > 0) \tag{6.6.65}$$

The soliton moves relative to an undisturbed medium with a velocity $\bar{C}_0 + \frac{1}{3}\tilde{p}_0$, i.e., the faster, the larger its compression amplitude $\tilde{p}_0$. The inherent thickness L of a soliton is

$$L = L_* \sqrt{\frac{12\beta_*}{\tilde{p}_0}} \tag{6.6.66}$$

If the thickness ($L_* = L$) of a soliton under consideration is taken as L_* (appearing in the expression for the parameter β_*) in such a manner that $u(1) \approx 0.42\tilde{p}_0$, then

$$\beta_* = \frac{\tilde{p}_0}{12}, \quad \sigma^2 = \frac{\tilde{p}_0}{\beta_*} = 12 \tag{6.6.67}$$

For each configuration of the initial impulse $\varphi(\xi)$, a critical value σ_c exists such that at $\sigma > \sigma_c$ (sufficiently strong and long disturbances); the original disturbance is divided into solitons, whose configurations resemble (6.6.65); the solitons of higher amplitude move ahead of all others because their velocity relative to the sonic wave is proportional to the amplitude. For the initial disturbances whose shape resembles

$$\varphi(\xi) = \exp(-\xi^2) \tag{6.6.68}$$

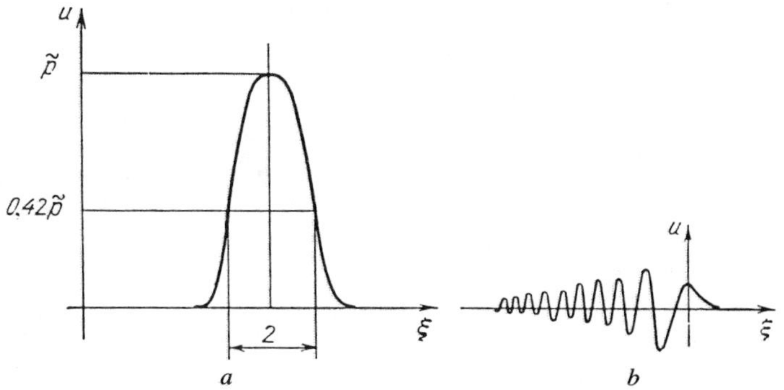

Figure 6.6.2 Soliton (*a*) and a wave packet (*b*) corresponding to Corteveg-de Vries equations.

we have $\sigma_c \approx \sqrt{12}$. The number of solitons is estimated as simple as

$$N \approx \sigma^2/12 \qquad (6.6.69)$$

At $\sigma \ll \sigma_c$ (weak and short impulses), the nonlinearity of the Corteveg-de Vries equation has an insignificant effect, and the solution has a configuration of a wave packet spreading with time (Fig. 6.6.2*b*); at $\sigma \to 0$, this solution approaches that of a linearized Corteveg-de Vries equation (without a nonlinear term). The solution of the latter may be written in an analytical form.

At $\sigma \sim \sigma_c$, the solution has a configuration consisting of both solitons and wave packet.

For an original disturbance of infinite duration of type (6.6.64), the Corteveg-de Vries equation does not yield (at $t_* \to \infty$) a stationary wave, since there is no dissipative mechanism in this equation.

For a negative disturbance (rarefaction waves, $\Delta p < 0$), the solution has only the configuration of a wave packet (Fig. 6.6.2*b*) spreading with time.

Properties of solutions of the Corteveg-de Vries equation. In the $\sigma/\mathbf{Re} \ll 1$ case, the dissipative term is weak, and the indicated properties of the CdV equation take place for the BCdV equation as well. However, with the growth of $\sigma/\mathbf{Re}$, the dissipation qualitatively alters the solution reducing the extent of manifestation of both oscillations and solitons; in this situation, the solitons decay due to dissipation in the process of their propagation.

The BCdV equation, as well as the Burgers equation, admits a solution in the form $u = u(\xi - w_0 t_*)$, which corresponds to a stationary compression wave propagating in a system of coordinates ξt_* with a fixed velocity with no change of configuration; this wave transforms the medium from an undisturbed state ($u = u_0 = 0$) into another state ($u = u_e > 0$). As distinct from the Burgers equation, this wave may have an oscillatory structure. Substituting $u = u(\xi - w_0 t_*)$ into Eq. (6.6.54), we obtain an ordinary differential equation (cf. Eq. (6.6.33)) for a wave structure $u(\xi)$ with the appropriate boundary conditions

$$\beta_* \frac{d^3u}{d\xi^3} - \mu_* \frac{d^2u}{d\xi^2} + (u - w_0)\frac{du}{d\xi} = 0$$

$$\xi \to \infty: \quad u = 0, \quad \frac{du}{d\xi} = \frac{d^2u}{d\xi^2} = 0 \qquad (6.6.70)$$

$$\xi \to -\infty: \; u = u_e, \quad \frac{du}{d\xi} = \frac{d^2u}{d\xi^2} = 0$$

Integrating this equation with given boundary conditions, we obtain

$$\beta_* \frac{d^2u}{d\xi^2} - \mu_* \frac{du}{d\xi} + \frac{u^2}{2} - w_0 u = 0, \qquad w_0 = \frac{u_e}{2} = \frac{\varkappa + 1}{4\sqrt{\varkappa}}\Delta p_e \qquad (6.6.71)$$

Thus, the stationary wave under consideration moves relative to the undisturbed medium with a velocity $D_0 = \bar{C}_0 + w_0$, whose expression in terms of $\Delta\bar{p}_e$ coincides with formula (6.3.10) obtained for weak waves ($\Delta p_e \ll 1$) (cf. also the remark concerning Eq. (6.6.33)).

Let us analyze the behavior of the solution $u(\xi)$ in the vicinity of state behind the wave ($\xi \to -\infty$, $u \to u_e$), when $u = u_e + y$, $|y| \ll u_e$. For the disturbance y, we obtain a linear equation

$$\beta_* \frac{d^2y}{d\xi^2} - \mu_* \frac{dy}{d\xi} + \frac{u_e}{2} y = 0 \tag{6.6.72}$$

The characteristic roots consistent with this equation are

$$K^{(1,2)} = \frac{\mu_*}{2\beta_*} \pm \sqrt{\frac{\mu_*^2}{4\beta_*^2} - \frac{u_e}{2\beta_*}} \tag{6.6.73}$$

In order that the stationary compression wave has an oscillatory structure, the following must hold

$$u_e > \frac{\mu_*^2}{2\beta_*} = \frac{4}{3\varkappa} \frac{\mu_{\text{эф}}}{a_0^2 \rho_0 p_0} \qquad \left(u_e = \frac{p_e - p_0}{p_0}, \ \frac{(\varkappa + 1)\, v_e}{2\alpha_{20} C_\alpha} \right) \tag{6.6.74}$$

which is in agreement with a similar condition (6.3.27) obtained from the complete system of equations based on the same physical scheme involving an effective viscosity as the BCdV approximation being analyzed.

Employment of the BCdV equation for investigations of weak waves in liquids with gas bubbles was proposed by V. E. Nakoryakov, V. V. Sobolev, I. R. Shraiber (1972) and by L. Wijngaarden (1972). Later on, this method was substantiated and developed in a number of theoretical and experimental investigations by B. E. Nakoryakov et al.

The theory of shock waves in bubbly liquids based on the BCdV equation, in spite of the limitations of this equation (unidirectional weak waves, absence of reflected waves, insufficient account of the heat exchange effects), nevertheless provided the following extremely important result. The evolution of an impulse of a predetermined configuration as a function of its amplitude and duration, and as a function of the original pressure and physical characteristics of a bubbly medium is defined only by two dimensionless parameters: **Re** and σ. The indicated theory permitted various types of disturbances to be singled out: wave packet (Fig. 6.6.2*b*), soliton (Fig. 6.6.2*a*), spread waves similar to thermal waves, triangular waves with a steep front (Fig. 6.6.1) whose realization is defined by parameters **Re** and σ. Presently, thanks to accumulated data, each type of wave may be related to a certain region on the **Re**, σ-diagram. Such a diagram (V. Kuznetsov et al., 1978; V. E. Nakoryakov, B. G. Pokusaev, and I. R. Shraiber, 1983) is depicted in Fig. 6.6.3. The indicated wave types are also found experimentally.

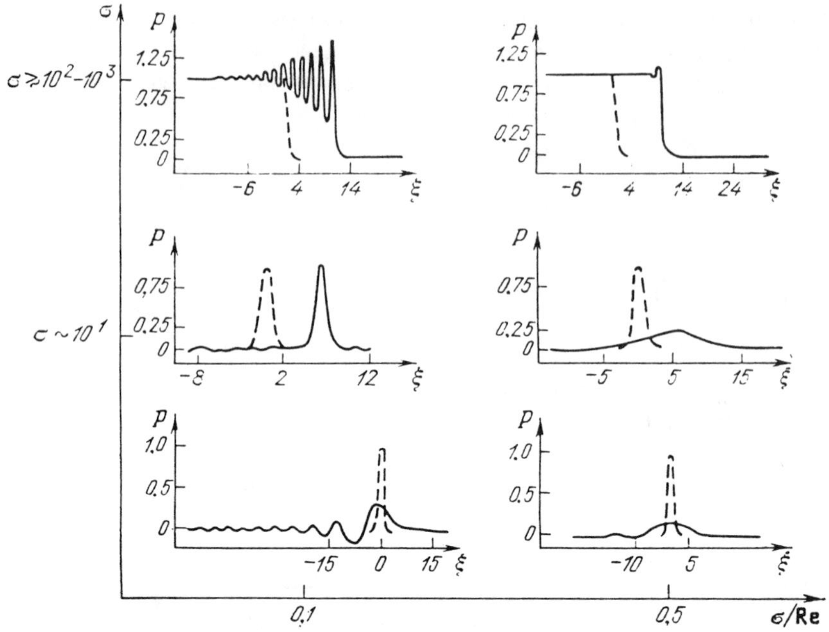

Figure 6.6.3 Generalization of solutions of the BCdV equation. Profiles of the original impulse are shown by dashed lines.

Bubbly media may be used for simulation of wave phenomena in other media (for instance, in plasma) where these phenomena are described by the BCdV equation.

By taking the interphase heat exchange into more consistent consideration than the effective viscosity scheme, V. E. Nakoryakov et al. (1983) obtained a more complicated equation for a quasi-simple wave for a particular case of low compression of gas; this equation contains, in addition to terms appearing in the BCdV equation, an integral, or hereditary, term (Duhamel integral of type (2.6.15) but for heat exchange within a bubble), which is determined at time t by the history of the bubble pressure variation for the entire period of the process time from 0 to t.

§6.7 NONSTATIONARY WAVE MOTIONS OF BUBBLY LIQUIDS

This section presents a more general (as compared to §6.6) approach to numerical modelling and investigation of one-dimensional nonstationary motions of bubbly media based on the two-temperature single-velocity scheme with an incompressible carrying liquid (see §1.5). Use of this method enables us to analyze significantly stronger (than those described in §6.6) disturbances, when $\Delta p/p_0 \gtrsim 1$, by giving more consistent consideration to thermal effects. The method described below together with a number of obtained results were first reported by A. I. Ivandaev, R. I. Nigmatulin (1976, 1978), and R. I. Nigmatulin (1981).

Equations of nonstationary motion of a bubbly liquid with an incompressible carrying (continuous) phase. In addition to the adopted assumptions mentioned in §1.5, we assume the absence of phase transitions, that velocities of macroscopic phase motion are equal to each other, and that the true density of carrying liquid is constant

$$j = 0, \quad v_1 = v_2 = v, \quad \overset{\circ}{\rho}_1 = \text{const} \tag{6.7.1}$$

Phase transitions in fairly weak shock waves propagating in "cold" liquids with bubbles of nonsoluble gas are absent (see discussion of (1.6.1)).

Admissibility of the phase-velocity coincidence assumption follows from the analysis[2] of stationary shock waves (see §6.4). The same analysis has proven the acceptability of the carrying-liquid incompressibility provided that the condition (6.4.17) is satisfied.

The above-indicated assumptions (6.7.1) significantly simplify the theoretical investigation of the nonstationary flow of bubbly mixtures.

The problem of a one-dimensional nonstationary flow of a single-velocity medium may be reasonably solved in Lagrangian coordinates (r, t), where r is the distance from a particle to the origin at initial time $t = 0$. The parameter values at $t = 0$ are denoted by a subscript 0. In particular, $\rho_0 = \rho(r, 0) = \rho_0(r)$ is the mixture density at $t = 0$. The current position of the medium's particle is characterized by the Eulerian coordinate $x(r, t)$, so that

$$\frac{\partial x}{\partial r} = \frac{\rho_0}{\rho}, \quad \frac{\partial x}{\partial t} = v \tag{6.7.2}$$

As a result of assumed simplifications, the system of equations (1.5.4) of motion for a monodisperse mixture of incompressible liquid with gas

[2]It should be noted that the relative motion of liquid and bubbles, in addition to effects already discussed, may cause intensification of the heat exchange within a bubble, distortion of the bubble sphericity, and, as an extreme consequence of the latter effect, bubble fragmentation. Nevertheless, there exists a considerable range of the wave-motion regime parameters when these effects do not exhibit themselves. At the same time, there is a range of regime parameters when these effects may have predominant significance. In particular, fragmentation of original bubbles into smaller bubbles—which generally occurs across the front of a sufficiently strong wave—results in the wave propagating through a medium with smaller (than in original state) bubbles, thereby reducing the wave relaxation-zone thickness by many times. Consequently, the equilibrium scheme of the mixture may become quite sufficient; this scheme is reduced to a model of a perfect compressible liquid with the predetermined equation of state (see §1.5).

bubbles in the absence of phase transitions in the framework of a single-velocity two-temperature approximation assumes the following form in Lagrangian variables

$$\frac{\partial \rho_1}{\partial t} + \frac{\rho_1 \rho}{\rho_0}\frac{\partial v}{\partial r} = 0, \quad \frac{\partial \rho_2}{\partial t} + \frac{\rho_2 \rho}{\rho_0}\frac{\partial v}{\partial r} = 0 \quad (\rho = \rho_1 + \rho_2)$$

$$\frac{\partial a}{\partial t} = w, \quad \frac{\partial}{\partial t}\left(\overset{\circ}{\rho}_2 a^3\right) = 0, \quad \alpha_2 = \frac{\rho_2}{\overset{\circ}{\rho}_2} = 1 - \frac{\rho_1}{\overset{\circ}{\rho}_1}, \quad \overset{\circ}{\rho}_1 = \text{const}$$

$$(1 - \varphi^{(1)})\, a\frac{\partial w}{\partial t} = \frac{p_2 - p_1 - 2\Sigma/a}{\overset{\circ}{\rho}_1} - \frac{4\nu_1^{(v)} w}{a} - (1 - \varphi^{(2)})\frac{3w^2}{2} \tag{6.7.3}$$

$$\frac{\partial v}{\partial t} + \frac{1}{\rho_0}\frac{\partial p_{1*}}{\partial r} = g, \quad p_{1*} = -\sigma_{1*} = p_1 + \alpha_2\left(p_2 - p_1 - \frac{2\Sigma}{a} - \overset{\circ}{\rho}_1 w^2\right)$$

$$\frac{\partial p_2}{\partial t} = \frac{3(\gamma_2 - 1)}{2a^2}\lambda_2 \mathrm{Nu}_2 (T_1 - T_2) - 3\gamma_2\frac{p_2 w}{a}$$

$$T_1 = T_0 = \text{const}, \quad T_2 = p_2/R_2\overset{\circ}{\rho}_2$$

The generalized Rayleigh-Lamb equation (the fifth differential equation) may be transformed identically to a form in which, instead of average pressure p_1, the reduced pressure p_{1*} appears

$$\left(1 - \varphi_*^{(1)}\right) a\frac{\partial w}{\partial t} = \frac{p_2 - p_{1*} - 2\Sigma/a}{\overset{\circ}{\rho}_1} - \frac{4\nu_1^{(v)} w}{a} - \left(1 - \varphi_*^{(2)}\right)\frac{3w^2}{2}$$

$$\varphi_*^{(1)} = \alpha_1\varphi^{(1)} + \alpha_2, \quad \varphi_*^{(2)} = \alpha_1\varphi^{(2)} + \alpha_2 \tag{6.7.4}$$

$$p_1 = p_{1*} - \alpha_2\left(p_2 - p_{1*} - 2\Sigma/a + \overset{\circ}{\rho}_1 w^2\right)$$

With the prescribed physical properties of liquid (density $\overset{0}{\rho}_1$, viscosity $\nu_1^{(v)}$, surface tension Σ) and those of gas (adiabatic index γ_2, gas constant R_2, coefficient of heat conductivity λ_2), and also with the given parameter of interphase heat exchange $\mathbf{Nu}_2$, and in the presence of initial and boundary conditions, the represented system of equations is closed.

In connection with the condition of incompressibility, we shall perform some identical transformations. The first two mass-conservation equations (6.7.3) of the system may be represented as

$$\frac{\partial \alpha_1}{\partial t} + \frac{\alpha_1 \rho}{\rho_0}\frac{\partial v}{\partial r} = 0, \quad \frac{\partial \alpha_2}{\partial t} + \frac{\alpha_2}{\overset{\circ}{\rho}_2}\frac{\partial \overset{\circ}{\rho}_2}{\partial t} + \frac{\alpha_2 \rho}{\rho_0}\frac{\partial v}{\partial r} = 0$$

whence, upon summing these two equations, we obtain

$$\alpha_2\frac{\partial \overset{\circ}{\rho}_2}{\partial t} + \frac{\overset{\circ}{\rho}_2 \rho}{\rho_0}\frac{\partial v}{\partial r} = 0$$

Using the mass-conservation equation for the bubble (the fourth equation (6.7.3)) in the form

$$a^3 \frac{\partial \overset{\circ}{\rho}_2}{\partial t} + 3a^2 \overset{\circ}{\rho}_2 w = 0$$

we obtain that the longitudinal deformation of mixture with an incompressible continuous phase takes place only due to the bubble radial deformation

$$\frac{\partial v}{\partial r} = \frac{3\rho_0}{\rho} \frac{\alpha_2 w}{a} \tag{6.7.5}$$

In this case, equations governing variation of both mixture density and bubble volumetric concentration may be written as

$$\frac{\partial \rho}{\partial t} = -\frac{\rho^2}{\rho_0} \frac{\partial v}{\partial r} = -3\rho \frac{\alpha_2 w}{a}, \quad \frac{\partial \alpha_2}{\partial t} = \frac{\alpha_1 \rho}{\rho_0} \frac{\partial v}{\partial r} = \frac{3\alpha_1 \alpha_2 w}{a} \tag{6.7.6}$$

The same equations directly follow from (1.3.9), given the relation between the Lagrangian and Eulerian variables

$$\frac{1}{\rho} \frac{\partial v}{\partial x} = \frac{1}{\rho_0} \frac{\partial v}{\partial r}, \quad \frac{d_2 \overset{\circ}{\rho}_2}{dt} = \left. \frac{\partial \overset{\circ}{\rho}_2}{\partial t} \right|_r \tag{6.7.7}$$

Let us differentiate both the momentum equation with respect to r (recall that, g = constant, $\rho_0 = \rho_0(r)$), and the continuity equation (6.7.5) with respect to t

$$\frac{\partial^2 v}{\partial r\, \partial t} = -\frac{1}{\rho_0} \frac{\partial^2 p_{1*}}{\partial r^2} + \frac{1}{\rho_0^2} \frac{d\rho_0}{dr} \frac{\partial p_{1*}}{\partial r}$$

$$\frac{\partial^2 v}{\partial t\, \partial r} = 3\rho_0 \left(\frac{w}{\rho a} \frac{\partial \alpha_2}{\partial t} + \frac{\alpha_2}{\rho a} \frac{\partial w}{\partial t} - \frac{\alpha_2 w}{\rho^2 a} \frac{\partial \rho}{\partial t} - \frac{\alpha_2 w}{\rho a^2} \frac{\partial a}{\partial t} \right) \tag{6.7.8}$$

These two expressions in the domain of continuous motion may be set equal to each other. If, in this case, the time derivatives in the last equation are replaced by their values in accordance with both (6.7.6) and equations of radial motion, we obtain

$$\frac{\partial^2 p_{1*}}{\partial r^2} + K \frac{\partial p_{1*}}{\partial r} + M = 0, \qquad K(r) = -\frac{1}{\rho_0} \frac{d\rho_0}{dr}$$

$$M(p_{1*}, a, w) = \frac{3\rho_0^2 \alpha_2}{\rho a^2 (1 - \varphi^{(1)})} \left\{ \frac{p_2 - p_{1*} - 2\Sigma/a}{\overset{\circ}{\rho}_1} - \frac{4\nu_1^{(v)} w}{a} \right. \tag{6.7.9}$$

$$\left. + \left(1 - 4\varphi_*^{(1)} + 3\varphi_*^{(2)}\right) \frac{w^2}{2} \right\}$$

Thus, the condition of incompressibility of continuous phase, as well as in hydrodynamics of a single-phase incompressible liquid, led to a second-order differential equation which contains derivatives with respect only to the spatial coordinate. This equation makes it possible to determine the pressure distribution at any point in time provided the remaining parameters (α_2, a, w, and p_2) are prescribed at the same point in time. The overall effect on the pressure distribution in a non-single-phase mixture (the presence of bubbles) is specified on the right-hand side of this equation. In a single-phase ($\alpha_2 = 0$), homogeneous (ρ_0 = constant) liquid, Eq. (6.7.9) always yields a rectilinear pressure-distribution profile

$$\frac{\partial^2 p_{1*}}{\partial r^2} = \frac{\partial^2 p_1}{\partial r^2} = 0$$

In order to solve the differential equation (6.7.9) for pressure p_{1*}, the boundary conditions on the tube ends $r_* = (0, L)$ must be prescribed; then, the pressure p_{1*} distribution at any point in time is determined from a solution of a boundary-value problem for Eq. (6.7.9), which contains derivatives with respect only to r. Boundary conditions of two kinds may be identified: a boundary condition of the first kind when pressure is prescribed on the boundary

$$r = r_*, \quad p_{1*} = p_{1*}(r_*, t) = \Pi(t) \tag{6.7.10}$$

and the boundary condition of the second kind, which may be obtained when a law of motion is prescribed on the boundary (e.g., speed of the "piston")

$$r = r_*, \quad v = v(r_*, t) = V(t) \tag{6.7.11}$$

Then, recalling the momentum equation, the latter condition predetermines the derivative of pressure with respect to coordinate

$$r = r_*, \quad \frac{\partial p_{1*}}{\partial r} = \rho_0(r)\left[g - \frac{\partial v(r_*, t)}{\partial t}\right] = \rho_0(r_*)\left[g - \frac{dV}{dt}\right] \tag{6.7.12}$$

In a particular case, if there is an immobile rigid wall on the boundary ($V(t) = 0$), we have

$$r = r_*, \quad \frac{\partial p_{1*}}{\partial r} = \rho_0(r_*)\, g \tag{6.7.13}$$

As to the initial conditions, they must prescribe distributions of α_2, a, w, and p_2, with respect to coordinate r at $t = 0$.

It should be kept in mind that, pressure p_{1*} distribution and velocity v distribution at each point in time, including $t = 0$, are not independent variables because of the incompressibility of continuous liquid. The p_{1*} distribution is determined from the above-indicated boundary-value problem in terms of distributions of α_2, a, w, and p_2, and also in terms of boundary conditions on the ends $r_* = (0, L)$, and the distribution of velocity (υ) is found from Eq. (6.7.5), or from the momentum equation. The variation of α_2, a, w, and p_2 in time is defined by differential equations (containing only time derivatives) following from Eqs. (6.7.3) and (6.7.6).

To facilitate both the analysis and solution of the system of equations being discussed, we shall use dimensionless variables (6.3.6) and (6.4.5) together with parameters (6.4.6) denoted, as before, by a bar ($\bar{p}_{1*}$, $\bar{\rho}$, $\bar{\rho}_2^0$, $\bar{a}$); the latter are determined as a ratio of parameters corresponding to their characteristic values (scales) designated by a subscript 0. Some fixed values (p_{10}, ρ_1^0, ρ_{20}^0, a_0), being characteristics for distributions of p_{1*}, ρ, ρ_2^0, and a at the initial time $t = 0$, are chosen as scale factors. The latter, for convenience, are selected so that they satisfy the conditions of equilibrium

$$p_{20} = p_{10} + 2\Sigma/a_0, \quad \overset{\circ}{\rho}_{20} = p_{20}/(R_2 T_0) \tag{6.7.14}$$

As before (see (6.4.5)), the equilibrium speed of sound $C_e = C_\alpha$ is chosen as the inherent velocity of the longitudinal motion, and the speed C_* is chosen as the inherent velocity of radial motion near bubbles. The linear scale L_α is adopted the same as in (6.3.6). Accordingly, the time scale factor t_0 is selected as

$$\bar{t} = \frac{t}{t_0}, \quad \bar{r} = \frac{r}{L_\alpha} \quad \left(L_\alpha = \frac{a_0}{\sqrt{\alpha_{10}\alpha_{20}}}, \quad t_0 = \frac{L_\alpha}{C_\alpha} = \frac{a_0}{C_*}\right) \tag{6.7.15}$$

As a consequence, the system of equations may be reduced to the following form

$$\frac{\partial^2 \bar{p}_{1*}}{\partial \bar{r}^2} - \frac{d\bar{\rho}_0}{d\bar{r}}\frac{\partial \bar{p}_{1*}}{\partial \bar{r}} = \frac{3\bar{\rho}_0^2}{\bar{\rho}\bar{a}^2} \frac{\alpha_2}{\alpha_{10}\alpha_{20}\left(1-\varphi_*^{(1)}\right)}$$

$$\times \left\{\left(1 - 4\varphi_*^{(1)} + 3\varphi_*^{(2)} - 12\alpha_2\right)\frac{\bar{w}^2}{2} - \frac{4\mu_1^* \bar{w}}{\bar{a}} + \bar{p}_2 - \bar{p}_{1*} - \frac{2\Sigma^*}{\bar{a}}\right\}$$

$$\frac{\partial \bar{\upsilon}}{\partial \bar{r}} = \frac{3\bar{\rho}_0}{\bar{\rho}}\frac{\alpha_2 \bar{w}}{\bar{a}}, \quad \frac{\partial \alpha_2}{\partial \bar{t}} = \frac{3\alpha_1\alpha_2\bar{w}}{\bar{a}}, \quad \frac{\partial \bar{a}}{\partial \bar{t}} = \bar{w} \quad (\tilde{\rho} \approx 1 - \alpha_2)$$

$$\left(1-\varphi_*^{(1)}\right)\bar{a}\frac{\partial \bar{w}}{\partial \bar{t}} = \bar{p}_2 - \bar{p}_{1*} - \frac{2\Sigma^*}{\bar{a}} - \frac{4\mu_1^*\bar{w}}{\bar{a}} - \left(1-\varphi_*^{(2)}\right)\frac{3\bar{w}^2}{2} \tag{6.7.16}$$

$$\frac{\partial \bar{p}_2}{\partial \bar{t}} = \frac{\beta_2}{\bar{a}^2}\left(\bar{p}_{20} - \bar{p}_2\bar{a}^3\right) - 3\gamma_2\frac{\bar{p}_2\bar{w}}{\bar{a}} \quad \left(\bar{p}_{20} = \frac{p_{20}}{p_{10}} = 1 + 2\Sigma^*\right)$$

$$\beta_2 = {}^3/_2\gamma_2\lambda_2^*\mathrm{Nu}_2$$

The above-discussed variants ((6.7.10)–(6.7.13)) are considered as the conditions over the boundaries of the selected volume of mixture ($\bar{r} = 0$ and $r = \bar{L}$)

$$\bar{r} = 0\colon\ \bar{p}_{1*} = \bar{\Pi}_0(\bar{t}) \quad \text{or} \quad \frac{\partial p_{1*}}{\partial \bar{r}} = \Gamma_0(\bar{t}) = \bar{\rho}_0\left(\bar{g} - \frac{d\bar{V}_0}{dt}\right)$$

$$\bar{r} = \bar{L} = \frac{L}{L_0}\colon\ \bar{p}_{1*} = \bar{\Pi}_L(\bar{t}) \quad \text{or} \quad \frac{\partial p_{1*}}{\partial \bar{r}} = \Gamma_L(t) = \bar{\rho}_0\left(\bar{g} - \frac{d\bar{V}_L}{d\bar{t}}\right) \qquad (6.7.17)$$

$$\left(\bar{g} = \frac{ga_0}{C_*^2}\sqrt{\alpha_{10}\alpha_{20}}\right)$$

Note that thermophysical properties of phases (surface tension, density and viscosity of liquid, heat conductivity, heat capacity and density of gas) affect the process through the four dimensionless parameters: Σ^*, μ_1^*, β_2 (see (6.4.6)) and γ_2. The external body forces are included in equation through the parameter $\bar{g}$.

Note that in the framework of the effective-viscosity model, a system of equations analogous to (6.7.16) is of a lower order: the polytropic equation $\bar{p}_2 = \bar{a}^{-3\varkappa}$ appears here instead of the last differential equation (heat-influx equation). In addition, in the next to the last equation (equation of radial motion), instead of the dimensionless coefficient of liquid viscosity there appears an analogous coefficient of effective viscosity.

The obtained system consists of 6 differential equations, each of which contains only a single derivative with respect to one of coordinates ($\bar{r}$ or $\bar{t}$). The first two equations of the system are intended for determining the reduced pressure and velocity of mixture at an arbitrary time using the known fields of the rest of the parameters; the remaining equations describe the laws of variation of Lagrangian particles of the medium in time.

For numerical integration of the obtained system of equations, we shall divide the selected volume of the medium by points $\bar{r} = \bar{r}_i (i = 1, 2, \ldots, n)$ into n material particles: the values of all sought functions we determine at points $\bar{r} = \bar{r}_i (i = 1, 2, \ldots, n)$. Then, the last four differential equations in partial time-derivatives of variables α_2, $\bar{a}$, $\bar{w}$, and $\bar{p}_2$ are transformed into $4n$ ordinary differential equations with respect to time, for whose numerical integration the modified Euler-Cauchy method may be conveniently used. To find the pressure $\bar{p}_{1*}$ values at points $\bar{r} = \bar{r}_i$ at each fixed point in time, a linear (for $\bar{p}_{1*}$) boundary-value problem must be solved for the first differential equation (with respect to $\bar{r}$) of the second order with boundary conditions (6.7.17).

To facilitate the solution of this problem, a trial-run approach is advisable. Note that in the above-indicated case of one-dimensional flow with plane waves, the velocity of the medium appears only in the second equation (6.7.16); therefore, to calculate it on each step of integration with respect to time is not necessary; and, in cases when the $\bar{v}$ distribution must be calculated, it is better to do it for each time interval of integration following the computation of p_{1*} distribution based upon the momentum equation which in dimensionless variables is written in the form

$$\frac{\partial \bar{v}}{\partial t} = \bar{g} - \frac{\alpha_{10}\alpha_{20}}{\bar{\rho}_0}\frac{\partial \bar{p}_{1*}}{\partial \bar{r}} \tag{6.7.18}$$

In all examples being discussed below, parameter $\bar{g} = ga_0\,\rho_1^0\sqrt{\alpha_{20}}/p_0$, defining the gravitational effects, is a small value of order of 10^{-4}. The smallness of $\bar{g}$ implies an insignificant influence of gravitational force upon the wave process, which may be analyzed disregarding this effect; afterward, the hydrostatic component $\Delta p_{st} = \rho_0 gr$ of pressure due to the superjacent column of liquid to predict pressure diagrams may be added.

The results of numerical calculations and analysis of nonstationary wave processes in water or in a solution of glycerine in water (1:1) containing gas bubbles (air, carbon dioxide, or helium) are presented below.

Evolution of nonstationary shock waves into stationary waves. The effects of bubble-gas properties. In studying the evolution of shock waves propagating along an infinite tube at fixed pressure p_e and $r = 0$, when the characteristic wave intensity is $\bar{p}_e = p_e/p_0$, for determining the heat transfer coefficient of a gas bubble, the evaluation (1.6.16) was used as the first approximation

$$\mathrm{Nu}_2 = \frac{2a_0}{\sqrt{\nu_{20}^{(T)}t_*}}, \quad \beta_2 = \frac{3\gamma_2}{2C_*}\sqrt{\frac{\nu_{20}^{(T)}}{t_*}} \tag{6.7.19}$$

where t_* is an inherent time of the bubble compression in the wave, or a period of their radial oscillations. For a preliminary evaluation of t_*, formulas (6.4.30)–(6.4.32), obtained from the analysis of stationary waves, were used; in conformity with these formulas, we have

$$\frac{2\pi a_0}{C_*\sqrt{3(\gamma_2\bar{p}_e - 1)}} \leqslant t_* \leqslant \frac{2\pi a_e}{C_*\sqrt{3(\bar{p}_e - 1)}} \qquad \left(t_* = \frac{2\pi L_\alpha}{D\,\mathrm{Im}\,\{k_e\}}\right) \tag{6.7.20}$$

For the purpose of comparison, we shall mention that the period of natural free oscillations of a single bubble equals $t_r = 2\pi a_0/(C_*\sqrt{3\gamma_2})$. Although in the process of evolution to a stationary wave configuration both oscillation period and bubble compression time change across the wave front (more specifically, they are smaller than in a stationary wave, and approach the corresponding characteristic values only as their limits), this change of t_* has a fairly weak impact, unless the fragmentation takes place.

In more general situations, the heat exchange of gas bubble has been prescribed in accordance with schemes (1.6.45) and (1.6.46).

The analysis showed that in bubbly mixtures of low viscosity of liquid ($\mu_1^* \ll 1$), e.g., in air-water mixtures with the bubble size of order of 1 mm ($p_0 \approx 0.1$ MPa), the structure of sufficiently weak waves ($p_e/p_0 < \gamma_2$) evolves from an oscillatory to monotonic pattern. The structure of stronger waves ($p_e/p_0 > \gamma_2$) approaches an ultimate oscillatory configuration in the process of evolution. The characteristic feature of the phenomenon is the growth of the number of oscillations in the process of wave evolution. The wave evolution in indicated mixtures stems from the effects of interphase heat exchange and also from the effects of transfer of kinetic energy of radial motion to neighboring volumes of mixture caused by pressure disturbance (but not on account of viscosity-effects during the relative phase motion). By way of example, Fig. 6.7.1 represents pressure profiles at various points in time following the pressure raise at $r = 0$ from p_0 to p_e, and also when it is maintained invariable. The limiting stationary configuration of a wave in a corresponding mixture has already been discussed above, in §§6.4 and 6.5 in connection with L. Noordzij's experiment (1971), and is depicted in Figs. 6.4.5 and 6.5.4. This stationary wave is monotonic and is established at distances of about four meters. At smaller distances, when the wave is still nonstationary, it has oscillations; this, in particular, was obtained in L. Noordzij's experiment (1971) for the indicated conditions (see §6.4). The pressure sensors in this experiment were arranged at distances of an order of 1 meter from the wave-initiation location, and, therefore, they recorded only nonstationary wave profiles.

The effects of heat exchange, in particular variations of a parameter $\mathbf{Nu}_2$, on the wave evolution process have also been investigated. Figure 6.7.2 represents the predicted pressure profiles realized at a fixed point in time after the shock but with various values of $\mathbf{Nu}_2$ (or β_2). It is evident that in conditions of almost adiabatic ($\mathbf{Nu}_2 = 3$, $\beta_2 \ll 1$) and almost isothermal ($\mathbf{Nu}_2 = 3000$, $\beta_2 \gg 1$) gas behavior, the dissipative effects are displayed far less, and oscillations are well-defined. The dissipation maximum is observed here at $\mathbf{Nu}_2 \approx 300$ (or $\beta_2 \sim 1$). Since the actual values of $\mathbf{Nu}_2$ (or β_2) depend on the gas diffusivity $\nu_2^{(T)}$, it follows that the wave structure evolution and attenuation of short impulses in a bubbly mixture must depend on the gas properties within a bubble, or, more specifically, on its diffusivity

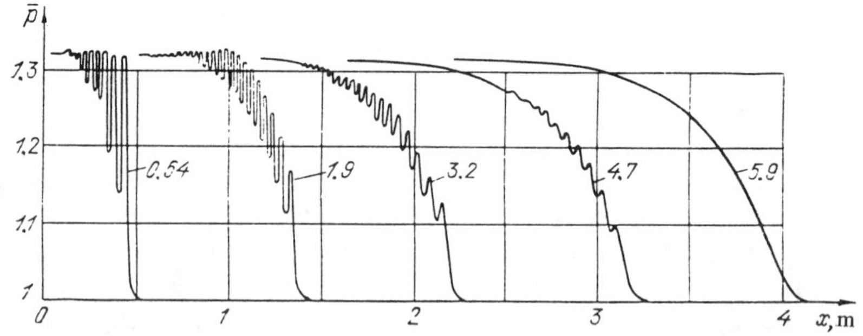

Figure 6.7.1 Evolution of a shock wave during stationary effects ($\bar{p}_e = p_e/p_0 = 1.32$) upon an air-water bubbly mixture ($p_0 = 0.09$ MPa, $\alpha_{20} = 0.025$, $a_0 = 1.4$ mm). The labels on curves indicate time in microseconds.

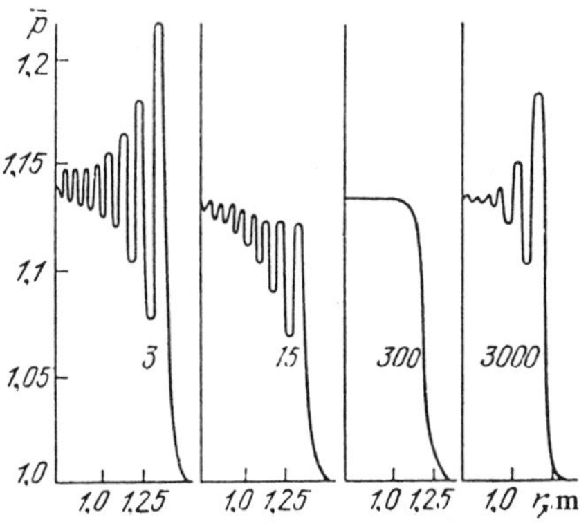

Figure 6.7.2 The effect of the heat exchange parameter $\mathbf{Nu}_2$ on the pressure profile in wave at time $t = 15$ microseconds measured from the beginning of a stationary pressure ($\bar{p}_e = p_e/p_0 = 1.13$) upon the air-water bubbly mixture ($p_0 = 0.1045$ MPa, $\alpha_{20} = 0.017$, $a_0 = 1.2$ mm). Labels on curves indicate values of $\mathbf{Nu}_2$.

$\nu_2^{(T)}$. In this connection, the comparison of three gases proves to be representative (see also the discussion of Fig. 1.6.2): carbon dioxide (CO_2), air and helium. At $p = 0.1$ MPa and $T = 293$ K, the diffusivity $\nu_2^{(T)}$ of helium is one order higher than that of air, and near 20 times higher than that of CO_2 (see Appendix).

The ratios of dimensionless heat exchange coefficients β_2 for bubbles of various gases, given the effect of $\nu_2^{(T)}$ on $\mathbf{Nu}_2$ (see (6.7.20)), equal $\gamma_2(\nu_2^{(T)})^{1/2}$. For the above-indicated gases—helium, nitrogen (air) and carbon dioxide—the values of β_2 are interrelated as 3.5:1:0.62.

In the regimes under consideration, β_2 are sufficiently small, and their increase must entail an intensification of dissipative effects. Therefore, attenuation of oscillations and formation of monotonic wave configuration in mixtures with helium bubbles, with all other conditions the same, must proceed faster than in mixture with air bubbles and CO_2 (cf. Fig. 1.6.2). The obtained analytical results and their comparison with experimental data confirm this conclusion. As an example, Fig. 6.7.3 represents in the same scale both predicted and experimental pressure oscillograms in shock waves propagating in mixtures of aqueous solution of glycerine with helium bubbles, air, and carbon dioxide. The significant impact of a kind of gas used upon the shock-wave structure in a bubbly medium is quite obvious. In particular, in mixtures with bubbles of carbon dioxide and air, the shock waves are of an explicitly oscillatory nature, and in a mixture with helium bubbles they are monotonic. The predicted oscillograms are in satisfactory agreement with experimental results.

The paradox of wave properties of bubbly mixtures with inert gas is that the wide variation of such thermophysical properties of liquid phase (which occupies nearly total volume of mixture) as heat-conduction and viscosity (that, for instance, may result from changing the glycerine concentration in a solution), does not affect wave propagation, its structure, and attenuation. Only when approaching a condition of pure glycerine does the liquid viscosity begin to have a noticeable effect. At the same time, altering the bub-

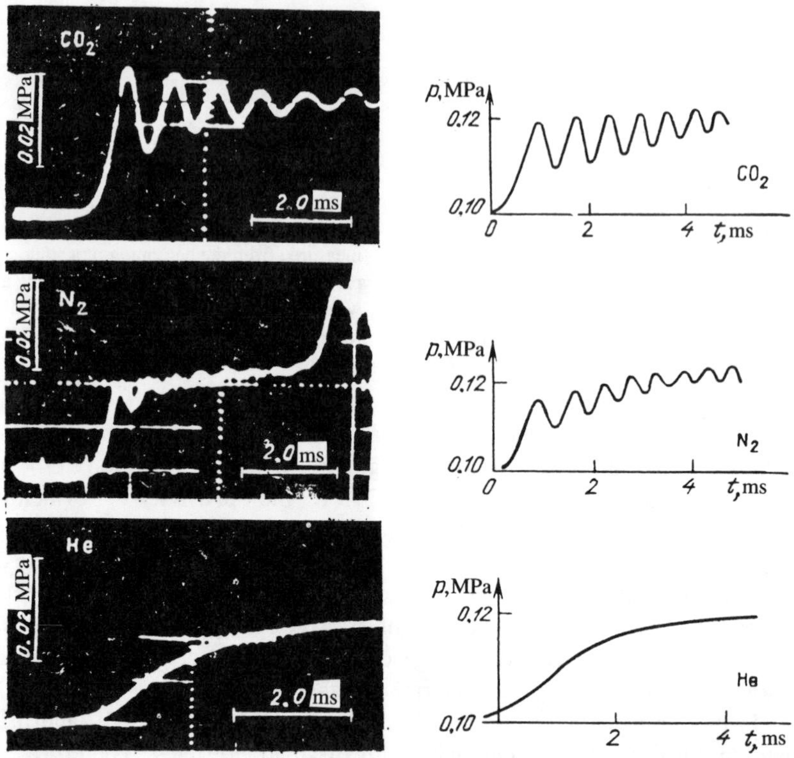

Figure 6.7.3 The predicted and experimental pressure oscillograms in shock waves of fixed intensity ($\bar{p}_e = 1.18$), propagating in a tube filled with a bubbly mixture with a fixed liquid (glycerine + water: 1:1, $p_0 = 0.118$ MPa), with fixed gas concentration ($\alpha_{20} = 0.95\%$) and bubble size ($a_0 = 1.0$–1.1 mm). Various kinds of gas within bubbles were used: bubbles of carbon dioxide, bubbles of nitrogen (air), and bubbles of helium. The oscillograms were taken at a fixed depth $x = 1.6$ meters. (The data are used from an experiment conducted by V. V. Kuznetsov et al., 1977).

ble-gas properties (adiabatic index and diffusivity) significantly affects the wave processes and their attenuation; it is understood that the bubble-gas occupies a very small volume of mixture, to say nothing of its negligible mass-concentration.

The evolution process of a shock wave under a stationary action in numerical modelling using a system of equations (6.7.4) or (6.7.17) is defined by only three dimensionless values: intensity $\bar{p}_e = p_e/p_0$, the gas adiabatic index γ_2, and the heat exchange coefficient β_2, because in low-viscosity liquids with fairly large bubbles and under fairly high pressures, the other three parameters (Σ^*, μ_1^*, $\bar{g}$) are very small and have little effect upon the process.

To study the impact of the initial pressure p_0 in an undisturbed mixture upon the shock-wave evolution process, waves of identical relative inten-

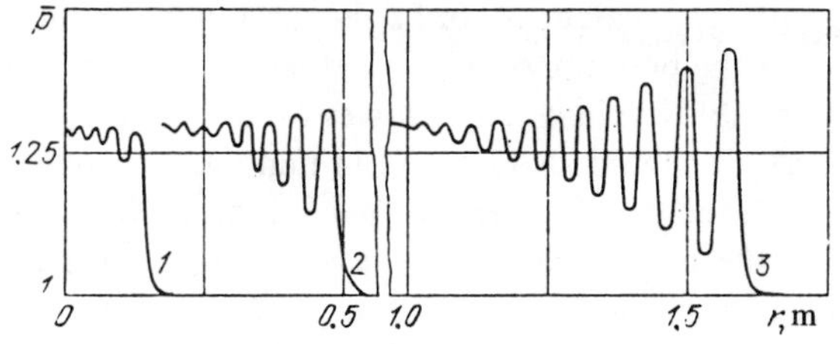

Figure 6.7.4 Pressure distributions in three kinds of shock waves with a fixed intensity $p_e/p_0 = 1.3$ but under various initial pressures p_0; the waves form their configuration in 6 microseconds after initiation by a fixed pressure p_e in water with air bubbles: $\alpha_{20} = 2.5\%$; $a_0 = 1.5$ mm; curve *1* is for $p_0 = 0.01$ MPa; curve *2* is for $p_0 = 0.1$ MPa; curve *3* is for $p_0 = 1.0$ MPa.

sities $\bar{p}_e$ were investigated in similar mixtures (mixtures with fixed a_0, α_{20}, $\overset{0}{\rho}_1$, λ_2, c_2, ...) but under various initial pressures p varied from 0.01–1.0 MPa (see Fig. 6.7.4). In this case, the gas original density $\overset{0}{\rho}_{20}$ was changed proportional to p_0 and the gas diffusivity $\nu_2^{(T)}$ was changed inversely proportional to pressure p_0. Analysis showed that the similarity of the relationships was distorted because of the effects of pressure upon β_2. An increase of pressure not only entails the increase of the bubble oscillation frequency proportional to $p_0^{1/2}$ and the increase of the wave velocity D (which follows from the similarity $\bar{p}(\bar{r}, \bar{t})$ when velocity D is proportional to $p_0^{1/2}$), but it also leads to an oscillation amplitude increase in the wave. The latter is due to the fact that the relative intensity of thermal dissipation in the process of heat exchange between phases reduces with the pressure increase because the characteristic dimensionless heat exchange parameter β_2 (see (6.7.19)) is proportional to $p_0^{-3/4}$ (the values $\overset{0}{\rho}_{20}$, $\nu_{20}^{(T)}$, C_*, t_* are, respectively, proportional to p_0, p_0^{-1}, $p_0^{1/2}$, $p_0^{-1/2}$). The value of $\mathbf{Nu}_2$ is proportional to $p_0^{3/4}$.

The experiments conducted in shock tubes (B. E. Gel'fand, E. I. Timofeev, and V. V. Stepanov, 1978) under various initial pressures p_0 in the LPCh (see (6.1.1)) showed no increase in the oscillation amplitudes and frequencies when the pressure was increased from $p_0 \approx 0.1$ MPa to $p_0 \approx 1.2$ MPa at fixed intensity $\bar{p}_e = 1.2$ and bubble size from 1 to 2 mm. In particular, with the indicated pressure p_0 rise, the oscillation frequencies have not altered and were equal to approximately 3 kcycles. The authors of these experiments indicated the nonuniformity of rising bubbles across the tube cross section and their accumulation near the axis or wall of the tube as possible causes of this situation. We shall note that the concentration of bubbles being ablated in indicated zones of the tube cross section may, in addition to distortion of parameter uniformity across the cross section and distortion of the unidirectional nature of the process, promote the coalescence of several bubbles into individual groups. Each such cluster of joint bubbles may be viewed as a unit "large" bubble divided by liquid films retaining the original interface for heat exchange. Such "grouping" results in an increase of the bubble effective size a and an increase of the effective diffusivity $\nu_2^{(T)}$ of gas, which, accordingly, accounts for the frequency diminishing and for the pressure oscillation attenuation in the wave. This may represent one possible explanation of lower oscillation frequencies than follows from theory. It should also be kept in mind that change of the original pressure p_0 may affect the bubble fragmentation. With a raise of pressure p_0 and fixed $\bar{p}_e$, the amplitude of the pressure-drop on the bubble wall increases proportional to p_0, but the action period of this pressure drop decreases. The first situation enhances the trend for fragmentation; the second, on the contrary, hampers it.

Evolution of a shock impulse of finite duration. Figure 6.7.5 illustrates the predicted evolution of a finite-duration shock impulse occurring due to

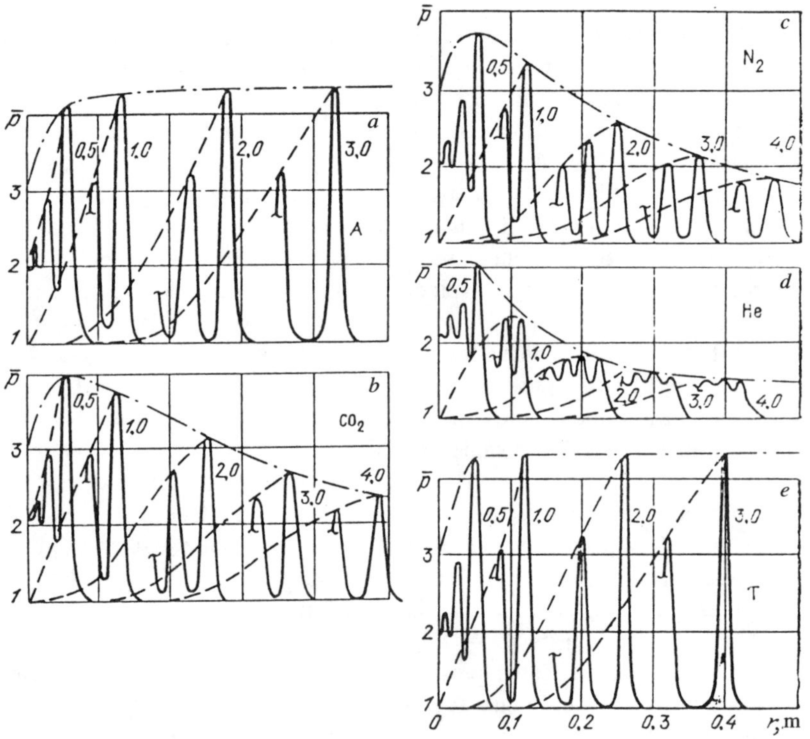

Figure 6.7.5 Evolution of a wave compression impulse of finite duration (in water with bubbles of hypothetical "thermally nonconductive" or adiabatic gas A(a, $\gamma_2 = 1.4$; $\lambda_2^* = 0$), "thermally superconductive" or isothermal gas T($e, \gamma_2 = 1$; $\lambda_2^* \to \infty$), and also in water with bubbles of carbon doxide CO_2 (b), air or nitrogen N_2 (c), helium He (d)). Mixture parameters: $p_0 = 0.1$ MPa, $T_0 = 293$ K, $\alpha_{20} = 0.02$, $a_0 = 1.0$ mm. The original pressure disturbance at $r = 0$ is predetermined by line K in Fig. 6.7.14d, and corresponds to a maximum pressure $\bar{p}_{max} = p_{max}/p_0 = 3$, and duration $\Delta t_0 = 1$ microsecond. Labels on curves indicate time t (microseconds) since the disturbance initiation. Dashed lines represent envelopes of maximum pressures at fixed points in time. Dot-dashed lines correspond to a variation of maximum pressure with depth r.

a pressure disturbance at $r = 0$; the "oscillogram" of this impulse $p(t, 0)$ has a triangular shape with duration $t_0 = 1$ microsecond and maximum pressure $p_{max} = 3p_0$. The results are shown for several variants for cold bubbly liquids made of water with a fixed volumetric concentration ($\alpha_{20} = 0.02$) of bubbles of a fixed size ($a_0 = 1$ mm), and fixed original pressure $p_0 =$ 0.1 MPa, but with different gases: hypothetical adiabatic gas (Fig. 6.7.5*a*; $\lambda_2^* = 0$) and isothermal gas (Fig. 6.7.5*e*; $\lambda_2^* \to \infty$), and real gases with a finite heat conductivity λ_2^*: carbon dioxide (Fig. 6.7.5*b*; $\lambda_2^* = 1.15 \times 10^{-3}$), air (Fig. 6.7.5*c*; $\lambda_2^* = 2.34 \times 10^{-3}$), helium (Fig. 6.7.5*d*; $\lambda_2^* = 20.3 \times 10^{-3}$). Comparison of indicated variants illustrates once more the strong effect of the nonequilibrium interphase heat exchange, which depends on the gas diffusivity, upon the shock-impulse evolution. In this situation, the two limiting schemes, one corresponding to an adiabatic ($\lambda_{20}^* = 0$) and another to isothermal ($\lambda_{20}^* \to \infty$) behavior of gas, do not cover the whole range (from "bottom to top") of the wave evolution characteristics in bubbly liquids containing real gas.

The represented data show that in mixtures with such gases (hypothetic adiabatic (*a*), isothermal (*e*), CO_3 (*b*), air (*c*)), the shock impulse breaks into solitons whose amplitude in the zone adjacent to the initiation plane $r = 0$ may even exceed the original impulse amplitude (wave intensification effect). For liquids with bubbles of CO_2, pressure across the front soliton reaches a noticeable higher value ($\bar{p} \approx 4$) than the maximum initiating pressure at $r = 0$ being equal $p_{max} = 3$. Following the indicated wave-intensification, the disturbance attenuation takes place for water with helium bubbles, the attenuation being accounted for only by thermal dissipation (it is, of course, understood that attenuation takes the place of both disturbances broken into solitons and those that were not (see Fig. 6.7.5*d*)).

The wave-intensification effect associated with the shock-impulse amplitude growth stems from the property of local deformation inertia of a bubbly liquid (see the discussion of (6.6.7) and (6.6.8)). Due to this property, the elementary volume of a bubbly liquid may undergo either contraction or expansion under its own momentum. The intensification effect strongly depends on both the duration and amplitude of the original signal, the volumetric concentration of bubbles and their size, and, astonishing as it is, on the properties of gas within bubbles (see also the conclusion of this section).

Figure 6.7.6 depicts, by way of example, a mixture of water with air bubbles with the same parameters ($\alpha_{20} = 0.02$; $a_0 = 1$ mm) and the same initial disturbance ($p_{max} = 3$; $t_0 = 1$ microseconds) as in variants represented in Fig. 6.7.5, but with different original pressures: 5 times higher ($p_0 = 0.5$ MPa) and 10 time smaller ($p_0 = 0.01$ MPa); this figure illustrates the effect of the original pressure upon the evolution and attenuation of the finite-duration disturbance. It is clearly seen that the lower the pressure is, the faster the attenuation of the shock impulse is. This is accounted for by the

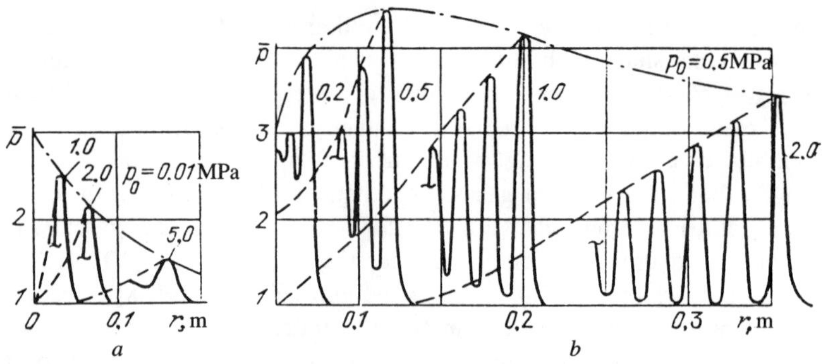

Figure 6.7.6 The same as in Fig. 6.7.5*c* (p_0 = 0.1 MPa), but with a different original pressure: p_0 = 0.01 MPa (*a*) and 0.5 MPa (*b*).

increase of the role of thermal dissipation (increase of β_2) with a decrease of pressure (see also the discussion of Fig. 6.7.4).

Similar to the demonstration of the effect of p_0, Fig. 6.7.7 shows the effect of a_0 upon the evolution of a shock impulse using the same mixture of water with air bubbles at the same original impulse as depicted in Fig. 6.7.5*c*, but with different bubble size a_0: 10 times smaller (a_0 = 0.1 mm) and 2 times smaller (a_0 = 0.5 mm). It is evident that a reduction of a_0

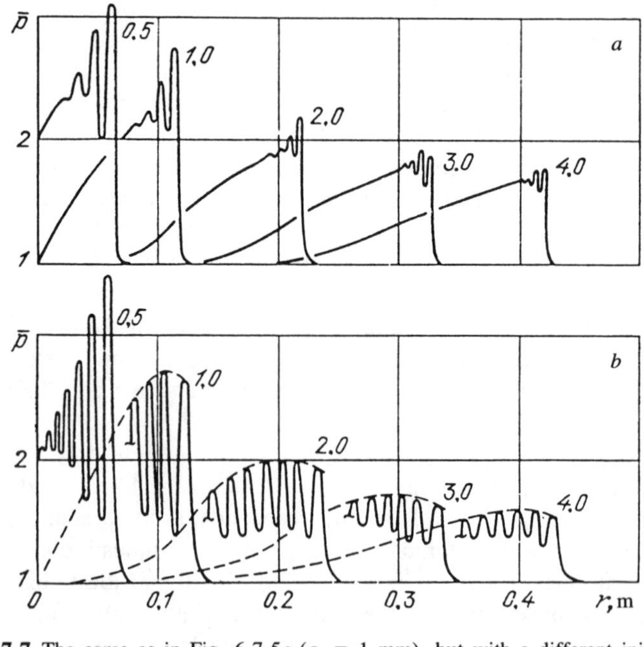

Figure 6.7.7 The same as in Fig. 6.7.5*c* (a_0 = 1 mm), but with a different initial bubble size: a_0 = 0.1 mm (*a*); a_0 = 0.5 mm (*b*).

increase the oscillation frequency, reduces the trend to develop oscillations, and accelerates the attenuation of the shock impulse under consideration.

Figure 6.7.8 illustrates the comparison of the predicted and experimental results of the investigation of the finite-duration impulse attenuation. Good agreement between the numerical and physical experimental data is obvious; this is indicative of the correctness of both the theory developed in this chapter and the results obtained based on this theory.

Interaction of oncoming solitons. Figure 6.7.9 depicts results of a numerical experiment concerned with analysis of a collision of two oncoming solitons of large (Fig. 6.7.9*a*, where $\Delta p^{(I)} \approx 4.3$, $\Delta p^{(II)} \approx 2.5$) and moderate (Fig. 6.7.9*b*), where $\Delta p^{(I)} \approx 0.6$, $\Delta p^{(II)} \approx 0.4$) intensities. In the case of moderate-intensity solitons (Fig. 6.7.9*b*), the interaction proceeds as an interaction of linear waves, namely: there is no energy interchange between solitons, and at the point in time of their collision (t = 1.2 microseconds), the pressure distribution is similar to their additive superposition; following the collision, solitons pass through each other, and then run separate ways having initial amplitudes (accounted for by attenuation due to dissipation) as if no interaction took place. Such interaction of low-amplitude solitons in bubbly liquids—when the soliton's "individuality" is retained—is reported in an experimental work by V. V. Kuznetsov et al. (1978).

For solitons of high amplitude, the linearity of their interaction may be perturbed, though only at the time of their collision. In the example depicted in Fig. 6.7.9*a,* the predicted results at the time of solitons collision exhibit

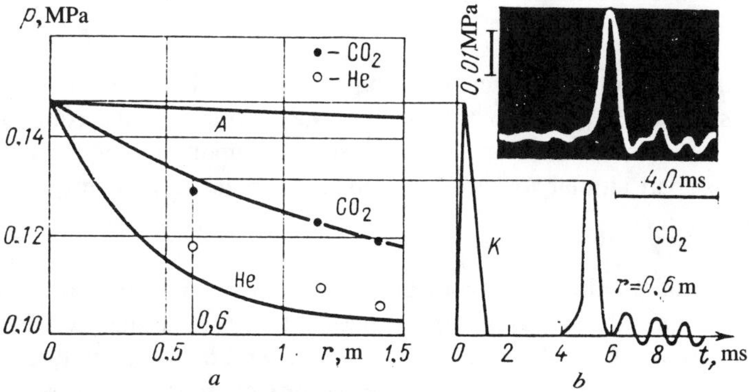

Figure 6.7.8 Predicted (A. A. Gubaidullin) and experimental (V. V. Kuznetsov et al., 1977) data on the attenuation of a triangular impulse whose reference oscillogram at r = 0 is shown by line K (intensity of the original impulse $\Delta \bar{p}_0 = \Delta p_0/p_0 = 0.48$, duration (length) $\Delta t_0 = 1.0$ microseconds) related to water with bubbles of carbon dioxide (CO_2), helium (He), and hypothetical adiabatic (A) gas. The mixture parameters: p_0 = 0.01 MPa, T_0 = 293 K, α_{20} = 0.01, a_0 = 1.4 mm. Figure 6.7.8*a* shows variation of maximum pressure with depth r; Fig. 6.7.8*b* shows both predicted and experimental pressure oscillograms $p(t)$ for the CO_2 bubble case on depth r = 0.6 m.

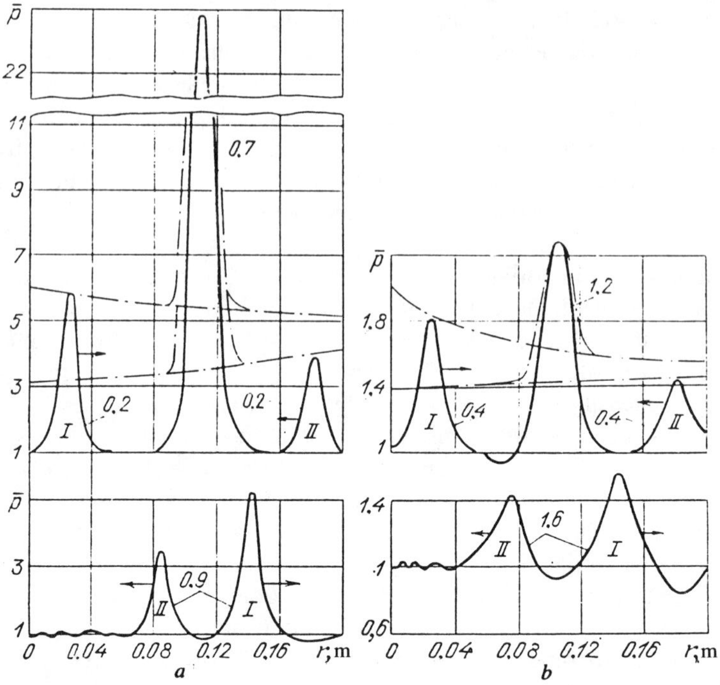

Figure 6.7.9 Interaction of solitons of large and moderate intensities when they move towards each other (*I*) moving to the right, (*II*) moving to the left) in liquid (water) with carbon-dioxide bubbles (p_0 = 0.1 MPa, T_0 = 293 K, α_{20} = 0.02, a_0 = 1.0 microseconds). Labels on curves indicate time in microseconds.

more than threefold amplification compared to an additive signal ($\Delta p_{\max} \approx$ 22.8 instead of $\Delta p_{\max} = \Delta p^{(I)} + \Delta p^{(II)} \approx 6.8$). Following the collision, the linearity of interaction is re-established: solitons, upon running through each other, go apart similar to waves of moderate intensity as if no interaction occurred.

Reflection of shock waves from plane stationary surfaces and effects of bubble fragmentation upon this process. Let us first analyze (using the equilibrium shock adiabats obtained in §§6.3 and 6.4) the mixture parameters which are realized resulting from the reflection of a shock wave from a rigid wall.

Let a stationary shock wave propagate through an equilibrium mixture (which is immobile relative to a wall) in the direction opposite to this wall (equilibrium states behind the wave are denoted by a superscript (1)). This wave strikes the wall, is reflected, and then runs in the opposite direction. The equilibrium state which is realized behind the reflected shock wave after its stationary regime is established is denoted by a superscript (2).

In accordance with the momentum conservation equation (6.3.5) across a shock wave, the velocity variation between two equilibrium states before and behind the stationary shock wave is determined by the pressure variation

$$|\Delta v| = \Delta p/\rho D \tag{6.7.21}$$

where ρ is the medium density before the wave, and D is the wave velocity relative to the medium before the wave. The medium velocity variations in both incident and reflected waves equal each other since the medium before the incident wave was at rest ($v_0 = 0$), behind the incident wave it acquires velocity $v^{(1)}$, and following the complete reflection from the wall it is again at rest relative to the wall ($v^{(2)} = 0$). Then, we obtain the reflection condition as

$$\frac{p^{(1)} - p_0}{\rho_0 D_0} = \frac{p^{(2)} - p^{(1)}}{\rho^{(1)} D^{(1)}} \qquad \left(D^{(1)} = D^* + v^{(1)}\right) \tag{6.7.22}$$

Here, it is taken into account that the reflected wave propagates through a medium in state (1) and that $D^{(1)}$ is the reflected wave velocity relative to the medium in state (1) induced by the incident wave: D^* is the velocity of the reflected wave relative to the wall.

Now, if the simplifications stemming from smallness of the bubble volumetric concentrations and incompressibility of the carrying liquid ($\alpha_C \ll \alpha_2 \ll 1$) are used, then, in conformity with the shock adiabats, we have

$$\begin{gathered} p^{(1)} = \overset{\circ}{\rho}_1 \alpha_{20} D_0^2, \quad p^{(2)} = \overset{\circ}{\rho}_1 \alpha_2 \left(D^{(1)}\right)^2, \quad \alpha_2^{(1)} = \alpha_{20} p_0/p^{(1)} \\ \rho_0 \approx \overset{\circ}{\rho}_1 \approx \rho^{(1)} \end{gathered} \tag{6.7.23}$$

Upon simple transformations, we obtain

$$\frac{p^{(2)}}{p_0} = \left(\frac{p^{(1)}}{p_0}\right)^2 \tag{6.7.24}$$

We shall emphasize that, because of the shock wave spreading, the pressure $p^{(2)}$ is established after a sufficiently long time. The process of the reflection of a fairly weak shock wave (being stationary to the point in time of reflection) from a wall is shown in Fig. 6.7.10, which demonstrates the results of analysis for two mixtures differing only by the bubble size. It is evident that, with the bubble size reduction, the relaxation zone lengths decrease, and with $a_0 = 0.5$ mm, waves are of monotonic structures.

In the shock tube experiments, pressure $p^{(2)}$ may not always be recorded because of the specific features of this kind of experiment. If the length of the incident-wave relaxation zone is great, the rarefaction wave from the high-pressure chamber of a shock tube may reach the opposite wall even before the establishment of an equilibrium pressure. In this case, the recorded maximum pressure p_* on the wall is smaller than its anticipated equilibrium magnitude. Thus, pressure p_* recorded in experiments may depend

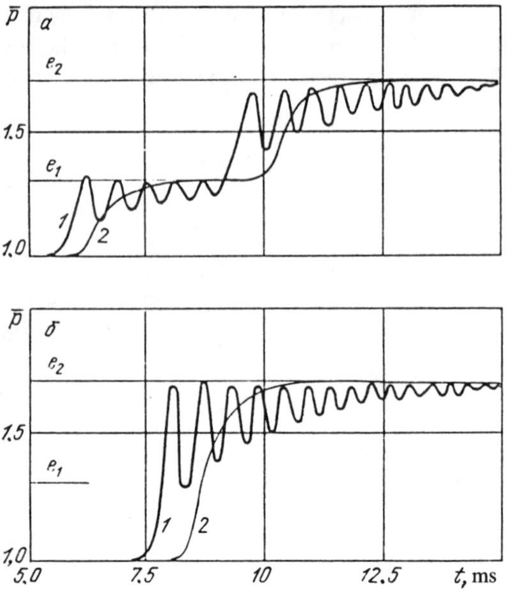

Figure 6.7.10 Reflection of a shock wave in an air-water bubbly mixture from a rigid wall. The incident wave in an equilibrium mixture was generated by means of an instantaneous rise of pressure from p_0 = 0.1 MPa to p_e = 0.13 MPa on the boundary $r = 0$. The wall is located at distance r = 1 m from the point of initiation. The mixture parameters: α_{20} = 0.01; a_0 = 1.5 mm (curves *1*) or a_0 = 0.5 mm (curves *2*); *a*) "oscillograms" for pressure at a distance 0.25 m from the wall (r = 0.75 m); *b*) oscillograms for pressure on the wall (r = 1 m).

on a number of factors: the original size of bubbles, thermophysical properties of the gaseous phase, wave intensity, presence or absence of the bubble fragmentation effects, etc. Under otherwise identical conditions, pressure p^* may also be affected by the high-pressure chamber length of the shock tube. Note that with sufficient incident-wave intensity, the bubbles break and the relaxation-zone length (profile spreading) sharply reduces. In this case, the equilibrium maximum pressures behind the reflected waves in general have sufficient time to be established before the arrival of rarefaction waves. With no bubble fragmentation in the wave, the situation sharply changes because the spreading of the incident-wave front strongly increases.

Examples illustrating the reflection of shock impulses of a finite (nonzero) length from a rigid wall are discussed below.

Figure 6.7.11 shows the process of a finite-length impulse propagation through a layer (of thickness $b = 0.2$ m and density ρ_ℓ^0), of pure (without bubbles) liquid which may be viewed as an incompressible body, the wave processes in which may be ignored. Then, on boundaries $r = r_0$ and $r = r_0 + b$ of this incompressible layer, the boundary conditions are

$$v(r_0,\ t) = v(r_0 + b,\ t) = V_b(t)$$
$$\left(\frac{\partial p}{\partial r}\right)_{r=r_0} = \left(\frac{\partial p}{\partial r}\right)_{r=r_0+b} = \frac{p(r_0+b,\ t) - p(r_0,\ t)}{b} = \rho_l^\circ \frac{dV_b}{dt} \tag{6.7.25}$$

The results of this analysis exhibit a noticeable attenuation of the propagated wave.

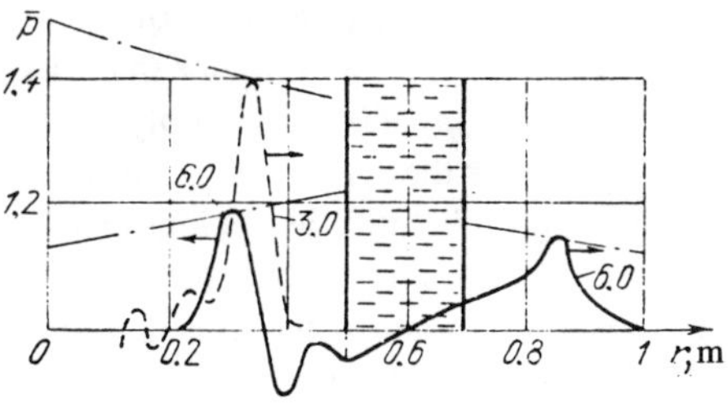

Figure 6.7.11 Passage and reflection of a nonzero length impulse through a layer of an incompressible liquid (water) of thickness b = 0.2 m ($0.5 \leqslant r \leqslant 0.7$ m) being found in a bubbly gas-liquid medium (water with bubbles of CO_2, p_0 = 0.1 MPa, T_0 = 293 K, α_{20} = 0.01, a_0 = 1.4 mm). Parameters of an original impulse at r = 0 are the same as in Fig. 6.7.8 ($\Delta\bar{p}_0$ = 0.48, Δt_0 = 1 microseconds; see line K in Fig. 6.7.8b). The dashed line represents the incident impulse (before reflection); the solid line represents the reflected and passed signal. Labels on curves indicate time t in microseconds.

Transit of a shock impulse through a contact boundary between bubbly and single-phase media. Solution of the above-indicated equations of bubbly liquids in conformity with the prescribed law of "piston" motion $\upsilon(0, t) = V(t)$ at the beginning (r = 0) of the bubbly layer determines the nonstationary wave flow in bubbly liquid and, also, the pressure $\Pi(t) = p(0, t)$ variation over the "piston." And, inversely: with a given pressure $\Pi(t)$, both the wave process and the law of "piston" motion (that generates this pressure) may be determined. Transit of wave impulses from a gas or liquid medium into a bubbly system is of particular interest.

Let the domain $r > 0$ be occupied by a bubbly liquid, and domain $r < 0$ by a single-phase medium, e.g., gas or liquid. We shall examine the compression shock-wave impulse transit from a single-phase medium to a bubbly liquid through the contact boundary $r = 0$ (see Fig. 6.7.12). Analysis of this process is generally associated with a joint solution of equations of motion in two media using the boundary condition of continuity of both normal stress (pressure) and velocity over the contact boundary.

If the domain $r < 0$ is occupied by gas, the wave R reflected from a "harder" (from the acoustic standpoint) bubbly liquid is a compression shock wave. Its parameters, together with the pressure variation $\Pi(t)$ over the contact boundary $r = 0$ may be independently determined sufficiently accurately from the solution of a problem concerning a wave S_0 reflection from a rigid

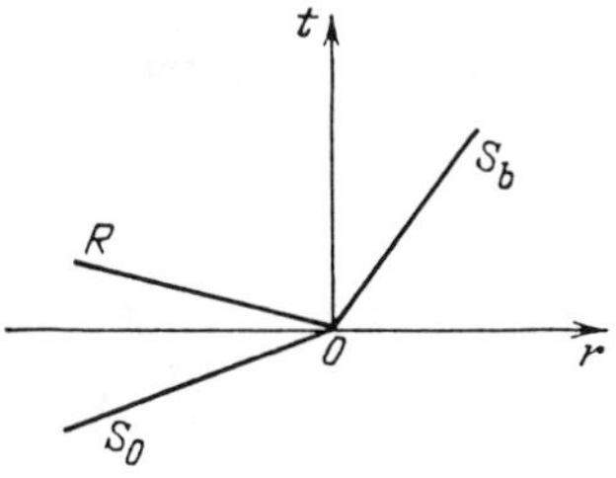

Figure 6.7.12 The "$r - t$" diagram illustrating the scheme of reflection of a shock wave S_0 (arrived from the domain $r < 0$) from the contact boundary r = 0 dividing the two media; S_b is a shock wave that entered the domain $r > 0$; R is a reflected wave (compression wave or rarefaction wave).

wall. This may be explained by the fact that acoustic hardness ρC of a bubbly medium, because of its high density ($\rho \gg \rho_g$), is much higher than that of gas ($\rho_g C_g$), and the velocity $V(t) = \upsilon(0, t) \sim \Delta p/\rho C$) of a bubbly liquid over the contact boundary, which is realized after the wave transit, is small compared to the gas bulk velocity $\upsilon_g \sim \Delta p_g/\rho_g C_g$ behind the incident shock wave S_0. Therfore, the reflected wave R in gas "does not sense" the yielding (pliancy) of the bubbly liquid boundary. Having obtained the pressure variation $\Pi(t)$ over the contact boundary, the equations of a bubbly liquid with a boundary condition of the first kind (6.7.10) can be solved separately by determining both the compression wave S_b evolution and the medium velocity variation $V(t)$ over the contact boundary.

If the domain $r < 0$ is occupied by a single-phase liquid (e.g., the same liquid as in domain $r > 0$ but without bubbles), then, as a result of interaction between the compression wave S_0 and the contact boundary dividing "pure" liquid from a softer (with regard to acoustic considerations) bubbly liquid, the reflected wave R will be a rarefaction wave. Let us discuss the issue of such wave interaction when a single-phase liquid in the domain $r < 0$ may be viewed as a linear acoustic medium with a density ρ_1^0 and speed of sound C_1. Motion of such medium is governed by linear wave equations for velocity υ and pressure p (see also §6.6). The general solution of these equations is represented in the form of a superposition of two simple waves propagating with velocity C_1 in positive (wave P) and negative (wave Q). directions of axis r

$$
\begin{aligned}
v(r, t)/C_1 &= P(r - C_1 t) + Q(r + C_1 t) \\
p(r, t)/(\rho_1^\circ C_1) &= P(r - C_1 t) - Q(r + C_1 t)
\end{aligned}
\tag{6.7.26}
$$

The inverse problem is solved in a simpler way: using prescribed functions $V(t)$ and $\Pi(t)$, which define variation of both velocity and pressure over the contact boundary $r = 0$ separating the bubbly liquid ($r > 0$) from a single-phase liquid ($r < 0$), the impulse may be determined in a single-phase liquid, which initiates the prescribed disturbances $V(t)$ and $\Pi(t)$; (the function $\Pi(t)$ being found using $V(t)$ (or vice versa) from the solution of equations of bubbly liquid at $r > 0$). Indeed, from the condition of continuity of pressure and velocity on the contract boundary $r = 0$, we have

$$
\begin{aligned}
V(t)/C_1 &= P(-C_1 t) + Q(C_1 t) \\
\Pi(t)/(\rho_1^\circ C_1^2) &= P(-C_1 t) + Q(C_1 t)
\end{aligned}
\tag{6.7.27}
$$

Solving these equations, we obtain, respectively, for incident and reflected waves in "pure" liquid

$$
P(-C_1 t) = \frac{V(t)/C_1 + \Pi(t)/(\rho_1^\circ C_1^2)}{2}
\tag{6.7.28}
$$

$$
Q(C_1 t) = \frac{V(t)/C_1 - \Pi(t)/(\rho_1^\circ C_1^2)}{2} \quad (t > 0), \qquad Q(C_1 t) = 0 \quad (t < 0)
$$

We will now set forth the solution of a direct problem: given the impulse $P(r)$ in a single-phase liquid at $t < 0$ (prior to its arrival on the boundary with the bubbly liquid), to determine both motion occurring in the bubbly liquid ($r > 0$) and the reflected signal $Q(r + C_1 t)$. Let the bubbly liquid at $t = 0$ be at rest, and the original signal $P(r - C_1 t)$ at $r < 0$ approach the contact boundary $r = 0$. Assume small time increments $t^{(1)} = \Delta t$ and velocity increments $\Delta\tilde{V}^{(1)} = \tilde{V}(t_1)$, Δt_1 being much smaller than the inherent duration of the signal, and $\Delta\tilde{V}$ is much smaller than the characteristic bulk velocity in the original signal. Solving equations of bubbly liquid, one can find the appropriate pressure $\tilde{\Pi}(t^{(1)})$, and then, using (6.7.28), find both $\tilde{P}(t^{(1)})$ and $\tilde{Q}(t^{(1)})$. Disagreement between $\tilde{P}(t^{(1)})$ and $P(t^{(1)})$ indicates a necessity of refinement of $\Delta V^{(1)}$, etc. The correction may be performed on the next step $t^{(2)} = t^{(1)} + \Delta t$. When choosing $\Delta\tilde{V}$, one must keep in mind that the maximum acoustic hardness of the domain $r > 0$ takes place when the domain is occupied by a pure liquid and $V(t) = C_1 P(C_1 t)$, and minimum when the constant pressure at $r = 0$ is realized, and $V(t) = 2C_1 P(C_1 t)$. Therefore

$$C_1 P(C_1 t) < V(t) < 2C_1 P(C_1 t) \tag{6.7.29}$$

If the bubble volumetric concentration is sufficient that the condition $C_e \ll C$ is satisfied (or, according to (6.4.14), $\alpha_2^{1/2} \gg \alpha_C^{1/2}$ which specifies the sufficient condition of insignificance of the continuous-liquid compressibility in a bubbly mixture), the solution of the above-indicated direct reflection-problem is simplified. Indeed, given that the evaluation $\Pi \sim \rho C_e V$ may be used for determining the pressure disturbance in a bubbly mixture, from Eq. (6.7.28), we have

$$P(-C_1 t) = \frac{V}{2C_1}\left[1 + O\left(\frac{C_e}{C_1}\right)\right], \quad Q(C_1 t) = \frac{V}{2C_1}\left[1 - O\left(\frac{C_e}{C_1}\right)\right] \tag{6.7.30}$$

Hence, it follows that in the case under consideration, the contact boundary velocity $V(t)$ during reflection of a shock impulse $P(r - C_1 t)$ arriving from a single-phase liquid may be determined using the scheme of reflection from a free surface where the pressure disturbances are zero. In accordance with this scheme, the bulk velocity after reflection becomes two times higher than that behind the incident wave

$$V(t) \approx 2C_1 P(-C_1 t) \tag{6.7.31}$$

Solving equations of bubbly liquids with a boundary condition of the second kind (6.7.11), both flow in the bubbly domain and the piston pressure $\Pi(t)$ may be found. If necessary, this solution may be made more accurate in the second approximation by determining the reflected impulse $Q(C_1 t)$ using $\Pi(t)$ and specifying $V(t)$

$$\begin{aligned} Q(C_1 t) &= P(-C_1 t) - \Pi(t)/(\overset{\circ}{\rho}_1 C_1^2) \\ V(t) &= C_1 [2P(-C_1 t) - \Pi(t)/(\overset{\circ}{\rho}_1 C_1^2)] \end{aligned} \tag{6.7.32}$$

It should be kept in mind that, if the incident compression impulse S_0 in a single-phase liquid is sufficiently strong ($\Delta p > p_0$) and has a nonzero (finite) duration (length), i.e., an unloading (decompression) wave follows the compression wave, then, pressure in the reflected rarefaction wave R after reflection from the contact boundary with a bubbly liquid may become negative; this situation—with a sufficient impulse length—may cause cavitation.

Bubbly screens for shock-wave damping and amplification. Bubbly, or void-fraction liquids, as any porous medium, may be used for damping or absorbing the impact impulses. Radial inertia of the carrying liquid and thermal processes within bubbles are responsible for some unusual effects. The fact is that the bubble or porous screens, as is shown below, sometimes enhance rather than damp the impulse effects upon adjacent structures.

Figure 6.7.13 illustrates an example of analysis of the wave process in

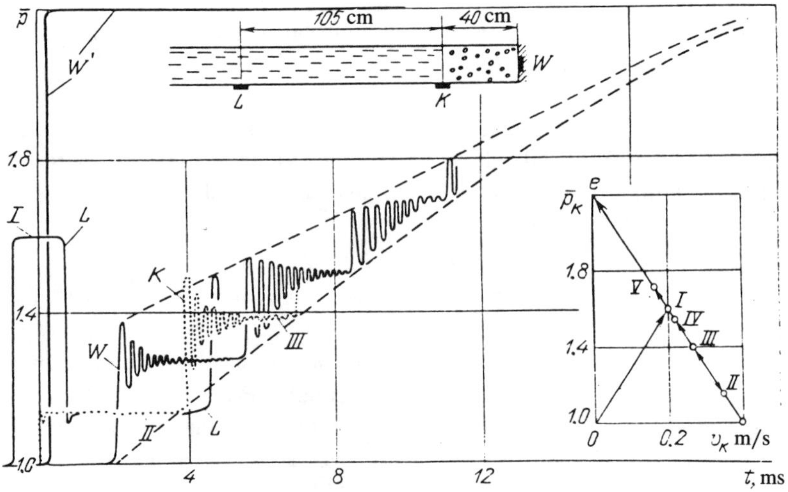

Fig. 6.7.13 Predicted "oscillograms" on sensors L, K, and W during propagation of an infinitely long (stationary) shock wave of intensity $\bar{p}_1 = p_1/p_0 = 1.6$ arriving from a single-phase liquid (water; $r < 0$, $p_0 = 0.5$ MPa) into a bubbly screen ($0 < r < 0.4$ m, water + air; $p_0 = 0.5$ MPa; $T_0 = 293$ K; $\alpha_{20} = 0.02$; $a_0 = 1.0$ mm), and its reflection from a rigid wall ($r = 0.4$ m). Letters L, K, and W correspond to "sensors" in liquid ($r = -1.05$ m), over the contact boundary ($r = 0$) and on the solid wall ($r = 0.4$ m). Letter W' indicate the "oscillogram" when there is "pure" (single-phase) water instead of a bubbly screen. The relationship between pressure and velocity on the contact boundary K upon propagation of the shock wave is shown in coordinates $\bar{p}_K v_K$. Label I indicates the state behind the incident wave, II indicates the state on the contact boundary K upon the shock wave entering into the bubbly screen, III indicates the state after the arrival of the wave reflected from the solid wall W onto the contact boundary. The readings of the "sensor" L, after the wave passes the contact boundary, repeat the readings of "sensor" K with a delay of $\Delta t = 1.05/1500 = 0.7$ microseconds.

a layer of bubbly liquid (water + air) of thickness $L_b = 0.4$ m directly adjacent to a stationary rigid wall, which separates it from a domain occupied by a bubbly-free single-phase liquid (water). The original state of the system is state of rest ($v = 0$, $p = p_0$). Three pressure-variation traces with respect to time are shown at three points: in a single-phase liquid (point L at distance 1.05 m from the contact boundary with a bubbly liquid), on the contact boundary (point K), and on the rigid wall W at the time of the infinite-length shock wave incident coming from the "pure" liquid (state of "pure" liquid behind the incident wave is denoted by I). Time $t = 0$ corresponds to the arrival of this wave front onto the contact boundary with the bubbly liquid, following which the rarefaction wave runs into a "pure" liquid, and the compression wave to the bubbly layer. The state of medium over the contact boundary after both compression in the incident shock wave and unloading is denoted by II. Since the pure liquid is unbounded from the left, and there are no, other than incident, waves propagating to the right, only simple waves propagating to the left (Q-waves) are feasible in pure liquid at $t > 0$

$$\begin{aligned} t>0:\ & p(r,t) = p_{\mathrm{I}} - \overset{\circ}{\rho}_1 C_1^2 Q(r + C_1 t) \\ & v(r,t) = v_{\mathrm{I}} + C_1 Q(r + C_1 t) \quad (p_{\mathrm{I}} - p_0 = \overset{\circ}{\rho}_1 C_1 v_{\mathrm{I}}) \end{aligned} \tag{6.7.33}$$

Thus, the relation $p(v)$ is relevant in a single-phase liquid and over the contact boundary

$$p(r,t) = p_0 + 2(p_{\mathrm{I}} - p_0) - \overset{\circ}{\rho}_1 C_1 v(r,t) \tag{6.7.34}$$

The inclined straight line in a coordinate system $\bar{p}_K(v_K)$ depicted in Fig. 6.7.13 corresponds to the indicated relation $p(v)$. The compression shock wave generated in a bubbly medium (state of the medium behind which is p_{II}, v_{II}) reaches the wall. Upon reflection, the shock wave returns to the contact boundary K from which it is again reflected. After this reflection and the decay of oscillations, pressure and velocity on the contact boundary become p_{III} and v_{III}. As a consequence of multiple reflections, the states on the contact boundary and in the bubbly screen (and also on the wall) tend to a state of complete reflection denoted by subscript e

$$p_e = p_0 + 2(p_{\mathrm{I}} - p_0), \quad v_e = 0 \tag{6.7.35}$$

In the variant under consideration, this state is reached by the time $t = t_b \approx 20$ microseconds. In a no-bubble case, the state of complete reflection e on the wall would have been reached almost instantaneously (see line W'). Thus, if duration of compression in the incident wave is shorter than t_b (t_b being proportional to L_b), the complete reflection with a pressure p_e does not have enough time to occur, and the bubbly screen damps the shock against wall.

The opposite effect may be realized in case the bubbly, or porous, screen "protects" the wall against sufficiently long shock waves of gas, or a medium being softer (regarding its acoustics behavior) than bubbly liquid.

Figure 6.7.14 demonstrates the results of a numerical experiment that illustrates the wave process in a layer of bubbly liquid, or, in other words, in a bubbly or porous screen ($0 \leqslant r \leqslant 0.4$ m) adjacent to a stationary wall W ($r = 0.4$ m) and separating it from a domain occupied by gas ($r < 0$).

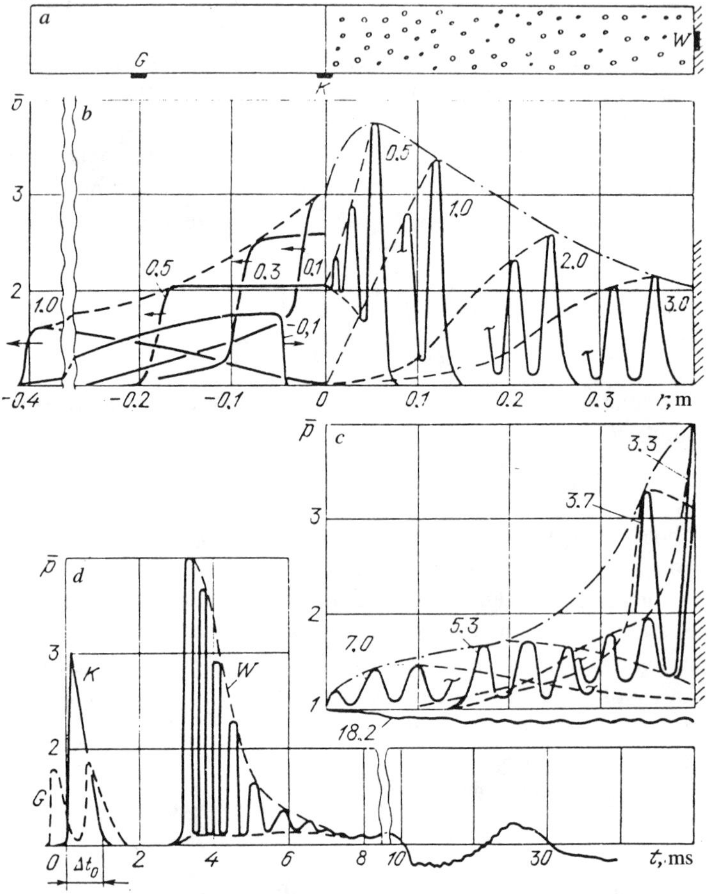

Fig. 6.7.14 The predicted evolution of the wave impulse arriving (b) from air ($r < 0$, $p_0 = 0.1$ MPa, $T_0 = 293$ K) into a layer of water with bubbles of air or nitrogen ($0 < r < 0.4$ m, $p_0 = 0.1$ MPa, $T_0 = 293$ K, $a_0 = 1$ mm, $\alpha_{20} = 0.02$) at time $t = 0$, and then reflected (c) from a rigid wall ($r = 0.4$ m) at time $t \approx 3.3$ microseconds. The process is shown in the form of pressure-diagrams $p(r)$ (b and c) at predetermined points in time t (μs) indicated by labels, and also in the form of pressure "oscillograms" $p(t)$ on three "sensors" G, K, and W (shown in Fig. 6.7.14a, namely: in air ("sensor G at $r = -0.2$ m), on the contact boundary ("sensor" K at $r = -0$), and on the rigid wall ("sensor" W at $r = 0.4$ m).

A shock impulse from gas strikes the contact boundary K ($r = 0$) between gas and bubbly liquid. The point in time when the front of this impulse reaches the boundary K is set $t = 0$. The pressure distribution over the original-impulse coordinate is shown in Fig. 6.7.14*b* at a time 0.1 microsecond before the impulse reaches the boundary K ($t = -0.1$ microsecond). At this time, the impulse length is $L_g \approx 0.35$ m. As a result of interaction between this impulse and contact boundary K, the shock wave is reflected into gas; the parameters and evolution of this wave are virtually the same as in reflection of this impulse from a stationary wall (see the discussion following Fig. 6.7.12). Simultaneously, the compression shock impulse will proceed into the bubbly layer. Figure 6.7.14 represents a variant when the bubbly liquid characteristics, trace of pressure $p(0, t)$ at $r = 0$ (shown by line K in Fig. 6.7.14*d*), and, consequently, the impulse arrived into the bubbly layer, are the same as in the above-discussed Fig. 6.7.5*c*. The time period before the point in time when the impulse reaches the wall W is shown in the form of pressure-diagrams in Fig. 6.7.14*b*. Upon reflection from the stationary wall W, the signal returns to the boundary K; a rarefaction wave similar to the wave on a free surface where $p = p_0$ is generated here. This wave may cause a pressure drop, as compared to the initial pressure. The pressure diagram at $t = 18.2$ microseconds resembles the maximum pressure reduction for the entire process when the bubbly screen expands due to gas elasticity and liquid inertia.

Figure 6.7.14*d* depicts pressure variation at three points: in gas (G), on the contact boundary (K), and on the stationary wall (W). The maximum pressure on the wall W equals $\bar{p} = 4$, although with no bubbly screen (even if the wall is located at $r = 0$ that makes the structure shorter), the pressure variation as a function of time is defined by the line K with maximum pressure $\bar{p} = 3$. Thus, as distinct from the bubbly screen which protects the wall from shocks in liquids, the bubbly screen between a wall and gas with the considered dimensions and parameters of both mixture and original impulse increases rather than reduces the maximum pressure against the wall. It is clear that extending the bubbly screen will finally account for damping the effect upon the wall because of the incident-signal decay due to thermal dissipation and erosion due to dispersion. In addition to the bubble volumetric concentration, their size, initial pressure, screen length, original impulse amplitude, the wall pressure variation, and the amplification or attenuation of the impulse depend to a great extent on both original impulse length and kind of gas. This is illustrated in Figs. 6.7.15 and 6.7.16. Impulses of greater length and reduced heat conductivity of gas within bubbles may cause much higher rise of pressure acting upon the wall accounted for by the bubbly screen due to the impact of the impulse arriving from gas.

Note that the "anomalous" pressure surges in the process of a shock impulse reflection from a rigid wall in a bubbly liquid were first reported in a theoretical treatise by V. K. Kedrinskii (1968 and 1980).

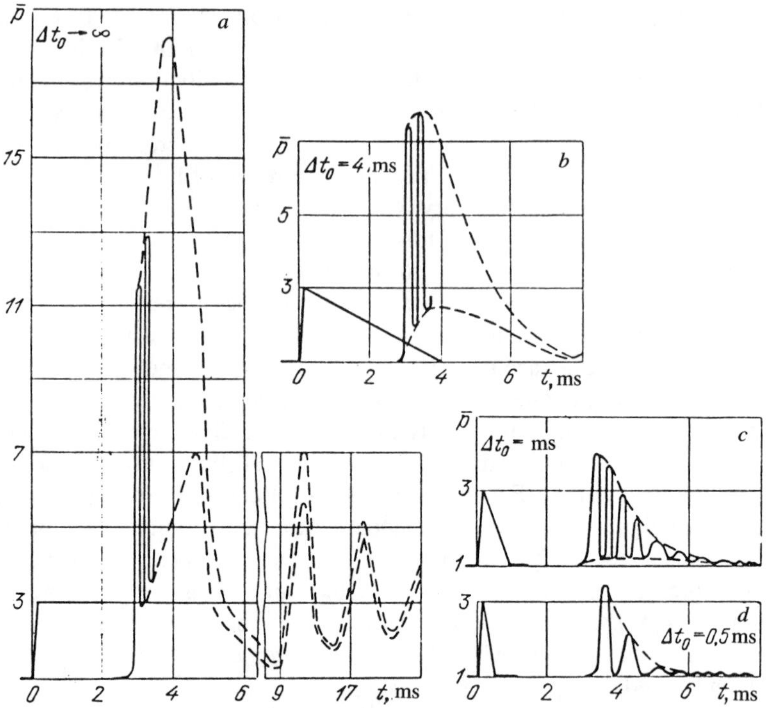

Fig. 6.7.15 Predicted pressure "oscillograms" related to the rigid wall (see "sensor" W in Fig. 6.7.14*a* at r = 0.4 m) as a consequence of passage of various-length wave impulses through a "screen" of bubbly liquid (water + air bubbles) of thickness L = 0.4 m adjacent to the wall. The screen parameters are the same as in Fig. 6.7.14. Lines K correspond to "oscillograms" of the original impulse pressures $p(t)$ at the entrance into the screen (r = 0). Figure 6.7.15*a*, *b*, *c*, and *d* correspond to various signal lengths: $\Delta t_0 \to \infty$; 4; 1; and 0.5 microseconds, respectively (Δt_0 = 1 microseconds (*c*) relates to Fig. 6.7.14*d*)

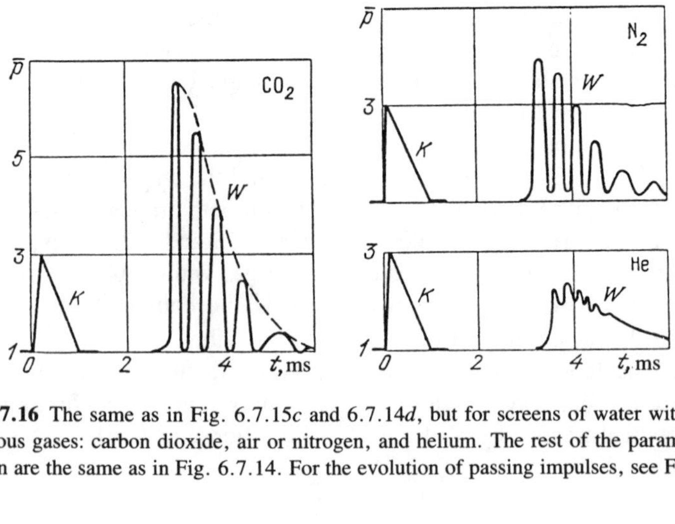

Fig. 6.7.16 The same as in Fig. 6.7.15*c* and 6.7.14*d*, but for screens of water with bubbles of various gases: carbon dioxide, air or nitrogen, and helium. The rest of the parameters and notation are the same as in Fig. 6.7.14. For the evolution of passing impulses, see Fig. 6.7.5.

Thus far, we have discussed gas-liquid systems with a homogeneous bubble distribution. Note that in addition to controlling the bubble distribution over the depth of the screen, regulation of thermophysical properties of gas within bubbles, bubble size, initial pressure in the medium, gas concentration etc., an additional, and sufficiently extensive, contribution to either damping or amplification of shock waves may be achieved. The latter is illustrated in Fig. 6.7.17. It can be clearly seen that maximum pressure on the rigid wall under the shock-wave effects may be reduced more than six-fold.

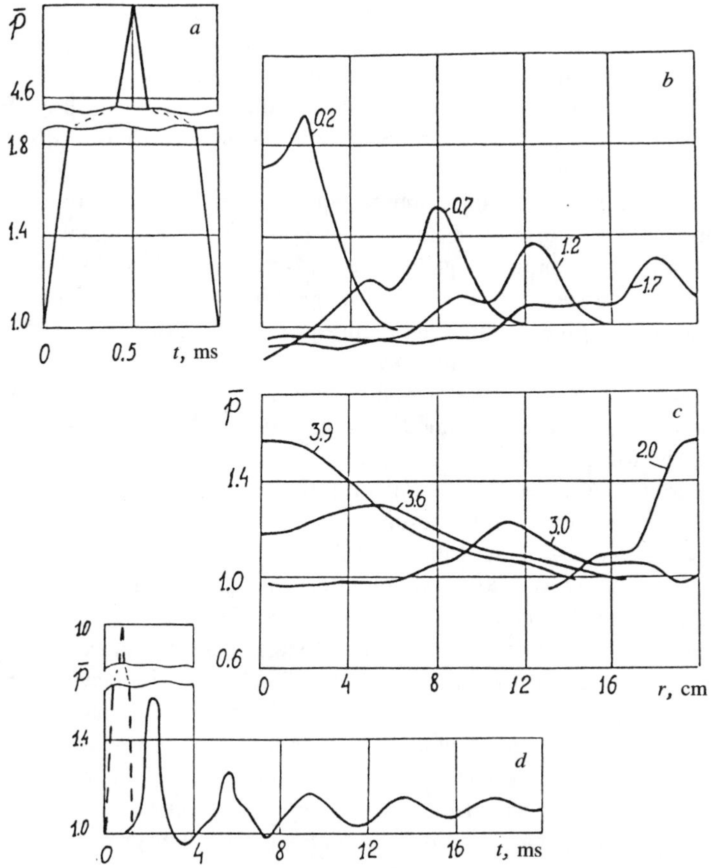

Fig. 6.7.17 Damping of a shock wave propagating through water with a bubbly screen (a_0 = 1 mm, hydrogen, p_0 = 0.1 MPa) characterized by a linear increase of gas concentration from 0.001–0.034.*a*) appearance of the original disturbance, *b*) evolution of attentuated impulse, *c*) pressure profiles in a shock wave reflected from a rigid wall. Numerical labels on curves indicate time measured in milliseconds from the beginning of the disturbance input into the bubbly screen. *d*) predicted pressure "oscillogram" for rigid wall (solid line); the same in the absence of the screen in shown by a dashed line.

We shall note that it is impossible to obtain the shown result with sufficient accuracy without using to the full an iteration algorithm for predicting the propagation of a shock impulse through the contact boundary between a bubbly and single-phase media, described above.

The possibility of amplification of effects upon a barrier from a shock wave propagating through gas is depicted in Fig. 6.7.18. Comparison of curves *1* and *2*(*c*) illustrates the influence of nonhomogeneous bubble distribution in a gas-liquid layer. It is seen that redistribution of the gas concentration itself in the bubbly screen may account for the qualitative modification of its property by producing a damper instead of an amplifier.

The shock-wave intensification effect in bentonite clays. The shock-wave intensification effect with manifestation of anomalously high pressure peaks has been discovered in aqueous suspensions of bentonite clay used as a drilling mud for carrying out the soil material in the process of drilling. Figure 6.7.19 shows pressure oscillograms obtained using two transducers placed in the percussion tube.

The oscillogram in a single-phase liquid (water) appears throughout the experiments in the form of an unvaried step with a pressure almost the same as the initiating magnitude equal to the pressure in HPCh; it displays a time spreading of the shock-wave front of about 150 microsecond, which is determined by the time of the diaphragm opening. As distinct from this, the oscillograms in a bentonite suspension reveal amplified pressure peaks up to 15 MPa throughout experimental study, although the initiating pressure did not vary and was maintained at a level of 3.5 MPa. The shock-wave front is significantly steeper—its spread is about $\leqslant$10 microseconds.

Figure 6.7.20 represents results of a model analysis, which are in good qualitative agreement with the results of the described experiments, and substantiate the validity of the following physical concepts concerning the nature of the shock-wave intensification effect in bentonite suspensions. When conducting experiments in a vertical shock tube (Fig. 6.1.1), the upper level of liquid in the low-pressure chamber is generally located closer to the diaphragm to ensure that the intensity of the shock wave initiated during the diaphragm rupture resembles pressure in the high-pressure chamber. In the process of the diaphragm rupture, in upper layers of bentonite suspension bubbles due to penetration of jets of gas from the high-pressure chamber may occur. Because of a gel-like consistency of suspension, the bubbles fail to rise and do not escape liquid; therefore, their amount grows with an increasing number of shocks, and the bubbly-layer height can increase, the gas concentration in the layer reducing with depth. The shock wave initiated in the upper bubbly layer following the diaphragm rupture has an intensity equal to the pressure in the high-pressure chamber. After passing through the bubbly layer, the shock wave enters and is reflected from a stiffer (from the acoustic standpoint), pure (bubble-free) suspension; its intensity in this

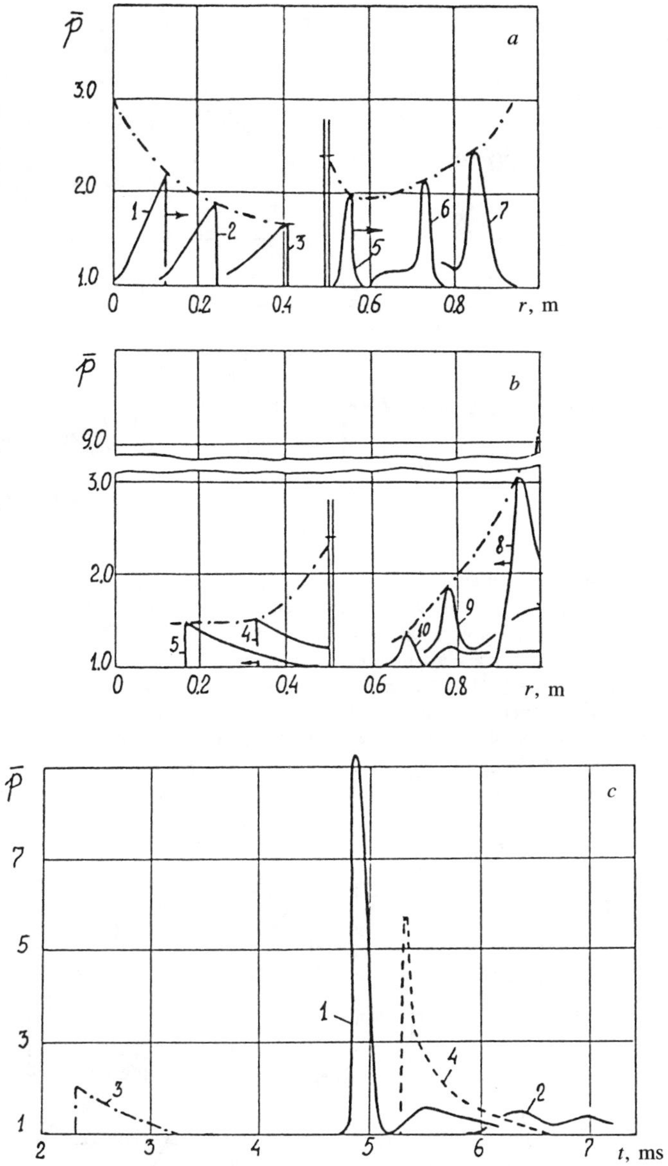

Fig. 6.7.18 Intensification of a dynamic action upon a barrier from a shock wave propagating (*a*) at time $t = 0$ from air into a layer of water with bubbles of carbon dioxide ($p_0 = 0.1$ MPa, $a_0 = 1$ mm) characterized by exponential decrease of gas concentration ($\alpha_{20} = (0.1) \exp(-x/b)$, $b = 0.1$ m), and reflected (*b*) from a rigid wall (P. K. Gazizov and A. A. Gubaidulin, 1989). Initial impulse length in air is 0.2 ms. $t_{1-10} = 0.2$; 0.5; 0.9; 1.5; 1.9; 3.9; 4.6; 4.8; 5.5; 6.3 milliseconds. *c*) predicted pressure "oscillograms" $\bar{p}(t)$ for a rigid wall. *1*) case (*a*), *2*) a uniform-distribution case for the same amount of bubbles, *3*) the bubbly-layer absence case, *4*) prediction of the profile *1* using a model of thermodynamic equilibrium.

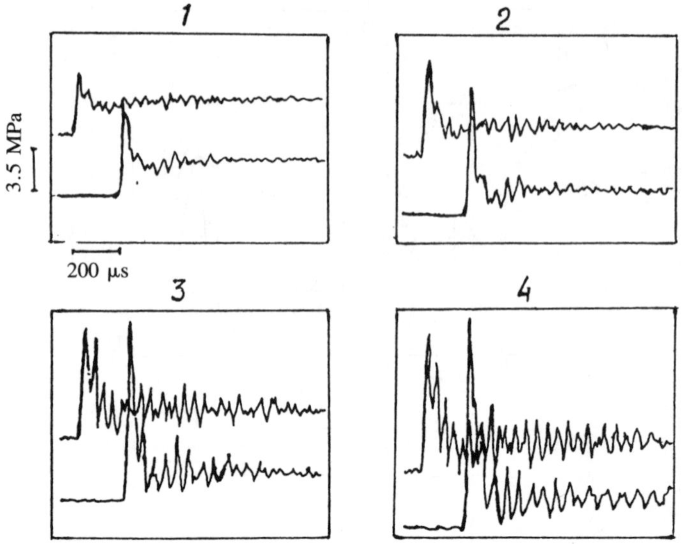

Fig. 6.7.19 Pressure oscillograms obtained from two piezoceramic pressure transducers in a percussion tube located approximately ~2 meters from the upper surface below the diaphragm, and spaced at 0.2 m between each other. The low-pressure chamber (LPCh) is filled with aqueous suspension of bentonite clay (mass concentration of clay is ~ 6%) containing particles of montmorillonite of size 0.01–0.1 micron. Oscillograms are shown for four successive experiments (shocks) conducted at intervals of ~ 5 minutes necessary for the diaphragm replacement and pumping-in the "striking" gas into high-pressure chamber (HPCh) to reach pressure up to p = 3.5 MPa.

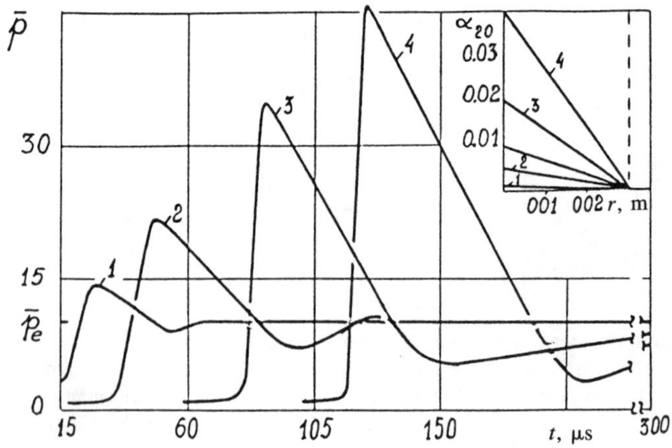

Fig. 6.7.20 Predicted pressure "oscillograms" in a liquid at a point located behind the bubbly layer for various distributions of the initial gas concentration (indicated by numerical labels *1–4* on curves). The air bubble size is a_0 = 10 microns, p_0 = 0.1 MPa, liquid-glycerine.

case increases. The higher the concentration of bubbles in the upper layer, the higher the intensity of the wave recorded by the sensors located deeper in suspension. The shock wave reflected from the boundary between a bubbly layer and suspension moves in an opposite direction, is reflected from the free surface of a bubbly layer, and propagates further as a rarefaction wave. As a consequence of interaction between the rarefaction wave and a shock wave moving ahead of it, a typical peak of amplification is produced, which first was detected experimentally; the nature of this peak gave rise to numerous discussions and arguments concerning the causes of its origination. Oscillations in the wave front are caused by a small-scale radial motion of liquid around bubbles. A steepening of the shock wave that was observed in Fig. 6.7.19 may also be explained by its passing through a layer of bubbly liquid which represents a strongly nonlinear medium.

Numerical experiments conducted by A. A. Gubaidulin and M. Turaev in the framework of a single-velocity, two-temperature approximation (6.7.3) and a model of a viscoelastic liquid (1.5.21) contributed to further comprehension of the nature of the shock-wave intensification effect in bentonite suspensions, and also revealed some specific features of this effect as a function of various parameters defining this process.

This intensification effect is a promising technique which may have long-term importance in applications for drilling and treatment of the near-well zone in oil-bearing beds.

Shock wave intensification in homogeneous gas-liquid bubbly media. The existence of shock waves with an oscillatory structure and the feasibility of the shock-wave intensification effect in bubbly liquids when shock waves are generated by a stationary source (e.g., by a constant-pressure piston) are manifestations of the combined effect of elasticity and local deformation inertia (i.e., that of dependence of p on the rate of density $\ddot{\rho}$ variation) of bubbly liquids upon the wave propagation process. The intensification effect consists of the fact that the pressure oscillations behind a shock wave propagating in a homogeneous medium may become significantly higher than pressure p_e behind the wave, or the maximum initiating pressure (on the piston).

Both the effect and rate of this intensification are characterized by a dimensionless coefficient of amplification

$$P_m = \frac{p_{\max} - (p_b)_{\max}}{(p_b)_{\max} - p_0} \qquad (6.7.36)$$

where $p_{\max}$ is the maximum pressure amplitude during the disturbance evolution, $(p_b)_{\max}$ is the maximum pressure at the initiating signal (on the piston when $r = 0$). For a stationary shock wave $(p_b)_{\max} = p_e$, and the amplification coefficient is

$$P_m = \frac{p_{\max} - p_e}{p_e - p_0} \tag{6.7.37}$$

For a stationary shock wave in a fixed-mass homogeneous mixture of liquid with gas bubbles, the shock-wave intensification is limited in accordance with (6.4.22): $p_{\max} - p_e < p_0$ that leads to the following limit for the coefficient of amplification

$$P_m < (\bar{p}_e - 1)^{-1} \tag{6.7.38}$$

In nonstationary waves propagating in homogeneous bubbly liquids, the rate of intensification may be many times higher than in stationary waves, and, therefore, may exceed the value specified by the latter formula. This is demonstrated in Fig. 6.7.21, which depicts the predicted results for a shock-wave evolution on the nonstationary initial stage (the wave is initiated by a "stationary piston" when $p_b = p_e =$ constant). The wave pressure exceeds not only p_e but also $p_e + p_0$. The amplification coefficient in represented variants is 0.9 and significantly exceeds the maximum value of 0.25 given by formula (6.7.38) for stationary waves.

It should be noted that the shock-wave propagation velocity D in represented variants equals approximately 250 m/s, i.e., $(D/C_1)^2 \ll 1$; there-

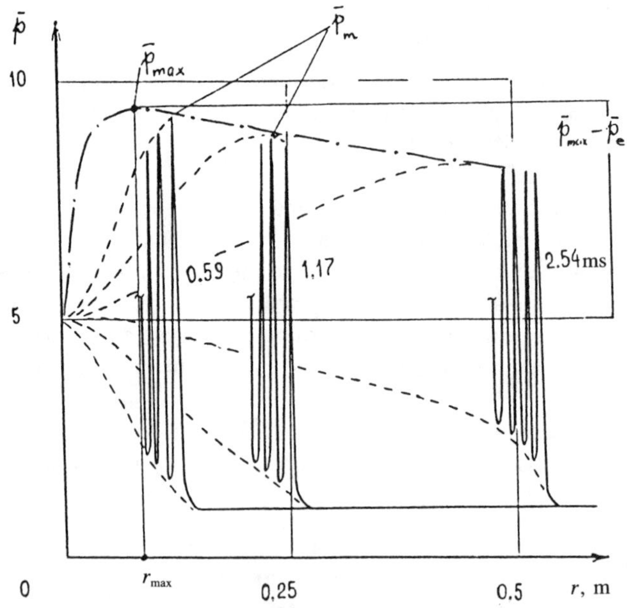

Fig. 6.7.21 Evolution of nonstationary shock waves of intensity $p_e = 5$ in water ($p_0 = 0.1$ MPa, $T_0 = 300$ K) with air bubbles ($\alpha_{20} = 0.02$, $a_0 = 0.5$ mm). Shock waves are generated by a step variation of pressure from p_0 to $p_e = \bar{p}_e p_0$ at time $t = 0$, and at section $r = 0$. Numerical labels on curves indicate time in milliseconds.

fore, the effects of the continuous-liquid compressibility on wave dynamics, as before, is negligible.

We shall note that bubble fragmentation may affect the process to a significant extent.

The bubble-radius a_0 reduction, with all other conditions (α_{20}, p_0, p_e, γ, etc.) being fixed, leads to the enhancement of the interphase heat-exchange intensity because of an increase of the specific surface of the phase boundary, and thereby to the reduction of the intensification rate within pressure waves. Rise of the heat exchange rate (on account of a reduction of a_0, an increase of $\nu_g^{(T)}$, etc.) increases, in represented conditions, the intensity of the dissipation of kinetic energy of small-scale radial motion of liquid around bubbles, which reduces the intensification effect.

Thermophysical properties of gas have a considerable impact upon the intensification effect. This becomes more obvious from Fig. 6.7.22. It can be seen that with a reduction of dissipation due to interphase heat exchange when helium is substituted by air, and, further, by carbon dioxide, the intensification effect exhibits itself much more strongly, and the zone L_m, where this intensification is displayed (i.e., pressure exceeds the maximum pressure in the initiating impulse: $p > (p_b)_{\max}$), expands.

The dependence of the rate of intensification P_m on the volumetric concentration of gas (void fraction) α_{20} in shock waves is not well-defined (Fig. 6.7.23): point $r_{\max}$ corresponding to P_m, with α_{20} increase, shifts in the direction towards the origin implying the reduction of zone L_m.

The effects of the initial pressure p_0 on the mode of the evolution of nonstationary shock waves (see, e.g., Fig. 6.7.4) have already been discussed. The predicted dependence of P_m, and $r_{\max}$ on p_0 for moderate values

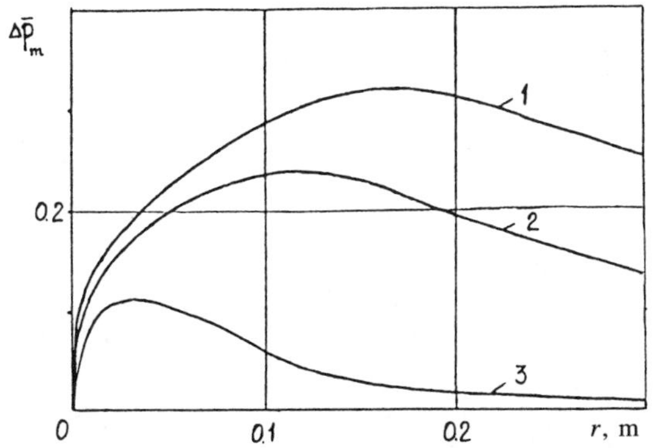

Fig. 6.7.22 Influence of kind of gas upon the shock-wave intensification effect in bubbly gas-liquid systems. Curves *1*, *2*, and *3* correspond to various gases (CO_2, air, and He) contained in bubbles of radius a_0 = 1 mm at p_0 = 0.1. MPa, $\bar{p}_e$ = 2, α_{20} = 0.01. ($\Delta\bar{p}_m = (p_m - \bar{p}_e)/\bar{p}_e$).

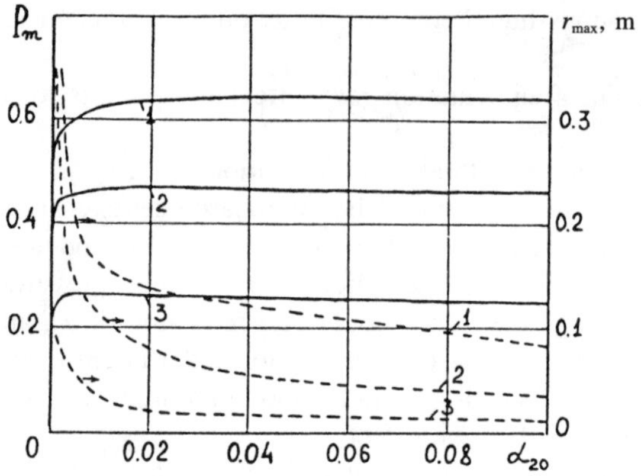

Fig. 6.7.23 Influence of the initial void fraction upon the rate of shock-wave intensity in homogeneous gas-liquid mixtures. The remaining parameters and nomenclature are the same as in Fig. 6.7.22.

$p_0 \leqslant 1$ MPa are depicted in Fig. 6.7.24. It is evident that with p_0 increase, the trend to the wave intensification grows, and zone L_m expands.

Analysis proved that the intensification effect manifests itself more strongly, the stronger the initiating impulse. It is most pronounced in substantially nonlinear waves, which cannot be studied in the framework of the theory of weakly nonlinear waves (described by equations of type of Burgers-Corteveg-de Vries or Boussinesq). The corresponding dependence of P_m and r_{max} on $\bar{p}_e$ is depicted in Fig. 6.7.25.

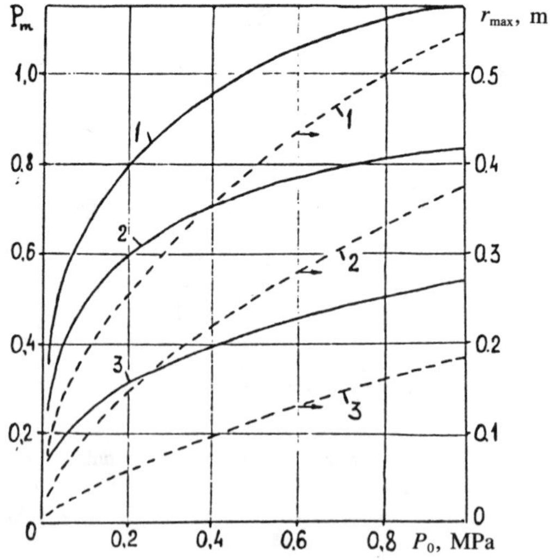

Fig. 6.7.24 Dependence of the amplification coefficient P_m on the initial pressure in bubbly liquid. The remaining parameters and nomenclature are the same as in Fig. 6.7.22.

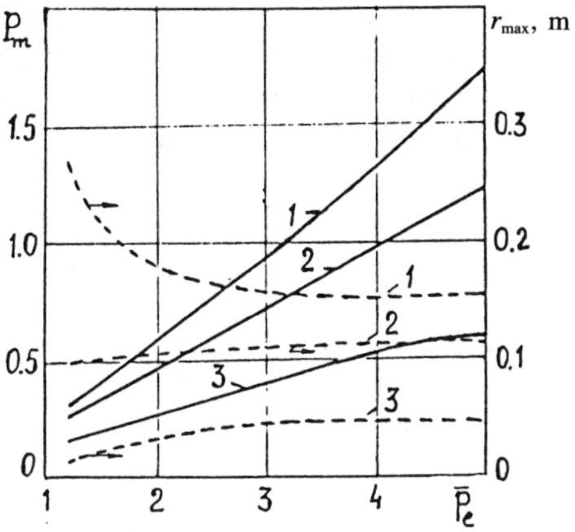

Fig. 6.7.25 Variation of amplification coefficient P_m as a function of a shock-wave intensity. The remaining parameters and nomenclature are the same as in Fig. 6.7.22.

Strong waves in high-viscosity liquids with fine gas bubbles. We shall review here the results of the numerical investigation of an evolution of nonstationary shock waves in bubbly liquid in the case when the viscoelastic model (1.5.21) holds good. A significant increase of the liquid viscosity promotes the growth of dissipation of kinetic energy of oscillatory motion of bubbles, which is responsible for a sharp reduction of the oscillation-peak amplitude in the shock-wave front. If, in addition to this, the bubble size is reduced, even strong waves will exhibit a monotonic structure. The latter is illustrated in Fig. 6.7.26, reflecting a pattern predicted in the framework of a single-velocity two-temperature model (6.7.3).

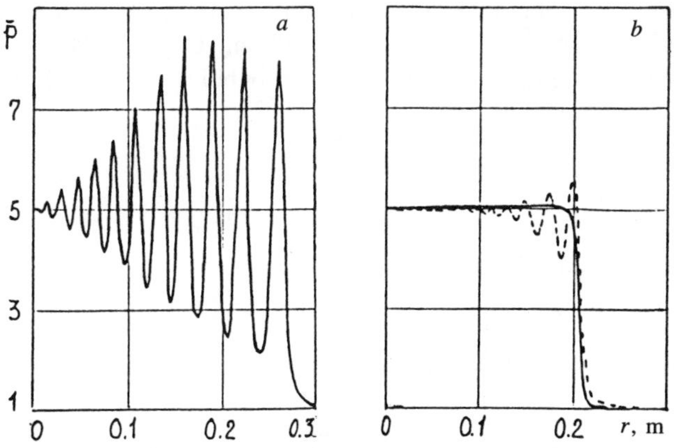

Fig. 6.7.26 Influence of both liquid viscosity and bubble size on the structure of a nonstationary shock wave in a gas-liquid bubbly mixture (p_0 = 0.01 MPa, α_{20} = 0.01). *a*) air-water mixture with bubbles of radius a_0 = 1 mm; *b*) air-glycerine mixture. Dashed line corresponds to a_0 = 1 mm, solid line to a_0 = 0.1 mm, t = 1 microsecond.

The most specific features of an evolution of strong disturbances in a high-viscosity liquid with extremely fine bubbles are: its monotonic structure, steepening of the front as the wave propagates, transformation of the compression wave to a shock, and rapid decay of the finite-length impulses. Figure 6.7.27 shows, by way of example, evolution of the finite-length impulse whose initial shape was a sinusoid half-period. It can be seen that even a small amount of bubbles ($\alpha_{20} = 0.001$) strongly affects the wave behavior in liquid.

To verify the adequacy of the model (1.5.21), analyses have been performed to compare it with available experimental results. Comparison between analytically predicted and experimental data is shown in Fig. 6.7.28.

§6.8 EFFECTS OF NONSPHERICITY, FRAGMENTATION, AND DISINTEGRATION OF BUBBLES ON WAVE PROPAGATION IN A BUBBLY LIQUID

Observations of the bubble behavior in shock waves conducted by means of high-speed filming (B. E. Gel'fand and S. A. Gubin et al., 1975; V. V. Kuznetsov and V. E. Nakoryakov et al., 1977) showed that bubbles may have a shape that significantly differs from spherical configuration; nevertheless, the theory based on an equation of the Rayleigh-Lamb type for radial motion around bubbles derived assuming the spherical shape is retained describes the wave evolution fairly well, unless fragmentation occurs. Apparently, the Rayleigh-Lamb equation correctly describes the most important consideration, namely, the bubble volume change in spite of the loss of spherical configuration. The bubble fragmentation is viewed as an extreme manifestation of the loss of sphericity. The realization of fragmentation has a major impact on a wave structure in a bubbly medium. In particular, intensive fragmentation of original bubbles into finer bubbles, occurring in

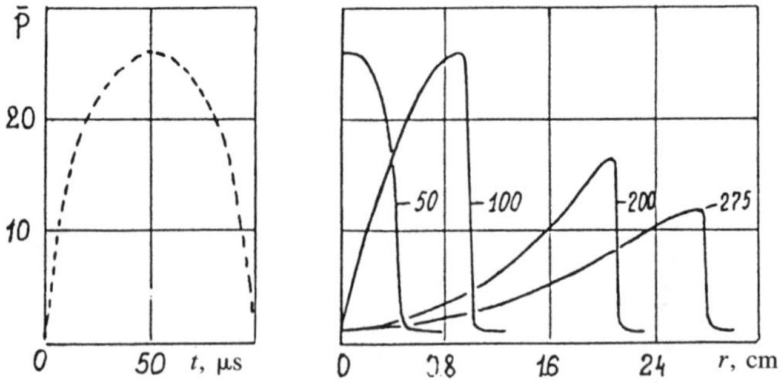

Fig. 6.7.27 Evolution of the finite-length impulse in a high-viscosity liquid (glycerine) with extremely small (a_0 = 10 micron) air bubbles (p_0 = 0.1 MPa, α_{20} = 0.001).

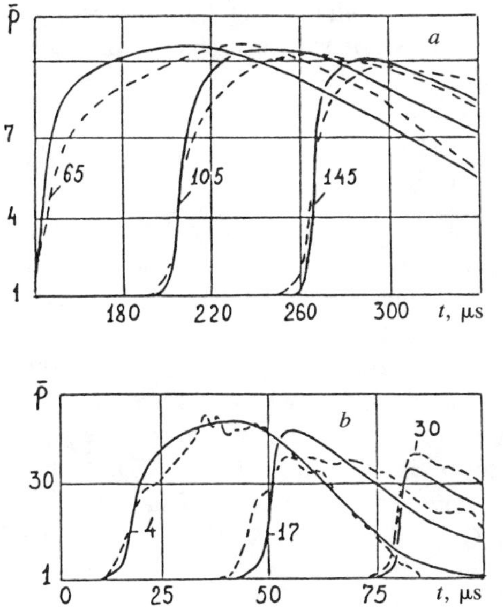

Figure 6.7.28 Comparison between predicted (A. A. Gubaidulin and M. Turaev) and experimental (G. A. Druzhinin) data on propagation of finite-length impulses in glycerine with air bubbles. Solid lines represent experimental oscillograms. Numerical labels on curves indicate distances in mm between pressure sensors and location of initiation of the original-shape impulses, which resemble the sinusoid half-period of duration θ at $p_0 = 0.1$ MPa. *a*) $\theta = 325$ microsecond, $a_0 = 20$ micron, $\alpha_{20} = 0.0017$; *b*) $\theta = 75$ microsecond, $a_0 = 20$ micron, $\alpha_{20} = 0.02$.

sufficiently strong waves generally leads to a situation when in the process of the very first compression of bubbles across the wave front, finer bubbles are found in the relaxation zone of the wave which are characterized by a much smaller pulsation period and cooling time compared to original bubbles. This reduces the wave relaxation-zone thickness by many times. As a consequence, the equilibrium model of the medium may become sufficient, which reduces the model to a perfect barotropic compressible liquid with a predetermined equation of state $p(\rho)$ (see (1.5.26)).

A model of a perfect barotropic and viscoelastic liquid for describing wave processes. The equation of state for a mixture of incompressible liquid (ρ_1^0 = constant) and gas with an assumption of negligibly small capillary effects ($2\Sigma/a \ll p$), in its equilibrium approximation ($p_1 = p_2 = p$, $T_1 = T_2 = T$) is written in the form which follows from Eq. (1.5.28)

$$p(\rho) = \frac{\alpha_{20} p_0}{\alpha_{10}} \frac{\rho}{\rho_1^\circ - \rho}, \quad C_e = \left(\frac{dp}{d\rho}\right)^{1/2} = \frac{\rho_1^\circ}{\rho_1^\circ - \rho} \sqrt{\frac{p_0 \alpha_{20}}{\rho_1^\circ \alpha_{10}}} \tag{6.8.1}$$

In this model, the elastic wave is described in the form of a shock, the analysis of which is based on well-known shock conservation laws

$$\rho_0 D = \rho_e (D - v_e), \quad \rho_0 D v_e = p_e - p_0, \; p_0 = p_0(\rho_0), \quad p_e = p(\rho_e) \tag{6.8.2}$$

where D is the shock velocity, v_e is the velocity "jump," or velocity of the medium behind the shock relative to that of the medium before the shock, ρ_0, p_0 and ρ_e, p_e are, respectively, the density and pressure before and behind the shock. As a result, the equilibrium shock adiabat may be easily obtained in an explicit form in accordance with (6.4.15); this adiabat is realized in the case of sufficiently small bubbles

$$D^2 = \frac{p_e}{\overset{\circ}{\rho}_1 \alpha_{10} \alpha_{20}}, \quad v_e = \frac{p_e - p_0}{\sqrt{p_e}} \sqrt{\frac{\alpha_{20}}{\overset{\circ}{\rho}_1 \alpha_{10}}} \tag{6.8.3}$$

Some solutions obtained based on the discussed equilibrium model are given below in §6.9. A model of a perfect compressible liquid based on Eq. (6.8.1) for predicting explosions in bubbly liquids were used in works by B. Parkin, F. Gilmore and H. Brode (1961).

In the case of very fine bubbles and very viscous liquid when $a\sqrt{p\rho_1^0}/\mu_1 \ll 1$, and the wave relaxation zone may be of a finite (nonzero) thickness, over which the phase pressures are equalized, the mixture is described by a system of equations for a viscoelastic liquid (1.5.21). A study of nonstationary flow of bubbly mixture in the framework of such model of viscoelastic liquid is described by G. M. Lyakhov (1982).

Effects on low density of gas on bubble fragmentation. Once the bubble fragmentation may have a significant impact upon the wave behavior structure, the fragmentation conditions and effects of phase parameters upon fragmentation process become essential; in particular, the effect of gas properties within bubbles on the shock-wave critical intensity above which the wave breaks the bubbles is of considerable importance.

A study of the process of flow past bubbles or drops with velocity w_{12} based, in particular, on a dimensional analysis, showed that the Weber number proves to be a reasonable criterion of bubble fragmentation

$$\mathrm{We} = \frac{\rho_* w_{12}^2 a}{\Sigma} \geqslant \mathrm{We}_* \sim 1 \tag{6.8.4}$$

where W_* is some critical value, ρ_* is the density; at first sight, this is the external-flow phase density, i.e., in the case of drops, it is the density of gas, and in the case of bubbles, it is the density of liquid.

The coefficient Σ of surface tension is defined by the liquid phase substance (the substance of gas phase has an insignificant effect on Σ), and depends on its temperature T on the interface, which (as was mentioned before, see §1.6), as distinct from the temperature of the bulk mass of gas, is invariable ($T_\Sigma = T_0$). The shock waves must be very strong ($p_e/p_0 > 10$) to be responsible for the rise of the liquid temperature on the bubble wall in the process of compression due to the gas temperature rise in the bubble core.

Now, the velocity of flow past bubbles $w_{12} = v_1 - v_2$ across the wave should also be independent of the gas-phase matter (see the momentum equations for υ_1 and υ_2 (in 1.5.4), since the bubble inertia is virtually defined by the associated mass of liquid which is many times higher than the bubble mass.

Thus, it can be concluded that the gas properties must not affect the bubble fragmentation conditions. Experimental data (B. E. Gel'fand, S. A. Gubin et al., 1977) demonstrated an opposite result. Under a fixed original pressure $p_0 \approx 0.1$ MPa and the bubble size $a_0 \approx 2$ mm, the air, or nitrogen, bubbles are broken by a wave of intensity $p_e/p_0 > 3$, and at $p_e/p_0 \approx 15$, the nitrogen bubbles break into "dust," i.e. into extremely fine bubbles ($a < 0.1$ mm). Helium bubbles are not broken by a wave of $p_e/p_0 = 8$, and hydrogen bubbles are not broken by a wave of intensity $p_e/p_0 = 50$. Thus, a qualitatively higher stability of helium and hydrogen bubbles was found against the shock wave effects compared to air, or nitrogen, bubbles.

To comprehend this effect, an analysis more sophisticated than that described above must be used; for example, analysis of both behavior and stability of a gas sphere with density ρ_g^0 in a liquid flow with density ρ_l^0 should be considered. Approximate analysis of this process is outlined in §2.2. The obtained formulas (2.2.8) and their discussion prove that, if a drop, or bubble, fragmentation is described by the Kelvin-Helmholtz mechanism (scheme *b* in (2.2.8) and in Fig. 2.2.2)), the condition of fragmentation is predetermined by the dynamic head in a gas phase, i.e., is defined by the gas density, as distinct from a speculative premise used in discussion of formula (6.8.4) and scheme *a* in (2.2.8) and in Fig. 2.2.2, where the break-up process is described by Rayleigh-Taylor mechanism, and is defined by liquid density.

Giving consideration to the above-indicated experimental fact of higher stability of hydrogen and helium bubbles in shock waves compared to air bubbles (of "higher density"), it can be concluded that the Kelvin-Helmholtz mechanism (scheme *b* in (2.2.8)) of bubble break-up in shock waves is a predominant mechanism governing the process.

The relative velocity of phases in shock waves (with regard to the order of magnitude) is

$$w_{12} \sim (p_e - p_0)/\rho D \qquad (\rho \approx \overset{\circ}{\rho}_l, \quad D \sim C_e \approx \sqrt{p_0/(\overset{\circ}{\rho}_l \alpha_{20})}) \tag{6.8.5}$$

Substituting this evaluation into formula (2.2.8b), we obtain the fragmentation condition in terms of phase parameters and shock-wave intensity

$$\frac{\overset{\circ}{\rho}_g}{\overset{\circ}{\rho}_l} \alpha_{20} \frac{2a}{\Sigma} \left(\frac{p_e - p_0}{p_0} \right)^2 \gtrsim 6 \tag{6.8.6}$$

This condition is supported by an experimental fact of higher "strength" of a lighter gas (of lower ρ_g^0) at all other conditions equal. For a more accurate account of bubble break-up in shock waves, the viscosity effects, initial pressure,[3] wave intensity, void fraction, and the initial nonsphericity of bubbles, further experimental and theoretical studies are imperative.

Note that the impact of the small parameter ρ_g^0/ρ_l^0 on the qualitative appearance of the motion of a dividing surface between gas and liquid was indicated by G. Birkhoff (1960) and Yu. L. Yakimov (1973).

The bubble fragmentation, resulting in reduction of their size, contributes also to a decrease of the shock-wave thickness, or the transient-zone thickness, in which the transition takes place from the initial state to a state behind the wave. Reduction of the wave thickness results in reduction of erosion, or the wave dispersion, that may lead to a delayed decay of the front wave because of a following from behind the unloading wave. In a medium with comminuted (because of fragmentation) bubbles, the wave reflection from a rigid wall may occur in a more rapid mode, resembling the mode of reflection of a perfect compressible liquid.

Thus, by stimulating and preventing the bubble fragmentation (e.g., by introducing surface-active agents affecting the coefficient of surface tension Σ, or by altering the composition of gas, or by changing the intensity and duration of perturbations), it becomes possible to affect the evolution, attenuation, and reflection of waves in bubbly liquids; it becomes possible, in particular, to affect the maximum pressure that acts upon structural elements in contact with a bubbly medium.

§6.9 SPHERICAL AND CYLINDRICAL WAVES IN BUBBLY LIQUIDS

In order to describe, in addition to the one-dimensional ($\nu = 1$) flow, the flows of bubbly liquids with cylindrical ($\nu = 2$) and spherical ($\nu = 3$) symmetries in Lagrangian variables r and t, it becomes necessary to use, instead of (6.7.2) and the first three equations (6.7.3), their generalizations following from Eqs. (1.10.14) or (1.10.15)

$$\begin{gathered}
\frac{\rho}{\rho_0}\left(\frac{x}{r}\right)^{\nu-1}\frac{\partial x}{\partial r} = 1, \quad \frac{\partial x}{\partial t} = v \\
\frac{\partial \rho_1}{\partial t} + \rho_1\left[\frac{\rho}{\rho_0}\left(\frac{x}{r}\right)^{\nu-1}\frac{\partial v}{\partial r} + \frac{(\nu-1)\,v}{r}\right] = 0 \\
\frac{\partial \rho_2}{\partial t} + \rho_2\left[\frac{\rho}{\rho_0}\left(\frac{x}{r}\right)^{\nu-1}\frac{\partial v}{\partial r} + \frac{(\nu-1)\,v}{r}\right] = 0 \\
\frac{\partial v}{\partial t} = -\frac{1}{\rho_0}\left(\frac{x}{r}\right)^{\nu-1}\frac{\partial p}{\partial r}
\end{gathered} \tag{6.9.1}$$

[3]Regarding the effects of initial pressure upon the bubble break-up in shock waves, see the discussion of Fig. 6.7.4.

The remaining equations have the same form as in (6.7.3). It can easily be proven that Eqs. (6.7.6) retain the same form, and instead of (6.7.5), we have

$$\frac{\partial v}{\partial r} = \frac{\rho_0}{\rho}\left(\frac{r}{x}\right)^{\nu-1}\left[\frac{3\alpha_2 w}{a} - \frac{(\nu-1)\,v}{x}\right] \tag{6.9.2}$$

Let us differentiate this equation with respect to t, and the momentum equation with respect to r, which leads to the generalization of Eq. (6.7.9)

$$\frac{\partial^2 p_{1*}}{\partial r^2} + K(r)\frac{\partial p_{1*}}{\partial r} + M(r) = 0$$

$$K(r) = -\frac{1}{\rho_0}\frac{d\rho_0}{dr} - \frac{\nu-1}{r}\left[1 - \frac{2\rho_0}{\rho}\left(\frac{r}{x}\right)^{\nu}\right]$$

$$M(r) = \frac{\rho_0^2}{\rho}\left(\frac{r}{x}\right)^{2(\nu-1)}\left\{\frac{3\alpha_2}{a^2\left(1-\varphi_*^{(1)}\right)}\left[\frac{p_2 - p_{1*} - 2\Sigma/a}{\overset{\circ}{\rho}_1} - \frac{4\mu_1 w}{\overset{\circ}{\rho}_1 a}\right.\right.$$
$$\left.\left. + \left(1 - 4\varphi_*^{(1)} + 3\varphi_*^{(2)}\right)\frac{w^2}{2}\right] + \frac{(\nu-1)\,v}{x}\left[\frac{\nu v}{x} - \frac{3\alpha_2 w}{a}\right]\right\} \tag{6.9.3}$$

The remaining equations governing the motions of bubbly liquid have the same form as in §6.7. The procedure of analysis is similar to that described in §6.7 with the only exception that with $\nu \neq 1$ ($\nu = 2$ and 3), the Eulerian coordinate $x(t, r)$ of particles must be calculated using the second equation of (6.9.1).

Nonequilibrium effects. Figure 6.9.1 depicts the results of solutions to problems with spherical ($\nu = 3$), cylindrical ($\nu = 2$) and plane ($\nu = 1$) waves initiated by a widening "piston" which produces a constant pressure during its expansion in accordance with a law $x_p(t)$ determined in the process of solving the problem under consideration. Attenuation of the wave front due to spreading leads to reduced oscillations, and, because of this fact, to a decrease of the thermal-dissipation and gas-property effects. As the wave propagates, the disturbance of every particle of the medium becomes more and more smooth, and both the difference between phase pressures and the impact of radial inertia of liquid reduce. The nonequilibrium effects, associated with the difference in phase pressure and temperatures, decrease for both cylindrical ($\nu = 2$) and spherical ($\nu = 3$) regimes (with the piston pressure or velocity being kept constant); smoothing of the parameter distribution which tends to the equilibrium distribution of gas-liquid medium (dashed lines for $\nu = 3$ in Fig. 6.9.1) is responsible for the indicated decrease of nonequilibrium effects.

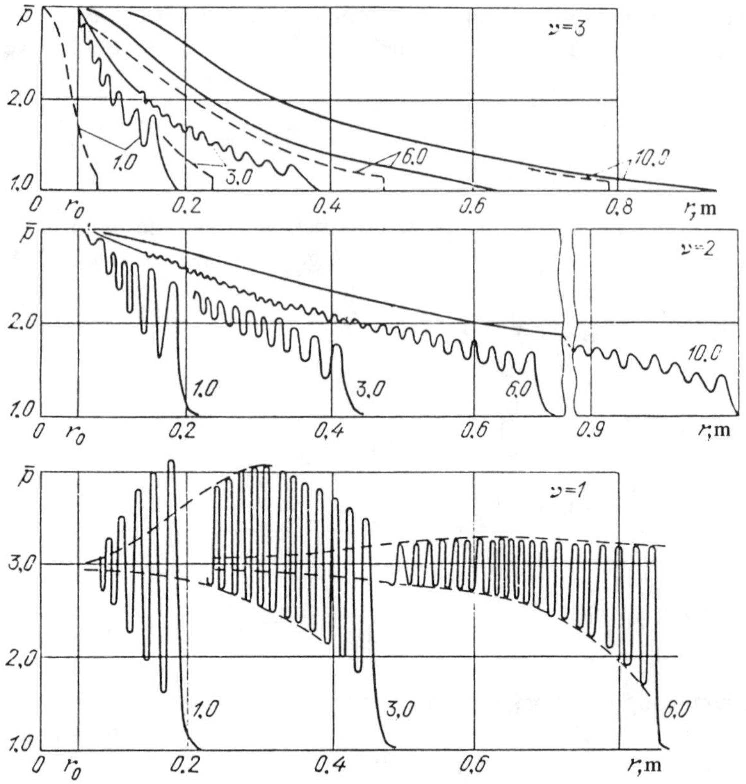

Figure 6.9.1 Comparison of pressure distributions at different times after the motion onset of plane ($\nu = 1$), cylindrical ($\nu = 2$), and spherical ($\nu = 3$) pistons (initial position of the piston: $t = 0$, $r_0 = 0.05$ m) producing the constant pressure $p_p = 3\ p_0$, and moved into the bubbly liquid (water + air: $p_0 = 0.1$ MPa, $T_0 = 293$ K, $a_0 = 1$ mm, $\alpha_{20} = 0.02$). Dashed lines for $\nu = 3$ relate to a self-similarity solution ($r_0 = 0$) for an equilibrium scheme ("$a = 0$") of gas-liquid mixture when v_p = constant in accordance with Eqs. (6.9.14) and (6.9.15), namely: for $\bar{p}_p = 3$, $\bar{v}_p = 0.172$ ($v_p = 12.2$ m/s) takes place. Labels on curves indicate time t in microseconds.

Self-similarity problem concerned with a piston in an equilibrium gas-liquid medium. We will now discuss a problem concerned with a plane or cylindrical or spherical piston uniformly widening following the law $x_p = v_p t$ in a gas-liquid mixture. For this particular problem we shall examine an equilibrium approximation for the behavior of the mixture as a perfect barotropic compressible liquid when the equation of state has the form (1.5.28). So far, we shall confine ourselves to situations when the continuous phase compressibility may be ignored ($\alpha_2 \gg \alpha_C$, ρ_1^0 = constant); then, equations of state (1.5.28) simplify and assume the form (6.8.1).

The mass and momentum equations of a spherically symmetrical motion in Eulerian coordinates x, t are

$$\frac{\partial\rho}{\partial t}+\frac{\partial\rho v}{\partial x}+\frac{(\nu-1)\rho v}{x}=0,\quad \frac{\partial v}{\partial t}+v\frac{\partial v}{\partial x}+\frac{1}{\rho}\frac{\partial p}{\partial x}=0 \tag{6.9.4}$$

It is natural to assume that the disturbance front propagates with velocity D in the form of a compression shock, the parameters behind which are denoted by subscript f. Mass and momentum conservation equations across the shock

$$\rho_0 D=\rho_f(D-v_f),\quad p_f-p_0=\rho_0 D v_f \tag{6.9.5}$$

together with equation of state (6.9.1) determine the bulk velocity v_f of a substance, and shock velocity D through the undisturbed medium in terms of pressure p_f behind the shock in the following form

$$v_f=\frac{p_f-p_0}{\sqrt{p_f}}\sqrt{\frac{\alpha_{20}}{\rho_0}},\quad D=\sqrt{\frac{p_f}{\rho_0\alpha_{20}}} \tag{6.9.6}$$

Let us introduce dimensionless variables

$$\bar{v}=\frac{v}{C_0},\quad \bar{p}=\frac{p}{p_0},\quad \bar{\rho}=\frac{\rho}{\rho_0},\quad \lambda=\frac{x}{C_0 t}\quad \left(C_0^2=\frac{p_0}{\rho_1^\circ\alpha_{10}\alpha_{20}}\right) \tag{6.9.7}$$

We will now find the similarity solution of the problem for a uniform motion of the piston when all dimensionless variables depend only on a single dimensionless variable λ

$$\bar{v}=\bar{v}(\lambda),\quad \bar{p}=\bar{p}(\lambda),\quad \bar{\rho}=\bar{\rho}(\lambda) \tag{6.9.8}$$

Then, Eqs. (6.9.4) are modified to ordinary differential equations

$$\bar{\rho}\frac{d\bar{v}}{d\lambda}+(\bar{v}-\lambda)\frac{d\bar{\rho}}{d\lambda}=-\frac{(\nu-1)\overline{\rho v}}{\lambda}$$

$$(\bar{v}-\lambda)\frac{d\bar{v}}{d\lambda}+\frac{\alpha_{20}}{\bar{\rho}}\frac{dp}{d\lambda}=0,\quad \bar{p}=\frac{\alpha_{20}}{\alpha_{10}}\frac{\bar{\rho}}{1-\bar{\rho}} \tag{6.9.9}$$

with boundary conditions on the piston ($\lambda = \lambda_p$), and across the shock ($\lambda = \lambda_f$)

$$\lambda=\lambda_p=v_p/C_0:\ \bar{v}=\bar{v}_p=\lambda_p\ (\bar{p}=\bar{p}_p,\quad \bar{\rho}=\bar{\rho}_p)$$

$$\lambda=\lambda_f=D/C_0:\ \bar{v}=\bar{v}_f,\quad \bar{p}=\bar{p}_f,\quad \bar{\rho}=\bar{\rho}_f \tag{6.9.10}$$

Note that, in accordance with (6.9.6), the shock coordinate and bulk velocity behind the shock are determined in terms of pressure

$$\lambda_f=\sqrt{\bar{p}_f},\quad \bar{v}_f=\alpha_{20}(\bar{p}_f-1)/\lambda_f \tag{6.9.11}$$

For plane waves ($\nu = 1$), a uniform (homogeneous) flow is realized behind the shock, and $\bar{v}_p = \bar{v}_f$, $\bar{p}_p = \bar{p}_f$. The solution of a boundary-value problem (6.9.9) and (6.9.10) for both cylindric and spherical waves may be found using a numerical method by variating the value $\bar{p}_f$, and by solving the Cauchy problem in the domain $\lambda_p < \lambda < \lambda_f$; the value $\bar{p}_f$ being chosen so that the known boundary condition on the piston ($\lambda = \lambda_p$, $\bar{\upsilon} = \lambda_p$) is satisfied; this boundary condition is defined by the piston speed υ_p. The described procedure permits the determination of both pressure $\bar{p}_p$ on the piston and the realized shock parameters (6.9.11).

Given that $1 \leqslant \bar{\rho} \leqslant \alpha_{10}^{-1}$ at small void fractions ($\alpha_{20} \ll 1$, but $\alpha_{20} \gg \alpha_C$), the solution may be found approximately provided that the density variability in differential equations is ignored, and assuming that $\bar{\rho} \approx 1$. Then, Eqs. (6.9.9) simplify, and both velocity and pressure distributions between the piston and shock become the same as in incompressible liquid. We will confine ourselves to a spherical-wave case ($\nu = 3$), and then

$$\frac{d\bar{v}}{d\lambda} = -\frac{2\bar{v}}{\lambda}, \quad (\bar{v} - \lambda)\frac{d\bar{v}}{d\lambda} + \alpha_{20}\frac{d\bar{p}}{d\lambda} = 0 \tag{6.9.12}$$

Upon integrating, using the boundary conditions on the piston, we readily obtain

$$\bar{v} = \bar{v}_p^3\lambda^{-2}, \quad \alpha_{20}(\bar{p} - \bar{p}_p) = -2\bar{v}_p^3(\lambda_p^{-1} - \lambda^{-1}) + {}^1/_2\bar{v}_p^6(\lambda_p^{-4} - \lambda^{-4}) \tag{6.9.13}$$

Here, $\bar{p}_p$ is still unknown. Taking into consideration both (6.9.11) and $\bar{\upsilon}_f = \bar{\upsilon}_p^3\lambda_f^{-2} = \bar{\upsilon}_p^3/\bar{p}_f$, we obtain the equation for determining $\bar{p}_f$.

$$\sqrt{\bar{p}_f}\,(\bar{p}_f - 1) = \bar{v}_p^3/\alpha_{20} \qquad \left(\frac{\bar{v}_p^3}{\alpha_{20}} = \frac{v_p^3\rho_0}{p_0}\sqrt{\frac{\rho_0\alpha_{20}}{p_0}}\right) \tag{6.9.14}$$

Having calculated values $\bar{p}_f$ (e.g., graphically) and λ_f, we obtain, from the second equation (6.9.13), an equation for determining pressure on the piston

$$\bar{p}_p = \bar{p}_f + \frac{3\rho_0 v_p^2}{2p_0} - 2(\bar{p}_f - 1) + \frac{\alpha_{20}(\bar{p}_f - 1)}{2\bar{p}_f} \tag{6.9.15}$$

The medium compressibility leads to $\bar{p}_f > 1$, and in the expression for $\bar{p}_p$ it is given consideration by both third and fourth addends. It is evident that, with a fixed velocity υ_p, compressibility of the medium leads to the reduction of pressure p_p on the piston.

If υ_p in the last equation is replaced by its value following from (6.9.14), we obtain an equation which may be used for determining p_f, and then, also, υ_p and λ_f, provided that pressure p_p on the piston is prescribed

$$\bar{p}_p = 2 - \bar{p}_f + \frac{3}{2\alpha_{20}^{1/3}}[\bar{p}_f(\bar{p}_f - 1)]^{1/3} + \frac{\alpha_{20}}{2}\frac{(\bar{p}_f - 1)^2}{\bar{p}_f} \tag{6.9.16}$$

The obtained equation can easily be generalized also for a compressible carrying liquid when the equation of state of the medium is written in the form (1.5.28). Such a generalization may be meaningful when the pressures are high but the density varies insignificantly ($\Delta\rho/\rho_0 < \alpha_{20} \ll 1$). Then, a more precise shock adiabat must be plotted in the form $D(p_f)$, from which a refined relation $\lambda_f(\bar{p}_f)$ follows instead of $\lambda_f = \sqrt{\bar{p}_f}$, and the pressure p_f should be found from the generalized equation (6.9.14)

$$(\bar{p}_f - 1)\,\lambda_f\,(p_f) = \bar{v}_p^3/\alpha_{20} \tag{6.9.17}$$

Now, v_f is readily found from the second formula (6.9.11), and the pressure p_f on the piston is found from Eq. (6.9.13).

Figure 6.9.2 depicts the results of a solution for spherical waves generated by a spherical piston which widens into an equilibrium bubbly medium which was originally at rest. The piston widens with a velocity v_p = constant from an initial radius $r_{p_0} = x_{p_0}$ at the original time t_0 (at $t < t_0$, state of rest), time t_0 being chosen so that with the piston-motion law $x_p = x_{p_0} + v_p(t - t_0)$, time $t = 0$ corresponds to $x_p = 0$. It is evident that the similarity solution corresponding to $x_{p_0} = 0$, $t_0 = 0$, and $x_p = v_p t$, analyzed here for an equilibrium model of a gas-liquid mixture ("$a = 0$"), is an asymptote of a solution of a piston-problem at $t \gg t_0$ (the piston begins its motion with a nonzero radius x_{p_0} into a bubbly liquid with consideration given to non-equilibrium effects).

§6.10 SHOCK WAVES IN A LIQUID WITH VAPOR (STEAM) BUBBLES

Thus far, shock waves propagating in a liquid with bubbles of noncondensing and insoluble gas in the absence of phase transitions were discussed in

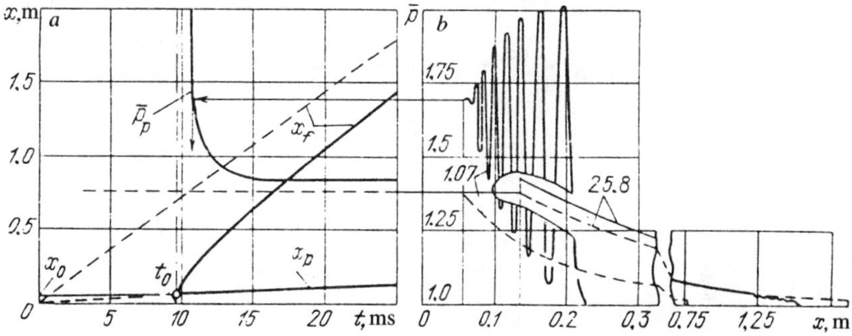

Figure 6.9.2 Variation of pressure $\bar{p}_p$ on a spherical piston as a function of time (*a*); the piston is widening with a constant velocity v_p = 5.2 m/s, beginning from a radius r_0 = 0.05 m ($x_p = r_0 + v_p\,(t - t_0)$ at $t > t_0 = r_0/v_p$ = 9.65 microseconds); law of propagation of the disturbance front $x_f(t)$ in a mixture of water with air bubbles (p_0 = 0.1 MPa, T_0 = 293 K, α_{20} = 0.02, a_0 = 1 mm) (*a*); and variation of pressure distribution (*b*) at two points in time t, in microseconds (indicated by labels on curves). The dashed lines correspond to a similar solution (r_0 = 0, t_0 = 0) for an equilibrium model of a bubbly liquid.

this chapter. We shall now be concerned with the study of effects arising when bubbles are filled with vapor of the carrying liquid and when the mixture is in equilibrium in its original state, i.e., at the saturation temperature.

Some experimental facts. The weak disturbances ($\Delta p/p_0 \ll 1$), as they propagate, undergo significant spreading due to strong dispersion when the speed of high-frequency perturbations is many times higher than that of low-frequency disturbances, since the low-frequency-perturbation velocity in such a mixture is very small (see §6.2), and, in particular, is many times smaller than in a mixture with gas bubbles. The latter may be explained by the fact that the vapor pressure at sufficiently slow compression of a bubble virtually does not increase because of condensation of "excessive" mass of vapor. If the initial effect is sufficiently strong ($\Delta p/p_0 \gtrsim 1$), the nature of the wave evolution, as is experimentally proven, alters.

Figure 6.10.1 schematically represents the pressure oscillogram exhibiting a strong effect of an anomalous pressure rise in shock waves propagating in both boiling nitrogen and water with steam bubbles (A. A. Borisov, B. E. Gel'fand, R. I. Nigmatulin et al., 1977; 1982; V. E. Nakoryakov, B. G. Pokusaev, et al., 1983). Waves have been produced in a vertical shock tube of diameter 50 mm resulting from rupture of the diaphragm separating the two-phase medium under study (original pressure $p_0 \approx 0.1$ MPa) from the high-pressure chamber (see §6.1). Boiling of the water-steam mixture was facilitated by a water-heater installed at the lower portion of the shock tube. For recording the shock waves, piezometric pressure sensors were employed whose natural frequency was about 30 kcycles. The volumetric concentration α_{20} of vapor bubbles in the boiling nitrogen near the bottom of a tube was about 0.90, and the mixture appeared to have a foamy structure. In the rest of the observed cases, $\alpha_{20} = 0.05$–0.2, and the mixture had a bubbly structure.

Disturbance of the original pressure p_0 consists of a front of the prop-

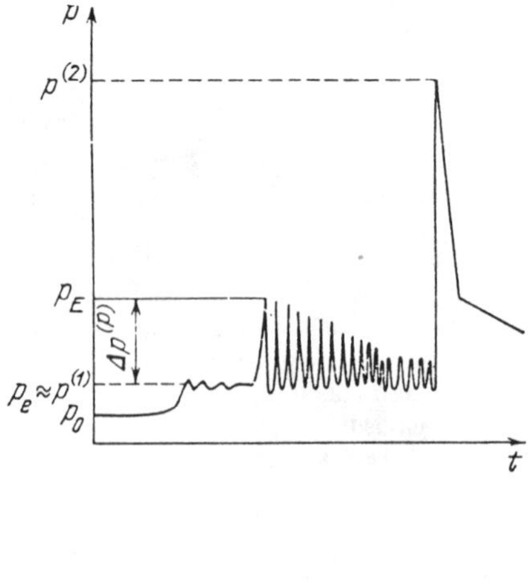

Figure 6.10.1 Schematic pressure oscillogram for shock waves propagating in boiling liquids with bubbles of either vapor or readily soluble gas.

agating wave with a pressure $p^{(1)}$, behind which pressure pulsations occur with an amplitude $\Delta p^{(1)}$ and frequency (approximately) 3 kcycles, and, finally, the pressure behind the wave reflected from the lower wall grows up to $p^{(2)}$. The pressure growth coefficients $p^{(2)}/p^{(1)}$ during reflection for boiling liquids reach values of 20–50, and they are many times higher than in a liquid with bubbles of low-solubility gases where $p^{(2)}/p^{(1)} = 4$–7. The strong rarefaction wave following the reflected wave is responsible for its attenuation.

In each of the three gas-liquid bubbly mixtures discussed above, the process in an incident wave proceeds in a peculiar mode. The disposition towards occurrence of high-frequency oscillation pressure-peaks in a steam-water mixture is much lower than in the rest of bubbly liquids (nitrogen, water + CO_2). Thus, in a wave $p^{(1)}/p_0 \approx 3.7$ ($p = 0.1$ MPa) propagating in boiling nitrogen, the amplitude of indicated peaks is $\Delta p^{(1)} = 0.1$–1.5 MPa, and in a similar wave propagating in a steam-water mixture there are no such peaks. This, as is shown below, is accounted for by different intensities of heat and mass exchange processes and those of bubble fragmentation processes. See also S. Bankoff (1978).

Analysis of wave amplification effects in vapor liquid media. The above-described effect of wave amplification in one-component mixtures of a liquid with vapor bubbles is associated with phase transitions. This is corroborated by the fact that an analogous effect has been observed in water with bubbles of carbon dioxide CO_2, which, compared to other gases, is of good solubility in water. However, the mechanism of the wave-amplification phenomenon described above is a nontrivial question.

Supposition of the wave amplification due to intensive vapor (steam) generation process cannot be criticized, since the adiabatic compression of mixture (because of small mass of gas) proceeds at a constant temperature of liquid, and condensation must take place in mixture under pressure rise rather than evaporation. The situation is similar as regards a supposition on an anomalous pressure rise in a mixture caused by the condensation-heat release since, due to small mass-concentration of vapor (steam) generated during complete condensation, this heat is insufficient even for a considerable rise of the liquid temperature, and it is all the more insufficient for raising the pressure.

Apparently, the mechanism of wave amplification in liquids with bubbles of vapor, or rapidly soluble gas, is associated with a rapid condensation-induced collapsing of bubbles. It is adviseable to discuss a specific feature of this collapsing using an example of a "trial" bubble (see §§1.6 and 2.6). Let us single out a spherical cell of fixed radius R around a trial bubble; this radius may be determined by the initial volumetric concentration of vapor (steam): $R^3 = a_0^3/\alpha_{20}$. In addition to the bubble-collapse law $a(t)$, we shall also look after the liquid mean-pressure variation $\langle p \rangle_1$ within the cell

$$\langle p\rangle_1 = \frac{1}{{}^4/_3\pi\,(R^3 - a^3)} \int_a^R 4\pi r^2 p'\,(r,\ t)\,dr \tag{6.10.1}$$

where the distribution $p'(r, t)$ is determined from the Cauchy-Lagrange integral (1.3.14) for an incompressible liquid in a spherically-symmetrical motion with a prescribed pressure $p_\infty(t)$ at a remote distance from the bubble, and in accordance with a law of its radius variation $a(t)$, which, in turn, is governed by the Rayleigh-Lamb equation. If it is assumed in this equation that p_2 = constant and p_∞ = constant, and also, that both capillary and viscosity effects are ignored ($p_{1\alpha} = p_2$, we obtain the regime of an accelerated inertial bubble-collapsing, referred to as the Rayleigh regime (2.6.52). Simple mathematical transformations prove that in this case, with $a \to 0$, the liquid average pressure within a cell approaches infinity

$$\langle p\rangle_1 \approx 0.5\alpha_{20}^{1/3}(a_0/a)^{2/3}p_\infty \tag{6.10.2}$$

The rate of an actual condensation process and the collapsing of a vapor (steam) bubble, as well as the mean-pressure $\langle p\rangle_1$ growth, are limited by the nonzero heat conductivity of liquid (see the discussion of (2.6.56)), owing to which the latter cannot remove all the heat needed for the realization of condition p_2 = constant (that provides the validity of (6.10.2)); therefore, on some stage of compression, the vapor pressure p_2 rises, causing the collapsing-process retardation and reducing the rate of the mean-pressure $\langle p\rangle_1$ growth. In order to investigate the potentials of the heat-removal properties of liquid (in particular, liquid nitrogen) for realization of a significant rise of the liquid mean-pressure $\langle p\rangle_1$ during vapor-bubble collapsing, computations have been carried out using the system of equations of a spherically-symmetrical motion, equations of nonstationary heat conduction in a compressible vapor and incompressible liquid, and boundary conditions over the interface $a(t)$ defined in the process of developing the solution (see §2.6).

Figure 6.10.2 shows, by way of example, the result of this computation for the case of a nitrogen bubble in liquid nitrogen when the pressure p_∞ abruptly increases at $t = 0$. It is evident from the graph that the mean pressure $\langle p\rangle_1$ reaches 1.3 MPa, which is much higher than pressure $p_\infty = 0.5$ MPa which initiates the process. But, as distinct from the Rayleigh regime (shown by a dashed line with a vertical asymptote and defined by the bursting-time t_R in accordance with (2.6.55)), many pulsations are realized due to the nonzero heat conduction of liquid. Bubble collapsing and disappearance may be enhanced by fragmentation, which is responsible for the condensation rate increase due to the interface-area increase.

Condensation wave. In connection with the above discussion, the statement of bubbly-liquid problems concerned with bubble-disappearance must envisage volumes, or zones $U^{(1)}$ and $U^{(2)}$, where single-phase and two-phase liquids are, respectively, realized, and must also specify the surfaces or

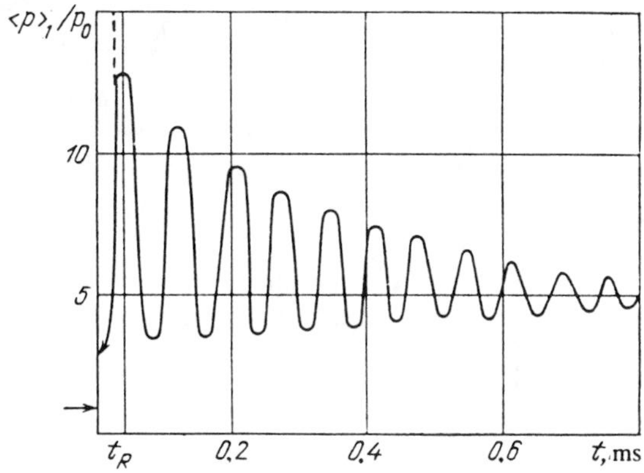

Figure 6.10.2 Variation of the mean pressure $\langle p\rangle_l$ within a cell with time for a case of a nitrogen bubble in liquid nitrogen at the following initial data: $a_0 = 1$ mm, $\alpha_{20} = 0.05$, $p_0 = 0.1$ MPa, $T_0 = T_S(p_0) = 77$ K; pressure p_∞ at a remote distance from the bubble undergoes a jump-like change from 0.1 MPa to 0.5 MPa. Dashed line corresponds to the Rayleigh inertial regime (2.6.52).

boundaries $F^{(12)}$ which separate these zones, and which may be referred to as the *condensation jumps;* the surfaces $F^{(12)}$ must be provided with boundary conditions analogous to those on dividing surfaces.

We shall now examine the indicated conditions in a coordinate system, in which the condensation surface $F^{(12)}$, or condensation wave, is at rest. The two-phase state (with bubbles) of the medium before this wave is denoted by subscript F, and state of the medium behind the wave (in the form of a single-phase liquid) by subscript e. Then, the mass, momentum, and energy conservation laws, provided that the bubble mass, momentum, and energy compared to similar parameters of liquid are neglected, assume the form

$$\begin{aligned}
&\alpha_{1F}\overset{\circ}{\rho}_{1F}v_F = \rho_{1e}v_e \\
&\alpha_{1F}\overset{\circ}{\rho}_{1F}\,v_F^2 + p_F = \overset{\circ}{\rho}_{1e}v_e^2 + p_e \qquad (6.10.3)\\
&\alpha_{1F}\overset{\circ}{\rho}_{1F}v_F\left({}^1\!/_2 v_F^2 + k_{1F} + u_{1F}\right) + p_F v_F = \overset{\circ}{\rho}_{1e}v_e\left({}^1\!/_2 v_e^2 + u_{1e}\right) + p_e v_e
\end{aligned}$$

Here, $\dot{k}_{1F}$ is the kinetic energy of a small-scale radial motion per unit mass of liquid before the shock, p_F is the average pressure of the medium before the shock, p_e is the liquid pressure behind the shock. We also recall that

$$p_F = \alpha_{1F}p_{1F} + \alpha_{2F}(p_{2F} - 2\Sigma/a), \quad p_e = p_{1e} \qquad (6.10.4)$$

Given that the void fraction is small ($\alpha_{2F} \ll 1$), we hereafter assume $p_F \approx p_{1F}$, and the linear equation (1.5.2) of state is used to define the continuous-liquid pressure. Then, from the mass equation across the shock (the first equation (6.10.3)), we have

$$\overset{\circ}{\rho}_{1e} = \overset{\circ}{\rho}_{1F}\left(1+\alpha_C(\bar{p}_e-1)\right), \quad v_F - v_e = \alpha_{2F}v_F\frac{1+(\alpha_C/\alpha_{2F})(\bar{p}_e-1)}{1+\alpha_C(\bar{p}_e-1)} \tag{6.10.5}$$

$$\left(\alpha_C = \frac{p_F}{\overset{\circ}{\rho}_{1F}C_1^2}, \quad \bar{p}_e = \frac{p_e}{p_F}\right)$$

From the momentum equation across the shock (the second equation (6.10.3)), we obtain an expression interrelating the shock intensity $\bar{p}_e$ and its velocity $D_F = -v_F$ relative to the medium before the front

$$D_F^2 = v_F^2 = \frac{p_F}{\alpha_{1F}\alpha_{2F}\overset{\circ}{\rho}_{1F}}(\bar{p}_e-1)\frac{1+\alpha_C(\bar{p}_e-1)}{1+(\alpha_C/\alpha_{2F})(\bar{p}_e-1)} \tag{6.10.6}$$

It is obvious that the effect of the carrying-liquid compressibility displays itself through the last multiplier; to ensure insignificance of this compressibility (i.e., when the overall compressibility of the mixture is due only to bubbles), the following is necessary and sufficient

$$\delta_C = \frac{\alpha_C}{\alpha_{2F}}(\bar{p}_e-1) = \frac{p_e-p_F}{\alpha_{2F}\overset{\circ}{\rho}_{1F}C_1^2} \ll 1 \tag{6.10.7}$$

This condition for $\alpha_{2F} \sim 10^{-2}$, $\rho_{1F}^0 \sim 10^3$ kg/m^3, $C_1 \sim 10^3$ m/s is satisfied if $p_e - p_F \leqslant 1.0$ MPa.

From (6.10.5) and (6.10.6), an expression is readily obtained that interrelates the variations of velocity and pressure across the condensation shock

$$\Delta v_e = v_F - v_e = \sqrt{\frac{\alpha_{2F}p_F}{\alpha_{1F}\overset{\circ}{\rho}_{1F}}\left(\frac{p_e}{p_F}-1\right)}\cdot\left(1+\frac{\alpha_{1F}}{2}\delta_C + O(\delta_C^2)\right) \tag{6.10.8}$$

From the energy equation across the shock (the third equation (6.10.3)), it follows

$$u_{1e} - u_{1F} = k_F - \left(\frac{p_e}{\overset{\circ}{\rho}_{1e}} - \frac{p_F}{\overset{\circ}{\rho}_{1F}\alpha_{1F}} + \frac{v_e^2-v_F^2}{2}\right) \tag{6.10.9}$$

In those cases when effects of the carrying-liquid compressibility are small, i.e. the (6.10.7) or $|\rho_{1e}^0 - \rho_{1F}^0| \ll \rho_{1F}^0$ hold true, the energy equation may be rewritten in the form

$$u_{1e} - u_{1F} = k_F - \frac{\alpha_{2F}}{\alpha_{1F}}\frac{p_e}{\overset{\circ}{\rho}_{1F}} + \frac{p_e-p_F}{\overset{\circ}{\rho}_{1F}}\delta_C \tag{6.10.10}$$

Reflection of the condensation shock wave from a solid wall. The equilibrium equation $p(V)$ of the state of a vapor-liquid bubbly mixture where $V = 1/\rho$ is schematically shown in Fig. 6.10.3. In the domain of original (two-phase) state, the $p(V)$-diagram is very gently sloping because of compressibility due to bubble condensation; then, following the bubble disap-

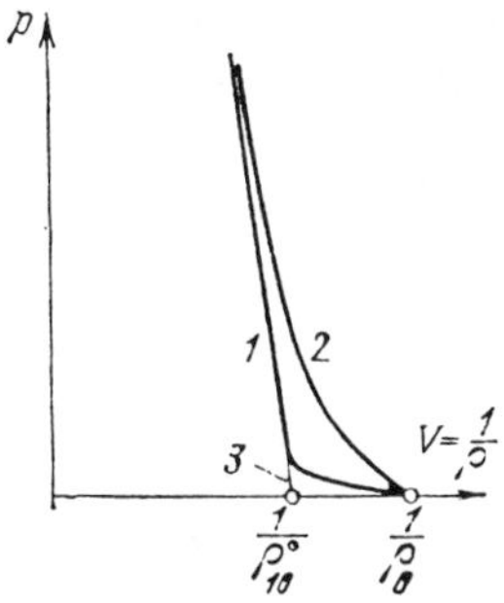

Figure 6.10.3 Schematic of an equilibrium shock adiabat of a bubbly vapor-liquid mixture (1), gas-liquid mixture (2), and a single-phase liquid (3).

pearance, when $V = 1/\rho_1^0$, compressibility of the medium equals that of liquid and the diagram experiences an angularity. Such a strong nonlinearity of the diagram leads to a very significant pressure rise during the shock-wave reflection from a rigid wall.

As in the derivation of formula (6.7.24), we shall use the subscript 0 and superscripts (1) and (2) to denote parameters, respectively, before the incident wave, behind the incident wave (before the reflected wave), and behind the reflected waves. Then, the condition of reflection from the immobile wall is written in the form (6.7.22). In this case, if the incident-wave intensity is sufficient to cause complete vapor (steam) condensation, D_0 is determined by formula (6.10.6), where the indices F and e should be repelaced by 0 and 1, respectively. Then

$$\rho^{(1)} = \overset{\circ}{\rho}_{10}, \quad D^{(1)} \approx C_1 \tag{6.10.11}$$

since the reflected wave propagates through a single-phase low-compressible liquid for which the $p(V)$-diagram below pressures 100–1000 MPa may be regarded as linear. Giving consideration to the above-indicated statements, we obtain an expression for pressure behind the reflected wave

$$\frac{p^{(2)}}{p^{(1)}} = 1 + \overline{C}_f \frac{p_0}{p^{(1)}} \sqrt{\frac{p^{(1)}}{p_0} - 1} \qquad \left(\overline{C}_f = \sqrt{\frac{\alpha_{20}}{\alpha_{10}} \frac{\rho_{10} C_1^2}{p_0}}\right) \tag{6.10.12}$$

This equation indicates the degree of intensification of a shock wave in a vapor-liquid mixture if reflected from a rigid wall when the incident wave causes complete condensation of vapor (steam), and the liquid compressibility exhibits itself only in a reflected wave. We will outline now, for comparison, the corresponding relation for a gas-liquid mixture of incompressible liquid with fixed-mass gas bubbles that follows from (6.7.24)

$$\frac{p^{(2)}}{p^{(1)}} = \frac{p^{(1)}}{p_0} \tag{6.10.13}$$

together with a relation for an acoustic medium of low compressibility obeying a linear law, which is represented by a bubble-free liquid ($\alpha_{20} = 0$)

$$\frac{p^{(1)} - p_0}{\rho_0 C_1} = \frac{p^{(2)} - p^{(1)}}{\rho_0 C_1}, \quad \text{т. е.} \quad \frac{p^{(2)}}{p^{(1)}} = 2 - \frac{p_0}{p^{(1)}} \tag{6.10.14}$$

Figure 6.10.4 depicts intensification diagrams for plane shock waves reflected from a rigid wall in steam-water mixture, gas-water mixture, and linear acoustic media, and also the relations (6.10.12)–(6.10.14).

As may be seen from formula (6.10.12), a considerable intensification of a shock wave after its reflection in a vapor-liquid mixture is possible. For instance, for a vapor-liquid mixture at $p_0 = 0.1$ MPa ($\rho_1^0 \approx 1000$ kg/m3, $C_1 \approx 1500$ m/s) with the void fraction $\alpha_{20} = 0.1$, we have $\bar{C}_f = 50$, and after reflection of waves with pressures $p^{(1)} = 0.2$ and 0.5 MPa, the wall pressures become $p^{(2)} = 5.2$ and 10.5 MPa, respectively, the indicated pressure increasing with increase of the vapor volumetric concentration.

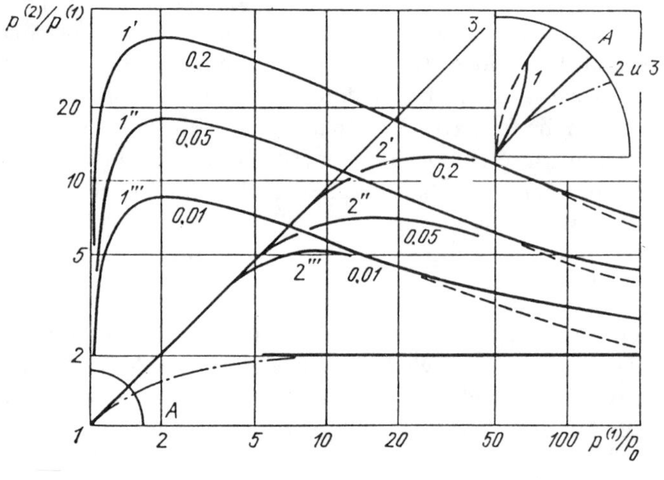

Figure 6.10.4 Intensification diagrams for waves reflected from a rigid wall in steam-water mixture (lines *1*, *1'*, *1''*, and *1'''*) ($p_0 = 0.1$ MPa, $T_0 = 373$ K), and in mixture of water with fixed-mass gas bubbles (lines *2*, *2'*, *2''*, and *2'''*) ($p_0 = 0.1$ MPa) as functions of the incident wave intensities and original void fraction (volumetric concentration of vapor or gas α_{20}). The numerical labels 0.01, 0.05, and 0.2 on curves indicate the values of α_{20}. The line *3* is related to formula (6.10.13), dashed lines to formula (6.10.12), dot-dashed line relates to acoustic medium represented by bubble-free liquid ($\alpha_{20} = 0$), in accordance with formula (6.10.14). Deviation of line *1* from a dashed line (which corresponds to formula (6.10.12)) at small $p^{(1)}/p_0$ (view *A*) is associated with the fact that very weak waves do not cause complete condensation of vapor (steam), and deviation of lines *1'*, *1''*, and *1'''* from dashed lines occurring at large $p^{(1)}/p_0$ is associated with the carrying-liquid compressibility-effects in an incident wave (finiteness of δ_C). Deviation of lines *2'*, *2''*, and *2'''* from line *3* at large $p^{(1)}/p_0$ is also affected by compressibility of the carrying liquid, which was not taken into account in the derivation of formula (6.10.13).

Structure of stationary shock waves in a liquid with vapor bubbles. We shall now be concerned with stationary waves in a liquid with vapor (steam) bubbles (condensating under compression) similar to the analysis of such waves in a liquid with gas bubbles. With this end in view, we shall use the equations described in §§1.5 and 1.6 in a single-velocity approximation (i.e., ignoring translatory motion of bubbles relative to liquid, and also assuming incompressibility of the continuous phase ($\delta_C \ll 1$)). The range of validity of these approximations has already been discussed for the case of gas bubbles, and within indicated limits these simplifications are acceptable for vapor bubbles, too. Since in the case of vapor bubbles, the impact of interphase heat and mass exchange increases to an even higher extent, the corresponding processes, analogous to §6.5, are being described with consideration given to nonhomogeneity of local temperatures around bubbles in the framework of a spherically-symmetrical model. Nonhomogeneity of both temperature and density within a vapor bubble only slightly affects its dynamics (see §§1.6 and 2.6). Therefore, pressure, temperature, and density within a bubble are said to be homogeneous and to satisfy the saturation condition ($T_2 = T_S(p_2)$, $\overset{0}{\rho}_2 = \overset{0}{\rho}_{2S}(p_2)$); the temperature nonhomogeneity is given consideration only as regards the continuous phase using the equation for radial heat conduction in liquid. Thus, all parameters, except for T_1', depend only on the longitudinal coordinate x, and the liquid local temperature is $T_1' = T_1'(x, r)$, where r is the distance to the bubble center. As a result, the system of equations for a one-dimensional stationary flow of liquid with vapor bubbles, similar to (6.5.2), assumes the form

$$\frac{d}{dx}(\rho v) = 0, \quad \frac{d}{dx}(nv) = 0, \quad v\frac{d}{dx}\left(\frac{4}{3}\pi a^3 \overset{\circ}{\rho}_2\right) = 4\pi a^2 \xi_{12}$$

$$\left(\rho = \rho_1 + \rho_2, \quad \rho_i = \overset{\circ}{\rho}_i \alpha_i \quad (i = 1, 2), \quad \alpha_1 + \alpha_2 = 1, \quad \alpha_2 = {}^4\!/_3\pi a^3 n\right)$$

$$\frac{d}{dx}(\rho v^2 + p_1) = 0$$

$$v\frac{da}{dx} = w_{1a} + \frac{\xi_{12}}{\overset{\circ}{\rho}_1}, \quad av\frac{dw_{1a}}{dx} + \frac{3w_{1a}^2}{2} + \frac{4\mu_1 w_{1a}}{\overset{\circ}{\rho}_1 a} = \frac{p_2 - p_1 - 2\Sigma/a}{\overset{\circ}{\rho}_1} \tag{6.10.15}$$

$$4\pi a^2 \xi_{12} l = -q_{\Sigma 1} - q_{\Sigma 2}$$

$$\overset{\circ}{\rho}_1 = \text{const}, \quad p_2 = R_2 \overset{\circ}{\rho}_2 T_2, \quad T_2 = T_S(p_2), \quad u_2 = c_{V2} T_2 + u_{20}$$

To close the represented system of equations, the interphase heat fluxes to both liquid and vapor phases, namely, $q_{\Sigma 1}$ and $q_{\Sigma 2}$, must be determined.

As in §6.5, we ignore both capillarity and viscosity effects, vapor mass concentration, and vapor density compared to liquid density, i.e., (see (6.4.6))

$$\Sigma^* \ll 1, \quad \mu^* \ll 1, \quad \rho_2/\rho_1 \ll 1, \quad \overset{\circ}{\rho}_2 \ll \overset{\circ}{\rho}_1 \tag{6.10.16}$$

If, as is specified above, we assume the model of a homogeneous bubble, its behavior (variation of $q_{\Sigma 1}$, $q_{\Sigma 2}$, p_g, and a along axis x) is described by Eqs. (2.6.13), where $dt = dx/v$ should be substituted. Before setting up the final closed system of equations to be written based on the indicated simplifications, we shall introduce dimensionless variables (6.3.6), understanding that $\bar{w}$ represents $\bar{w}_{1a}$, (6.4.5), and also the following variables (see (2.7.4))

$$\eta = \frac{r}{a(x)}, \quad \bar{T}_1' = \frac{T_1'}{T_0}, \quad \bar{\xi}_{12} = \frac{\xi_{12}}{\overset{\circ}{\rho}_{20} C_*}, \quad \bar{n} = \frac{n}{n_0} \tag{6.10.17}$$

The system of equations (6.10.15) has first integrals which, resulting from simplifications in dimensionless variables, may be written as

$$\alpha_1 \bar{v} = \alpha_{10} \bar{v}_0, \quad \bar{n}\bar{v} = \bar{v}_0 \quad (\alpha_2 = \alpha_{20} \bar{a}^3 \bar{n})$$
$$\frac{\bar{v}_0 \bar{v}}{\alpha_{20}} + \bar{p}_1 = \frac{v_0^2}{\alpha_{20}} + 1 \tag{6.10.18}$$

The differential equations (6.10.15) and (2.6.13) are rewritten in a similar manner

$$\bar{v}\frac{d\bar{a}}{d\bar{x}} = \bar{w} + \bar{\xi}_{12}\frac{\overset{\circ}{\rho}_{20}}{\overset{\circ}{\rho}_1} \approx \bar{w}, \quad \bar{a}\bar{v}\frac{d\bar{w}}{d\bar{x}} = \bar{p}_2 - \bar{p}_1 - \frac{3\bar{w}^2}{2}$$

$$\bar{v}\frac{d\bar{p}_2}{d\bar{x}} = -\frac{3\varkappa_{gS}\bar{p}_2}{\bar{a}}\left(\bar{w} + \bar{q}_{\Sigma 1}\right)$$

$$\bar{v}\frac{\partial \bar{T}_1'}{\partial \bar{x}} = \frac{\bar{w}(\eta^3 - 1)}{\bar{a}\eta^2}\frac{\partial \bar{T}_1'}{\partial \eta} + \frac{\lambda_{10}^*}{\bar{a}^2\eta^2}\frac{\partial}{\partial \eta}\left(\eta^2\frac{\partial \bar{T}_1'}{\partial \eta}\right) \qquad (\eta \geqslant 1)$$

$$\eta = 1: \; \bar{T}_1' = \bar{T}_S(\bar{p}_2) \qquad \left(\bar{T}_S = T_S/T_0\right) \tag{6.10.19}$$

$$\eta \to \infty: \; \bar{T}_1' = 1$$

$$\left(\frac{\varkappa_{gS}}{\gamma_2} = \left[1 + (\gamma_2 - 1)\left(1 - \frac{\bar{T}_S}{(\gamma_2 - 1)\, l_0^*}\right)^2\right]^{-1}\right.$$

$$\left.\bar{q}_{\Sigma 1} = \frac{q_{\Sigma 1}'}{\overset{\circ}{\rho}_2 C_* l} = -\frac{(\gamma_2 - 1)^{-1}\lambda_{10}^*}{l_0^* \tilde{c}_0 \bar{a}\bar{\overset{\circ}{\rho}}_2}\left(\frac{\partial \bar{T}_1}{\partial \eta}\right)_{\eta=1}\right)$$

The remaining variables—which do not appear under the sign of derivative—are determined from descrete relations following from the first integrals (6.10.18) and the equations of state of phases

$$\bar{\overset{\circ}{\rho}}_2 = \bar{p}_2/\bar{T}_S(\bar{p}_2)$$
$$\bar{v} = \bar{v}_0\left(\alpha_{10} + \alpha_{20}\bar{a}^3\right), \quad \bar{p}_1 = 1 + \bar{v}_0^2\left(1 - \bar{a}^3\right) \tag{6.10.20}$$

The dimensionless parameters l_0^*, $\tilde{c}$, and λ_1^* are determined, in this case, by formulas

$$l_0^* = \frac{l(p)}{\gamma_2 R_2 T_0}, \quad \tilde{c}_0 = \frac{\overset{\circ}{\rho}_{20} c_2}{\overset{\circ}{\rho}_1 c_1} = \frac{\tilde{\overset{\circ}{\rho}}}{c_l(\gamma_2 - 1)}, \quad \lambda_{10}^* = \frac{\lambda_1}{\overset{\circ}{\rho}_1 c_1 a_0 \sqrt{p_0/\overset{\circ}{\rho}_1}} \tag{6.10.21}$$

Thus, the thermophysical properties of phases appear in the obtained equations through four dimensionless parameters: γ_2, l_0^*, $\tilde{c}_0$, and λ_1^*. With the above-indicated assumptions γ_2, $\tilde{c}_0$, and λ_{10}^* are constant, and l_0^* is a slowly decreasing function of pressure p_2 so that with fairly small pressure drops, l_0^* may also be regarded as practically constant. The relation $T_S(p_2)$ appearing in the equation satisfies, in conformity with the Clausius-Clapeyron equation, the equation

$$\frac{\bar{p}_2}{\bar{T}_S^2} \frac{d\bar{T}_S}{d\bar{p}_2} = \frac{1}{\gamma_2 l_0^*}, \qquad \bar{T}_S(1) = 1 \tag{6.10.22}$$

If l_0^* = constant (that is consistent with l = constant (see also (1.3.78)), the solution of this equation becomes

$$\bar{T}_S = \frac{\bar{T}^\circ}{\ln(\bar{p}^\circ/\bar{p}_2)} \qquad (\bar{T}^\circ = -\ln\bar{p}^\circ = \gamma_2 l_0^*) \tag{6.10.23}$$

This formula describes well the actual relationship $T_S(p_2)$ for various substances within fairly wide pressure ranges.

The stationary-wave structure is defined by boundary conditions before and behind the wave. The initial equilibrium state before the wave is determined by parameters with subscript 0

$$\bar{v} = \bar{v}_0, \quad \bar{p}_1 = \bar{p}_2 = 1, \quad \bar{T}_1' = \bar{T}_2 = 1, \quad \bar{w} = 0, \quad \bar{a} = 1 \tag{6.10.24}$$

and the final state e behind the wave in the form of liquid phase is defined by parameters with subscript e, parameters in this state being found from the finite relations (6.10.18) and (6.10.20)

$$\bar{v}_e = \alpha_{10}\bar{v}_0, \quad \bar{p}_e = 1 + \bar{v}_0^2, \quad \bar{T}_{1e} = 1, \quad \bar{a}_e = 0 \tag{6.10.25}$$

As distinct from the mixture of incompressible liquid with gas bubbles where $\bar{a}_e^3 = \bar{p}_e^{-1}$, and the interrelationship between the wave intensity and velocity v_0 has a form (6.4.18): $\bar{p}_e = \bar{v}_0^2$, for the mixture of incompressible liquid with vapor bubbles, where $\bar{a}_e = 0$, this relationship has a form $\bar{p}_e = 1 + \bar{v}_0^2$, which indicates that with the same wave intensity $\bar{p}_e$, a wave in the mixture with vapor bubbles propagates slower than that in the mixture with gas bubbles.

Note that, from the lasts formula (6.10.20), it follows

$$\bar{p}_1 = \bar{p}_e - (\bar{p}_e - 1)\bar{a}^3 \leqslant \bar{p}_e \tag{6.10.26}$$

i.e., in the stationary wave propagating in an incompressible liquid with condensating vapor bubbles, the liquid pressure $\bar{p}_1$ cannot exceed pressure $\bar{p}_e$ behind the wave. In an incompressible liquid with constant-mass gas bubbles, a similar evaluation has the form (6.4.25): $\bar{p}_1 < \bar{p}_e + 1$, and rise of liquid pressure $\bar{p}_1$ within a stationary shock wave above the pressure $\bar{p}_e$ behind the wave cannot be excluded, although analyses show that such a pressure-rise in a stationary wave may not be significant. Pressure rise above the pressure behind the wave is more typical for nonstationary waves within a zone of their initiation (see §6.7).

For a limiting case of infinitely great heat conduction of liquid ($\lambda_{10}^* = \infty$), temperature on the bubble wall, analogous to the Rayleigh inertial regime (2.6.52) (with accelerating collapse of a single bubble), must be constant ($T_S(p_2) = T_0$). Hence, the vapor (steam) pressure $p_2 = p_0$ must also be constant. For this case, the system (6.10.19) simplifies and is reduced to the following equations

$$\bar{v}\frac{d\bar{a}}{d\bar{x}} \approx \bar{w}, \quad \bar{a}\bar{v}\frac{d\bar{w}}{d\bar{x}} = 1 - \bar{p}_1 - \frac{3\bar{w}^2}{2}$$

$$\bar{p}_1 = 1 + (\bar{p}_e - 1)(1 - \bar{a}^3), \quad \bar{v} = \bar{v}_0(\alpha_{10} + \alpha_{20}\bar{a}^3) \tag{6.10.27}$$

As distinct from the Rayleigh inertial regime (2.6.52), the collapsing of a single bubble—when, in addition to the condition $p_2 = p_0$, it is assumed that $p_1 = p_\infty = p_0 + \Delta p_0 =$ constant—here, consideration is given to the variability of the liquid pressure p_1 due to its longitudinal retardation during reduction of the trial bubble radius a, which leads to an increase of α_1 ($\alpha_1 \to 1$), and decrease of v ($v = \alpha_{10}v/\alpha_1$).

From Eqs. (6.10.27), we have

$$\bar{a}\frac{d}{d\bar{a}}\left(\frac{\bar{w}^2}{2}\right) + 3\frac{\bar{w}^2}{2} = -(\bar{p}_e - 1)(1 - \bar{a}^3)$$

The solution of this equation, satisfying the condition $\bar{w} = 0$ at $\bar{a} = 1$, is written in the form (cf. (2.6.52))

$$\bar{w}^2 = \frac{\bar{p}_e - 1}{3}\left(\bar{a}^{-3/2} - \bar{a}^{3/2}\right)^2 \quad \text{or} \quad w_{1a} = -\sqrt{\frac{p_e - p_0}{3\rho_1^\circ}\left\{1 - \left(\frac{a}{a_0}\right)^3\right\}\left(\frac{a_0}{a}\right)^3} \tag{6.10.28}$$

and from the first equation (6.10.19), we obtain

$$\int_0^{\bar{a}} \frac{\alpha_{10} + \alpha_{20}\bar{a}^3}{\bar{a}^{-3/2} - \bar{a}^{3/2}}\, d\bar{a} = \frac{\bar{x}}{\sqrt{3}}, \quad \text{or} \quad \frac{\bar{x}}{2\sqrt{3}} = \int_0^{\sqrt{\bar{a}}} \frac{y^4\,dy}{1 - y^6} - \frac{\alpha_{20}}{10}\bar{a}^{-5/2} \tag{6.10.29}$$

Here, the constant of integration is chosen such that $\bar{a} = 0$ at $\bar{x} = 0$. The obtained solution defines $\bar{x}(\bar{a})$, which, together with the closed-form expressions for $\bar{p}_1(\bar{a})$, $\bar{v}(a)$, and $\bar{w}(\bar{a})$, determines the structure of the stationary shock-wave front in a parametric form with respect to $\bar{a}$.[4] It can be easily calculated that with $\bar{a} = 0.95$, the inherent wave thickness becomes $\bar{x} = 1.3$ or $x = 1.3a_0\sqrt{\alpha_{20}}$. Both the kinetic energy of radial motion of liquid per its unit mass ($k_1 = 3/2w_{1a}^2\alpha_2/\alpha_1$), and per bubble ($k_1/n = 2\pi a^3\alpha_1^{-1}w_{1a}^2$), in a stationary shock wave at the time of bursting ($\alpha_2 \to 0$, to which the subscript F corresponds), in conformity with this solution are

$$k_{1F*} = \frac{1}{2}\frac{\alpha_{20}}{\alpha_{10}}\frac{p_e - p_0}{\overset{\circ}{\rho}_1}, \quad \frac{k_{1F*}}{n} = \frac{2\pi a_0^3\,(p_e - p_0)}{3\alpha_{10}\overset{\circ}{\rho}_1} \tag{6.10.30}$$

These values of kinetic energy (denoted by *) are the maximum possible for a prescribed intensity of a stationary wave. Because of a finite heat conduction of liquid, the vapor pressure within a bubble during its collapse increases; this, in turn, will lead to reduced values of realized kinetic energy of the radial motion of liquid.

Reverting to a more general case, described by the system of equations (6.10.19), we shall, similar to the approach used in §§6.3–6.5, investigate the asymptote of the behavior of a solution of this system of equations in the vicinity of an initial state before the wave. The solution of a linearized system is sought in a form of an exponential of type (6.3.16) decaying at $\bar{x} \to \infty$, this solution for a function of two variables T_1' $(\bar{x}, \eta)$ being sought in the form

$$T_1' = 1 + A_1^{(T)}(\eta)\exp k_0\bar{x}, \quad \mathrm{Re}\,\{k_0\} < 0, \quad \bar{v}_0 < 0 \tag{6.10.31}$$

For constants $A^{(a)}$, $A^{(w)}$, $A_1^{(p)}$, and $A_2^{(p)}$, we obtain the following algebraic equations

$$\begin{aligned} &k_0\bar{v}_0A^{(a)} = A^{(w)} \\ &k_0\bar{v}_0A^{(w)} = A_2^{(p)} - A_1^{(p)}, \qquad A_1^{(p)} = -3\bar{v}_0^2A^{(a)} \\ &k_0\bar{v}_0A_2^{(p)} = 3\varkappa_{gS}\left\{\frac{\lambda_{10}^*}{(\gamma_2 - 1)\,l_0^*\tilde{c}_0}\left(\frac{dA_1^{(T)}}{d\eta}\right)_{\eta=1} - A^{(w)}\right\} \end{aligned} \tag{6.10.32}$$

and for the function $A_1^{(T)}(\eta)$, a differential equation with pertinent boundary conditions is obtained as

$$\begin{aligned} &k_0\bar{v}_0A_1^{(T)} = \frac{\lambda_1^*}{\eta^2}\frac{d}{d\eta}\left(\eta^2\frac{dA_1^{(T)}}{d\eta}\right) \\ &\eta = 1: A_1^{(T)} = \frac{dT_S}{d\bar{p}_2}A_2^{(p)} = \frac{1}{\gamma_2 l_0^*}A_2^{(p)}; \quad \eta \to \infty: A_1^{(T)} = 0 \end{aligned} \tag{6.10.33}$$

Solution of this boundary-value problem is of the form

[4]The outlined solution for the case $\lambda_{10}^* = \infty$ has been obtained by V. Sh. Shagapov.

$$A_1^{(T)} = \frac{A_2^{(p)}}{\gamma_2 l_0^* \eta} \exp\left\{\frac{K}{\sqrt{\lambda_1^*}}(1-\eta)\right\} \qquad \left(K = \sqrt{k_0 \bar{v}_0}\right)$$

To satisfy the boundary condition at $\eta \to \infty$, it is required that Re $K > 0$, and the derivative appearing in (6.10.32) equals

$$\left(\frac{dA_1^{(T)}}{d\eta}\right)_{\eta=1} = -\frac{A_2^{(p)}}{\gamma_2 l_0^*}\left(1 + \frac{K}{\sqrt{\lambda_1^*}}\right)$$

The condition of existence of nonzero solutions of a system of linear algebraic equations (6.10.39) with respect to $A^{(a)}$, $A^{(w)}$, $A_1^{(p)}$, and $A_2^{(p)}$ leads to an algebraic characteristic equation of the sixth order with respect to K

$$z(K) = \sum_{j=0}^{6} a_j K^j = 0$$

$$a_6 = \gamma_2/\varkappa_{gS}, \quad a_5 = c_* \sqrt{\lambda_{10}^*}, \quad a_4 = c_* \lambda_{10}^*, \quad a_3 = 0 \qquad (6.10.34)$$

$$a_2 = 3\gamma_2\left(1 - \varkappa_{gS}^{-1}\right)\bar{v}_0^2, \quad a_1 = -3c_* \sqrt{\lambda_{10}^*}\,\bar{v}_0^2, \quad a_0 = -3c_* \lambda_{10}^* \bar{v}_0^2$$

$$\left(c_* = 3/[(\gamma_2 - 1)\,\tilde{c}_0 l_0^{*2}]\right)$$

Two conditions for a characteristic root:

$$\mathrm{Re}\{K\} > 0, \quad \mathrm{Re}\{k_0 \bar{v}_0\} = \mathrm{Re}\{K_0^2\} > 0$$

result in a situation that for the solution of type (6.10.31), only those roots which lie in the following sector on an imaginary plane are suitable

$$-{}^1/_4\pi < \arg K < {}^1/_4\pi \qquad (6.10.35)$$

Let us analyze the mapping of the line $OCAB$ bounding the indicated sector at a sufficiently large radius R of the arc CAB (see Fig. 6.10.5), by a polynomial of the sixth degree $z(K)$ ($z = x + iy$, $K = K^{(1)} + iK^{(2)}$). Since $z(0) = a < 0$, point K is mapped into point O' lying on the negative half-axis x. Because of $a_0 > 0$, we have $z(A) > 0$ at sufficiently great R, i.e., point A is mapped into point A' lying on the positive half-axis Ox. The image of the ray $OB = r + ir$ ($0 \leqslant r \leqslant R$) becomes

$$z(r + ir) = (a_0 + a_1 r - 2a_3 r^3 - 4a_4 r^4 - 4a_5 r^5)$$

$$+ i(a_1 r + 2a_2 r^2 + 2a_3 r^3 - 4a_5 r^5 - 8a_6 r^6)$$

Since $a_0 < 0$, $a_1 < 0$, $a_3 = 0$, $a_4 > 0$, and $a_5 > 0$, we have Re$\{z(r + ir)\}$ < 0, and with sufficiently great R, we obtain Im$\{z(R + iR)\} < 0$; consequently, mapping of the ray OB schematically shown in Fig. 6.10.5 as line $O'p'B'$ lies entirely in the left half-plane, and with sufficiently great R, point B', representing the image of point B, lies in the third quadrant. The ray OC, conjugate to ray OB, is mapped into line $O'q'C'$, conjugate with line $O'p'B'$, or symmetric to this line relative to axis Ox, because the coefficients of polynomial $z(K)$ are real values. Thus the image of ray OC is located in

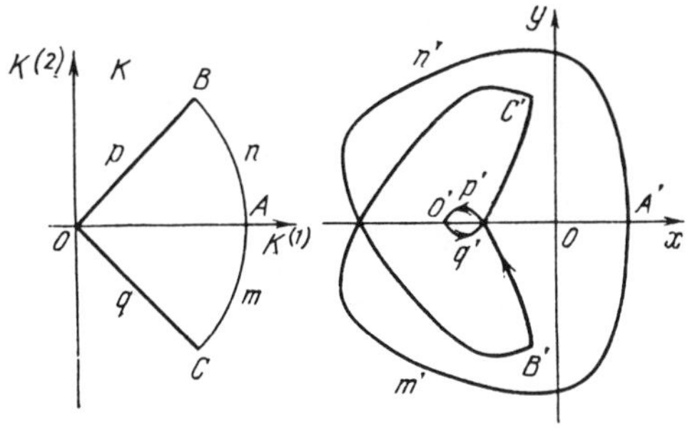

Figure 6.10.5 Schematic illustrating both the existence and uniqueness of a characteristic root of Eq. (6.10.34).

the left half-plane. The arc AnB, whose equation is $K = R \exp i\varphi$ $(0 < \varphi < 1/4\pi)$ is mapped into line $A'n'B'$ because $z(K)$ is a sixth degree polynomial, and $a_6 > 0$. With sufficiently large R, all points of this line satisfy the condition $0 < \arg z(AB) < 3/2\pi$, i.e., line $A'n'B'$ successively passes only through the first, second, and third quadrants. The arc AmC, conjugate with arc AnB, is mapped into line $A'm'C'$, which is symmetrical to line $A'n'B'$ relative to axis Ox, i.e., the image of the arc AmC successively passes only through the fourth, third, and second quadrants. Thus, with sufficiently large R, the image of the boundary of the sector (6.10.35), where the characteristic roots of Eq. (6.10.34) are sought, envelopes the point $z = 0$ only once. Hence (see M. A. Lavrent'ev and B. V. Shabat, 1973), the required root does exist and is unique. Since $z(K)$ is a polynomial with real coefficients, and for each of its complex roots, the conjugate number is also a root, the unique root K in the sector (6.10.35) is a real, positive number which can be calculated.

In the adiabatic case $(\lambda_1^* = 0)$, the characteristic equation (6.10.34) modifies[5] into a quadratic equation in terms of $k_0\bar{v}_0$

$$(k_0\bar{v}_0)^2 = 3(\bar{p}_e - 1 - \varkappa_{gS}) \tag{6.10.36}$$

In this case, the required root exists only at $\bar{p}_e > 1 + \varkappa_{gS}$. For water, at $p_0 = 0.1$ MPa, $T_0 = 373$ K, we have $\varkappa_{gS} = 1.14$. In the presence of heat exchange $(\lambda_1^* > 0)$, the required root exists only for compression waves when $\bar{p}_e > 1$, i.e., when $\bar{D}_0^2 = \bar{v}_0^2 > 0$. Thus, the discussed model of a bubbly vapor-liquid mixture yields the equilibrium speed of sound $C_e = 0$, which is based on neglect of the mass fraction of vapor; as a consequence, any compression wave of infinitesimal intensity (in conformity with this model) accounts for a complete condensation of vapor.

Figure 6.10.6 depicts the predicted dependence of the root on the wave

[5]It should be kept in mind that, at small λ_1^*, both the vapor homogeneity condition and its saturation within a bubble is violated, and the bubble behavior approaches that of fixed-mass gas bubbles.

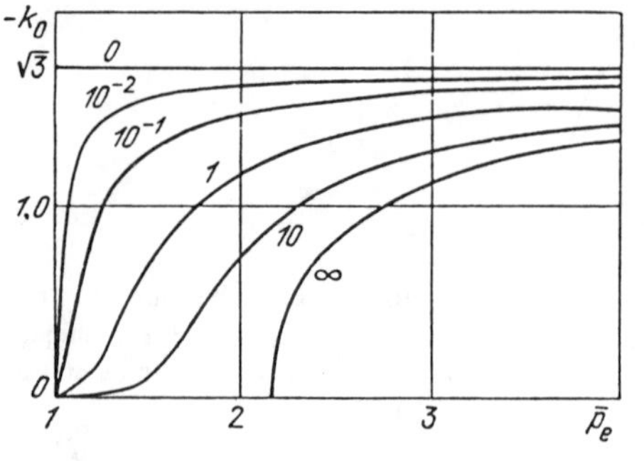

Figure 6.10.6 Relationship between the root of a characteristic equation (6.10.34), which defines steepness of the stationary wave front in a steam-water bubbly mixture at $p_0 = 0.1$ MPa, $T_0 = 373$ K, and the wave intensity $\bar{p}_e$ for various bubble sizes a_0 (mm) indicated by labels on curves.

intensity $\bar{p}_e$ for a stream-water bubbly mixture at various values of the bubble radius a_0. Different values of a_0 correspond to variation of $\lambda_{10}^* = a_\lambda/a_0$ where, for indicated conditions, $a_\lambda = \nu_1^{(T)}/C_* = 1.67 \times 10^{-5}$ mm.

The values of basic parameters characterizing thermophysical properties of some single-component vapor-liquid bubbly systems are represented in Table 6.10.1. It may be seen that magnitudes of parameters γ_2 and l_0^* for substances under consideration are close to each other. Therefore, the principal criteria of similarity, which may significantly differ in the wave processes simulation in various single-component vapor-liquid (two-phase) media, in addition to wave intensity $\bar{p}_e$ and void fraction α_{20}, are $\tilde{c}_0$ and $\lambda_{10}^*(\lambda_{10}^* = a_\lambda/a_0)$. It is noteworthy that parameter $\tilde{c}_0$ changes significantly due to $\tilde{\rho}_{20}^0 = \rho_{20}^0/\rho_1^0$ when the initial pressure p_0 in a system varies. With values of other parameters identical, the wave patterns and, in particular,

Table 6.10.1. Magnitudes of principal parameters characterizing thermophysical properties of phases for some single-component vapor-liquid bubbly systems ($a_\lambda = \nu_1^{(T)}/C_*$)

Medium	p_0, MPa	$T_0 = T_S(p_0)$, K	γ_2	l^*	$\tilde{c}_0 \cdot 10$	$a_\lambda = \lambda_{10}^* a_0$, 10^{-6} mm
1. Water (H_2O)	0.05	354	1.31	10.75	0.14	22.73
	0.1	373	1.30	9.92	0.29	16.79
	1.0	453	1.29	5.86	3.44	5.15
	2.0	486	1.29	4.34	8.05	3.48
	5.0	537	1.27	2.68	28.39	1.90
	6.5	554	1.25	2.20	42.68	1.54
2. Freon-21 ($CHFCl_2$)	0.185	299	1.19	8.75	2.62	6.25
3. Nitrogen (N_2)	0.1	77.3	1.32	8.81	3.36	5.29
4. Freon-12 (CF_2Cl_2)	0.7	301	1.18	7.19	11.90	2.33
	1.3	326	1.24	5.76	15.23	1.38

evolution of nonstationary shock waves and structure of stationary shock waves in vapor-liquid bubbly media with equal $\tilde{c}_0$, λ_{10}^*, γ_2, and l_0^* are similar. This permrits the use of one vapor-liquid bubbly system for simulating wave processes occurring in another system in which direct experiments are not feasible or expensive. From Table 6.10.1, it is seen that at other parameter (α_{20}, a_0, and $\bar{p}_e$) values identical, the wave pattern in boiling water at $p_0 \approx 1.0$ MPa must approach a similar pattern in boiling nitrogen at $p_0 \approx 0.1$ MPa.

The analysis of a stationary shock-wave structure is similar to the analysis desribed in §6.5. The difference consists of the fact that the thermal problem here is dealt with outside the bubble, i.e., in the liquid phase ($\eta > 1$; see also §§1.6, 2.4, and 2.6). The variants of a solution concerned with dynamics of a single vapor bubble at predetermined pressure at remote distance are used as test cases for debugging computer-programs developed for verification of the thermal-problem solution.

Structure of stationary shock waves under smooth transition of the medium to a single-phase state. We shall first discuss the variants when the significant values of $k_{1F}(k_{1F} \ll k_{1F*})$ are not realized. There are no oscillatory pressure-peaks across the wave tail. Such regimes are realized at small and moderate values $B(B \lesssim 1$, see (2.6.56)). The results obtained by Zuong Ngok Hai, R. I. Nigmatulin, and N. S. Khabeev (1982, 1989) are discussed below.

Figure 6.10.7 represents structures of stationary shock waves of various intensities $\bar{p}_e = 1.4$–3.0 in a steam-water bubbly mixture. It is obvious that,

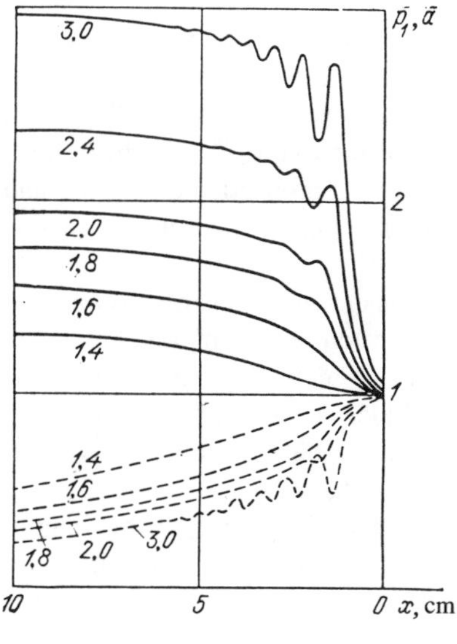

Figure 6.10.7 Structures of stationary shock waves (variation of pressure in liquid $\bar{p}$ (solid lines) and that of bubble radii $\bar{a}$ (dashed lines)) in a steam-water bubbly mixture ($p_0 = 0.1$ MPa, $T_0 = T_S = 373$ K, $\alpha_{20} = 0.05$, $a_0 = 1$ mm) at various intensities $\bar{p}_e$, indicated by labels on curves. The dot-dashed line corresponds to the pressure profile in a wave for $\alpha_{20} = 0.01$.

as distinct from shock waves in water with air bubbles (see Fig. 6.4.8) when stationary waves with intensity $\bar{p}_e \geqslant \gamma_2$ exhibited an oscillatory structure, the presence of the condensation phase-transitions in a steam-water mixture enhances the trend toward a monotonic structure and widens the range of intensities at which such structure is realized. This range to a great extent depends on the degree of dispersion of the steam phase (bubble radius a_0), but the intensity $1 + \varkappa_{gS}$ is a specific threshold above which the oscillatory structure is displayed in a wave with any size of bubbles. This intensity exceeds the corresponding value, equal to γ_2, in the gas-bubble case (see the discussion of Fig. 6.4.2). A significant distinction of shock waves in a liquid with insoluble-gas bubbles consists of the fact that the thickness of such waves in liquids with vapor bubbles is considerably smaller because of bubble contraction (coalescence) and the low velocity of the wave. So, at pressure $p_0 \sim 0.1$ MPa, the stationary wave thickness is only 5–10 cm for a mixture of steam bubbles of radius $a_0 \sim 0.1.$ mm in water, whereas this thickness amounts to approximately 1 m for a mixture of gas bubbles of the same size in water.

In the case of weak waves having a monotonic structure, the phase pressures almost coincide: $p_2 \approx p_1 \approx p$, and the dimensionless thermal flux (the Nusselt number $\mathbf{Nu}_1$, see (1.3.56)) varies monotonically (curves *1* in Fig. 6.10.8). The vapor-phase pressure in oscillatory waves fluctuates synchronously (simultaneously) with pressure in liquid, but with a significantly higher amplitude (Fig. 6.10.9). In this case, both Nusselt number $\mathbf{Nu}_1$ and kinetic energy k_1 of a small-scale motion of liquid around a bubble fluctuate (curves *2* in Figs. 6.10.8 and 6.10.10).

Analysis has proven that the influence of corrections $\varphi^{(1)}$ and $\varphi^{(2)}$ upon the bubble multiplicity in the Rayleigh-Lamb equation for a wave of a monotonic structure is negligible, and in the case of a wave with an oscillatory structure, it results only in a slight increase of the oscillation frequency.

Reduction of the vapor volumetric concentration α_{20} with the remaining mixture parameters kept unvarying, and also at fixed intensity of the shock wave leads to the wave velocity growth proportional to $\alpha_{20}^{-1/2}$, which, in turn, promotes increase of the oscillatory-wave lengths proportional to $\alpha_{20}^{-1/2}$, and also to some increase of the wave-structure thickness (cf. dot-dashed line and solid line 3.0 in Fig. 6.10.7).

Figure 6.10.11 depicts the liquid local-temperature T_1' distribution around the bubbles at various distances x along the structure of a monotonic shock wave, the bubbles in which contraction is exhibited. The spreading of a thermal boundary layer in a liquid around bubbles accounted for by the radial heat conduction of liquid is clearly observed (if the bubble radius were fixed, or subjected to a very slow change, this would have led to $\mathbf{Nu}_1 \to 2$).

In addition, Fig. 6.10.8 displays a convective thickening of this boundary layer because of its radial convergence to the bubble center. The latter effect contributes to the further reduction of both heat removal and conden-

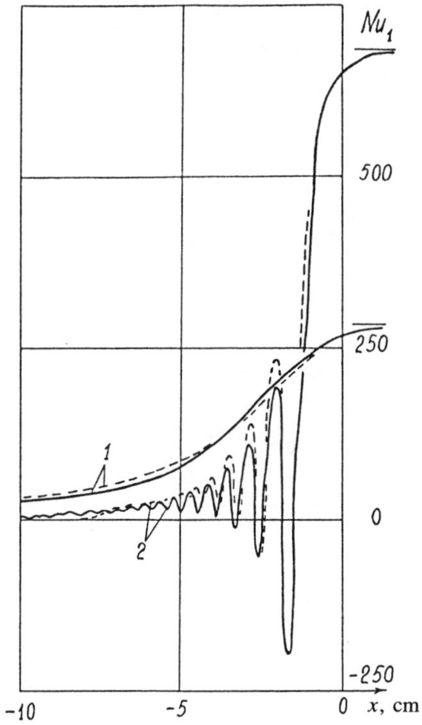

Figure 6.10.8 Variation of a dimensionless heat flux (Nusselt numbers $\mathbf{Nu}_1$) in shock waves $\bar{p}_e = 1.4$ (curves *1*) and $\bar{p}_e = 3.0$ (curves *2*) represented in Fig. 6.10.7 Dashed lines correspond to values obtained analytically using the Duhamel integral (2.6.48*b*)

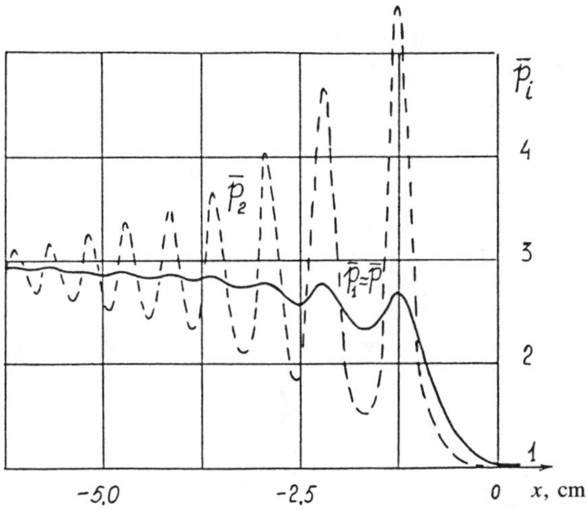

Figure 6.10.9 Structure of a stationary shock wave (variation of pressure in liquid (solid line) and in vapor (dashed line)) in a steam-water mixture (p_0 = 0.1 MPa, $T_0 = T_S$ = 373 K, α_{20} = 0.05, a_0 = 1 mm, $\bar{p}_e$ = 3.0).

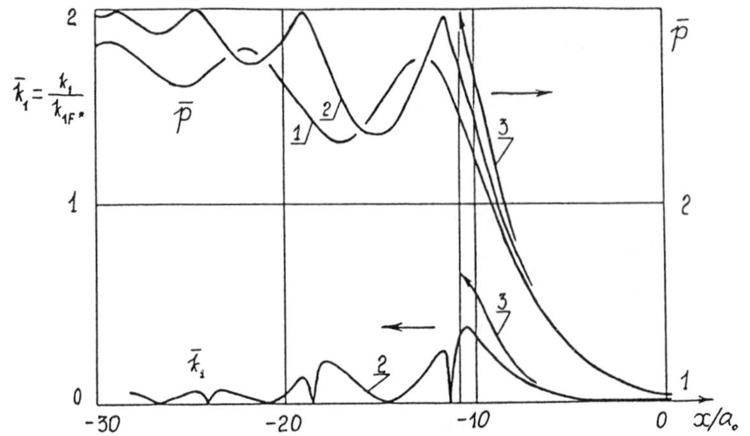

Figure 6.10.10 Structures of stationary shock waves (variation of both pressure $\bar{p}$ in liquid and kinetic energy k_1 of a small-scale motion of liquid around bubbles) in a steam-water bubbly mixture (p_0 = 0.1 MPa, $T_0 = T_S$ = 373 K, α_{20} = 0.05, $\bar{p}_e$ = 3.0) as a function of the steam-phase dispersion degree. Curves *1–3* correspond to a_0 = 1, 0.1, and 0.01 mm, respectively.

sation per unit surface of a bubble; (these are characterized by the Nusselt number $\mathbf{Nu}_1$, and cause its decrease (see discussion of (2.6.42)). We shall assume that the stationary wave thickness ΔL is measured from a point where the medium pressure begins to differ from the initial equilibrium pressure p_0 by a certain small value (for instance, when $|p - p_0|/|p_e - p_0| = 0.03$), to a point where it asymptotically approaches the final pressure p_e (from which it also differs by a small value, e.g., $|p_e - p|/|p_e - p_0| = 0.03$. We may also introduce an additional linear characteristic, namely, the shock-wave front thickness

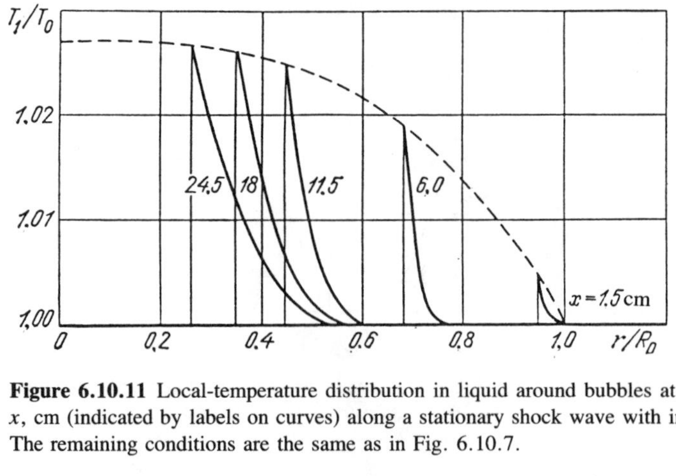

Figure 6.10.11 Local-temperature distribution in liquid around bubbles at various distances x, cm (indicated by labels on curves) along a stationary shock wave with intensity $\bar{p}_e$ = 1.4. The remaining conditions are the same as in Fig. 6.10.7.

$$\Delta L_f = \frac{p_e - p_0}{(\partial p/\partial x)_{\max}}$$

Figure 6.10.12 depicts thicknesses of stationary (both monotonic and oscillatory) shock waves in a liquid with vapor bubbles versus their intensity $\Delta p_e = (p_e - p_0)/p_0$. It is clearly seen that these relationships are substantially nonlinear. With the wave intensity growth, the thicknesses ΔL and ΔL_f reduce. The minimum wave-front thickness is realized when bubbles collapse in a Rayleigh regime at which p_2 = constant, and when ΔL is determined in accordance with (6.10.29). With the shock-effect intensity decrease, the wave-front thickness increases. For instance, for conditions shown in Fig. 6.10.12, the stationary shock wave with an intensity $\Delta \bar{p}_e = 0.2$ may have a front thickness ΔL_f about 0.2 m, whereas the entire wave thickness ΔL amounts to about 1 m. The pressure oscillograms for weak waves in vapor-liquid mixtures tested in shock tubes of lengths 1–2 m, as well as for a liquid with bubbles of noncondensing gas (see §6.4), correspond, mainly, to nonstationary regimes.

Reduction of the initial bubble radius a_0, which is responsible for the intensification of interphase heat and mass exchange due to increase of the specific surface of interface, with the volumetric concentration α_{20} of the vapor phase retained unvarying (see Fig. 6.10.10), leads to the evolution of shock waves from monotonic to oscillatory (curves *1* and *2*); in a limiting process $a_0 \to 0$, it leads to a structure which is realized during an inertial burst of bubbles (curves *3*). In this case, the reference wave thickness reduces significantly. The value of kinetic energy k_1 of a small-scale radial motion of liquid around a bubble undergoes, as well as pressure, attenuating oscillations, becoming very small at the final stage of the bubble contraction. In this case, both compression and pressure-rise phases are characterized by noticeable peaking. The specific kinetic energy k_1 in the process of collapse

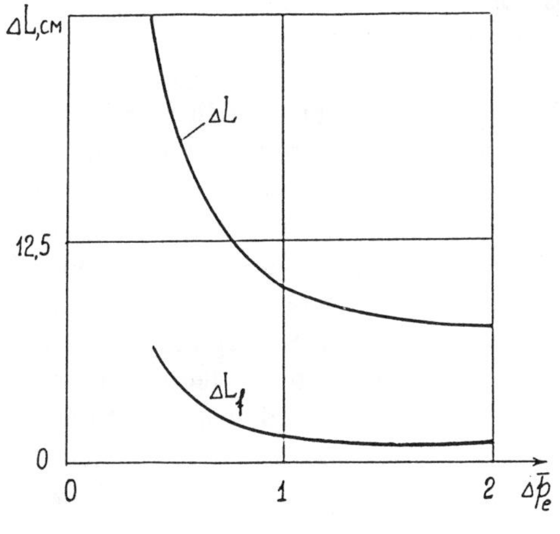

Figure 6.10.12 Dependence of the stationary-wave thickness ΔL in a steam-water bubbly mixture and that of thickness ΔL_f of their fronts on intensity $\bar{p}_e$ ($p_0 = 0.1$ MPa, $T_0 = T_S = 373$ K, $\alpha_{20} = 0.05$, $a_0 = 1$ mm).

of a bubble of radius $a_0 = 0.01$ mm reaches a value of 0.6 of its limit magnitude k_{1F*}, defined by formula (6.10.30) for the inertial Rayleigh regime. The maximum value of a similar quantity during collapse of a bubble of radius $a_0 = 0.1$ mm does not exceed 0.4.

The qualitative difference of shock-wave structures in a liquid with vapor bubbles becomes clear if we refer to the dimensionless parameter B determined by formula (2.6.56) and characterizing the relative contributions of the liquid inertia and interphase heat and mass exchange during this process. With a step variation of pressure, the large values of B ($B > B_*^{(p)} \approx 10$) correspond to a situation of a predominant influence of inertia of the radial motion of liquid, whereas small B ($B < B_*^{(T)} \approx 0.01$) are related to a thermal regime. The values $B = 0.012$; 0.12; and 1.2 correspond to curves *1–3* represented in Fig. 6.10.10 ($B = 0.0036$ corresponds to $\bar{p}_e = 1.4$ represented in Fig. 6.10.7).

In order to investigate the influence of thermophysical properties of phases on the shock-wave structure, analytical computations of the stationary shock-wave structures in liquid nitrogen containing nitrogen-vapor bubbles (liquid nitrogen being on a saturation line at atmospheric pressure) (Fig. 6.10.13) have been carried out. It is seen that, in this case, the wave thickness compared to stationary shock waves in water with bubbles of steam is several times larger than in boiling water with steam bubbles (see Fig. 6.10.7), and the inertial effects are less pronounced; both retardation of the interphase heat and mass exchange processes (because of reduction of the dimensionless parameter λ_{10}^*) and mass increase of each bubble are responsible for this phenomenon.

An increase of the initial static pressure p_0 leads (in the same mixture) not only to a wave velocity D increase, but also to growth of the initial mass of vapor in bubbles at fixed original radius a_0, and thereby, to a significant increase of the inherent time of bubble contraction and to an increase of the shock wave thicknesses ΔL and ΔL_f (cf. Figs. 6.10.14 and 6.10.7).

We shall note that phase transitions reduce the relaxation-zone thickness and increase the tendency to a monotonic structure.

On using the Duhamel integral for predicting the heat flux around vapor bubbles in shock waves. In order to predetermine the heat flux $q_{\Sigma 1}$ transferred to liquid from the interface, thereby closing the system of hydrodynamic equations (6.10.15) when investigating the structure of stationary shock waves in vapor-liquid bubbly media, Zuong Ngok Khai and N. S. Khabeev (1983) examined the possibility of employing a model with a quasi-stationary bubble radius leading to the Duhamel integral (2.6.48c) instead of solving an exact, nonlinear, and nonstationary thermal problem (2.6.13) for a liquid around a bubble. The results of comparison of the stationary shock-wave structures obtained by solving the system of equations (6.10.15), (2.6.13), and system of equations (6.10.15) with prescribing $q_{\Sigma 1}$

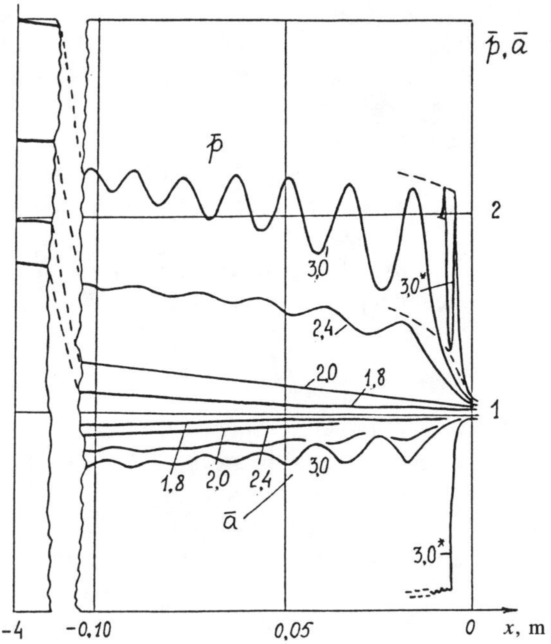

Figure 6.10.13 Structure of stationary shock waves in a mixture of liquid nitrogen with bubbles of nitrogen vapor (p_0 = 0.1 MPa, $T_0 = T_S$ = 77.3 K, α_{20} = 0.05, a_0 = 1 mm). Numerical labels on curves indicate the wave intensities $\bar{p}_e$. Curves with a label 3.0* correspond to a wave predicted at $\bar{p}_e$ = 3.0 with consideration given to bubble fragmentation into fine bubbles of a radius $a_* \approx$ 0.1 mm at $\mathbf{We}_* \approx 6$.

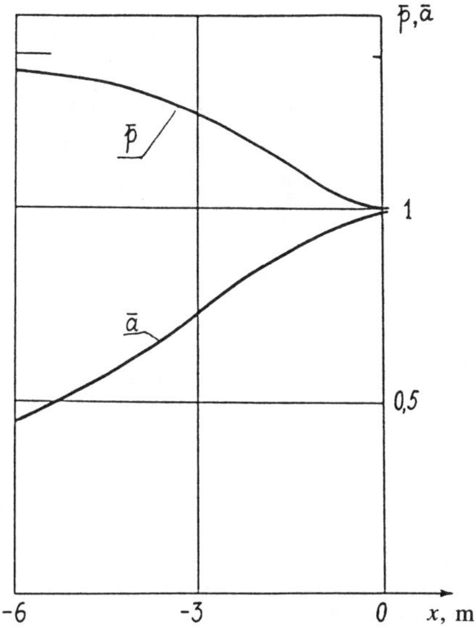

Figure 6.10.14 Structure of a stationary shock wave of intensity $\bar{p}_e$ = 1.4 in a steam-water bubbly mixture (a_0 = 1.4 mm, α_{20} = 0.018) at a higher initial pressure (p_0 = 0.5 MPa, $T_0 = T_S$ = 424 K) than in Fig. 6.10.7.

using (2.6.48b), showed good agreement only in the vicinity of their front (i.e., in the vicinity of the original state defined by point O) where one of conditions (2.6.13a) is always satisfied (the integral expression (2.6.48B) is obtained based on these conditions). The Duhamel integral yields a significant error in the zone of bubble contraction.

Note that when the Duhamel integral (2.6.48b) is used in analyses of shock waves, it is imperative that information concerning dp_2/dt throughout the entire prehistory of the process ranging from $-\infty$ to $x(dx = vdt)$ (x is the current coordinate) is retained; this, of course, requires an extensive computer memory, and more arithmetic operations than with the above-described solution of the nonlinear heat-conduction equation (2.6.13) obtained by the finite-difference method.

Shock waves with a wave packet of oscillatory pressure-peaks. Let a great kinetic energy k_1 of radial motion be realized in the process of a bubble contraction, or with decrease of its size by many times in an inertial regime (at $B > 10$) at points in time when $\alpha_2 \approx d_c \equiv (C_l^*)^{-3} = p_0/\rho_1^0 C_1^2$. Then, even upon the bubble disappearance, liquid continues to be compressed due to inertia, and the indicated kinetic energy may convert to the elastic-compression energy (state E in Fig. 6.10.15), and the relation (6.10.28) for $\bar{p}_1(\bar{a})$ is violated because of liquid compressibility, and, as is shown below, the wave nonstationarity. The final heat conduction of liquid, its elasticity, and that of residual vapor (steam) may lead to a situation when, upon collapsing due to compression-wave reflection, liquid begins to diverge from the bubble center, again forming a bubble; the latter will again collapse following its expansion, leading to the next pressure peak exceeding pressure p_e behind the wave (Fig. 6.10.6). These nonlinear oscillations must gradually decay because of dissipative processes accounted for by liquid viscosity, heat conduction, and nonlinear compressibility.

The typical magnitudes of pressure peaks p_E due to bubble collapse in the case of compressible liquid may be determined from a condition of transformation of the kinetic energy k_{1F} of radial motion to the energy $u_1^{(p)}$ of

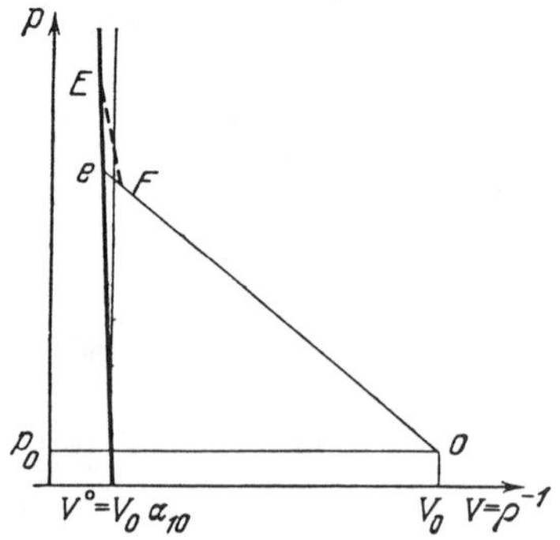

Figure 6.10.15 The pV-diagram for waves in a vapor-liquid medium.

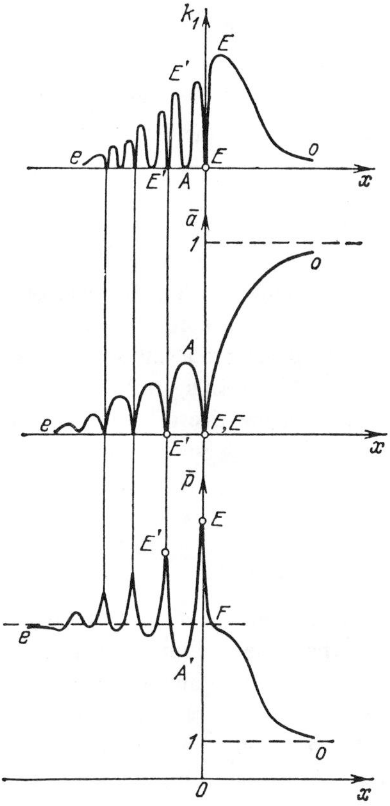

Figure 6.10.16 Structure of a nonstationary shock wave in a vapor-liquid mixture with oscillatory pressure peaks due to bubble collapse.

elastic compression of liquid at the point in time when the bubble radius $a_{\min}$ is minimum (in particular, $a_{\min}$ may equal zero); under this condition, the kinetic energy of radial motion becomes equal to zero

$$k_{1F} = u_1^{(p)}(p_E) - u_1^{(p)}(p_e)$$
$$= \int_{p_e}^{p_E} \frac{p}{(\overset{\circ}{\rho}_1)^2} d\overset{\circ}{\rho}_1 \approx \frac{p_E^2 - p_e^2}{2(\overset{\circ}{\rho}_1 C_1)^2} \tag{6.10.37}$$

Here, in calculating the integral defining work of internal compressive forces, the low compressibility of liquid was taken into consideration ($\rho_1^0 - \rho_{10}^0 < \rho_{10}^0$), and also, its acoustic equation of state. Solving the equation for p_E, we obtain

$$p_E = p_e \sqrt{1 + \bar{k}_{1F}} \qquad \left(\bar{k}_{1F} = k_{1F} \frac{\overset{\circ}{\rho}_1}{p_e} \frac{2\overset{\circ}{\rho}_1 C_1^2}{p_e}\right) \tag{6.10.38}$$

The maximum possible values of $\bar{k}_{1F} = \bar{k}_{1F}^*$ are determined using formula (6.10.31), which is consistent with $\lambda_1^* \to \infty$. Then, we obtain

$$(p_E)_{\max} = p_e \sqrt{1 + \frac{p_e - p_0}{p_e} \frac{\overset{\circ}{\rho}_1 C_1^2}{p_e} \frac{\alpha_{20}}{1-\alpha_{20}}} \tag{6.10.39}$$

If, for example, $p_e = 2p_0 = 0.2$ MPa, $\alpha_{20} = 0.05$, $C_1 = 1400$ m/s, $\rho_1^0 = 1000$ kg/m^3, then, $p_E \approx 3.3$ MPa, which corresponds to the oscillatory pressure-peaks in shock waves observed in experiments by A. A. Borisov, B. E. Gel'fand, et al., (1977).

It should be kept in mind that transition *OFEe* (see Fig. 6.10.15) cannot be realized in a stationary wave, since all states on the pV-diagram in a stationary wave must lie on line *oe* (Rayleigh-Michaelson line). Therefore, the anomalous pressure surges p_E may occur only on a nonstationary stage of the process, and they must decay as the wave turns into a stationary mode when the kinetic energy of a small-scale motion transforms, following some oscillations, into heat. The characteristic temperature rise of liquid due to dissipation k_{1F} is

$$\Delta T \approx \frac{k_{1F}}{c_1} \lesssim \frac{1}{2} \frac{\alpha_{20}}{\alpha_{10}} \frac{p_e - p_0}{\overset{\circ}{\rho}_1 c_1} \tag{6.10.40}$$

and it is generally very low.

A detailed theoretical and quantitive description of a structure of an oscillatory packet, or a shock-wave "tail," where the medium transits from a nonequilibrium state (resembling $F(k_{1F} \leqslant k_{1F}^*, \bar{a} \ll 1)$) into a single-phase state e ($k_1 = 0$, $a_e = 0$), realizing many oscillations of type *FEAE'* with the dissipation of kinetic energy accompanied by the strong reduction (and even disappearance) of bubbles, with the display of liquid compressibility and viscosity, and with bubble fragmentation, is a task of enormous complexity. The amplitude of oscillations is evaluated by formula (6.10.38), and it is the prediction of the duration of the oscillation existence (or the oscillation "tail" thickness) that is the most complex issue. This problem may be essential because in the process of vapor-bubble collapse, a great many oscillations are generated, and the wave "tail" may become considerably long.

Figure 6.10.17 demonstrates the predicted structure of a shock wave with significantly smaller bubbles than in variants represented in Figs. 6.10.7 and 6.10.8. In this case, the interface, where phase transitions occur, is sufficiently large, and bubbles exhibit the collapsing-regime rather rapidly. A coefficient $\eta = 0.6$ is used in the represented analysis; this coefficient determines the regeneration of kinetic energy upon reflection of the collapsed bubble from the center and its transition to expansion.

Figure 6.10.18 illustrates the predicted evolution of a nonstationary shock wave into a stationary wave at fixed pressure disturbance in water with steam bubbles. Here, the generalized equations of §6.7 for the phase-transition case were used for describing the mechanical parameters p, v, ρ, w_1, and a, which are functions of x and t; to describe the radial heat and mass exchange, and, also, the local-temperature distribution $T_1'(x, r, t)$, Eqs. (1.6.4),

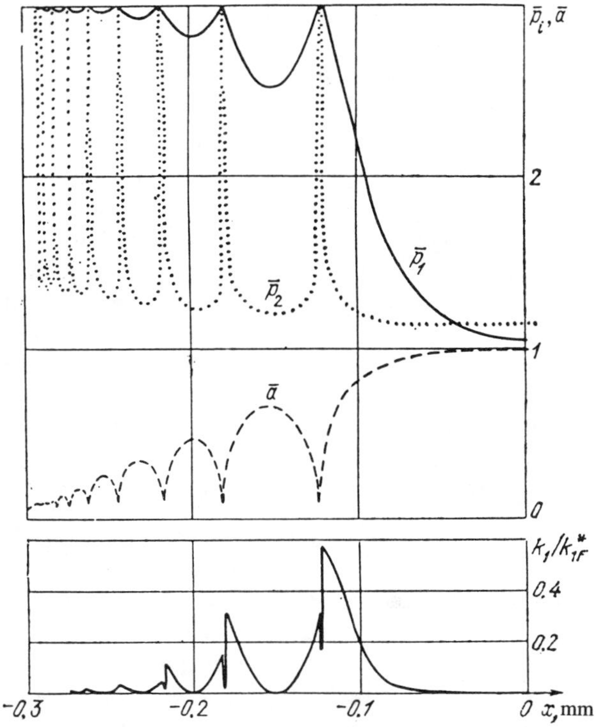

Figure 6.10.17 Structure of a stationary shock wave with intensity $\bar{p}_e = 3.0$ in a steam-water mixture ($p_0 = 0.1$ MPa, $T_0 = 373$ K, $\alpha_{20} = 0.05$) with very small bubbles ($a_0 = 0.01$ mm) (variation of pressure $\bar{p}_1$ in liquid and $\bar{p}_2$ in steam; variation of bubble radius a, and variation of kinetic energy k_1 of radial motion). The coefficient of the regeneration of kinetic energy upon the bubble contraction is $\eta = 0.6$).

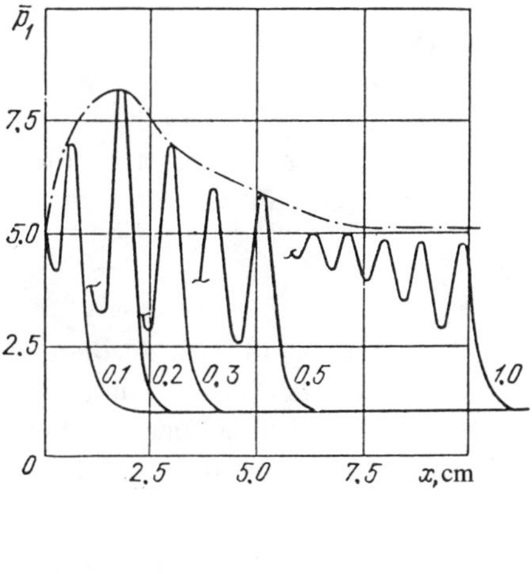

Figure 6.10.18 Evolution of a shock wave generated by a constant-pressure disturbance $\bar{p}_e = p_e/p_0 = 5$ at $t > 0$ in water with steam bubbles ($p_0 = 0.1$ MPa, $\alpha_{20} = 0.05$, $a_0 = 1.0$ mm). Labels on curves indicate time t in microseconds.

or generalizations of Eqs. (6.10.19) were used in case of a nonstationary longitudinal flow when $\upsilon = \upsilon(x, t)$. The analysis identifies a significant enhancement of the disturbance on the initial stage of the process following the wave origination.

Realization of the above-described regime with anomalous pressure surges may contribute to the bubble fragmentation when, because of a considerable decrease of the bubble radius a and increase of their number n, the interface, and the heat-rejection capacity of liquid associated with it, are significantly augmented. This results in an increase of the parameter B, which is indicative of the acceleration of the bubble-contraction process up to the finite values of $\bar{k}_{1F}$. As is shown in §6.8, the condition for the bubble fragmentation in shock waves is qualitatively determined by the gas density ρ_2^0, the coefficient of surface tension Σ, and the velocity w_{12} of the flow past bubbles; this condition is written using the Weber criterion

$$\mathrm{We} = \frac{2a\overset{\circ}{\rho}_g w_{12}^2}{\Sigma} \geqslant \mathrm{We}_* \approx 6 \tag{6.10.41}$$

Fragmentation proceeds until the bubble size a becomes so small that **We** reduces to $\mathbf{We}_*$. Under fixed original pressure p_0, wave intensity p_e/p_0, and void fraction α_2, which determine the velocity w_{12} for the low-viscosity liquids under consideration, the trend towards fragmentation is promoted with an increase of the parameter ρ_g^0/Σ. Based on conditions depicted in Fig. 6.10.1, this parameter for boiling nitrogen, water with carbon dioxide, and boiling water is 500, 25, and 10 s^2/m^3, respectively. Thus, from analysis of all the above-discussed mixtures, bubbles in steam-water mixture are the most "stable," and they are the least "stable" in the boiling nitrogen; therefore, an intensive bubble fragmentation takes place in the boiling nitrogen that promotes bubble collapse in a regime approaching self-acceleration of the process when the "anomalous" oscillatory pressure peaks manifest themselves.

In the analysis made by Zuong Ngok Hai (1982), it was assumed that bubbles break into smaller sizes (of diameter $a_* = 0.1$ mm), provided that condition (6.10.41) holds. The results are depicted in Fig. 6.10.13 (curves with the label 3.0*). In this case, in order to determine the velocity υ_2 of motion of the second phase, an additional equation was used (see the seventh equation in (1.5.4), or the ninth equation in (6.4.1)), and for υ_1, $\upsilon_1 = \upsilon$ was used. Thus, a steeper oscillatory wave structure is realized with a higher oscillation frequency.

Equations of nonstationary motion of a bubbly liquid with account of nonstationary interphase heat and mass exchange. With assumptions specified before (6.10.15), we obtain, instead of the first four equations (6.10.15), the following equations for motion of liquid with vapor bubbles described in a Lagrangian coordinate system Xt, which is moving[5] together with the medium with a velocity $\upsilon(X, t)$ (see also (6.7.3))

$$\frac{\partial \rho_1}{\partial t} + \frac{\rho_1 \rho}{\rho_0}\frac{\partial \upsilon}{\partial X} = -4\pi a^2 n \xi_{12}, \qquad \frac{\partial \rho_2}{\partial t} + \frac{\rho_2 \rho}{\partial_0}\frac{\partial \upsilon}{\partial X} = 4\pi a^2 n \xi_{12}$$

$$\frac{\partial}{\partial t}\left(\frac{4}{3}\pi a^3 \rho_2^0\right) = 4\pi a^2 \xi_{12}, \qquad \frac{\partial \upsilon}{\partial t} + \frac{1}{\rho_0}\frac{\partial p}{\partial X} = g \tag{6.10.42}$$

The rest of the equations and relations (6.10.15) remain unvaried, with the only difference being that in the equation for the bubble radius variation, as well as in the Rayleigh-Lamb equation for radial velocity (the fifth and sixth equations in (6.10.15)), the substantial derivative with respect to time should be given consideration instead of differential operator $(\upsilon d/dx)$

$$\upsilon \frac{d}{dx} \rightarrow \left.\frac{\partial}{\partial t}\right|_x + \upsilon \frac{\partial}{\partial x} = \left.\frac{\partial}{\partial t}\right|_X$$

A system of hydrodynamic equations (6.10.42) supplemented by equations (6.10.15) is closed, provided that Eqs. (2.6.9) and (2.6.13) for determining the interphase heat and mass intensities (namely, the quantities $q_{\Sigma 2}$ and $q_{\Sigma 1}$ in accordance with §1.6) are attached to them.

From Eqs. (6.10.42) and the Rayleigh-Lamb equation, the same equations follow which were used for investigating shock waves in a liquid with constant-mass bubbles, viz., Eqs. (6.7.5), the first equation in (6.7.6), and (6.7.9). The form of writing the differential equation for p_2 changes and assumes the form (2.6.12).

The complication of analyses, compared to §6.7, takes place only due to the determination of the microfield of temperatures T_1' (t, r', X) around trial bubbles (see (2.6.13)) at every "macropoint" with coordinate X.

The obtained system of differential equations with corresponding boundary conditions is numerically solved using a computer technique based on a combination of the modified Euler-Cauchy approach for solving ordinary first-order differential equations with the factorization method (trial runs) for determining both pressure field in a mixture $p(X, t)$ and microfields of temperatures T_1' (t, r', X).

In this case, the computational field with respect to coordinate X is split into cells—Lagrangian particles, and with respect to coordinate $\eta = r'/a$, liquid around the trial bubble is split into spherical layers. Values of parameters ρ, T_2, p_2, w_{1a}, a, and p are obtained similar to the solution (6.7.16) at each point in time and in each Lagrangian cell. Using these values, we ob-

[5]Here, as distinct from §6.7, the longitudinal Lagrangian coordinate is denoted by X instead of r. Notations r and r' are used in this section to denote the radial microcoordinate around a trial bubble.

tain the temperature field T_1' (t, η, X) around the bubble at time t and at each fixed spatial point X by integrating a nonlinear, nonstationary equation (2.6.13) using trial runs with respect to coordinate η.

The results of numerical investigation obtained by Zuong Ngoc Hai and N. S. Khabeev (1983), Zuong Ngoc Hai, R. I. Nigmatulin, and N. S. Khabeev (1984), and R. I. Nigmatulin, N. Khabeev, and Zuong Ngoc Hai (1988) are discussed below.

Evolution of nonstationary shock waves into stationary waves. We shall now analyze the evolution of nonstationary waves in liquids containing vapor bubbles under stationary (continuous and sustained) or sufficiently prolonged effects upon mixture (see also (6.7.17))

$$\begin{aligned} &t = 0 : p = p_0, \quad p_2 = p_0 + 2\Sigma/a_0, \quad a = a_0, \quad T_{10}' = T_{20} = T_0 \\ &\upsilon = w_{1a} = 0, \quad \rho = \rho_0 \\ &X = 0 : \bar{p} = p/p_0 = \bar{\Pi}_0(t) = \bar{p}_e \equiv p_e/p_0 = \text{const} \\ &X \to \infty : \bar{p} \to 1 \end{aligned} \tag{6.10.43}$$

By varying the shock-wave intensity $\bar{p}_e$, the medium structure (through parameters α_{20} and a_0), initial state of the system (through p_0), and physical properties of phases (mixture of boiling water with bubbles of steam, mixture of liquid nitrogen with bubbles of nitrogen vapor, etc.), the effects of basic dimensionless criteria ($\bar{p}_e$, α_{20}, $\tilde{c}_0$, λ_{10}^*, γ_2, and l_0^*) on evolution of waves has been investigated; the indicated criteria characterize the influence of nonlinearity ($\bar{p}_e - 1$), structure of the mixture (α_{20}, a_0) and thermophysical phase parameters (l_0^*, λ_{10}^*, γ_2, and $\tilde{c}_0$) on the process under consideration.

The initial stage of evolution of finite-intensity waves during shock effects (6.10.43) is always of an oscillatory nature (see Fig. 6.10.19). Then, the wave assumes a stationary configuration, which in cases represented in Fig. 6.10.19 is monotonic for waves of intensities $\Delta\bar{p}_e = (\bar{p}_e - 1) \lesssim 1$. An increase of the shock-wave intensity leads to its velocity growth, and to an increase of both amplitude and frequency of the medium-parameter fluctuation behind the wave front. Both duration t_{st} of the process of evolution to a stationary regime and distance L_{st} over which the shock waves assume the stationary configuration in this case reduce. Figure 6.10.19 represents the shock-wave configurations (the last solid lines in each case) which do not differ from their limiting stationary configurations. It is seen that in the case under discussion, they build up over a distance $L_{st} \sim 0.2$–0.4 m for a time period t_{st} of the order of 4–10 microseconds.

To illustrate the effect of phase transition upon wave dynamics of bubbly liquids, the evolution of shock waves, in the same scale, in water with bubbles of insoluble and noncondensing gas (air) (*a*) and in boiling water with bubbles of steam (*b*) is shown in Fig. 6.10.20. The strong impact of

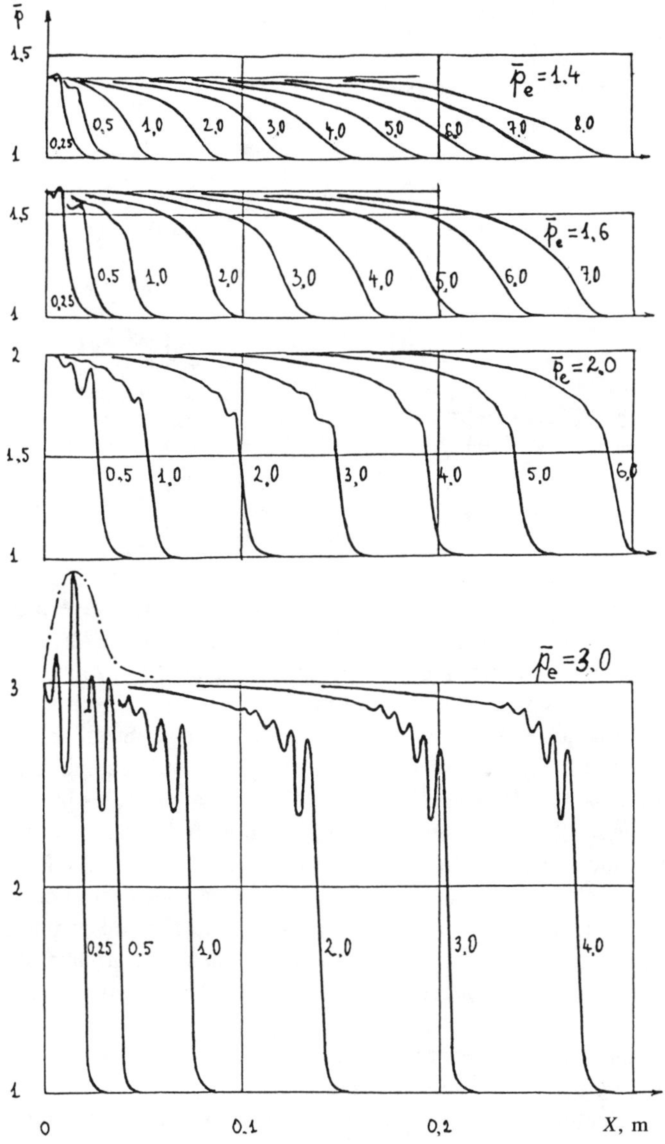

Figure 6.10.19 Evolution of shock waves (variation of pressure p) in a steam-water bubbly mixture (p_0 = 0.1 MPa, $T_0 = T_S$ = 373 K, α_{20} = 0.05, a_0 = 1 mm) of various intensity $\bar{p}_e$. Numerical labels on curves indicate time in milliseconds. The dot-dashed line is an envelope of maximum pressure peaks.

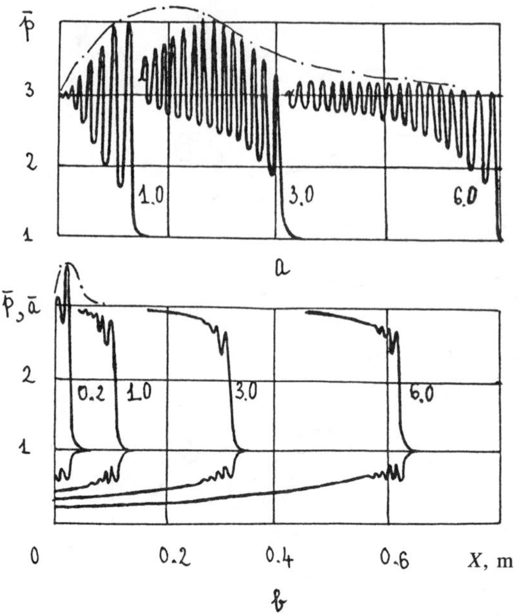

Figure 6.10.20 Evolution of shock waves (variation of both pressure p and bubble radius a) at intensity $\bar{p}_e = 3.0$ in water with air bubbles ($p_0 = 0.1$ MPa, $T_0 = 293$ K, $\alpha_{20} = 0.02$, $a_0 = 1$ mm) (a), and in a steam-water bubbly mixture ($p_0 = 0.1$ MPa, $T_0 = T_S = 373$ K, $\alpha_{20} = 0.02$, $a_0 = 1$ mm) (b). Numerical labels on curves indicate time in microseconds.

the interphase mass exchange on the process of pressure-wave propagation in bubbly liquids is evident. The pertinent characteristics of shock waves in liquids with vapor bubbles are much smaller compared to evolution of shock waves in liquids with gas bubbles of radius $a_0 \sim 1$ mm. Both the time and distance for which these waves assume a stationary configuration are, depending on kind of gas, usually as high as several milliseconds and meters, respectively (see §6.7). The stationary configuration of waves in liquids with vapor bubbles is built up over a distance of about 0.4 meters for time of the order of 4 microseconds. In this case, both oscillation amplitude and wave propagation velocity reduce. A compression wave causes both compression and vapor condensation within bubbles. Condensation takes place due to vapor pressure growth, which becomes higher than saturation pressure corresponding to initial temperature T_0, and the bubble radius is gradually diminishing to zero.

A comparison of curves related to $p_e = 3.0$ in Figs. 6.10.19 and 6.10.20 shows that with an increase of the initial concentration α_{20} of vapor phase, the wave velocity reduces. The length of oscillatory waves behind front and the distance over which waves assume a stationary regime varies as their velocity varies. In this case, the time for which waves acquire the stationary regime remains constant, approximately equal to 4 microseconds; bubble size, initial pressure, and thermophysical properties of phases are responsible for this.

In order to investigate the impact of dimensionless parameters γ_2, $\tilde{c}_0$,

l_0^*, and λ_{10}^* on the shock-wave evolution in a liquid containing vapor bubbles with all other conditions being fixed, computations of shock-wave evolution in water with steam bubbles in systems with initial static pressures p_0 = 0.05 MPa (Fig. 6.10.21*a*) and 0.5 MPa (Fig. 6.10.21*b*) have been conducted. From Fig. 6.10.21, it is seen that with an increase of the initial static pressure in a system, the distance over which a wave acquires a stationary regime increases. At initial pressure p_0 = 0.5 MPa (Fig. 6.10.21*b*), this distance amounts to several meters, whereas for shock waves of intensity $\bar{p}_e = 2$ at an initial pressure in a system p_0 = 0.05 MPa (Fig. 6.10.21*a*), the distance is only 0.10–0.15 m, and at p_0 = 0.1 MPa (Fig. 6.10.19), it is approximately 0.3 m. An increase of initial static pressure p_0 in a system not only leads to an increase of the distance over which a wave assumes a stationary regime, but also to an increase of both length and amplitude of oscillation parameters behind the wave front. The latter is not only associated with an increase of the shock-wave propagation speed D proportional to $p_0^{1/2}$ with growth of p_0, but also with the relative retardation of the interphase heat and mass exchange (because of the λ_{10}^* reduction) and the growth of vapor mass within bubbles (with an increase of $\bar{\rho}_2^0$ and $\bar{c}_0$) at fixed initial radius. An increase of vapor mass within bubbles leads to an increase of both condensation time and time of bubble contraction. With an increase of the initial static pressure p_0, the elasticity of the system enhances, and, consequently, its tendency toward realization of oscillatory-structure shock waves augments. It is noteworthy that in the case of p_0 = 5 MPa represented in Fig. 6.10.21*c*, a soliton-wave breaks away from the main wave, which gradually decays, because of dissipation, as it propagates through the mixture. Therefore, when studying the structure of stationary shock waves in bubbly liquids of a huge thickness of about 10 m, such soliton cannot be detected. The possibility of emanation of a soliton from the main signal at a nonstationary stage of propagation of long shock waves in vapor-liquid bubbly media is associated with the interrelationship between two kinds of velocities: rate of transfer of energy delivered by a shock wave into a medium before the wave front (which is determined by velocity of propagation of the first pressure peak (by its amplitude)), and rate of transfer of energy of the main wave to its front, which is associated with velocity of motion of subsequent pressure peaks (by their amplitude).

Note that in cases when $\alpha_2 \leqslant \alpha_c = p/\rho_1^0 C_1^2$, or when the shock-wave velocity D approaches speed of sound C_1 in a liquid phase (e.g., due to a reduction of void fraction α_2, or a pressure p increase), the investigation of wave processes becomes more complicated because it is necessary to take into account the liquid phase compressibility and finiteness of the perturbation-transfer velocity through continuous medium. Moreover, with the system initial pressure p_0, and, consequently, temperature $T_0 = T_s(p_0)$ growth, the interphase surface tension Σ reduces, and vapor density within bubbles (at fixed radius a_0) increases. All these may become responsible for wave-

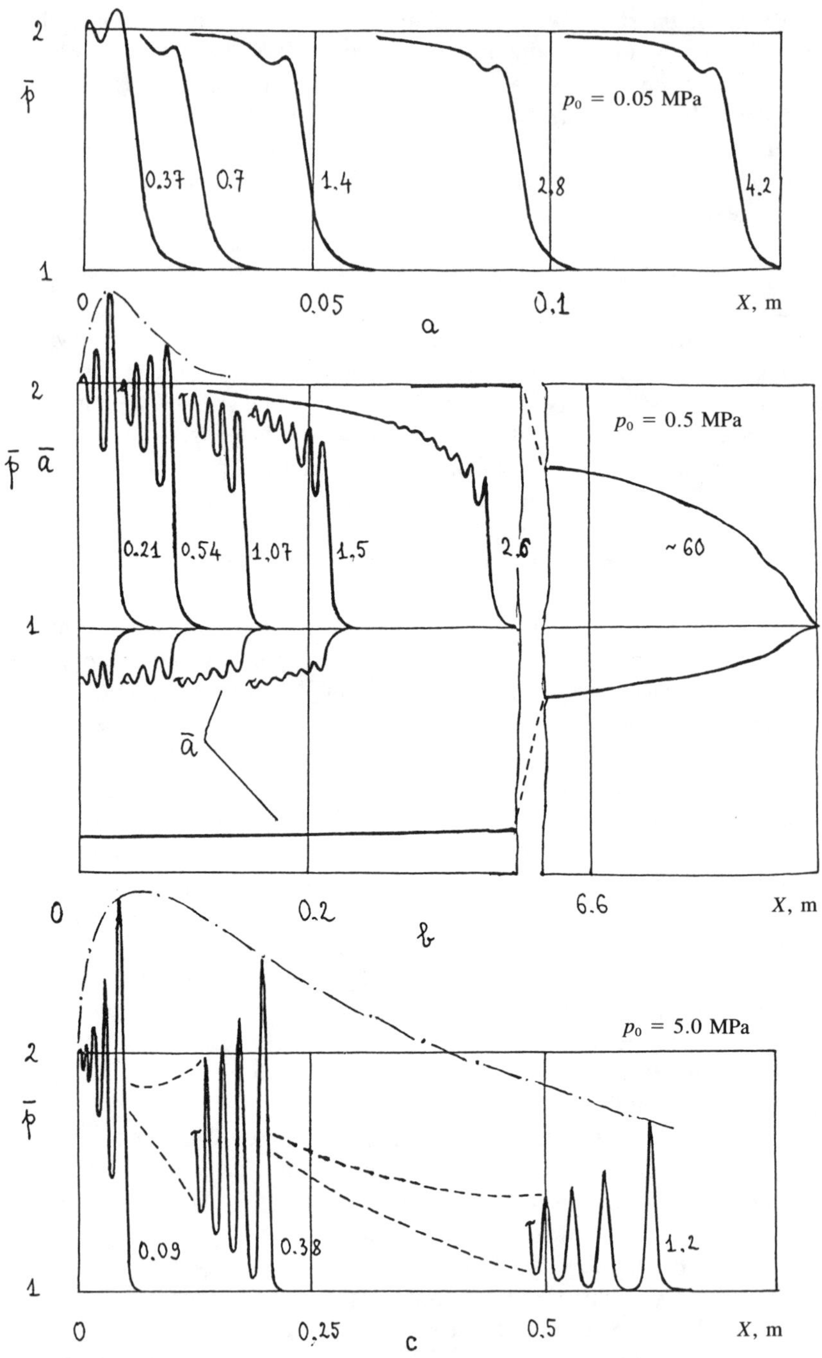
2
$\bar{p}$
1
p_0 = 0.05 MPa
0.37
0.7
1.4
2.8
4.2
0
0.05
0.1
X, m
a
2
$\bar{p}$ $\bar{a}$
1
p_0 = 0.5 MPa
0.21
0.54
1.07
1.5
2.6
~60
$\bar{a}$
0
0.2
6.6
X, m
b
p_0 = 5.0 MPa
2
$\bar{p}$
1
0.09
0.38
1.2
0
0.25
0.5
X, m
c

induced bubble fragmentation which—as is proven by experiments—occurs at the first pressure peak. Bubble fragmentation results in a significant modification of both medium structure and wave picture in the vapor-liquid mixture (see, e.g., curve 3.0* in Fig. 6.10.13).

Varying the bubble initial radius is adequate for varying the dimensionless criterion λ_{10}^* proportional to a_0^{-1}. The λ_{10}^* criterion strongly affects the nature of the shock wave evolution in a liquid with vapor bubbles. With all other conditions being fixed, the reduction of λ_{10}^* (e.g., by increasing the initial bubble size a_0), leads to interphase heat and mass exchange retardation and thereby, to an increase of both amplitude of parameter oscillation and wavelength of oscillatory waves behind the front; it also accounts for an increase of both time and distances of establishment of stationary configurations. In a small bubble-radius case ($a_0 \sim 0.01$ mm, and, more precisely, when $B \gtrsim 10$ (see (2.6.56)), the bubbles condense almost in the Rayleigh regime, in which the liquid inertia, not thermal effect, is predominant. With the bubble compression reaching sufficiently high rates, the vapor condensation does not have sufficient time to compensate for compression, and vapor pressure within bubbles grows, causing an inverse motion—bubble expansion. The wave in this case may have a tapered oscillatory structure.

To demonstrate the effects of thermophysical properties of liquid, Fig. 6.10.22 represents both predicted evolution of a shock wave in liquid nitrogen with nitrogen-vapor bubbles and its limiting stationary configuration. Comparison of curves depicted in Figs. 6.10.19 ($\bar{p}_e = 2$), 6.10.21 and 6.10.22 once more substantiates the predominant effect of interphase heat and mass exchange (of thermophysical properties of a liquid and that of thermal dissipation) upon evolution of shock waves in a liquid containing vapor bubbles. An increase of $\tilde{\rho}_2^0$ and a reduction of λ_{10}^*, as was already indicated, signify the increase of the vapor mass within bubbles and retardation of interphase heat and mass exchange. This leads to an increase of both duration and distance over which waves establish the limiting stationary configuration, to an increase of their thickness, and to the enhancement of the tendency for oscillations, which are characteristic for waves in a liquid with bubbles of nonsoluble and noncondensing gas. The nonstationary wave in liquid nitrogen containing bubbles of nitrogen vapor, depicted in Fig. 6.10.22, has a well-defined oscillatory structure. The wave establishes a monotonic stationary configuration over a distance of approximately 10 m, and the wave thickness, in this case, amounts to about 3 meters.

Figures 6.10.23 and 6.10.24 illustrate the results of a comparison of

---

Figure 6.10.21 Evolution of shock waves (variation of pressure p and bubble radius a) in saturated steam-water bubbly medium ($\alpha_{20} = 0.05$, $a_0 = 1$ mm, $\bar{p}_e = 2$) at various initial pressures: $p_0 = 0.05$ MPa, $T_0 = T_S = 354$ K (a); $p_0 = 0.5$ MPa, $T_0 = T_S = 424$ K (b); $p = 5.0$ MPa, $T_0 = T_S = 537$ K (c); The variant for $p_0 = 0.1$ MPa, $T_0 = T_S = 373$ K, see Fig. 6.10.18. Numerical labels on curves indicate time in microseconds.

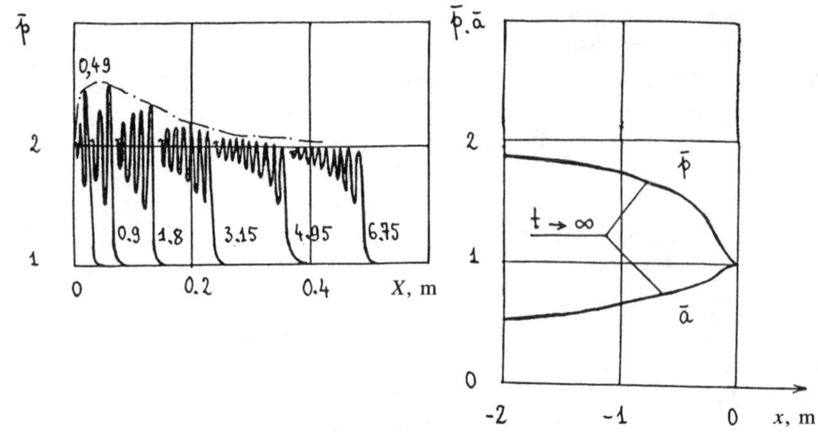

Figure 6.10.22 Evolution of a shock wave (variation of pressure p) of intensity $\bar{p}_e = 2$ in liquid nitrogen containing nitrogen-vapor bubbles ($p_0 = 0.1$ MPa, $T_0 = T_S = 77.3$ K, $\alpha_{20} = 0.05$, $a_0 = 1$ mm). Curves $t \to \infty$ represent pressure p configurations and variation of the bubble radius a in a corresponding stationary wave. Numerical labels on curves indicate time in milliseconds.

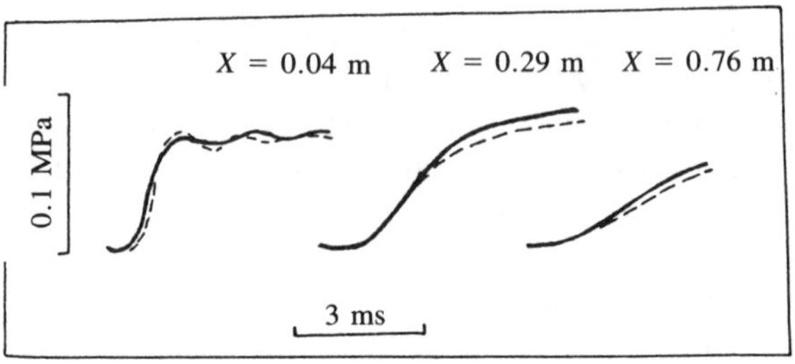

Figure 6.10.23 Predicted (dashed lines) and experimental (solid lines, B. G. Pokusaev (1979)) pressure oscillograms in the process of propagation of a shock wave produced by a stationary disturbance ($\bar{P}_e = 1.1$) in steam-water bubbly mixture ($p_0 = 0.5$ MPa, $T_0 = T_S = 424$ K, $\alpha_{20} = 0.015$, $a_0 = 1.4$ mm). Oscillograms are recorded at various points of a shock tube; coordinates of the test points are measured from the point of initiation and are indicated by numerical labels on curves.

numerical solutions with experimental pressure oscillograms at various points of a shock tube. Good agreement between theory and experiments is obvious.

It should be noted that in spite of the available experimental size-distribution of bubbles and the observed significant departure of bubble shape from spherical configuration, the supposition of the bubble sphericity does not impede good agreement between theory and experiments. This is indicative of the predominant impact of the volume rather than the shape of bubbles on the radial motion process.

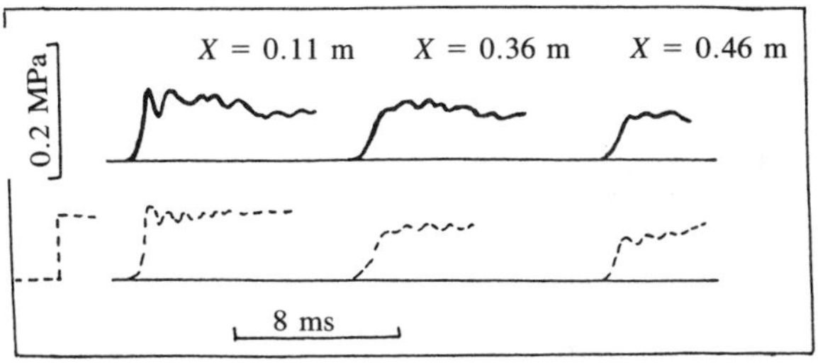

Figure 6.10.24 The same as in Fig. 6.10.23, but for liquid freon-21 with bubbles of freon-21 vapor (p_0 = 0.18 MPa, $T_0 = T_S$ = 299 K, α_{20} = 0.01, a_0 = 1.2 mm, $\bar{p}_e$ = 1.4). (The experimental data were obtained by V. E. Nakoryakov et al. (1984).)

Shock wave intensification in a homogeneous vapor-liquid bubbly mixture. Although in stationary shock waves propagating in a liquid with vapor bubbles we have $p < p_e$ and pressure increase is impossible, pressure increase ($p > p_e$) in nonstationary waves does occur (see Figs. 6.10.19–6.10.25); in this case, the intensification parameter $P_m = (p_{\max} - p_e)/(p_e - p_0)$ ranges from 0.3–0.7. Analysis proves that reduction of the interphase heat and mass exchange enhances wave intensification, and in a liquid with fixed-mass bubbles (see §6.7), intensification is significantly higher.

An increase of both initial pressure p_0 and disturbance intensification augment the intensification parameter P_m. At the same time, the intensification parameter depends to a smaller extent on the volumetric concentration of bubbles in the range $\alpha_c \ll \alpha_{20} \lesssim 0.1$; the indicated void fraction affects the size X_m of a zone where the wave intensification is observed (see Fig. 6.10.26). The size X_m is proportional to $\alpha_{20}^{-1/2}$ and a_0.

An even more significant wave amplification ($P_m \approx 5$) of higher intensity ($\Delta\bar{p}_e \sim$ 10–20) that is promoted by both bubble fragmentation and ac-

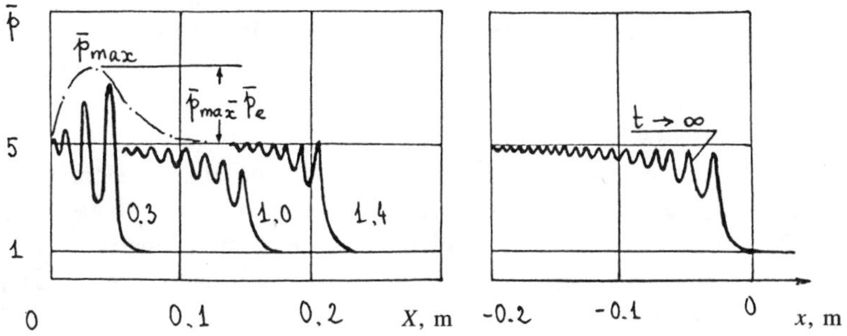

Figure 6.10.25 A shock wave evolution (variation of pressure p) in steam-water bubbly mixture (p_0 = 0.1 MPa, $T_0 = T_S$ = 373 K, α_{20} = 0.02) of intensity $\bar{p}_e$ = 5. The dot-dashed line envelopes the maximum pressure peaks. Curve $t \to \infty$ represents a pressure p configuration in a corresponding stationary wave. Numerical labels on curves indicate time in microseconds.

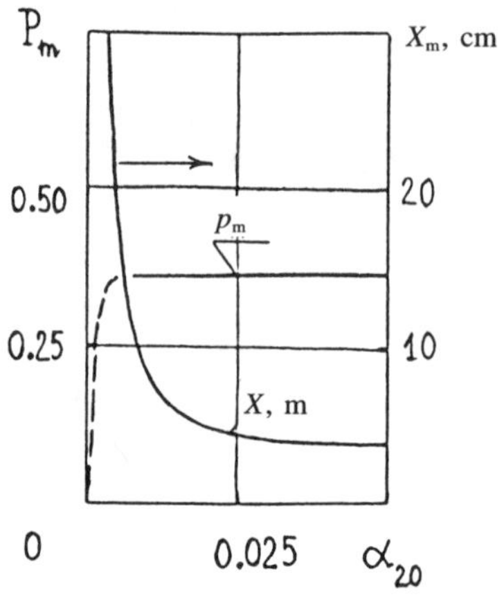

Figure 6.10.26 Dependence of intensity P_m and the size X_m of the intensification zone (prediction) within pressure shock waves in a vapor-liquid bubbly mixture (p_0 = 0.1 MPa, $T_0 = T_S$ = 373 K, a_0 = 1 mm, $\bar{p}_e$ = 3) on the volumetric concentration (void fraction) α_{20} of the second vapor phase.

celeration of vapor condensation has been recorded in the above-described experiments by A. A. Borisov, B. E. Gel'fand, R. I. Nigmatulin, et al. (1977).

The size of a zone where pressure increase is observed in shock waves under normal conditions (p_0 = 0.1 MPa) is generally fairly small. As the rate of interphase heat and mass exchange reduces and the wave velocity increases, the dimension of this zone increases. The value X_m in water with steam bubbles of radius a_0 = 1 mm for waves of intensity $\bar{p}_e$ = 2 ranges from 0.01–0.1 m when the initial pressure in a system grows from 0.1–0.5 MPa. The width of this zone in liquid nitrogen containing bubbles of nitrogen vapor under the above-indicated parameters amounts to 0.4 m.

Figure 6.10.27 proves that both wave intensity $\bar{p}_e$ (factor of nonlinearity) and thermophysical properties of phases affect the rate of pressure increase P_m to a considerable extent. The rate of pressure P_m increase within waves of moderate intensities ($\bar{p}_e \leqslant 5$) in boiling water with steam bubbles (p_0 = 0.1 MPa) is not higher than 0.8, and in boiling nitrogen (p_0 = 0.1 MPa), the analogous magnitude may reach ~1.2. It is of interest that dependence of P_m on the wave intensity $\bar{p}_e$ (Fig. 6.10.27) and on the initial static pressure p_0 in a system (Fig. 6.10.28) is of a nonlinear and nonmonotonic nature. This is associated with the nonlinear character of the division of energy delivered by a shock wave to a medium: part of the energy is transmitted to a medium in the form of kinetic energy of microscopic motion determined by $\rho v^2/2$, and part of the energy which, because of the property of a local deformation inertia inherent in bubbly liquids, is transformed to energy ($\sim\rho_1^0 w_{10}^2 \alpha_2$) of oscillatory motion of liquid around bubbles. It is the second part of energy delivered to a medium by a shock wave and the process of its transformation that cause both the existence of shock waves with pulsatory structure and the feasibility of pressure magnification within waves.

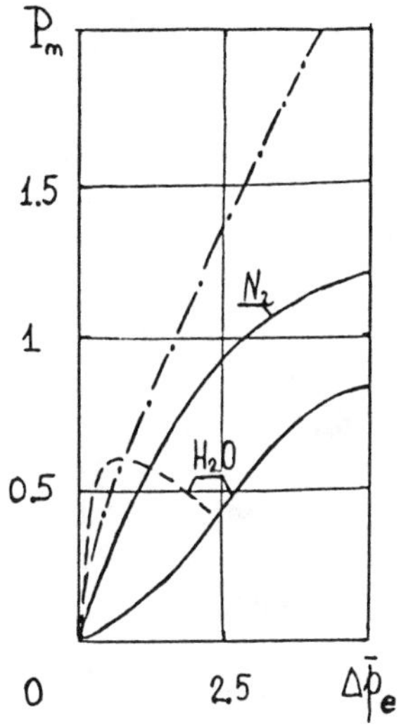

Figure 6.10.27 Relationship between the rate of increase of a shock wave pressure P_m (prediction) and its intensity (pressure on a "stationary" piston) $\bar{p}_e$ in several bubbly liquids (α_{20} = 0.05, a_0 = 1 mm): in boiling water with steam bubbles (p_0 = 0.1 MPa, $T_0 = T_S$ = 373 K (solid line) and p_0 = 6.5 MPa, $T_0 = T_S$ = 554 K (dashed line) (label H_2O), in water with air bubbles (p_0 = 0.1 MPa, T_0 = 293 K, dot-dashed line), and boiling nitrogen with bubbles of nitrogen vapor (label N_2) (p_0 = 0.1 MPa, $T_0 = T_S$ = 77.3 K).

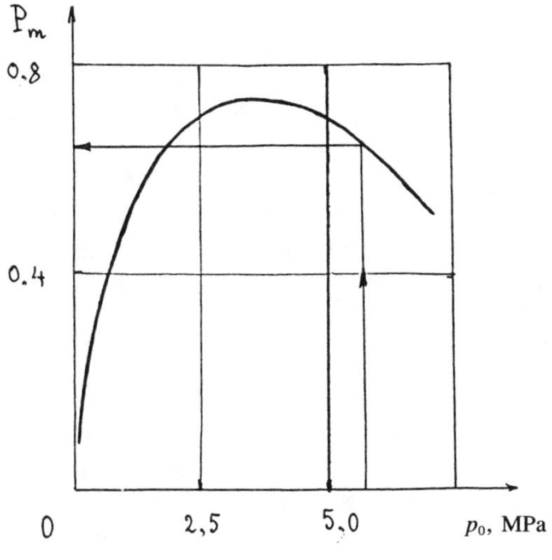

Figure 6.10.28 Relationship between rate of P_m increase in shock waves with a fixed intensity $\bar{p}_e$ = 2 (prediction) in steam-water bubbly mixture (α_{20} = 0.05, a_0 = 1 mm) and the initial static pressure p_0 in the mixture ($T_0 = T_S(p_0)$).

An increase of the shock-wave front propagation velocity, e.g., by augmenting their intensity or by an increasing the initial pressure in a system, may lead to an increase of energy delivered by a wave to a medium before the wave front, and, thereby, to the reduction of the part of energy delivered by the wave through its transformation of oscillatory motion around bubbles to energy, i.e., to the reduction of the rate of intensification of pressure P_m within shock waves (see Figs. 6.10.27 and 6.10.28).

Evolution of finite-length impulses in a liquid containing vapor bubbles. We shall now examine impulses initiated by short-duration disturbances whose pressure oscillogram at the location of their initiation is of a triangular shape: there is, first, a uniform (linear) pressure growth during time $t_b^{(1)}$ up to a maximum value equal $(p_b)_{\max} = (\bar{p}_b)_{\max} p_0$, and then linear drop during time $t_b^{(2)}$.

The evolution of a finite-length impulse depends on the "shape" of the original disturbance. In addition to triangular configuration, a Π-like shape, sinusoid half-wave, etc. may be indicated. But its principal characteristics include the maximum pressure $(p_b)_{\max}$ and length t_b (for a triangular disturbance under discussion, $t_b = t_b^{(1)} + t_b^{(2)}$).

Thus, besides four thermophysical parameters of vapor-liquid bubbly medium (γ_2, l_0^*, $\tilde{c}_0$, and λ_{10}^*), an evolution of a signal depends on the shape of original signal and its two major characteristics: amplitude $(\bar{p}_b)_{\max}$ and length $\bar{t}_b = t_b/t_0$ (t_0, see (6.7.15)).

For waves of a low but finite intensity ($(\Delta\bar{p}_b)_{\max}) \ll 1$), the following dispersion parameter may be introduced (V. Nakoryakov et al., 1984)

$$d = \frac{C_* t_b}{a_0}\sqrt{(\Delta\bar{p}_b)_{\max}}\left((\Delta\bar{p}_b)_{\max} = (p_b)_{\max} - 1,\ C_* = \sqrt{\frac{p_0}{\rho_1^0}}\right)$$

And, there exists a critical value of this parameter, which is denoted by d_*, such that at $d > d_*$ the disturbance is transformed to a single or several solitons which gradually decay; at $d < d_*$, the disturbance assumes transformation to a wave packet (see §6.6). In vapor-liquid media where dissipation due to heat and mass exchange is very extensive, value d_* depends on thermophysical parameters and impulse intensity

$$d_* = d_*((\Delta p_b)_{\max}, \tilde{c}_0, \lambda_{10}^*, l_0^*, \gamma_2)$$

It should be remembered that impulse may have an oscillatory structure at the starting stage of evolution, but subsequently the oscillations decay. As distinct from infinite-length waves ($t_d \to \infty$), the bubbles, following propagation of the disturbance, retain their reduced size (if they had insufficient time to condense), and the impulse dissipates and gradually vanishes.

The pattern of attenuation of the impulse effect depends on its configuration and initial length. The higher the intensity and length of impulse, the longer the duration of their decay. In the process of evolution of finite-length impulse effects as well as during propagation of nonstationary shock waves, a pressure-rise effect in a mixture compared to maximum initiating pressure $(p_b)_{\max}$ is observed. In the case depicted in Fig. 6.10.28 ($(\bar{p}_b)_{\max} = 5$), $P_m = 0.27$ takes place.

Figure 6.10.30 represents a pattern of evolution for propagation of a wave initiated by a triangular impulse in liquid nitrogen on a saturation line containing nitrogen-vapor bubbles. The wave is of oscillatory structure. The results of analysis showed that a triangular impulse propagating in liquid nitrogen with nitrogen-vapor bubbles may have an oscillatory structure even at $(p_b)_{\max} = 1.4$, which is indicative of a less intensive pattern of interphase heat and mass exchange (thermal dissipation) in it compared to water under similar conditions ($p_0 = 0.1$ MPa, $T_0 = T_S(p_0)$). However, as has been indicated before (see discussion of Table 6.10.1), the wave picture in boiling water at $p_0 = 0.1$ MPa must approach a similar pattern in boiling nitrogen under an increase of the initial static pressure to $p_0 = 1$ MPa in the first system. The predicted results confirm this statement. Note that at the same initial pressure $p_0 = 0.1$ MPa and initial impulse (cf. Figs. 6.10.29 and

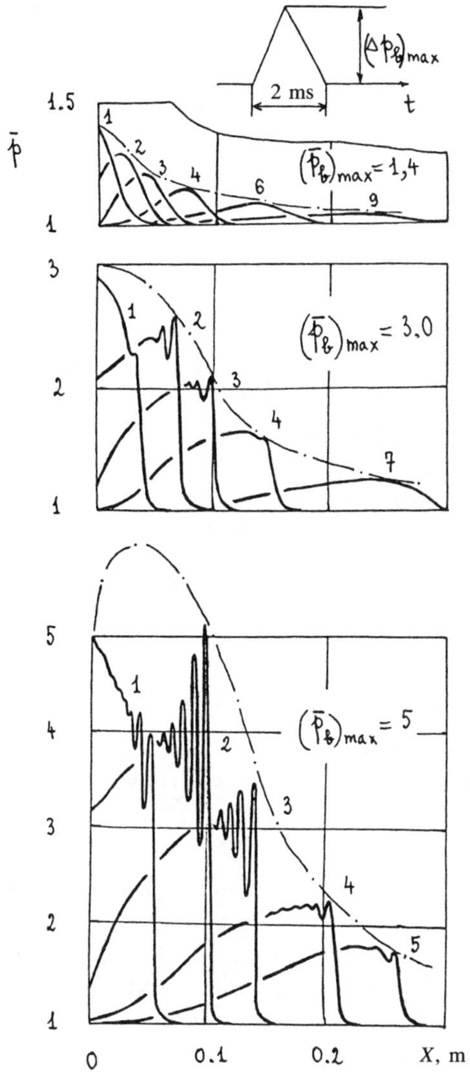

Figure 6.10.29 Evolution of triangular impulses (variation of pressure $\bar{p}$) in a steam-water bubbly mixture ($p_0 = 0.1$ MPa, $T_0 = T_S = 373$ K, $\alpha_{20} = 0.05$, $a_0 = 1$ mm, $t_b^{(2)} = 2\ t_b^{(1)} = 2\ \mu$s). Curves *1–9* correspond to points in time: t = 1; 1.5; 2; 3; 4; 4.9; 6.5; 7.8; 8.8 microseconds.

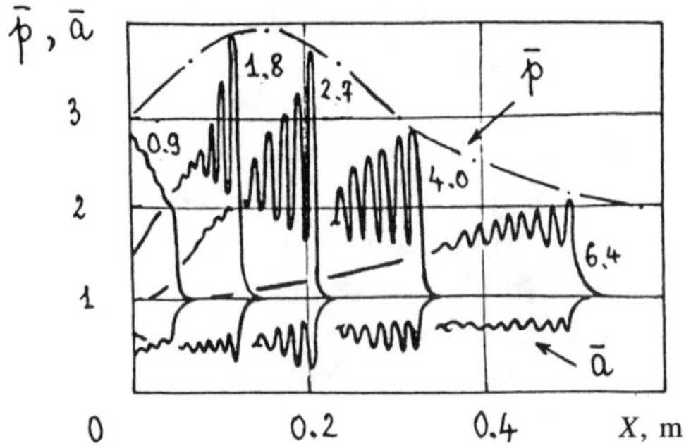

Figure 6.10.30 The same as in Fig. 6.10.29, but for variation of pressure $\bar{p}$ and bubble radius $\bar{a}$ in a liquid nitrogen with nitrogen-vapor bubbles (p_0 = 0.1 MPa, T_0 = T_S = 77.3 K, α_{20} = 0.05, a_0 = 1 mm, $(\bar{p}_b)_{max}$ = 3, $t_b^{(2)}$ = 2 $t_b^{(1)}$ = 2 μs). Numerical labels on curves indicate time t in microseconds.

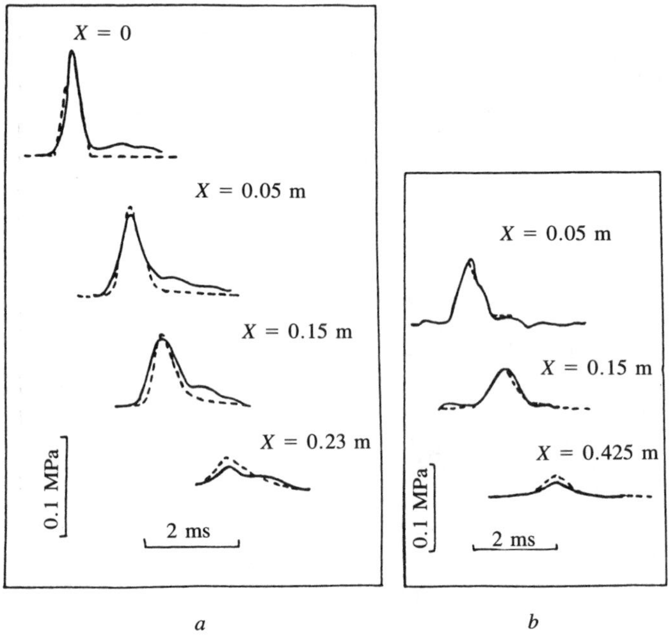

Figure 6.10.31 Comparison of analytically predicted results (dashed lines) on the evolution of a triangular impulse (variation of pressure $\bar{p}$) in a steam-water bubbly mixture ($\bar{p}_0$ = 0.5 MPa, T_0 = T_S = 424 K, α_{20} = 0.015, a_0 = 1.4 mm, $(\Delta\bar{p}_b)_{max}$ = 0.2 (*a*), and a_0 = 1.4 mm, $(\Delta\bar{p}_b)_{max}$ = 0.14 (*b*)) with experimental data obtained by B. G. Pokusaev (1979) (solid lines) at various points of a shock tube indicated by labels on curves.

6.10.30), the tendency towards amplification in boiling nitrogen is higher than that in boiling water.

Figure 6.10.31 shows the comparison of numerical and experimental results for finite and decaying impulses in boiling water. Good agreement between data on both signal-amplitude attenuation and signal expansion is evident.

§6.11 NONSTATIONARY OUTFLOW AND RAREFACTION WAVES IN FLASHING LIQUID

In a channel of length L with a constant, or slow-varying, cross section $S(x)$ and with both ends (bottoms) closed, let there be contained a homogeneous subcooled or saturated water under pressure p_0 and at temperature $T_0 \leqslant T_S$ (p). At time $t = 0$, the bottom on one of the closed ends of the tube is removed (emergency situation), and the liquid outflow is initiated together with steam generated as a consequence of water flashing; both liquid and steam outflow directly into the environment under pressure $p_\infty < p_0$. It is necessary to describe the occurring flow.

The formulated physical problem is one of the key issues arising in studies of a nonstationary outflow of flashing liquids out of high-pressure vessels. The problem of a theoretical explanation of this process is urgent from the standpoint of the analysis of emergency situations on nuclear (atomic) power plants, in chemical engineering, oil-and-gas pipelines, and other installations of modern technology where liquid or two-phase substances are used.

Specific features of flow of flashing liquid in a rarefaction wave in the condition of high pressure differences. When describing the theoretical aspects of an outflow of flashing liquid, two essential circumstances should be kept in mind.

1. High pressure-differences (0.1–10 MPa) realized in these processes entail more detailed, and, consequently, more intricate equations of state of phases and phase-transition conditions than equations of state of vapor and liquid used in the above discussion (see, e.g., §6.10). In particular, the compressibility of a liquid phase must be taken into consideration for analysis of wave processes in a single-phase subcooled or overheated metastable liquid, and also in a mixture with small volumetric concentrations of vapor.

2. The nonequilibrium effects are possible because of the high velocity of outflow due to high pressure drop, and, in particular, the "delay" of flashing also becomes feasible. The flashing process causes the expansion of the medium and of the rate of the vessel evacuating. Treatment of these effects is associated with an account of the structure of a two-phase mixture (bubble formation and growth, transition of a bubbly structure of the flow into a disperse-annular or drop structure, etc.), and, also, with equations of kinetics of nonequilibrium processes.

Outflow of subcooled or saturated liquid begins as a flow of a single-phase liquid, in which, due to high pressure drop below the saturation-pressure p_S (T_0), formation and growth of bubbles initiate, and the flow, for some period of time, proceeds in a bubbly regime. Therefore, to describe this stage of the process, we shall use equations of two-phase bubbly media giving consideration to the assumptions indicated in §1.5, except for fixed temperature of liquid, since significant mass concentrations of vapor are possible. In this case, we shall employ the mechanism of a heterogeneous nuclei-formation (see §1.7), supposing that there are uniformly distributed microparticles with $a_0 \sim 10^{-6}$ m, and number concentration $n_0 \sim 10^{11}$–10^{13} m^{-3} in the original single-phase liquid; these microparticles are believed to become centers of nucleation.

We shall also note that the channel-wall roughness and the presence of various particles on it may also become centers of vapor (steam) formation; therefore, with sufficiently small tube diameters (of an order of 1 cm and less), the surface flashing over the walls may also play a certain role.

The formation of vapor on ready nuclei limits the possible superheating ΔT of liquid during its outflow; this superheating (ΔT) does not generally exceed 10 K. Under such conditions, the homogeneous nucleation (thermofluctuation) whose intensity exponentially grows with increase of superheating (see §1.7), during the time (0.01–0.1 seconds) available in the outflow process, cannot exhibit itself.

Note that at the initial temperatures of the outflowing liquid that are close to critical temperature T_{er} ($T_0/T_{cr} \gtrsim 0.9$, the superheating ΔT at which the sufficiently intensive homogeneous nucleation becomes possible is about 10 K (see also §1.7 and the discussion of (6.11.38) that follows below).

Analysis of the bubble-growth process under a sudden drop of pressure from p_0 to $p_0 - \Delta p$ shows (see §2.6) that two stages of the process are observed: first, a dynamic stage during which for the time $t^{(p)} \sim a_0$ $(\rho_l^0/\Delta p)$, the vapor pressure within a bubble differs from the pressure in liquid because of the liquid radial inertia; the second, a thermal stage when the pressure in vapor differs from the pressure in liquid due only to the surface tension, and a bubble monotonically grows in a superheated liquid; the rate of this growth is determined by the capacity of liquid to supply heat for evaporation. The evaluations show that the duration of the dynamic stage is as low as 1 microsecond, and, therefore, this stage need not be taken into account. In addition, because of high pressure, in spite of the bubble smallness, the capillary pressure may be neglected ($2\Sigma/a_0 p_0 \ll 1$). The inherent time of equalization of the temperature within a bubble is $t_2^{(T)} \sim a^2/\nu_2^{(T)}$, and, if the characteristic radial velocities w of the bubble growth are such that $t_2^{(T)} w/a \ll 1$, both temperature and density within a bubble, as in §6.10, may be considered uniform, and defined by conditions of saturation (see §2.6). As in the analysis of shock waves, we ignore the translatory motion of bubbles relative to liquid. Generalizing the above-indicated statements, we assume the following simplifications

$$v_1 = v_2 = v, \quad p_1 = p_2 = p, \quad T_2 = T_S(p), \quad \overset{\circ}{\rho}_2 = \overset{\circ}{\rho}_{2S}(p) \tag{6.11.1}$$

Also, let both heat exchange and friction on the channel walls be neglected.

Mass, momentum, and energy conservation equations; equations of state of phases; interphase heat and mass exchange equations. In a quasi-one-dimensional approach, the mass and momentum conservation equations, the equations of conservation of the number of bubbles (see (1.5.11)), and also the heat-influx equations for each phase (similar to (1.4.8)). together with the heat-influx equation on the interface, are written (using the above-indicated assumptions and the variability of a cross section along the longitudinal coordinate) in the form

$$\frac{\partial(\rho S)}{\partial t}+\frac{\partial(\rho v S)}{\partial x}=0, \quad \frac{n}{n_0}=\frac{\rho}{\rho_0}$$

$$\frac{\partial(\rho v S)}{\partial t}+\frac{\partial(\rho v^2 S)}{\partial x}=-S\frac{\partial p}{\partial x}$$

$$\overset{\circ}{\rho}_1\alpha_1\frac{di_1}{dt}-\alpha_1\frac{dp}{dt}=nq_{\Sigma 1}-j_{12}(i_{1S}-i_1) \tag{6.11.2}$$

$$\overset{\circ}{\rho}_2\alpha_2\frac{di_2}{dt}-\alpha_2\frac{dp}{dt}=nq_{\Sigma 2} \qquad (i_2=i_{2S}(p)=i_2(p,\ T_S(p)))$$

$$q_{\Sigma 1}+q_{\Sigma 2}=-j_{12}l$$

The state of a single-velocity, two-temperature medium under consideration is determined, in addition to its velocity v, by independent parameters ρ, p, T_1, and n. Using the equations of state, the following quantities are also determined

$$\overset{\circ}{\rho}_1=\overset{\circ}{\rho}_1(p,\ T_1), \quad T_2=T_S(p), \quad \overset{\circ}{\rho}_2=\overset{\circ}{\rho}_{2S}(p)$$
$$\alpha_2=\frac{\overset{\circ}{\rho}_1-\rho}{\overset{\circ}{\rho}_1-\overset{\circ}{\rho}_2}, \quad a^3=\frac{\varepsilon\alpha_2}{\pi n} \tag{6.11.3}$$

We will also later use the phase-mass conservation equations written in the form

$$\frac{d\rho_1}{dt}=-\frac{\rho_1}{S}\frac{\partial(vS)}{\partial x}-nj_{12}, \quad \frac{d\rho_2}{dt}=-\frac{\rho_2}{S}\frac{\partial(vS)}{\partial x}+nj_{12} \tag{6.11.4}$$

The sum of these equations yields the mixture-mass conservation equation.

Analysis of wave motions of the mixture by the method of finite differences implies calculations of phase properties in every node of the difference grid at each time step. Accordingly, the employed equations of state, in addition to the high accuracy of calculation of the thermodynamic functions (6.11.2), must also ensure high accuracy of calculation of their derivatives, in terms of which both speed of sound and wave velocity are determined. At the same time, they must be sufficiently simple that the amount of calculations is not increased. In order to satisfy these requirements, the use of the equation of state which approximates the phase properties on the finite intervals of the parameter variation is advisable. The thermodynamic approximations applied to waves in the range from $p = 0.1$ MPa to $p = 10$ MPa are set forth below.

As was indicated, the phase equilibrium curve in the form of relation "saturation temperature vs. pressure" is well approximated by expression (1.3.78) (see also (6.10.23)), which yields a relative error for water in the indicated range of pressure p at $T^0 = 4640$ K, $p^0 = \exp 10.26 = 28{,}570$ MPa, not exceeding 1%.

The phase-transition heat, enthalpy, and specific volume of liquid in the state of saturation are approximated by polynomials (A. I. Ivandaev and A. A. Gubaidulin, 1977), which are written (for water) as

$$\begin{aligned} l(p) &= 2199 - 170{,}7p + 15.22p^2 - 0.707p^3 \\ i_{1S}(T) &= i_1(p_S(T),\ T) \\ &= -1299 + 5.663T - 0.446\cdot 10^{-2}T^2 + 0.49\cdot 10^{-5}T^3 \\ \overset{\circ}{V}_{1S}(p) &= (\overset{\circ}{\rho}_1(p,\ T_S(p)))^{-1} \\ &= 0.1046\cdot 10^{-2} + 0.789\cdot 10^{-4}p - 0.776\cdot 10^{-5}p^2 + 0.389\cdot 10^{-6}p^3 \end{aligned} \tag{6.11.5}$$

where i_{1S}, and l are expressed in kJ/kg, p in MPa, T_S in K, V_{1S} in m^3/kg. Both the saturation enthalpy and specific volume of vapor are expressed in terms of these approximations using the Clausius-Clapeyron equation

$$i_{2S}(p) = i_{1S}(p) + l(p), \quad \overset{\circ}{V}_{2S}(p) = \frac{1}{\overset{\circ}{\rho}_{2S}} = \overset{\circ}{V}_{1S} + \frac{l(p)}{T_S(p)}\frac{dT_S}{dp} \tag{6.11.6}$$

Experiments concerned with determining the specific volume $V_1^0 = 1/\rho_1^0$ of a subcooled liquid (V. P. Skripov, 1972) as a function of pressure and temperature proved that the relationship between V_1^0 and p is linear and continuous during transition through the state of saturation

$$\overset{\circ}{V}_1(p,\ T) = \overset{\circ}{V}_{1S}(T) + \beta(T)\cdot(p - p_S(T)) \tag{6.11.7}$$

where $\beta = (\partial V_1/\partial p)_T$ is an isothermal coefficient of volumetric expansion. It is well-known that the relations $V_1\ (p,T)$ and $i_1\ (p,T)$ must satisfy the condition of integrability of the entropy differential

$$T\,ds_1 = di_1 - \overset{\circ}{V}_1 dp \tag{6.11.8}$$

whence

$$T\left(\frac{\partial \overset{\circ}{V}_1}{\partial T}\right)_p - \overset{\circ}{V}_1 = -\left(\frac{\partial i_1}{\partial p}\right)_T \tag{6.11.9}$$

and upon integration, we obtain the energy equation

$$i_1(p, T) = i_{1S}(T) + \int\limits_{p_S}^{p} \left[V_1^\circ - T \left(\frac{\partial V_1^\circ}{\partial T} \right)_p \right] dp$$
$$= i_{1S}(T) + I^{(1)}(T) \cdot [p - p_S(T)] + I^{(2)}(T) \cdot [p - p_S(T)]^2 \qquad (6.11.10)$$
$$I^{(1)} = V_{1S}^\circ(T) - T \frac{dV_{1S}^\circ}{dT} + \beta(T) \cdot T \frac{dp_S}{dT}, \quad I^{(2)} = \frac{1}{2} \beta(T) - T \frac{d\beta}{dT}$$

At states relatively far from critical, the first terms in expressions for coefficients $I^{(1)}$ and $I^{(2)}$ are the most significant

$$I^{(1)} \approx V_{1S}^\circ(T_1), \quad I^{(2)} \approx {}^1\!/_2\, \beta(T_1) \qquad (6.11.11)$$

The values $\beta(T_1)$ may be obtained from thermophysical data for both subcooled and superheated liquids using the experimental data of the speed of sound in liquid on the saturation line $p_S(T_1)$. The corresponding relation may be obtained based on assumed equations of state

$$C_1^2 = -(V_1^\circ)^2 \left(\frac{\partial p_1}{\partial V_1^\circ} \right)_s = -(V_1^\circ)^2 \left(\frac{\partial i_1}{\partial T_1} \right)_p \left\{ \beta(T_1) \left(\frac{\partial i_1}{\partial T_1} \right)_p + \left(\frac{\partial V_1^\circ}{\partial T_1} \right)_p \left[V_1^\circ - \left(\frac{\partial i_1}{\partial p} \right)_T \right] \right\}^{-1} \qquad (6.11.12)$$

Figure 6.11.1 depicts the relationships between the speed of sound C_1 on temperature and between isothermal compressibility β and temperature for saturated water ($P = p_S(T_1)$). The values of the so-called "isothermal" speed of sound C_{1T} is also shown for the sake of comparison

$$C_{1T}^2 = -(V_1^\circ)^2 \left(\frac{\partial p_1}{\partial V_1^\circ} \right)_T = -\frac{(V_1^\circ)^2}{\beta} \qquad (6.11.13)$$

The intensity of vapor formation within a volume of superheated liquid is defined by the number of nuclei on which bubbles occur and also by the rate of their growth. Analysis of experimental and theoretical data related to formation and growth of single vapor (steam) bubbles within a volume of overheated liquid (see the references in §1.6) show that a bubble undergoes several stages of its development (growth). During a sudden pressure drop in liquid, as soon as the bubble-nucleus size exceeds the critical diameter, a slow bubble-growth stage ensues, which is defined by the liquid surface tension ("capillary stage"). A rapid increase of the bubble-growth rate now takes place. In this stage, the bubble-growth rate—defined by velocity w_{1a} of liquid on its wall—is hampered by the liquid radial inertia, defined by the liquid density ρ_l^0; the radial inertia is overcome by the pressure difference $p_2 - 2\Sigma/a - \rho_1$ in conformity with the Rayleigh-Lamb equation. The radial velocity increases in the inertia stage of the bubble-growth until it is limited by the intensity of vapor-formation defined by the heat influx

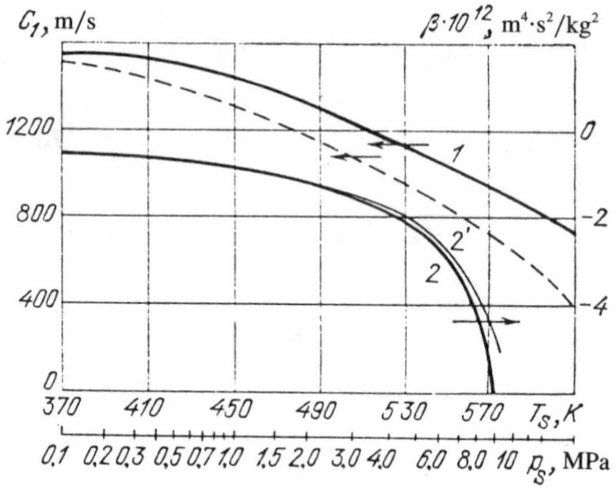

Figure 6.11.1 The speed of sound C_1 (line 1), the isothermal speed of sound C_{1T} (dashed line), and the coefficient of isothermal compressibility β (lines *2* and *2'*) of saturated water, as functions of temperature or pressure. Line *1* represents experimental data obtained by V. V. Sychev (1961). Line *2* relates to β calculated using speed of sound (formula (6.11.12)); line *2'* corresponds to experimental data obtained by V. P. Evstefeev et al. (1977) in accordance with relation $V_1^0(p,T)$.

to the bubble-walls from superheated liquid; the inertia stage is succeeded by the thermal stage of the bubble-growth defined by the heat influx $q_{1\Sigma}$. Evaluations of the duration of both capillary and inertia stages preceding the thermal regime indicate that, for a bubble of diameter of order 10^{-6} m under p being about 7.0 MPa, it is $t_b \approx 10^{-6}$–10^{-5} seconds. Under seal failure, bubbles are generated resulting from a sharp pressure drop in rarefaction waves. The inherent velocity of wave propagation in a single-phase liquid amounts to 1000 m/s; the characteristic scale for the channel length is 1 meter; then, the characteristic time of the wave-propagation process equals 10^{-3} seconds, and the characteristic time of flashing and that of vapor-volume growth (which defines the outflow) is many times greater, and amounts to $t_0 \approx 10^{-2}$–10^{-1} seconds. Therefore, during the first two stages t_b, preceding the thermal stage of flashing, there is no time needed for the sufficient amount of vapor (steam) to be generated in order to affect the outflow process; this permits the first two stages of the bubble-growth to be ignored, and it can be suggested that the thermal stage of the bubble-growth initiates no sooner than the size of the nucleus-bubble or particle exceeds the critical diameter.

We have already mentioned, in §1.6, the analytical solution of the Scriven problem on the self-similarity growth of a steam (vapor) bubble in superheated liquid in a thermal regime under fixed both pressure p and over heating ΔT. The inherent time of the temperature-distribution "adjustment" for a similarity time under constants p and ΔT for a bubble of radius a equals

$t_1^{(T)} \sim a^2/\nu_1^{(T)}$. A more accurate analysis shows that this "adjustment" time is, in fact, ten times smaller. If the inherent time t_0 of: the pressure variation $p(t)$, superheating $\Delta T(t)$, and the bubble growth is many times greater than $t_1^{(T)}$ ($t_0 \gg t_1^{(T)} \approx a^2/(10\nu_1^{(T)})$), then the bubble-growth may be said as proceeding in a thermal (see (2.6.48) where $p_\infty = p$) quasi-similarity regime; i.e., the bubble-growth rate, or the intensity of the heat influx $q_{1\Sigma}$ at each point in time t, are determined by formulas (1.3.56), (1.6.19), and (1.6.20) (or by their generalizations (2.6.42) and (2.6.44)), which follow from the similarity solution, and in which both pressure $p(t)$ and overheating $\Delta T(t) = T_1(t) - T_S(p(t))$ appear now as dependent on time as a parameter.

Note that during a loss of sealing in vessels, channels, etc., the Jacob number is generally Ja $\gg 1$, which corresponds to both the thin thermal boundary layer around a bubble and only a slight impact of neighboring bubbles on this layer. Therefore, there is no need to give any consideration to the effect of bubble interaction upon the interphase heat and mass exchange. The appropriate relation that takes into account the bubble interaction effects at Ja $\ll 1$ ($\mathbf{Nu}_1 \approx 2$) was obtained by B. I. Nigmatulin et al., (1979).

Thus, the following model for water flashing is assumed. As soon as the liquid pressure drops below the saturation pressure ($p < p_S$), its flashing begins immediately on a fixed number n_0 of nuclei particles of radius a_0 in a liquid. The further growth of bubbles proceeds in accordance with the similarity solution of a problem concerning the thermal growth of a single bubble in an infinite volume of liquid.

The initial radius a_0 of bubbles and their number n_0 per unit volume of liquid are determined by the reached level of superheating ($T - T_S$), and by the quality of the original liquid; they must be prescribed *a priori*.

The heat influx $q_{\Sigma 2}$ to vapor (steam) is responsible for the change of its state along the saturation line and, therefore, is unequivocally determined by the pressure variation in conformity with (2.6.9).

The system of differential equations (6.11.2) may be modified to a quasi-divergent form

$$\frac{\partial \rho}{\partial t} + \frac{\partial (\rho v)}{\partial x} = -\rho v \frac{1}{S}\frac{\partial S}{\partial x}, \quad \frac{n}{n_0} = \frac{\rho}{\rho_0}$$

$$\frac{\partial (\rho v)}{\partial t} + \frac{\partial}{\partial x}(p + \rho v^2) = -\rho v^2 \frac{1}{S}\frac{\partial S}{\partial x}$$

$$\frac{\partial p}{\partial t} + \frac{\partial (pv)}{\partial x} = (p - \rho C_f^2)\frac{\partial v}{\partial x} + \psi_p \tag{6.11.14}$$

$$\frac{\partial T_1}{\partial t} + \frac{\partial T_1 v}{\partial x} = \left[T_1 + \rho C_f^2 \frac{(\partial i_1/\partial p)_T - V_1^\circ}{(\partial i_1/\partial T_1)_p}\right]\frac{\partial v}{\partial x} - \psi_T$$

where the values ψ_p, ψ_T, and C_f appearing on the right-hand side of equations are determined by the following formulas

$$\psi_p = \rho C_f^2\left\{\left[\frac{V_{2S}^\circ - V_1^\circ}{l} - \frac{(\partial V_1^\circ/\partial T_1)_p}{(\partial i_1/\partial T_1)_p}(1-\delta)\right] nq_{1\Sigma} - \frac{v}{S}\frac{\partial S}{\partial x}\right\}$$

$$\psi_T = \frac{V_1^\circ (1-\delta)\, nq_{1\Sigma}}{\alpha_1 (\partial i_1/\partial T_1)_p} + \frac{(\partial i_1/\partial p)_T - V_1^\circ}{(\partial i_1/\partial T_1)_p}\psi_p$$

$$\frac{1}{C_f^2} = \rho\left[\alpha_2\left(\rho_{2S}^\circ \frac{di_{2S}}{dp} - 1\right)\frac{V_{2S}^\circ - V_{1S}^\circ}{i_{2S} - i_{1S}} + \frac{\alpha_2 V_{2S}^\circ}{C_{2S}^2} + \frac{\alpha_1 V_1^\circ}{C_1^2}\right] \qquad (6.11.15)$$

$$C_{2S}^2 = -\left(V_{2S}^\circ\right)^2\frac{dp}{dV_{2S}^\circ}, \quad \delta = \frac{i_1 - i_{1S}}{l}$$

Note that C_{2S} is not the speed of sound in vapor. It can be shown that C_f is an inherent velocity of the obtained system of equations, and a frozen speed of sound of a vapor-liquid medium which is a relaxing, or nonequilibrium, medium. Nonequilibrium here is accounted for only by the nonequilibrium of heat and mass exchange between vapor (steam) and liquid. The frozen speed of sound is a speed of propagation of small disturbances of an infinite frequency ($\omega \to \infty$), when the relaxation processes (in this case, the interphase heat and mass exchange) do not have enough time to occur, and each phase behaves isentropically

$$\frac{1}{\rho C_f^2} = -\frac{1}{V}\left(\frac{\partial V}{\partial p}\right)_{s_1, s_2} = \frac{\alpha_1}{V_1^\circ}\left(\frac{\partial V_1^\circ}{\partial p}\right)_{s_1} + \frac{\alpha_2}{V_2^\circ}\left(\frac{\partial V_2^\circ}{\partial p}\right)_{s_2} \qquad (6.11.16)$$

Let us examine the heat-influx equation, or equation for the liquid temperature (the fourth equation (6.11.14)), in more detail. The use of simplifications (6.11.11) leads to $(\partial i_1/\partial p)_T - V_1^0 = 0$, and the equation for T_1 assumes the form

$$\rho_1 c_1 \frac{dT_1}{dt} = -nq_{1\Sigma} \qquad (6.11.17)$$

Thus, the indicated simplifications are consistent with disregard of the variation of both enthalpy and the temperature of the liquid phase when the latter undergoes either compression or expansion. In the pressure range under consideration ($0.1 < p < 10$ MPa), this condition is satisfied with sufficient accuracy. At the same time, the use of the above-indicated simplifications (6.11.11) in the pressure-equation (the third equation (6.11.14)) where the medium compressibility appears (which is defined by the frozen isentropic speed of sound C_f) may lead to a significant error. The latter may be explained by the fact that the isothermal speed of sound C_f, obtained instead of C_{1T}, is significantly different from C_f in the pressure range under consideration (see Fig. 6.11.1).

We shall now discuss the behavior of the system of Eqs. (6.11.14) in the absence, or disappearance, of the vapor phase ($\alpha_2 \to 0$). In this situation, $\rho \to \rho_1^0$, $C_f \to C_1$, $a \to 0$, $q_{1\Sigma} \to 0$, and the bubble number-concentration n, appearing in the equation together with $q_{1\Sigma}$, cease to affect the parameter distribution, and the system of equations (6.11.14) transforms into a system describing the motion of a compressible two-parameter medium. Thus, the system of equations (6.11.14) enables us to predict all stages of the flashing liquid outflow using a single approach, without it being necessary to isolate domains of a single-phase or two-phase flows.

If equations of state of thermally perfect gas ($p = \rho_1^0 R_1 T_1$, $i_1 = c_1 T_1$, R, c_1 = constant) are employed in Eqs. (6.11.14), they reduce to regular equations of gas dynamics; this may be used in experimental research and analyses.

Initial and boundary conditions for loss-of-sealing problems. For a problem whose physical statement is formulated at the beginning of this section, the initial conditions for a homogeneous liquid are

$$t = 0,\ 0 \leqslant x \leqslant L:\ v(0, x) = 0, \quad p(0, x) = p_0$$
$$T_1(0, x) = T_0, \quad n(0, x) = n_0, \quad \rho(0, x) = \overset{\circ}{\rho}_1(p_0, T_0) \tag{6.11.18}$$
$$(p_S(T_0) \leqslant p_0)$$

The boundary conditions are based on a requirement that on the closed end of a tube ($x = 0$), nonleaking is ensured, and on the open end of a tube ($x = L$), pressure equals the ambient pressure

$$x = 0:\ v(t, 0) = 0$$
$$x = L:\ p(t, L) = p_\infty \qquad (p_\infty < p_0) \tag{6.11.19}$$

Thermodynamic equilibrium and polytropic approximations. First, we shall consider a limiting case when the liquid heat conductivity is so high that there is no delay in flashing, and the liquid does not experience superheating, i.e.

$$T_1 = T_2 = T_S(p) \quad \text{or} \quad i_1 = i_{1S}(p), \quad i_2 = i_{2S}(p) \tag{6.11.20}$$

This is of course consistent with an equilibrium model of a saturated two-phase mixture.

Adding the phase heat-influx equations (6.11.2), we obtain an equation for mixture heat-influx which, given Eqs. (6.11.3) and the fact that in equilibrium, $i_2 - i_1 = i_{2S} - i_{1S} = l$, is modified to a form reflecting the isentropic nature of the process for a material particle

$$\rho \frac{di}{dt} = \frac{dp}{dt} \quad \text{or} \quad \frac{ds}{dt} = 0 \tag{6.11.21}$$

We will now demonstrate that for each particle, the process of its expansion is barotropic, i.e., all parameters depend on pressure, in particular, $\rho = \rho(p)$; these relations are determined by functions $T_S(p)$, $i^0_{1S}(p)$, $i_{1S}(p)$ for which appropriate tables are available, or their approximations of types (6.11.5) and (6.11.6) are possible. Indeed

$$dV = r_1 dV^\circ_{1S} + r_2 dV_{2S} + \Delta V^\circ_S dr_2$$
$$(r_i = \rho_i/\rho \quad (i = 1, 2,\ r_1 + r_2 = 1),\quad \Delta V^\circ_S = V^\circ_{2S} - V^\circ_{1S}) \tag{6.11.22}$$
$$dp = \rho\, di = \rho\,(r_1 di_{1S} + r_2 di_{2S} + l dr_2)$$

whence

$$l\frac{dr_2}{dp} = r_1 V^\circ_{1S} + r_2 V_{2S} - r_1\frac{di_{1S}}{dp} - r_2\frac{di_{2S}}{dp} \tag{6.11.23}$$

Upon integrating, this equation allows $r_2(p)$ to be obtained, and, further, we obtain the equation of barotropy independent of both mixture structure (whether it is a bubbly, or drop mixture, etc.) and bubble (or drop) size

$$\rho(p) = \frac{1}{V(p)}\frac{1}{(1 - r_2)V^\circ_{1S} + r_2 V^\circ_{2S}} \tag{6.11.24}$$

Now, the equilibrium speed of sound C_e in the equilibrium vapor-liquid mixture under consideration is

$$C_e^2(p) = \frac{dp}{d\rho} = -\frac{V^2}{dV/dp} = \frac{V^2 l}{(A - V)\Delta V^\circ_S - Bl}$$
$$\left(A(p) = r_1\frac{di_{1S}}{dp} + r_2\frac{di_{2S}}{dp},\quad B(p) = r_1\frac{dV^\circ_{1S}}{dp} + r_2\frac{dV^\circ_{2S}}{dp}\right) \tag{6.11.25}$$

This expression determines speed of sound C_{1e} on the side of the two-phase domain for a saturated single-phase liquid ($\rho_2 = 0$)

$$C_{1e}^2(p) = V_{1S}^2 l \Big/ \left[\left(\frac{di_{1S}}{dp} - V_{1S}\right)\Delta V^\circ_S - \frac{dV_{1S}}{dp} l\right] \tag{6.11.26}$$

If liquid is subcooled in its initial state i.e., is in a single-phase state ($T_0 < T_S(p_0)$, or $p_0 > P_S(T_0)$), the speed of sound in it is a speed of sound C_1 in a pure liquid (see (6.11.15)); this speed of sound (C_1), because of the linear relationship between p and V_1, is independent of subcooling and, since the liquid temperature variation on the stage of a single-phase outflow are negligible, the speed C_1 may be regarded as a fixed quantity, and, also, may be prescribed based on the initial temperature T_0, i.e.

$$C_1 = C_{10} = C_1(T_0)$$

Thus, the medium under consideration in its equilibrium approximation is a perfect compressible liquid whose motion is governed by Euler equations involving a more complicated equation of state or compressibility. For a purely one-dimensional motion (S = constant), we have

$$\frac{\partial \rho}{\partial t} + \frac{\partial (\rho v)}{\partial x} = 0, \quad \frac{\partial (\rho v)}{\partial t} + \frac{\partial (\rho v^2 + p)}{\partial x} = 0$$

$$\frac{dp}{d\rho} = C^2 = \begin{cases} C_{10}^2, & p > p_S(T) \\ (C_e(p))^2, & p < p_S \end{cases} \tag{6.11.27}$$

It should be kept in mind that the relation $p(V)$ has a change of slope at the point of transition from a single-phase to a two-phase state (see line OSE in Fig. 6.11.2), and the speed of sound experiences a step change from C_1 to C_{1e}. Figure 6.11.3 shows, by way of example, the relations $C_e(p)$ and $p(V)$ (the latter is schematically shown by line SE in Fig. 6.11.2) along several isentropic lines in a two-phase equilibrium state for water, methane, and propane.

The presence of a change of slope in $p(V)$-diagram leads to a separation of the rarefaction wave in a subcooled liquid into two waves, and the process of outflow of subcooled liquid during an emergency loss of sealing proceeds as follows. To start with, the first, or "rapid," rarefaction wave arrives in the liquid with a speed $C_1 \approx 1000$ m/s (OS in Fig. 6.11.4). Pressure drops to the saturation pressure $p_{S0} = p_S(T_0)$, and liquid behind the wave begins to move with velocity

$$v_{S0} = \frac{p_0 - p_{S0}}{\overset{\circ}{\rho}_{10} C_1} \tag{6.11.28}$$

Then, the second, or "slow," wave comes ($S'C$ in Fig. 6.11.4); the speed of the front of this wave equals the equilibrium speed of sound $C_{1e} = C_{1e}(p_{S0})$ in liquid on the side of the two-phase region. The qualitative investigation of the motion of an equilibrium vapor-liquid mixture in this wave may be carried out with an assumption of a polytropic barotropy by approximating the relation $p(\rho)$ as

$$\frac{p}{p_{S0}} = \left(\frac{\rho}{\rho_{S0}}\right)^{\varkappa} \quad \left(p_{S0} = p_S(T_0), \quad \rho_{S0} = \overset{\circ}{\rho}_{1S}(T_0) = \frac{1}{\overset{\circ}{V}_{1S}(p_{S0})}\right) \tag{6.11.29}$$

with the polytropic index $0 < \varkappa < 1$. This relation is shown by straight lines on a diagram depicted in Fig. 6.11.3. For a vapor-liquid mixture, $\varkappa < 1$; this is accounted for by the fact that at the same degree of rarefaction p/p_{S0}, this type of mixture—because of the formation of vapor of a higher-density liquid—exhibits a much higher increase of specific volume than gas (see Fig. 6.11.3).

For the first stage of outflow of a homogeneous ($p = p_0$), quiescent ($v = 0$) liquid (see (6.11.18)) in its initial state ($t = 0$), up to a point in time

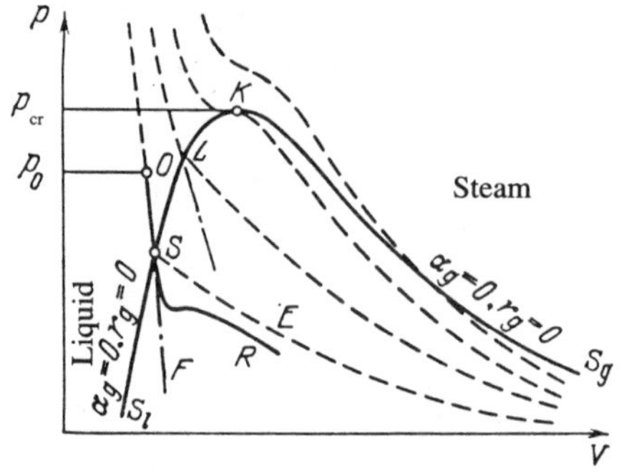

Figure 6.11.2 Process of expansion of flashing vapor-liquid mixture in coordinates pV; OS is a subcooled liquid, S is a saturated liquid, SE is an equilibrium vapor-liquid mixture, SF is a metastable (superheated) liquid, SR is an actual non-equilibrium process. Dashed lines relate to isentropic lines; KS_l and KS_g represent the limiting lines for, respectively, saturated liquid and saturated vapor; point K indicate a critical state.

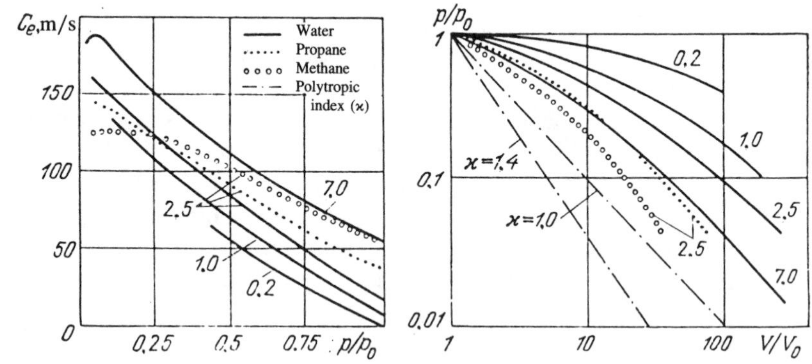

Figure 6.11.3 Equilibrium speed of sound C_e and specific volume V of a vapor-liquid mixture as functions of pressure p in the process of an equilibrium flashing due to pressure drop when the medium in its original state is a saturated liquid under pressure p_0 = 0.2; 1.0; 2.5 and 7.0 MPa (for water), and p_0 = 2.5 MPa (for methane and propane). The labels on curves indicate the original pressure p_0, MPa. Dot-dashed lines correspond to a polytropic law $p(V)$ with a polytropic index $\varkappa$ = 1.0 and 1.4. All represented lines are analytically predicted based on formulas (6.11.26) and (6.11.25) using approximations (6.11.5) and (1.3.78).

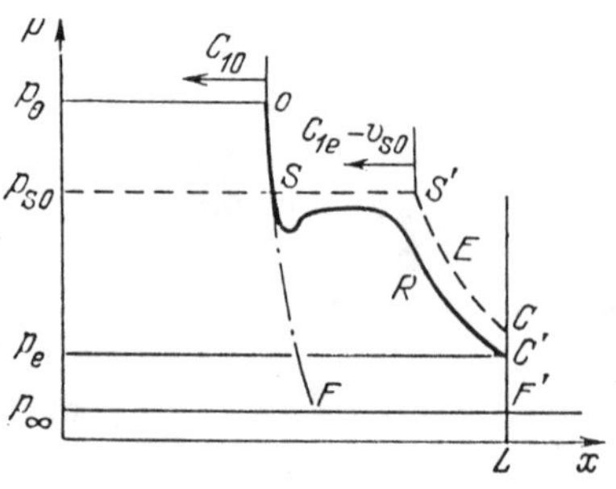

Figure 6.11.4 Schematic of a two-front rarefaction wave in a subcooled liquid flashing during pressure drop; $OSS'EC$ is the equilibrium boiling (OSE in Fig. 6.11.2); $OSFF'$ is the absence of flashing with a transition of liquid to a metastable state (OSF in Fig. 6.11.2); $OSRC'$ is the rarefaction wave in the presence of a non-equilibrium boiling (OSR in Fig. 6.11.2). The rarefaction-wave front velocity (points o) equals C_{10}, and the velocity of a wave-front of the equilibrium boiling relative to the channel walls (points S') equals $C_{1e} - v_{S0}$.

of reflection of a "rapid" rarefaction wave from a closed end, the solution of equations (6.11.27) and (6.11.29) is a similarity solution where independent variables x and t determine the solution only through the combination x/t; this solution is written in the form of a simple wave (see §6.6) for which the Reimann integral takes place

$$-v(p) = \int_{p_0}^{p} \frac{dp}{\rho C} = \frac{p_0 - p_{S0}}{\overset{\circ}{\rho}_{10} C_{10}} + \int_{p_{S0}}^{p} \frac{dp}{\rho C_e} \tag{6.11.30}$$

where in the polytropic approximation

$$\rho = \overset{\circ}{\rho}_{S0} \left(\frac{p}{p_{S0}}\right)^{1/\varkappa}, \quad C_e = C_{1e} \left(\frac{p}{p_{S0}}\right)^{\frac{\varkappa-1}{2\varkappa}}, \quad C_{1e} = \sqrt{\frac{\varkappa p_{S0}}{\rho_{S0}}} \tag{6.11.31}$$

If $\upsilon(p_\infty) < C_e(p_\infty)$, rarefaction to pressure p_∞, propagating relative to the tube with velocity $\upsilon(p_\infty) - C_e(p_\infty) < 0$, moves in direction $x < L$, i.e., into the tube, and the ambient pressure p_∞ is established at the exit section of a tube.

Otherwise, if $\upsilon(p_\infty) \geqslant C_e(p_\infty)$, rarefaction to pressure p_∞ inside the tube becomes impossible since the correspondent disturbance is displaced by the medium out of the way $x > L(\upsilon(p_\infty)) - C(p_R) > 0$, and a critical pressure p_c is established at the exit section of the tube; this pressure realizes $\upsilon(p_c) = C_e(p_c)$. The further rarefaction from p_e to p_∞ on the first stage of the process under discussion proceeds out of the tube. From the Reimann integral, we obtain the following expression for p_c

$$p_c = p_{S0} \left(\frac{2}{\varkappa + 1}\right)^{\frac{2\varkappa}{\varkappa-1}} \left[1 + \frac{\varkappa - 1}{2} \frac{p_0 - p_{S0}}{\overset{\circ}{\rho}_{10} C_{10} C_{1e}}\right]^{\frac{2\varkappa}{\varkappa-1}} \tag{6.11.32}$$

and a "slow" rarefaction wave becomes centered (see Fig. 6.11.4). Its front moves relative to the tube with velocity $C_{1e} - \upsilon_{S0}$.

If subcooling is fairly low, the second addend in the brackets is much lower than unity. Then, the effect of subcooling on the magnitude of critical pressure p_c may be taken into account by means of a simpler formula

$$p_c = p_{S0} \left(\frac{2}{\varkappa + 1}\right)^{\frac{2\varkappa}{\varkappa-1}} \left[1 + \varkappa \frac{p_0 - p_{S0}}{\overset{\circ}{\rho}_1 C_1 C_{1e}}\right] \tag{6.11.33}$$

The polytropic index $\varkappa$ depends on the initial temperature T_0, or on $p_{S0} = p_S(T_0)$. For water at $p_{S0} = 7.0$ MPa, $\varkappa \approx 5$, and at $p_{S0} = 2.5$ MPa, $\varkappa \approx 0.33$.

The pressure $p_c \geqslant p_\infty$ at the tube exit is retained constant up to the arrival of the rarefaction wave reflected from the closed end of the tube. To predict the outflow stages of the process following the rarefaction-wave reflection, numerical methods should be used; all the more, if a higher-degree accuracy approximation of the equations of state, which is based on (6.11.5), (6.11.6), and (6.11.24) is used; and, also, if the nonequilibrium effects due to flashing delay are taken into consideration.

The results of numerical investigation of the issue of a flashing liquid flow out of a tube of a finite length in an equilibrium approximation are reported by N. G. Rassokhin, V. S. Kuzevanov, and G. V. Tsiklauri (1974), A. I. Ivandaev and A. A. Gubaidulin (1977, 1978), A. I. Ivandaev (1978); these data are discussed below in regard to experimental and theoretical studies of the effects of a nonequilibrium interphase heat and mass exchange characteristic for a vapor-liquid medium.

On the numerical integration technique. The theoretical results discussed below were obtained by B. I. Nigmatulin and K. I. Soplenkov (1980) using the numerical integration of Eqs. (6.11.14) with boundary and initial conditions (6.11.18) and (6.11.19); the integration was performed by means of an explicit two-step difference model proposed by Lax-Vendroff (see R. Richtmyer and K. Morton (1967)). The scheme is characterized by second-order accuracy. The Courant condition concerning the integrating step with respect to time Δt is a necessary requirement for its stability (similar to difference schemes of equations of gas dynamics in Eulerian variables)

$$\Delta t < \Delta t_C = \frac{\Delta x}{|v| + C_f} \tag{6.11.34}$$

Since in the system of equations under consideration there are source terms ψ_p and ψ_T, associated with relaxation, or nonequilibrium processes of heat and mass exchange, the integrating step with respect to time Δt with the chosen explicit scheme must be much smaller than either kinetic or relaxation time t_ψ

$$\Delta t \ll t_\psi, \quad \text{or} \quad \Delta t \lesssim \frac{\Delta x}{L} t_\psi, \text{where} t_\psi = \min\left\{\frac{p_0}{\psi_p}, \frac{T_0}{\psi_T}\right\} \tag{6.11.35}$$

so that increments of required functions accounted for by the kinetic process were small on each step ($\psi_p \Delta t \ll p_0$, $\psi_T t \ll T_0$). This condition with rapid relaxation processes, i.e., when $t_\psi < \Delta t_C$, makes the explicit finite-difference method of calculations inefficient when nonequilibrium processes are given consideration. In such occasions, either implicit models or equations of equilibrium approximations should be used. In addition, the analysis involving rapid relaxation processes may be made more efficient (i.e., to use fairly large integration steps with respect to time Δt), if the solution is to be represented in the form of piecewise exponential relations

$$p - p^{(m)} = (\partial p/\partial t)^{(m)} t_\psi \exp(t/t_\psi), \qquad t \in [t^{(m)};\ t^{(m)} + \Delta t] \tag{6.11.36}$$

rather than in the form of piecewise linear relations, as is generally done

$$p - p^{(m)} = (\partial p/\partial t)^{(m)} (t - t^{(m)}), \qquad t \in [t^{(m)};\ t^{(m)} + \Delta t] \tag{6.11.37}$$

In both expressions, the superscript m denotes the m-th time step.

In the above-mentioned work by B. I. Nigmatulin and K. I. Soplenkov (1980) (their results are discussed below), there were indications concerning numerical oscillations, especially near the exit section of the tube. To suppress them, a procedure of step-by-step flattening (smoothing) which used a "differential analyzer" proposed by A. I. Ivandaev (1975) was employed.

Investigation of an outflow in the presence of thermodynamic nonequilibrium. Note that an adiabatic but thermodynamically nonequilibrium two-phase medium described by Eqs. (6.11.2)–(6.11.5) is not barotropic, and the process curve in pV-coordinates depends on the actual rate of the phase transition defined by the interphase heat exchange with the liquid phase; i.e., it depends on the rate (speed) of the process (see line SR in Fig. 6.11.2, located between line SE of the equilibrium process and line SF of the metastable process).

The flow of a liquid, flashing in a nonequilibrium regime, out of a tube of finite size with initial parameters consistent with "subcooled" or saturated state of water $p_S(T_{10}) \leqslant p_0$ may be conveniently studied by considering two specific time periods: $t \leqslant t_f$ and $t \gg t_f$, where $t_f = L/C_f$. During the first period, a rarefaction wave propagates in the channel (the elastic preindicator) through pure liquid with a speed of $C_1 \sim 1000$ m/s; a metastable condition is generated behind this wave, and the liquid flashing initiates. This flashing leads to the decay of the elastic preindicator to pressure p_S in conformity with (6.2.42). The second period is characterized by an outflow of the two-phase mixture with nonequilibrium, or quasi-equilibrium, heat and mass exchange over the entire zone of flow.

Initial stage of outflow. We shall now discuss an analytical solution concerned with attenuation of the first wave front in the form of a rarefaction shock propagating with a speed of sound C_1 in "pure" liquid with its amplitude approaching $p_0 - p_S(T_0)$. We assume that at the channel outlet ($z = L$), a rarefaction shock is produced with an amplitude $p_0 - p_\infty$. We shall determine the rarefaction-shock amplitude variation in the process of its propagation along a channel filled with liquid characterized by parameters p_0, and T_0, and containing n_0 of nucleation centers of radius a_0; the analysis is based on the premise that this shock moves along a characteristic through a resting liquid ($v_0 = 0$), and the following characteristic conditions must be satisfied for a system of equations (6.11.14)

$$\frac{dx}{dt} = -C_1, \qquad \frac{dp^{(1)}}{dx} = -\frac{\overset{0}{\rho}_1 C_1 n_0 j}{2\overset{0}{\rho}_{2S}(p^{(1)})}, \qquad Z = L: p^{(1)} = p_\infty$$

This equation describes the pressure variation in the elastic preindicator during its motion along the channel. Substituting the expression for liquid vaporization rate within a bubble, we obtain

$$j = 4\pi a_0^2 \lambda_1 \mathbf{Nu}_1 \frac{T_0 - T_S(p)}{2a_0 l(p)}$$

Let us now assume that $\mathbf{Nu}_1$, in accordance with considerations discussed before (6.11.14), is determined from (2.9.33) for $\mathbf{Ja} \gg 1$, i.e., $\mathbf{Nu}_1 = (12/\pi)\mathbf{Ja}$. Then, we obtain differential equation for $p^{(1)}$

$$\frac{dp^{(1)}}{dx} = -12a_0 n_0 \rho_1^0 C_1 c_1 \lambda_1 \frac{[T_0 - T_S(p^{(1)})]^2}{l^2(p^{(1)})\rho_{2S}^{02}(p^{(1)})}, \quad z = L: p^{(1)} = p_\infty$$

This equation permits the prediction of the evolution of elastic preindicator $p_0 - p^{(1)}$ with respect to coordinate x. It is seen that decay of elastic preindicator ($p^{(1)} \rightarrow p_S(T_0)$) strongly depends, in addition to thermophysical properties of liquid and vapor density, on $a_0 n_0$, i.e., on both the amount and concentration of admixtures.

Figure 6.11.5 represents an envelope of the elastic preindicator amplitudes, which corresponds to relationship $p^{(1)}(x)$. The rate with which $p^{(1)}$ approaches $p_S(T_0)$ is proportional to a square of superheatings occurring in the process of its motion. The specific features of rarefaction wave propagation in a tube ($dS/dz = 0$) of length 4 meters containing a substantially subcooled (near the saturation parameters) water were investigated: $p_0 = 6.9$ MPa, $T_{10} = 515$ K, $V_{10}^0 = V_1(p_0, T_{10})$. It has been assumed that the pressure at the tube exit becomes instantaneously equal to the ambient pressure.

Figure 6.11.5 represents the pressure diagrams in the rarefaction wave. At a high initial number of bubbles (dotted lines correspond to $n_0 = 10^{12}$ m^{-3}), the solution resembles that for an equilibrium model (dashed lines): there are a "rapid" rarefaction wave, pressure behind which only slightly differs from the saturation pressure, and a "slow" wave whose displacement

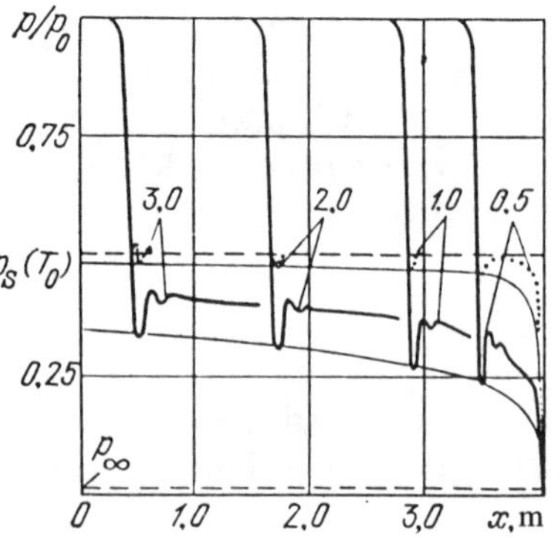

Figure 6.11.5 Evolution of distributions of both pressure p and volumetric vapor-concentration α_g on the initial stage ($t < L/C_{10}$) immediately after an instantaneous loss of sealing ($p_\infty = 0.1$ MPa) of a channel of $L = 4$ meters at $t = 0$ ($x = 0$ is the closed end, $x = 4$ m is the open end of the channel) containing underheated water under high pressure: $p_0 = 6.9$ MPa, $T_0 = 515$ K ($p = 3.5$ MPa (dashed line), $C_{10} = 1200$ m/s); $n_0 = 0.5 \times 10^9$ m^{-3}, $a_0 = 10$ microns (heavy solid lines); $n_0 = 10^{12}$ m^{-3}, $a_0 = 1$ micron (dotted lines). The line enveloping the "kink" bases (maximum p after minimum p), as distinct from the line enveloping the "kink" peaks ($p_{\min}$ is the thin solid line) is independent of the nuclei radii, and is determined only by their number n_0 per unit volume. Labels on curves indicate time in microseconds.

for the time period t_f, compared to the tube length L, is negligible. The qualitative distinction from equilibrium solution for this stage of the process consists of the fact that to the point in time $t = t_f$, in accordance with the nonequilibrium model, the liquid flashing occurs along the entire tube, whereas in the equilibrium model, it takes place only at a distance $C_e t_f$ from the exit section, where C_e is the equilibrium speed of sound in the mixture.

As distinct from the most "rapid" wave, the flow behind it depends to a significant extent on the initial concentration n_0 of the nuclei bubbles: the lower n_0, the higher the liquid overexpansion, or overheating (cf. the dashed, dotted, and solid lines in Fig. 6.11.5). In other words, the lower n_0, the lower the pressure compared to saturation pressure (in this case, $p_S/p_0 \approx 0.5$) following the propagation of the rapid wave, or the greater the departure from the thermodynamic equilibrium.

Reflection of a rarefaction wave. The process of reflection of a "rapid" rarefaction wave in a flashing liquid from the tube closed end (wall), which depends on the possibility of this liquid existing in a metastable state ($p < p_S(T)$), and, also, on the intensity of flashing in this state, is depicted in Fig. 6.11.6 in the form of diagrams of both pressure and volumetric vapor-concentration near the closed end of the tube ($x = 0$) at times $t = 3.5$; 4.0, and 5.0 microseconds. When a "rapid" rarefaction wave reaches the tube closed end, pressure in this zone rapidly drops and becomes significantly lower than the saturation pressure. In this, the abnormally strong (compared to a condition before the wave reflection) nonequilibrium state of liquid when superheating reaches about 45 K, and the pressure is 2 MPa lower than the saturation pressure, an intensive flashing takes place. The rapid growth of α_2 (and, consequently, of the mixture specific volume) causes reduction of

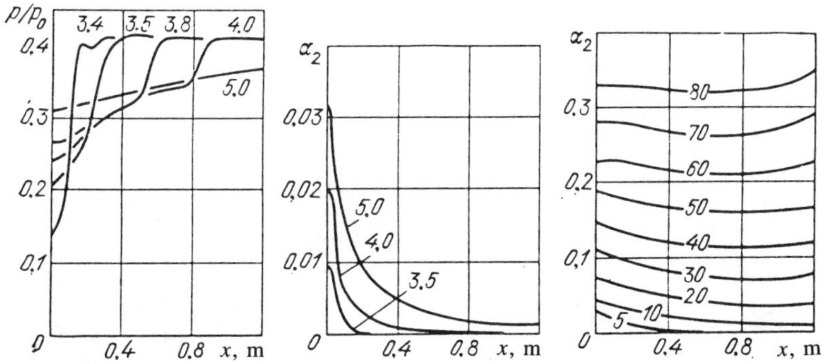

Figure 6.11.6 Distributions of pressure p and volumetric concentration α_2 during reflection ($t \geqslant L/C_{10}$) from the closed end, or a wall ($x = 0$), of a rarefaction wave resulting from loss of sealing of a channel (the conditions are the same as for Fig. 6.11.5, $n_0 = 0.5 \times 10^9$ m^{-3}). Labels on curves indicate time t in microseconds.

the reflected wave intensity and its rapid attenuation along the length of tube. It can be seen that at a distance of about 0.5 m from the tube wall, the reflected wave is indiscernible.

Quasi-static stage of the outflow process. With the loss of sealing of a channel carrying a high-temperature heat-transfer agent when the flashing wave intensity is sufficiently high ($p_{S0}/p_\infty \geqslant 3$), the flow, in accordance with a thermodynamically equilibrium model, during the time period $t_f < t < t_e$ ($t_e = L/C_{1e}$), is blocked, i.e., at the tube exit $p = p_c$ (see (6.11.32)), and $\upsilon = C_{1e}$, and a "slow" rarefaction wave propagates along the tube, with its front moving with a velocity $C_{1e} - \upsilon_{S0}$. Note that, at time $t > t_e$, the tube discharging (evacuation) is of a quasi-static, or homobaric, nature when pressure is uniform along the tube (the pressure gradient is noticeable in a region 1 m from the exit), and depends only on time; the effect of wave processes becomes insignificant as early as with the just mentioned decay of the reflected wave. On this homobaric stage of outflow, a time-scale of variation of pressure and other parameters is two orders larger. The pressure variations at four points of the tube at $t \gg t_e$, for the same regime as shown in Fig. 6.11.5, are represented in Fig. 6.11.7*a*. It is seen that pressure at fixed points, including the exit section of the tube, slightly decreasing towards the open end of the tube, is retained on an almost constant level during 50–100 microseconds (the pressure at the exit section is much higher than p_∞); then, it monotonically decreases up to a complete evacuation of the tube. Simultaneously, Fig. 6.11.7*b* depicts the behavior of the remaining parameters at the exit section of the tube. It can be seen that the mixture velocity continually increases because of the vapor-concentration growth, following the increase of C_f, but it never reaches its magnitude. With an almost complete evacuation of the tube, the outflow velocity υ begins to decrease sharply.

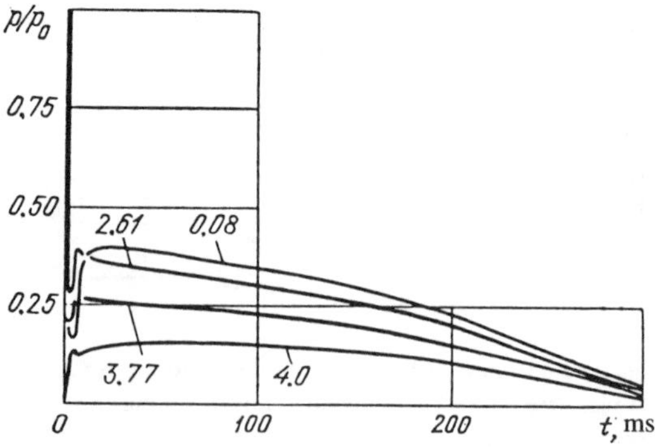

Figure 6.11.7 Variation of pressure with time at several points with coordinates $x = 0.08$; 2.61; 3.77; and 4.0 meters (indicated by labels on curves) for a sealing-failure regime represented in Fig. 6.11.6.

Effects of initial parameters of the flashing nuclei. In the used model of a flow of liquid flashing in a nonequilibrium regime, both the initial number of bubbles n_0 and their original nucleus radius a_0 are said to be insufficiently known empirical parameters. A study of the effects of these parameters on the mixture outflow revealed the following. The initial diameter of a bubble does not affect the mode of flow as early as at time $t > 10^{-4}$–10^{-5} seconds. This generally follows from the adopted model of liquid flashing. The law of the bubble growth is written in the form $a \approx A\sqrt{t + t_*}$, where $A \approx$ constant, t_* is the time needed for the growth of original radius a_0. For bubbles with $a_0 \approx 10^{-2}$–10^{-3} mm, time $t_* \approx 10^{-6}$–10^{-5} s, and as late as at $t \approx 10^{-5}$–10^{-4} s, the effect of t_* (and, consequently, the effect of a_0) on bubble growth becomes insignificant. In the computations presented here, it was assumed that $2a_0 = 1$ micron, which is consistent with initial vapor concentrations $\alpha_{20} \approx 10^{-10}$–$10^{-5}$; the latter is known as having no impact on the liquid properties in its original state. Recall that, in the adopted model, vapor formation begins at $p \leqslant p_S\,(T_0)$.

The initial number n_0 affects the mixture flow to a considerable extent: reduction of n_0 leads to enhancement of the flow nonequilibrium (see Fig. 6.11.5, where results are represented for $n_0 = 10^{12}$ and 0.5×10^9 m^{-3}).

Comparison of predicted analytical results with experimental data. The most comprehensive experimental data concerning the process of nonstationary outflow are represented in works by A. Edwards and T. O'Brien (1970); N. G. Rassokhin et al., (1977). Experiments reported in the first of these works were conducted in a tube 4.1 m long and with a 7.3 cm diameter filled with water with a subsequent increase of both pressure and temperature. The glass disk (diaphragm) which sealed one end of the tube was broken by a special striker. The breaking time was as low as 1 microsecond. The remnants of the broken disk were retained in the fixing block, thereby reducing the exit area by 10–15%. The water temperature T (in K) could be maintained approximately fixed ($\pm 3\%$) along the entire tube by using sectional heating devices covering 70% of the outer surface of the tube.

Pressure in the tube was measured at seven points by quick-response pressure sensors. A device was installed in one of sections for measuring the average volumetric vapor concentration by the method of gamma-radioscopy. The accuracy of this device turned out to be fairly low since the measuring results depend to a significant extent on the flow structure. The actual flow may have a structure significantly different from the "calibration" structure, which accounts for the errors. Nevertheless, such measurements represent clear qualitative information about inherent times of vapor formation.

The axial forces acting upon the tube fixtures have also been measured. The force-sensor pick-up time was about 10^{-4} s, which enabled the experimenters to identify the oscillation and wave processes at the earlier stage of outflow.

We will now describe the general scheme of the transient process of sealing-failure in a tube. Breaking the glass disk results in the occurrence of two rarefaction (unloading) waves in a tube. The first wave propagates within the tube walls (an elastic preindicator within tube walls) with a velocity of about 400 m/s, which causes a reduction of longitudinal stresses, a slight radial expansion of the tube, and a hardly noticeable pressure drop of about 1%. The second wave (elastic preindicator in liquid) moves through liquid with a velocity of $C_1 \approx 1000$ m/s causing its flashing. The reflected wave, as predicted (see Fig. 6.11.6), can only be seen on an oscillogram taken in the direct neighborhood of the closed end of the tube.

From the experimental oscillograms, as well as from the above-mentioned analysis, it follows that at times $t \gg t_e$, the outflow is almost of a homobaric nature up to a complete evacuation of the tube. The pressure begins to drop rapidly when the volumetric vapor concentration in the tube approaches unity.

The qualitative analysis of experimental studies leads to an important conclusion: the degree of nonequilibrium ($T_1 - T_S(p)$, or $p - p_S(T_0)$), which is realized in a flashing flow behind the rapid rarefaction wave (elastic preindicator), grows as fast as the absolute values of the initial temperature T_0 of liquid increase. Variation of the initial degree of subcooling $T_0 - T_S(p_0)$, or $p_0 - p_S(T_0)$, of liquid due to pressure variation p_0 at a fixed temperature T_0 does not, in this case, significantly affect the nonequilibrium. Figure 6.11.8 schematically shows the oscillograms of two rarefaction waves obtained (N. G. Rassokhin et al., 1977) at a point near the exit diaphragm. It is evident that a decrease of the initial temperature T_0, although contributing to an increase of the amplitude of the "rapid" unloading wave, nevertheless results in the reduction of the maximum nonequilibrium of flow. An analogous result has also been obtained in an experimental study (A. Edwards and T. O'Brien, 1970), where at $T_0 = 559$ K, $p_0 = 10.5$ MPa, the liquid over-

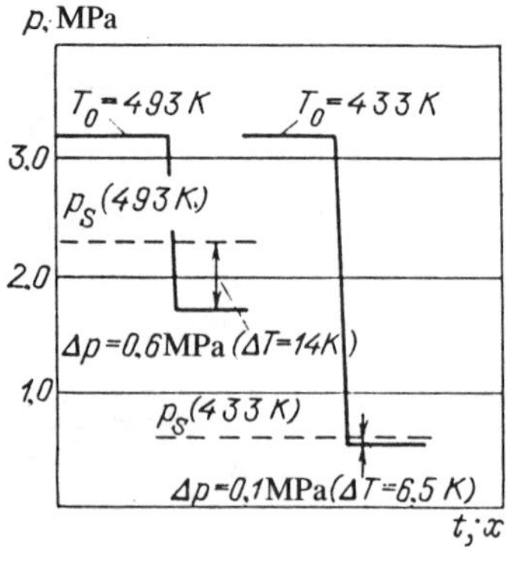

Figure 6.11.8 Schematic oscillograms for "rapid" rarefaction waves in subcooled water obtained in experiments by N. G. Rassokhin et al., (1977) under a fixed initial pressure ($p_0 = 3.2$ MPa), but at various initial temperatures ($T_0 = 493$ K and 433 K).

heating behind the rarefaction wave in the middle part of the tube was $\Delta T = T_0 - T_S(p) \approx 20$ K, and at $T_0 = 515$ K, $p_0 = 7.0$ MPa, $\Delta T \approx 13$ K.

For better comprehension of this effect, we shall note that the greater the volume of vapor produced behind the "rapid" rarefaction wave, the closer the pressure p_f behind it approaches (from below) the saturation pressure $p_S(T_0)$, the lower the water superheating $T_0 - T_S(p_f)$. If the number density n of bubbles be suggested as approximately the same for all regimes of flow (and is defined by the degree of water purification), the volumetric vapor concentration α_2 is proportional to a^3. For liquid flashing during the outflow, large Jacob numbers **Ja** are characteristic (see (1.6.20)), and, therefore, at a fixed superheating ΔT, we may, analogous to (1.6.18), write

$$a^3 = 8\zeta_T^3 \Delta T^3 t^{3/2}, \qquad \zeta_T = \frac{\sqrt{\lambda_l \overset{\circ}{\rho}_l c_l}}{\overset{\circ}{\rho}_g l} \tag{6.11.38}$$

$$\Delta T = T_0 - T_S(p_f), \quad \overset{\circ}{\rho}_g = \overset{\circ}{\rho}_g(T_0, p_S(T_0)) = \overset{\circ}{\rho}_{gS}(T_0), \quad l = l(T_0)$$

Figure 6.11.9 illustrates a relationship for the parameter $\zeta(T)$, which determines the rate of vapor formation on bubbles in superheated water. It can be seen that, at $T_0 \approx 620$ K $\approx 0.95\, T_{cr}$ (where $T_{cr} \approx 647$ K is the critical temperature), the value $\zeta = 1.6 \times 10^{-5}$ m/K·s$^{1/2}$ is minimum. Therefore, the maximum nonequilibrium, with other conditions equal, is at temperatures near $0.95\, T_{cr}$, namely, at $T_0 \approx (0.85\text{–}0.97)\, T_{cr}$; and at water temperatures $T_0 < 0.95\, T_{cr} \approx 620$ K, the realized nonequilibrium is greater, the higher the water temperature T_0. In particular, for the example depicted in Fig. 6.11.8, at the same superheating ΔT, the volumetric vapor concentration α_2 grows proportional to ζ^2, and at initial temperature $T = 493$ K, α_2 grows 34 times faster than at temperature $T_0 = 433$ K. This difference in the rate of vapor formation behind the rarefaction wave results in a situation whereby at temperatures closer to $T_0 \approx 620$ K, the value of the volumetric vapor content at which the pressure diagram behind the wave sharply flattens is reached at higher superheatings.

Thus, to describe a flow out of channels with a size of an order of 1 m, with relatively low initial parameters $T_0 \leqslant 530$ K, $p \leqslant 5.0$ MPa, when

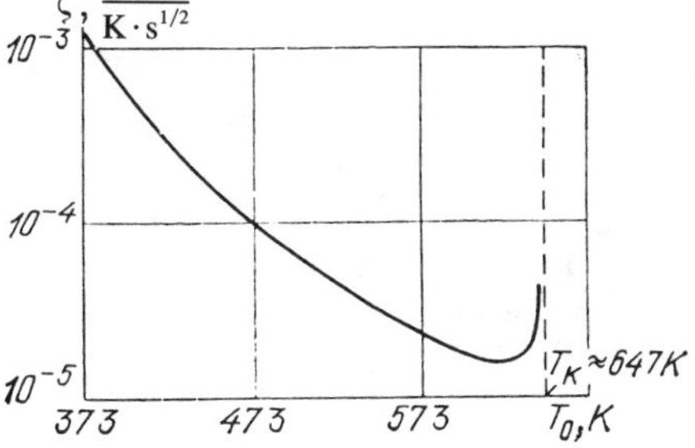

Figure 6.11.9 The relationship between the parameter $\zeta = (\lambda_l \rho_l^0 c_l)^{1/2}/(\rho_g^0 l)$, which determines the bubble growth, and the temperature T_0.

the degree of nonequilibrium in real flows is small, the equilibrium model may well be used. However, at temperatures corresponding to saturation pressure $p_S(T_0) \geqslant 7$ MPa, employment of an equilibrium model may cause a substantial error. In this case, a nonequilibrium model with the flashing delay must be used. Moreover, at temperatures near critical values ($T_0 \approx$ 0.9–0.97 T_{cr}), when vapor formation on the available nuclei bubbles proceeds very slowly and significant superheatings occur in waves, one may expect (see O. A. Isaev and P. A. Pavlov, 1980) that a certain contribution may also be made by the homogeneous nuclei-formation which (see §1.7) is facilitated under conditions resembling the critical state.

The discussed experiments have been conducted in conditions when about 10–15% of the exit section remained obstructed by remnants of the broken diaphragm. In order to take this situation into account in the framework of the quasi-one-dimensional model, the sections $S(x)$ of the exit segment of the tube $x - \Delta L < x < L$ ($\Delta L/L \approx 0.05$) were approximated by a linear function so that $S(x - \Delta L) = S_0$, and $S(L) = (0.85\text{–}0.9)S_0$.

The noninstantaneous rupture of the diaphragm leads to an additional spreading ("erosion") of the rarefaction wave. Therefore, for more accurate agreement between analytically predicted and experimental results, the boundary conditions at the tube exit were prescribed as

$$p(L) = p_\infty + (p_0 - p_\infty)\exp(-t/t_b) \tag{6.11.39}$$

where t_b is the inherent time of the diaphragm rupture and its impact on the process. Varying time t_b (being of order of 0.1 microseconds) in analytical calculations, the spreading of the "rapid" rarefaction wave may be obtained the same as in experiments.

Figure 6.11.10 represents both analytical (predicted) (A. I. Ivandaev and A. A. Gubaidulin, 1978; B. I. Nigmatulin and K. I. Soplenkov, 1980) and experimental (A. Edwards and T. O'Brien, 1970) oscillograms of variation of pressure and volumetric vapor content with time at a fixed point of the tube for the same conditions that are depicted in Fig. 6.11.5.

An equilibrium model yields both overrated pressure magnitudes at $t \leqslant$ 0.15 s and volumetric vapor concentration at $t > 0.1$ s. A thermally nonequilibrium model improves agreement with an experiment at $t \leqslant 0.1$ s, since it takes into consideration the possibility of metastable state.

At time $t > 0.15$ s, the above-described model with vaporization of vapor on a fixed number of bubbles exhibits a more rapid pressure drop than in experiments, implying an underrated intensity of vapor generation at $t >$ 0.15 s. Accordingly, the suggestion that intensification of vapor generation may be associated with an additional and significant increase of phase boundary where vaporization takes place due to bubble fragmentation arises. This fragmentation may be attributed to a "slip" of bubbles with a velocity w_{12} relative to liquid, which is calculated in accordance with a model described

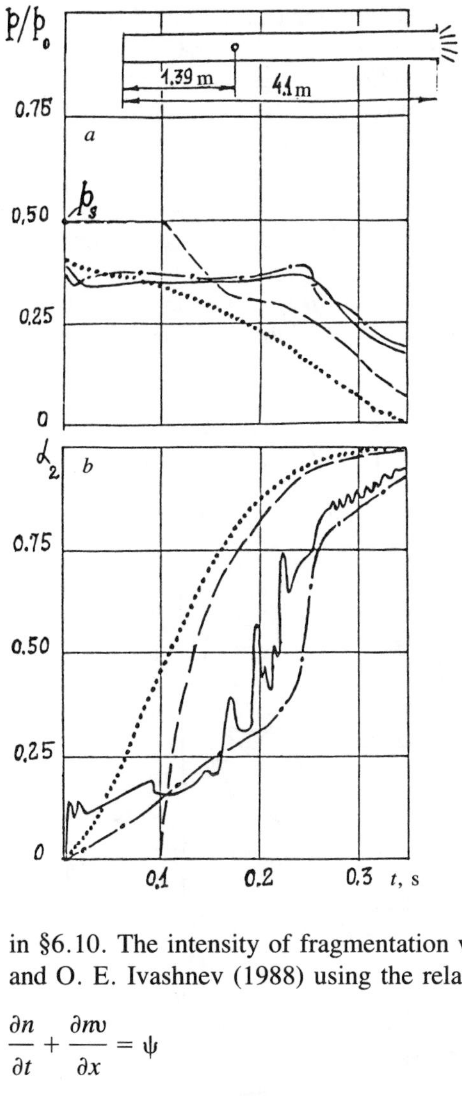

Figure 6.11.10 Experimental (solid lines) and predicted (dashed, dot-dashed lines and sequence of dots) pressure oscillograms (*a*) and oscillograms of volumetric vapor concentration (*b*) at a tube section (x = 1.39 m from the closed end of the tube) during loss of sealing in a tube of length L = 4.1 m filled with water (p_0 = 6.9 MPa, T_0 = 515 K); dashed lines relate to an equilibrium scheme, sequence of dots represents prediction with consideration given to thermal nonequilibrium ($n_0 = 0.5 \cdot 10^9$ m^{-3}), dot-dashed lines represent prediction with consideration given to bubble fragmentation.

in §6.10. The intensity of fragmentation was described by K. I. Soplenkov and O. E. Ivashnev (1988) using the relaxation approximation

$$\frac{\partial n}{\partial t} + \frac{\partial n v}{\partial x} = \psi$$

$$\psi = \begin{cases} 0, & \mathbf{We} < \mathbf{We}^* \\ (n^* - n)/t^*, & \mathbf{We} \geqslant \mathbf{We}^* \ (\mathbf{We}^* \approx 2\pi) \end{cases} \tag{6.11.40}$$

where n^* is the number of bubbles per unit volume immediately after breakup. It has been assumed in the analysis that a bubble breaks into two parts ($n^*/n = 2$). The inherent bubble-fragmentation time was estimated using (2.2.11)

$$t^* \approx (\rho_1^0 a^3/\Sigma)^{1/2} (\mathbf{We}^*/\mathbf{We})^{3/2}$$

Analysis showed that the onset of a weak relative motion of phases ($\upsilon_{12} \ll \upsilon$), well before the flow inversion, when $\alpha_2 \leqslant 0.8$ may lead to an intensive bubble fragmentation which significantly augments the phase boundary (interface), promotes vapor generation and delays pressure drop at $t >$ 0.1 s. This particular consideration leads to satisfactory agreement between analytical and experimental data on $p(t,x)$ and $\sigma_2(t,x)$ up to a point in time when outflow of the major part of liquid occurs.

Parameter distribution (number of bubbles, volumetric concentration of vapor and pressure) along the tube at various points in time in the process of a "slow" rarefaction wave propagation are represented in Fig. 6.11.11. A moving front of bubble fragmentation before the wave is a distinctive

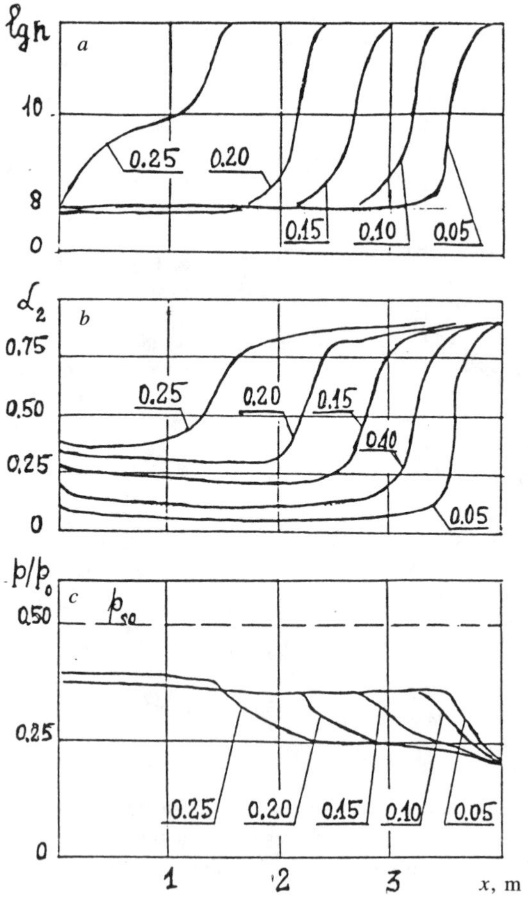

Figure 6.11.11 Distributions of the number of bubbles (a), volumetric vapor concentration (b), and pressure (c) in the process of a "slow" rarefaction wave propagation ($p_0 = 6.9$ MPa, $T_0 = 515$ K). Numerical labels on curves indicate time in seconds.

feature of the "flashing" wave. Bubble fragmentation causes an intensive boiling of liquid, drop of pressure, and flow acceleration, which, in turn, lead to an increase of the phase velocity difference, thereby generating conditions for bubble fragmentation upstream the flow. The mixture in a "slow" wave transforms into a near-equilibrium state.

The velocity of the bubble-fragmentation front motion and, consequently, the "flashing" wave propagation velocity, continuously grow. This is associated with the fact that the flashing wave moves through a medium with continuously increasing vapor concentration, and that the growth of vapor concentration in a mixture facilitates the bubble fragmentation process.

If the tube were of an unlimited length, the mixture at a sufficiently remote distance from the outlet would have transformed to a near-equilibrium state with no collapse of bubbles. The "flashing" wave, in this case, degenerates into a centered rarefaction wave propagating through an equilibrium mixture.

Let us compare the data of numerical computations with experimental results. The pressure oscillograms at five various sections of the channel, obtained by means of a model (giving consideration to bubble fragmentation), are represented in Fig. 6.11.12. Their comparison with experimental oscillograms recorded at the same sections of the channel (Fig. 6.11.11*a*) and with oscillograms predicted using a model with no consideration given

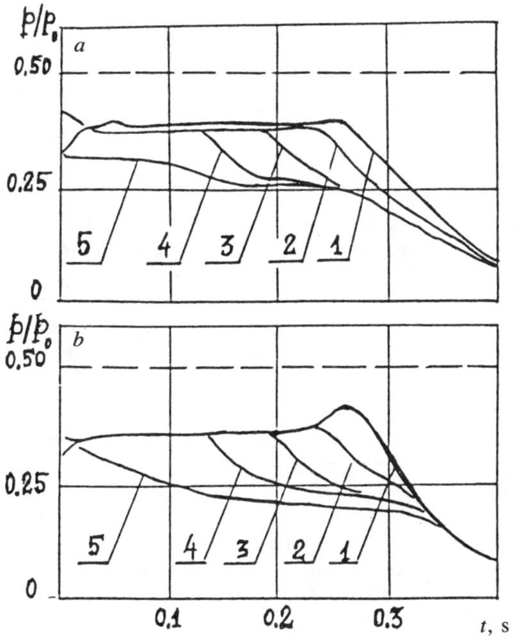

Figure 6.11.12 Pressure oscillograms at various sections of the tube: *1*) x = 0.08 m, *2*) x = 1.39 m, *3*) x = 2.02 m, *4*) x = 2.94 m, *5*) x = 3.8 m. (*a*) experiment (A. Edwards and T. O'Brien (1970)); (*b*) prediction based on the bubble-fragmentation model.

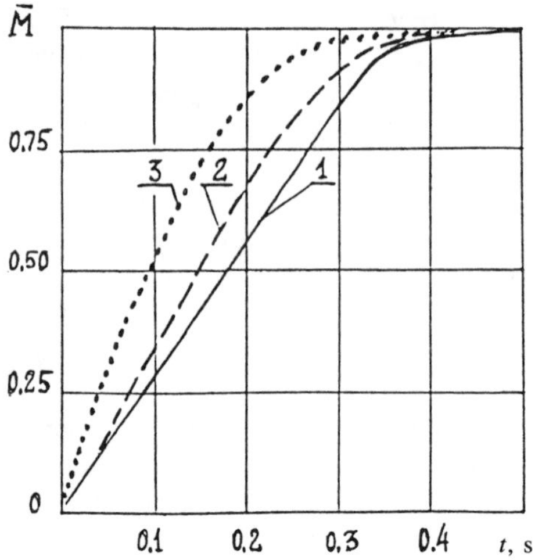

Figure 6.11.13 The relative mass of liquid discharged from a tube versus time for various models: *1*) nonequilibrium model with bubble fragmentation; *2*) equilibrium model *3*) nonequilibrium model with no consideration given to bubble fragmentation.

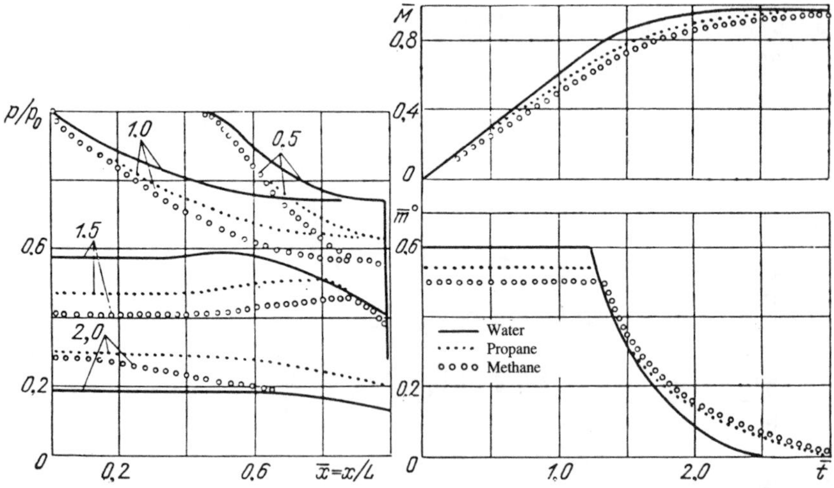

Figure 6.11.14 Evolution of distributions of pressure p, discharge $\bar{m}^0$, and mass $\bar{M}$ of a liquid discharged through an exit section $x = L$ of a tube of length L with a closed end ($x = 0$), during a nonstationary outflow of three liquids into the atmosphere ($p_\infty = 0.1$ MPa): methane ($T_0 = 173$ K, $C_{1e} = 53.3$ m/s, $\rho_0 = 304$ kg/m^3), propane ($T_0 = 341$ K, $C_{1e} = 39.5$ m/s, $\rho_0 = 412$ kg/m^3), water ($T_0 = 496$ K, $C_{1e} = 20.1$ m/s, $\rho_0 = 829$ kg/m^3); at the initial point in time, all liquids were in a single-phase (liquid) state of rest and saturation ($T_0 = T_S(p_0)$) under a pressure of $p_0 = 2.5$ MPa. Both outflow and flashing of all liquids occur following an instantaneous and complete opening of the exit section at time $t = 0$. The results of numerical calculations are shown in an equilibrium approximation ($T_1 = T_2 = T_S(p)$). Labels on curves indicate various dimensionless times $\bar{t} = t/t_*(t_* = L/C_{1e})$.

to bubble fragmentation (Fig. 6.11.7) shows that prediction based on the bubble-fragmentation model describes the seal-failure dynamic at late stages of outflow more accurately.

The mass of liquid escaping from the tube as a function of time is of great practical importance. The intensity of the vessel evacuation is determined by the specific discharge at the tube outlet ($x = L$)

$$m(t) = \rho(t,L)v(t,L) = \bar{m}\rho_0 C_{1e} \tag{6.11.41}$$

In this case, the relative mass of liquid discharged from the tube is

$$\bar{M} = \frac{1}{\rho_0 L}\int_0^t m(t)dt = \int_0^{\bar{t}} \bar{m}(\bar{t})d\bar{t}\left(\bar{t} = \frac{t}{t_e} = \frac{tC_{1e}}{L}\right) \tag{6.11.42}$$

Figure 6.11.13 represents analytical results obtained using various models of variation with time of the ratio between mass of liquid discharged from a vessel and the original mass of liquid in it at initial parameters, $p_0 = 6.9$ MPa, $T_0 = 515$ K. From the analysis, it follows that the nonequilibrium model which takes into account bubble fragmentation predicts the slowest evacuation of the system even when compared to an equilibrium model.

Figure 6.11.14 depicts the analytical results concerning the equilibrium mode of outflow of originally-liquid methane, propane, and water, which demonstrate, in addition to pressure-distribution evolution, the characteristic rate of the tube evacuation in the form of relations $\bar{m}(\bar{t})$ and $\bar{M}(\bar{t})$.

§6.12 DYNAMICS OF GAS BUBBLES UNDER VIBRATIONAL EFFECTS

Using methods presented in §4.6, it becomes possible to analyze motions of not only incompressible particles, but also of compressible gas bubbles of small void fraction in liquids at various predetermined periodic motions.

As distinct from gas suspensions with a fixed radius of particles, the bubble radius varies. This is described by the Rayleigh-Lamb equation. Let us set the bubble radius $a(t) = a_0(1 + \mu a')$, where μ is a small parameter equal by order to the disturbance intensity ε (see (4.6.7)). Then, similar to (4.6.8), the equations for motion of bubbles with a polytropic gas (polytropic index $\varkappa$) of constant mass in a predetermined nonhomogeneous field of flow of continuous liquid may be modified to

$$\ddot{\mathbf{r}} = f_{(r)}(\mathbf{r}, \dot{\mathbf{r}}, a', \dot{a}'), \quad \ddot{a}' + \omega_r^2 a' = f_{(a)}(\mathbf{r}, \dot{\mathbf{r}}, a', \dot{a}')$$

$$\omega_r^2 = \frac{3\varkappa}{\overset{\circ}{\rho}_1 a_0^2}\left(p_0 + \frac{3\varkappa - 1}{3\varkappa}\cdot\frac{2\Sigma}{a_0}\right), \qquad 1 \leqslant \varkappa \leqslant \gamma_g \tag{6.12.1}$$

where ω_r is the natural frequency of the bubble radial oscillations in the presence of capillary pressure.

Results of investigations (see R. F. Ganiev et al., 1978) are described below for various regimes of oscillatory motion of continuous liquid with sufficiently high frequency when

$$\tau_0^2 \sim \varepsilon \quad \left(\tau_0 = 9\mu_1/(\rho_0 a_0^2 \omega)\right), \qquad \bar{g} = g/(C_{10}\omega) \sim \varepsilon^2 \tag{6.12.2}$$

although, in this case, $C_{10}/a_0\omega \gg 1$.

Bubble motions in a standing wave. For a standing wave of type (4.6.17) and (4.6.19), these equations for a continuous (carrying) liquid, similar to (4.6.23), are modified upon averaging to a form which reveals the vibrating (vibromotive) force

$$\ddot{\xi} + \mu\tau_0\dot{\xi} = -\mu^2\bar{F}, \qquad \bar{F} = \frac{3}{2}\varepsilon^2\left(1 + \frac{C_{10}^2}{a_0^2(\omega_r^2 - \omega^2)}\right)\frac{\sin 2(\xi - \bar{L})}{\cos^2 \bar{L}} \tag{6.12.3}$$

Compared to a solid particle (see (4.6.24)), the vibromotive force for bubbles—because of $C_{10}/(a_0\omega) \gg 1$—may become much greater. The obtained averaged equation allows stationary solutions

$$\xi = \bar{L} - {}^1/_2 n\pi \tag{6.12.4}$$

Part of these solutions, for which the following conditions hold

$$(-1)^n(\omega_r - \omega) > 0 \tag{6.12.5}$$

are stable. In this case, the larger bubbles, having a natural frequency less than resonance frequency ($\omega_r > \omega$), migrate into the velocity nodes of liquid, and the smaller bubbles, having a natural frequency larger than resonance frequency ($\omega_r < \omega$), migrate into antinode that is consistent with experimental data (see L. Bergman, 1954). In a case of resonance ($\omega = \omega_r$), there are no stable stationary solutions.

Bubble motion in a moving wave. The averaged equation of the bubble motion in a moving wave of type (4.6.30) and (4.6.31) is written in the form

$$\ddot{\xi} + \mu\tau_0\dot{\xi} = \mu^3\bar{F}$$
$$\bar{F} = \varepsilon^2\frac{9\mu_1}{\rho_0\omega a_0^2}\left[\frac{11}{2} + \frac{C_{10}^2}{a_0^2(\omega_r^2 - \omega^2)} + \frac{4}{3}\frac{C_{10}^2\omega^2}{(\omega_r^2 - \omega^2)^2 a_0^2}\right] \tag{6.12.6}$$

Similar to (4.6.34), this equation reveals a stationary bubble-drift velocity if $\omega \neq \omega_r$. The obtained vibromotive force F is directed from the source of oscillations and is associated with the liquid viscosity. The formula for the vibromotive force accounted for by the radial pressure on the bubble in the perfect-liquid case is analogous to (4.6.35), and is written in the form (see V. A. Krasil'nikov and V. V. Krylov, 1984)

$$\frac{F_{(rp)}}{F_0} = \overline{F}_{(rp)} = \varepsilon^2 \frac{3\overline{\omega}^3}{\overline{\omega}^6 + (\overline{\omega}^2 - \omega_r^2)^2} \qquad \left(\overline{\omega} = \frac{\omega a}{C_{10}}, \quad \overline{\omega}_r = \frac{\omega_r a}{C_{10}}\right) \tag{6.12.7}$$

Motion of gas bubbles in a vibrating vessel. We will now discuss the motion of bubbles in a liquid placed in a rigid vertical cylindrical vessel with a hard bottom, when the vessel is subjected to vertical vibrations with angular frequency ω and the displacement amplitude Δ in the field of gravitation g. The liquid has a free surface, and the column of liquid is L ($\Delta \ll L$, $\omega \ll \omega_r$). This process was investigated and described by S. Zwick (1959), S. S. Grigoryan et al. (1965), and R. F. Ganiev et al. (1976).

Because $\omega \ll \omega_r$, the distributions of pressure and velocity of the carrying liquid may be described by an equilibrium model of a bubbly medium characterized by both initial density ρ_0 and equilibrium speed of sound C_0; this model, at small disturbances of both density and pressure, yields a linear wave equation (§6.6) with a linear boundary condition ($\Delta \ll L$) on the free surface $z = L$

$$\frac{\partial^2 u}{\partial t^2} - C_0 \frac{\partial^2 u}{\partial z^2} = 0 \quad (u = (p, \rho, v)), \qquad \rho_0 \frac{\partial v}{\partial t} = -\frac{\partial p}{\partial z} + \rho_0 g$$

$$z = L:\ p = p_{00}; \qquad z = 0:\ v = \Delta \cdot \omega \cdot \cos \omega t \tag{6.12.8}$$

where the coordinate z is measured upward from the bottom, and p_{00} = constant is pressure on the free surface. The solution of this problem in the form of steady sine-shaped oscillations (standing wave, $v = \psi(z) \cos \omega t$) is of the form

$$p = p_{st}(x) + \rho_0 C_0 \Delta_{00} \omega \sin \omega t \cdot \sin(\omega x / C_0) \qquad (p_{st} = p_{00} + \rho_0 g x)$$

$$v = \Delta_{00}\omega \cdot \cos \omega t \cdot \cos(\omega x / C_0) \tag{6.12.9}$$

$$(\Delta_{00} = \Delta / \cos(\omega L / C_0), \quad x = L - z)$$

For an incompressible liquid ($C_0 = \infty$), this solution simplifies ($\sin(\omega x/C_0) \to \omega x/C_0$, $\cos(\omega x/C_0) \to 1$), and assumes the form

$$p = p_{00} + \rho_0 g x + \rho_0 \Delta \cdot \omega^2 \cdot x \cdot \sin \omega t, \qquad v = \Delta \cdot \omega \cos \omega t \tag{6.12.9a}$$

In the field of velocities and pressure (6.12.9) or (6.12.9a), we may, similar to §4.6, investigate the field of bubble velocities v_2 in accordance with equations for longitudinal and radial motions (see (1.3.49))

$$\frac{d_2 v_2}{dt} = -\frac{3}{\rho_1} \frac{\partial p_1}{\partial z} + g + F_\mu^* + F_a^*$$

$$p_{00} \left(\frac{a_0}{a}\right)^{3\varkappa} = p_1 + \frac{2\Sigma}{a} + \overset{\circ}{\rho}_1 \left(a \frac{d_2^2 a}{dt^2} - \frac{3}{2} \left(\frac{d_2 a}{dt}\right)^2 \right) + \frac{4\mu_1}{a} \frac{d_2 a}{dt} \tag{6.12.10}$$

$$\left(F_\mu^* = \frac{9\mu_1}{\rho_1 a^2} (v_1 - v_2), \quad F_a^* = \frac{3(v_1 - v_2)}{a} \frac{d_2 a}{dt} \right)$$

For the case when $a \sim 1$ mm, $\omega \sim 100\ \mathrm{s}^{-1}$, the radial inertia of liquid in the second equation (terms defined by d_2^2a/dt^2 and $(d_2a/dt)^2$) may be ignored.

The state of equilibrium may be found either from the stationary solution (see §4.6) or from a condition of zero-displacement of the bubble during the time period (H. Bleich, 1956; S. S. Griogoryan et al., 1965)

$$\int_0^{2\pi/\omega} v_2 dt = 0 \tag{6.12.11}$$

H. Bleich (1956) showed that in the approximation of a perfect ($\mu = 0$), incompressible ($C_0 \to \infty$) liquid (see (6.12.9a)), the unstable stationary level z_*, measured from the bottom of the vessel, is determined by the formula

$$\frac{z_*}{L} = 1 - \frac{g}{\omega^2\Delta}\frac{2\varkappa p_0}{\overset{\circ}{\rho}_1\omega^2 L\Delta}\left(1 + \frac{3\varkappa - 1}{3\varkappa}\cdot\frac{2\Sigma}{a_0 p_0}\right)\left(1 - \frac{\omega^2}{\omega_r^2}\right) \tag{6.12.12}$$

A three-dimensional problem for a perfect, incompressible liquid, together with an investigation of viscosity effects, has been discussed in R. F. Ganiev and V. F. Lapchinskii (1978), and Yu. L. Yakimov (1978) reported the effects of compressibility when determining the stationary level z_*.

The physical implications of this unstable stationary level z_*, if $z_* > 0$, consist of the fact that the bubbles located below this level ($0 < z < z_*$), in their averaged motion, descend onto the vessel bottom, and those above the level z_* ($z > z_*$) rise to the surface. The averaged motion of light bubbles downward, in spite of the uplifting action of the static buoyancy force, is accounted for by the nonlinear interaction of radial and longitudinal motions of liquid around bubbles, or, in other words, by nonlinear inertial and viscous forces characterized by values F_a and F_μ.

The condition of existence of a domain with a descending motion of bubbles is deduced from the requirement $z_* > 0$, which, at $a > 10^{-1}$ mm (when there are no effects of surface tension) and at $\omega^2 \ll \omega_r^2$, leads to

$$\frac{g}{\omega^2\Delta}\frac{2\varkappa p_0}{\overset{\circ}{\rho}_1\omega^2 L\Delta} < 1 \tag{6.12.13}$$

Changing the amplitude Δ, vibration frequency ω, and acceleration g due to gravity (the latter may be changed in conditions of space flight), the level z_* can be monitored. With growth of ω, Δ, and reduction of g, this level moves upward towards the free surface.

Formation of a gas-liquid system during vibration. Following the work by S. S. Grigoryan, Yu. L. Yakimov, and E. Z. Apshtein (1965), we shall describe the process of formation of a gas-liquid system, as well as its evolution under vertical oscillations of a cylindrical vessel with liquid[6] at a frequency 20–200 s^{-1}. At low frequencies, liquid is at rest relative to the tube. At higher frequencies (of about 50 cycles), the cavitation arises in layers nearest to the bottom. The cavitation nuclei in the bottommost zone (20–30 cm from the bottom) grow in the rarefaction phase and collapse in the phase of pressure rise. The bubble collapse leads to shocks. Splashes occur on the free surface of liquid, and the produced air bubbles swoop down. As was shown above, the phenomenon may be explained by the fact that under sufficiently intensive oscillations, the level of neutral stability coincides with the free surface ($z_* \approx L$), and bubbles that happen to be below the level of neutral stability ($z < z_*$) descend to the bottom of the vessel. Accumulation of bubbles on the bottom leads to a restructuring of the disperse mixture, and a foamy, or gas, cushion is formed. Under some conditions, several cushions may be produced; they may undergo a slow displacement.

The presence of a cushion generates an oscillatory gas-liquid system qualitatively different from the disperse mixture; the role of an elastic element in this system is played by a gas localized in the cushion of a variable volume and mass; the column of liquid above the cushion represents the inertial element. In this case, the gas cushion has two degrees of freedom: translatory displacement and pulsating motion accounted for by variation of its volume characterized by natural frequency Ω of the gas-cushion pulsations. This frequency may be determined from a simplified one-dimensional scheme of motion (S. S. Grigoryan et al., 1965), according to which the cushion is a single bubble with a cylindrical lateral surface coinciding with the surface and flat ends of the tube. Only the height y of the cushion varies under oscillations, and its cross section remains equal to the section of the tube. We shall assume that only that part of liquid which is located above the cushion with a fixed height H is in motion; both liquid under the cushion and the tube are at rest. The latter is accounted for by the fact that the amplitude of oscillations of the liquid column above the cushion is significantly higher than that of oscillations of the vibration table. Neglecting friction, and assuming a polytropic scheme for the change of volume of the cushion, we may write the equation for the liquid motion above the cushion as well as the gas-polytrope equation

$$\overset{\circ}{\rho}_1 H \frac{d^2y}{dt^2} = -\overset{\circ}{\rho}_1 gH + p - p_{00}, \quad \frac{p}{p_0} = \left(\frac{y_0}{y}\right)^{\varkappa}, \quad p_0 = p_{00} + \overset{\circ}{\rho}_1 gH \tag{6.12.14}$$

Here, y_0 and p_0 are, respectively, the height of the cushion and the pressure in it in a state of equilibrium, p_{00} is the constant pressure above the column

[6]In addition to this article, the process is investigated in works by S. Zwick (1959), E. Z. Apshtein et al. (1969), and R. F. Ganiev et al. (1976).

of liquid. Upon linearization, we obtain an expression for the natural oscillation frequency

$$\Omega = \sqrt{\frac{\varkappa p_0}{\overset{\circ}{\rho}_1 y_0 H}} \qquad (6.12.15)$$

Effect of vibrating mixing of gas and liquid. Based on the work by R. F. Ganiev et al., (1976), we shall now describe the resonance model of the bubble motion in liquid, and the enhancement of oscillations accounted for by both gas cushion and vibrating mixing of liquid in a vessel subjected to vertical vibrations.

In the process of its growth due to descending bubbles, the gas cushion undergoes forced pulsations in the prior-to-resonance regime ($\omega < \Omega$). As the cushion volume increases, the natural frequency of its pulsations Ω decreases and approaches the frequency of exterior periodic actions ω, thereby leading to the resonance effect. In this case, the amplitude Δ of liquid pulsations above the cushion sharply increases, and the level of neutral stability of a bubble rises towards the free surface. With a sufficiently high intensity of vibrations, the level z_* of neutral stability may be made as close to the free surface as desired. Because of the cavitational rupture of the latter, the gas bubbles enter liquid from the atmosphere over its own surface; virtually all these bubbles rush downward to the vessel bottom. Thus, under a resonance regime of the gas cushion oscillations, conditions are provided for entrainment of a great number of gas bubbles, which in a few seconds results in the saturation of liquid with bubbles, part of which joins the cushion. The volume of cushion increases, thereby reducing its natural frequency Ω even to a higher extent, and takes the cushion out of resonance ($\Omega < \omega$). The liquid oscillation decreases, and the equilibrium level of neutral stability descends, possibly down to the very cushion. In this situation, part of gas becomes detached from the cushion in the form of bubbles and moves towards the free surface, thereby returning the system into the resonance regime $\Omega \approx \omega$. Such periodic displacements of the level of neutral stability of bubbles in a vertical direction set out a mechanism of dynamic stabilization of the cushion volume and the amount of gas bubbles in liquid, or, in other words, set out a dynamic regime of automatic adjustment.

By virtue of vibrating (in addition to the averaged) bubble displacement, the nonstationary regime of the level of neutral stability due to variation of vibrating accelerations $\omega^2\Delta$ in liquid, the complicated spatial configuration of the gas cushion, and oscillations of the free surface, the bubble motion may resemble a random regime contributing to an intensive mixing of liquid, which may be utilized in technology (R. F. Ganiev et al., 1978; 1980). The mixing intensification may also be promoted by the occurrence of a variety of motions in different directions of very fine, or cavitational, bubbles ($a < 10^{-2}$ mm) affected by surface tension; the size of these bubbles affects the level of neutral stability of bubbles.

It should be remembered that, affecting the gas cushion (for instance, removing or feeding gas from outside), the process of vibrating mixing can be controlled. In particular, to perform a rapid mixing of several liquid media, a resonance volume of gas may be fed all at once from below and outside, thereby eliminating the process of the cushion formation by the bubble entrainment.

Thus, the resonance effect of vibrating mixing consists of the fact that, in the process of vibrating action on a two-phase gas-liquid bubbly mixture, the produced gas cushion, or a system of several cushions, constitutes an amplifier (resonator) of oscillations for the bubbly mixture, which provides vibrating accelerations needed for the cavitational disruption of liquid near the free surface, and the entrainment and retention of a great amount of gas bubbles involved in an intensive periodic motion which is responsible for an intensive mixing of liquid.

CHAPTER
SEVEN

HYDRODYNAMICS AND THERMOPHYSICS OF STATIONARY ONE-DIMENSIONAL GAS-LIQUID AND VAPOR-LIQUID FLOWS IN CHANNELS

In flows in tubes and other channels, interaction takes place between the flow and exterior body (the channel wall); namely, a force interaction occurs due to friction and pressure upon the wall, and, also, thermal interaction occurs due to heat exchange with the wall. The intensity of this process for two-phase gas-liquid flows depends on the realized structure of the flow, in particular, on the presence of either liquid or vapor film on the channel wall, on phase distribution over the cross section of the channel, and, also, on the intrinsic processes in the flow.

§7.1 SPECIFIC FEATURES OF GAS-LIQUID FLOWS. BASIC METHODS OF THEIR IDENTIFICATION

The amount of experimental data needed to establish a complete theory of gas-liquid turbulent flows in channels is much more extensive than the information needed for single-phase flows. In this regard, we shall indicate some specific features of experiments aimed at a study of stationary gas-liquids flows in long tubes. Such experiments are generally conducted on stands with either a horizontal or vertical experimental section of a tube with ID = 10–100 mm. There are both "cold" and "hot" stands. Usually, the experiments with air-water flows at normal temperature ($T \approx 300$ K) and under moderate pressures ($p \sim 0.1$ MPa) are conducted on "cold" installations. The steam-water flows under various pressures ($p = 0.1$–20 MPa) and at temperatures near saturation temperature, are most frequently investigated on "hot" stands.

The experimental passage consists of two sections: the flow-stabilizaton section of about 200 diameters in length, and a measurement section. These sections represent a single channel of a constant diameter. The flow in the stabilization section undergoes stabilization of all its major parameters (temperature, phase distribution over the cross section of the tube, spectrum of the drop or bubble sizes, film thickness on the tube wall, etc.) before it enters the measurement section. Because of stabilization at adiabatic conditions along the measurement section, the indicated characteristics of the structure of gas-liquid flow do not change. The liquid mass flow rate m_l and gas mass flow rate m_g, pressure p, temperature T, and some other flow characteristics (such as volumetric concentration of phases α_l and α_g), liquid flow rate within the wall film, and the film thickness, drop or bubble sizes, phase velocity distribution over the cross-section area, velocity-fluctuation spectrum, etc. are measured at several points along the measurement section of the installation. Principles of these measurements are described below.

The most essential characteristics of a stationary two-phase flow in a channel are both mass and volumetric flow rates of each phase in mass and volumetric flow rate of the mixture, respectively. The gas (vapor) and liquid parts of the mass flow rate of mixture are referred to as mass concentration of gas flow (vapor concentration) x_g, and mass concentration of liquid flow x_l, respectively

$$x_g = m_g/m, \quad x_l = m_l/m \quad (m = m_g + m_l, \quad x_g + x_l = 1) \tag{7.1.1}$$

In describing the one-dimensional (hydraulic) two-phase flow in a channel, in addition to the average values of volumetric concentrations of phases α_g and α_l (the quantity α_g is often denoted in literature by φ) over a cross section or a segment of the channel, and to related phase velocities v_g and v_l, the modified phase velocities W_g and W_l, gas concentration β of the flow rate (which equals the part of volumetric gas flow rate in the volumetric flow rate of mixture), and the specific mass flow rate of mixture (m^0) and that of phases (m_g^0, m_l^0) are also used

$$\begin{gathered} W_g = \alpha_g v_g, \quad W_l = \alpha_l v_l, \quad \beta = W_g/(W_g + W_l) \\ m^\circ = m/(^1/_4\pi D^2), \quad \overset{\circ}{m}_g = m_g/(^1/_4\pi D^2), \quad \overset{\circ}{m}_l = m_l/(^1/_4\pi D^2) \end{gathered} \tag{7.1.2}$$

In stationary adiabatic flows (in the absence of heat removal and heat influx over the experimental segment of an installation), the mass flow rates of phases along the experimental segment do not change; then, nor do the corresponding relative values x_g and x_l. In cold air-water, or analogous binary (two-component) flows, this is ensured by the absence of phase transitions between air and water, and in "hot" steam-water flows by the absence of heat removal or heat influx, and also by small pressure difference Δp (and, consequently, temperature difference ΔT) in the measurement section ($\Delta p \ll p$, $c_l \Delta T \ll l(p)$).

If the true densities of phases ρ_l^0 and ρ_g^0 are known (at small pressure and temperature differences they experience no change along the measurement section), the specific mass and volumetric flow rates are expressed in terms of each other

$$\overset{\circ}{m}_l = \overset{\circ}{\rho}_l W_l, \quad \overset{\circ}{m}_g = \overset{\circ}{\rho}_g W_g, \quad \beta = [1 + (\overset{\circ}{\rho}_g / \overset{\circ}{\rho}_l)(m_l/m_g)]^{-1} \tag{7.1.3}$$

The void fraction $\alpha_g \equiv \varphi$ differs from the flow-rate void fraction β because of relative motion (slipping) of phases ($\upsilon_g \neq \upsilon_l$). The above-indicated parameter φ is important, in particular, for evaluation of the neutron absorption by the two-phase heat-transfer agent in nuclear reactors, since the volumes occupied by liquid and steam (vapor), having substantially different densities ρ_l^0 and ρ_g^0, absorb neutrons in different ways. Furthermore, the value φ is essential for determining the weight of a vertical column of a two-phase liquid required for analysis of force interaction between the flow and the tube walls (see §7.3 below). Numerous data are available on the relationship $\varphi(\beta)$ for horizontal and vertical downward and upward flows at various parameters (D, p, m, ρ_g^0, ρ_l^0) of flow regimes; these data are obtained by various methods,[1] but most of all, by the cut-off technique (for air-water and other "cold" flows) and gamma-radioscopy.

For steam-water upward flows in vertical round tubes, the pressure effects together with the effects of variation of thermophysical properties of phases ρ_g^0, ρ_l^0, μ_g, μ_l, Σ, tube diameter D and the mixture flow rate m upon the relation $\varphi(\beta)$ is described by the following approximation, based on a large amount of experimental data (Z. L. Mitropol'skii et al., 1965)

$$\frac{1}{\varphi} = 1 + \frac{1-\beta}{\beta}\frac{v_g}{v_l}, \quad \frac{v_g}{v_l} = 1 + \frac{13.5\,(1 - p/p_K)}{\mathrm{Fr}_0^{5/12}\mathrm{Re}_0^{1/6}}$$

$$\left(\mathrm{Fr}_0 = \frac{w_0^2}{gD}, \quad \mathrm{Re}_0 = \frac{\overset{\circ}{\rho}_l w_0 D}{\mu_l}, \quad w_0 = \frac{m}{{}^1/_4 \pi D^2 \overset{\circ}{\rho}_l}\right. \tag{7.1.4}$$

$$g = 9.8\ \mathrm{m/s^2}, \quad p_K = 22.1\ \mathrm{MPa}$$

$$\left.\overset{\circ}{\rho}_l w_0 = (0.3 \div 3)\, 10^3\ \mathrm{kg/(m^2 \cdot s)}, \quad D = 5 - 35\,\mathrm{mm}\right)$$

Dependence of the relation between φ and β on the Froude number $\mathbf{Fr}_0$ indicates that this relation does generally depend on the flow orientation relative to the gravity force.

[1]A brief description of these methods is given later in this section. See also S. Banarjee, R. Lahey (1981).

Regimes (structures) of flow. The practically encountered regimes of flow of gas-liquid mixtures in channels present a noticeable diversity. They are defined by a great number of factors such as volumetric concentration of phases, density, viscosity, surface tension, and other physical characteristics of phase constituents, and also by phase velocity, presence of phase transitions and chemical reactions, diameter of the channel and its spatial position, methods of the phase feeding to the channel, distance from the channel entrance, etc. For the flow-regime identification, the so-called *regime diagrams* are generally used (S. S. Kutateladze, and M. A. Styrikovich, 1976; S. Tong, 1979). There are various types of these diagrams presently available. They all are plane diagrams, i.e., they identify the regime by two dimensionless parameters obtained based on some of the above-indicated parameters.

In spite of the variety of flow regimes and the complexity of their identification, it is still possible to evolve the following major regimes which replace each other as the void fraction or the flow rate of gas (vapor) increase: bubbly, slug, dispersed-film, and dispersion (bubble, drop) regimes.

A *bubbly* regime usually exists at a volumetric concentration of gas phase (void fraction) of $\varphi \equiv \alpha_g < 0.2–0.3$. At higher volumetric concentratons, the bubble coalescence takes place with the formation of large bubbles of a slug-like shape; such bubbles occupy almost the entire cross section of the channel, and the bubbly regime turns into a slug regime.

If the velocity of the gas phase is sufficiently high ($\upsilon_g > 5–10$ m/s in the case of flow of air-water mixture under normal conditions), the slug structure of the flow becomes unstable. A transient flow regime is produced, which is sometimes identified as an individual regime referred to as a *foamy,* or *semiannular,* regime of flow. This regime is characterized by a flow of liquid film on the channel wall and on the gas-liquid core which has a foamy structure. The slug, or semiannular, regimes generally exist at $0.2–0.3 < \alpha_g < 0.6–0.8$.

With a further increase of the volumetric concentration of the gas phase, at $\alpha_g > 0.6–0.8$, a *film,* or *annular,* flow regime is realized at which the liquid phase forms a continuous film flowing along the channel wall, and the vapor (steam) phase constitutes the core of flow. Because of the dynamic interaction between the gas core and liquid film, waves are generated on the surface of the latter; drops may break away (separate) from the wave crest and be carried away by the flow core. A *dispersed-film* regime is realized in this case, which is referred to in literature as a *dispersed-annular* regime of flow.

In heated channels, the film may vaporize, and the dispersed-annular regime turns into purely *dispersed* (*drop*) regime, or flow of mixture of vapor (steam), and drops. This regime represents inversion relative to the bubbly regime of flow.

Similarly, a so-called *inversed dispersed-annular* regime exists relative to the dispersed-annular regime, which is realized during a high-velocity flow of a boiling liquid out of the channel, when the boiling mostly takes place on the channel wall. Then, a vapor (steam) film is built-up in the near-wall zone, and liquid with bubbles flows in the core.

During a flow of mixture in vertical channels, an axisymmetrical distribution of phase concentrations and velocities over the cross-section area takes place. In flows through horizontal and inclined tubes, the axial symmetry in phase distribution over the cross-section area is distorted due to gravity. An increased gas, or vapor concentration, is found in the upper part of the tube cross section; this concentration is higher the smaller the angle of the tube inclination to a horizontal line, and the smaller the mixture velocity. The phase symmetry distortion may become indiscernible at sufficiently large Froude numbers $\mathbf{Fr} = v_g^2/(gD) \gtrsim 100$, where $g = 9.81\ \mathrm{m/s^2}$ is acceleration due to gravity, and D is the channel diameter.

Methods of measurement of the gas-liquid flow parameters. The analysis of local structure in two-phase flows is made considerably more complicated. This is associated with both the necessity to perform complicated measurements of such values as local phase velocities and shear (viscous) stresses (that are also measured in single-phase flows) and the need to develop measurement methods for such values as void fraction, film thickness and film flow rates, their wave characteristics, size of drops and bubbles, which are inherent only for two-phase flows.

Let us review the methods of measurement of void fraciton (volumetric concentration of gas) $\varphi \equiv \alpha_g$. The *cut-off approach* consists of rapid simultaneous shutting-down of two valves over a certain segment of a channel with subsequent measuring of the phase volumes. The *dynamic-balance method* is based on measuring pressure in the gas-liquid flow when it is exerted upon a disk installed normally to the flow. The *gamma-* and *beta-radioscopy* enable us to measure density of the gas-liquid mixture flowing through the channel; the gas concentration is then found based on determined phase densities. The *electric-conductivity* method permits the measurement of the local gas concentrations and liquid-film thickness in flows with a conductive liquid phase. The *optical* method utilizes the effects of a total internal reflection when passing through the interface, and proves to be a promising approach especially for nonconducting liquids. The *photographic* method is based on photography using a special lens with small depth of focus. Both *microthermocouple* and *hot-wire anemometer* methods in applications for measuring the local void fraction are based on the difference in heat removal when using sensors in both gas and liquid.

We shall now review the methods of measuring local phase velocities. The *dynamic-head* method (of the Pitot tube type) is used in separated phase motions, at sufficiently high velocities. The *isokinetic probing* method is based on taking samples with a speed equal the velocity of the medium at a given point; this method is applicable only to disperse and bubbly regimes of flow. The *hot-wire anemometer* method is analogous to measuring the local velocity in single-phase flows.

We will now turn to measurements of both shear stresses on the channel wall and their pulsations. In stabilized flows with unvarying parameters along the channel, the average shear stress on the channel wall may be found based on the pressure difference. But for flows experiencing acceleration along the channel (for instance, due to boiling), for nonstabilized flows (for instance, at small distances from the channel entrance), and in vertical flows, to determine friction, the direct measuring of shear stresses is needed; all the more if fluctuations of their local values is required, e.g., for the flow regime identification. To this effect, a group of authors (S. Kutateladze, A. Burdukov et al., 1969; V. Nakoryakov, A. Burdukov et al., 1973) have developed an electrodiffusion method (previously used for investigating mass exchange (V. Popov and N. Pokryvailo, 1966). The method employs a redox reaction in a weak solution of electrolyte on a cathode in a diffusion regime when the diffusion boundary layer is contained inside a viscous boundary layer; by virtue of the latter, a linear velocity profile is realized in the diffusion boundary layer. The corresponding velocity gradient which defines the shear stress on the sensor when viscosity is known is calculated in accordance with the convective-diffusion equation (which involves the linear velocity profile) based on the measured saturation current between anode and cathode of an electrochemical cell. Based on this principle, not only shear-stress sensors but also the local-velocity (end cylindrical sensor of size 30–50 micrometers for which the mass exchange coefficient is known as a function of the external-flow velocity) and local void-fraction sensors (based on contact time between the sensing element with gas and liquid) are developed.

Employment of this method for vapor-liquid flows and for investigating the boiling process is limited by fairly low temperatures, since at temperatures higher than 80°–90°C the electrolytes suitable for this method decompose.

Using this method, distributions of local phase velocities and void fractions over the cross section and along the channel together with spectral characteristics of shear stresses on the channel wall were investigated for both bubbly and projectile regimes of flow of air-water mixtures at velocities below 5 m/s. Some of the most important principal results of this study are discussed below.

When evaluating the potentials of the above-indicated methods, it should be remembered that high pressure ($p \sim 5$–10 MPa) and temperatures (300°–500° C), which are of the most interest for power engineering, significantly complicate any kind of measurements. The related diagnostics for the disperse-film flows is discussed in §7.4.

Hydrodynamic effects of both bubbly and slug regimes. Following the works by V. E. Nakoryakov, A. P. Burdukov et al. (1973), A. P. Burdukov, B. K. Koz'menko et al. (1975), V. Nakoryakov, O. Kashinsky et al., (1981), we shall now review the specific features of both bubbly and slug regimes discovered by means of the electrodiffusion diagnostics of vertical stationary air-water upward flows in tubes with $ID = 15–86$ mm at small velocities $v_0 = 0.1–5.0$ m/s and flow-rate gas concentrations $\beta = 0–0.9$.

It was found (Fig. 7.1.1) that, for the bubbly and slug regimes of flow with $W_l > 1.2–1.5$ m/s, the hydraulic resistance $\Delta p = \pi D \tau_W$ may be determined using a relation proposed by A. A. Armand (1946)

$$\frac{\tau_W}{\overset{\circ}{\tau}_W} = \frac{A}{(1-\varphi)^n}, \quad \varphi = 0.833\beta \qquad (A = 0.5 - 1.7; \quad n = 1.4 - 2.2) \tag{7.1.5}$$

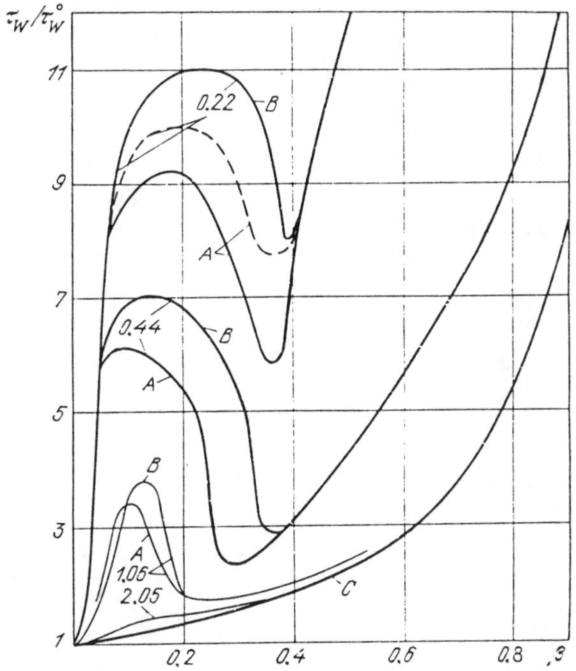

Figure 7.1.1 Experimental relationship (V. Nakoryakov et al., 1981) between friction over the tube wall ($D = 86.4$ mm) and modified velocity W_l of liquid, and the flow-rate gas concentration β in both bubbly and slug regimes of an air-water upward flow ($p_0 = 0.1$ MPa, $T_0 = 297$ K). Numerical labels on curves indicate the values of W_l, m/s. Labels A and B indicate two subregimes of a bubbly flow. The line C corresponds to the A. A. Armand relation (7.1.5) with an index $n = 1.53$.

where τ_W^0 is the friction stress on the tube wall determined by formulas for a single-phase liquid with the same volumetric flow rate W_l.

The friction losses under the bubbly regime at small velocities $W_l < 1.2$ m/s are many times higher than those obtained from (7.1.5). In this case, a strong dependence of τ_W on the distance from the mixture entrance into the tube, and a considerable data scatter are also observed. With a liquid Reynolds number $\mathbf{Re}_l = \rho_l^0 W_l D/\mu_l$ increase, the zone of such regime with anomalous high friction, shifts towards the smaller β, and the value τ_W/τ_W^0 drops. The boundary between bubbly and slug flows is indefinite, and generally two subregimes *A* and *B* of the bubbly flow are observed, and their realization depends on the method of gas feeding, and distance from the entrance into the channel; it is worthwhile to indicate that, upon a sufficiently long time, the sudden transitions from one regime to another are possible (these regimes are characterized by a substantial difference (by a factor of several times) in hydraulic resistance (see Fig. 7.1.1).

Figure 7.1.2 depicts the results of measurements of local structure of a turbulent gas-liquid vertical upward flow. It is evident that in a bubbly regime ($\beta = 0.044$ and 0.091), related to an anomalously high friction, the gas concentration profile near the wall has a distinctive peak which exceeds the gas concentration (void fraction) in the main flow by many times, and the liquid velocity profile's configuration is more spread, and is characterized by a greater velocity gradient near the wall compared to a single-phase motion. At $\beta = 0.45$, corresponding to a slug regime, the gas concentration and liquid velocity profiles are of parabolic shape with maxima on the chan-

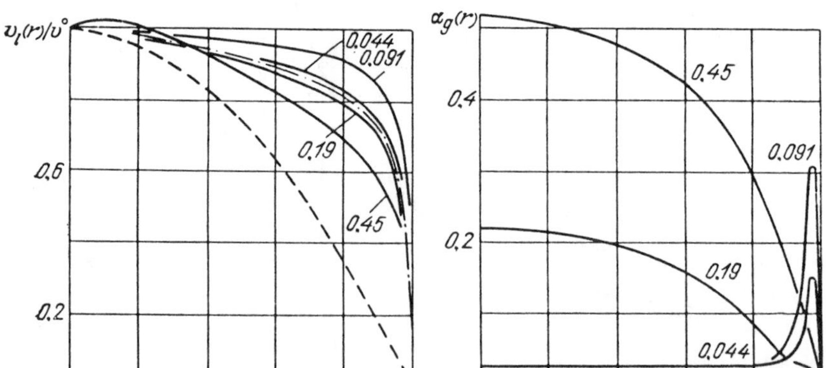

Figure 7.1.2 Radial distribution (experimental data) of both volumetric concentration of gas (void fraction) and liquid velocity in a vertical upward air-water flow in a tube of radius R = 43 mm at $W_l = 1.17$ m/s ($\mathbf{Re}_l = 107{,}400$). Numerical labels on solid curves indicate the flow-rate gas concentration β. The dot-dashed line corresponds to a single-phase flow of liquid ($\beta = 0$), the dashed line corresponds to the velocity distribution at laminar regime.

nel center-line, the velocity profile being of lesser "spread" than in a single-phase flow. In the regime of transition from a bubbly to a slug regime, two gas-concentration maxima may be encountered: at the wall where fine bubbles are accumulated, and in the center part of the tube where the gas bubbles of greater-size pass.

Investigation of spectra characteristics of shear stress over the wall, e.g., those of relative dispersion of friction pulsations

$$d_\tau = \frac{\sqrt{\langle(\tau'_W - \tau_W)^2\rangle}}{\tau_W}, \qquad \tau_W = \langle \tau'_W \rangle \tag{7.1.6}$$

showed that they depend to a great extent on the flow regime, and, also, that they may be used for identification of bubbly, slug, and annular regimes of flow. This is especially important when the visual methods are used for the regime identification (even when transparent tubes may be utilized under conditions of fairly low pressures and temperatures), which often lead to errors. The approach based on employment of analysis of spectra characteristics of pressure fluctuation was first used by M. Habbard and A. Duckler (1966).

Investigation of upward bubbly flows ($\beta = 0.01$–0.2) in a vertical tube ($D = 15$ mm) with smaller bubbles ($a < 0.1$–0.5 mm), produced in a flow by means of a special generator at small Reynolds numbers $\mathbf{Re}_l < 5000$, at which a laminar regime is realized in a single-phase liquid ($\beta = 0$), revealed a "microturbulent" flow instead of a laminar regime. This flow, in addition to abnormal friction many times exceeding the values predicted by formula (7.1.15), strong gas-concentration peaks in the near-wall zone, and a significantly more "spread" velocity profile than in a turbulent flow of a single-phase liquid, is characterized by high values of relative dispersion d_τ of the friction pulsations which grow with decrease of $\mathbf{Re}_l$, and whose spectrum is continuous. Thus, at $\mathbf{Re}_l = 2000$, an addition of a small amount of gas $\beta = 0.01$ increases shear stress by one order, and the values d_τ become equal to 0.35–0.38, which is higher than in turbulent flow of a single-phase liquid. The indicated effects depend on the bubble size.

In turbulent flows at $\mathbf{Re}_l > 5000$ and low concentrations of gas (appearing in a form of fine bubbles), the liquid velocity profile, wall friction, and relative dispersion ($d_\tau \approx 0.3$) are the same as for a single-phase flow. And, in this case, there is no distinctive boundary in the transition from a "microturbulent" to turbulent regimes of flow.

The gas velocity profiles for indicated regimes are almost similar to those for liquid, and differ by a value resembling the velocity of rise of bubbles of a given size in liquid at rest.

A possible reason for the "microturbulent" flow is the realization of transverse pulsations of the liquid velocity due to the flow lines deflection in the flow past bubbles, which, similar to turbulence, leads to a lateral transfer of momentum, i.e., to an increase of effective viscosity of the wall layer. This effect is intensified by an increased bubble concentration in the wall layer, and by the transverse random motion of bubbles. Accordingly, the characteristic "transfer length" in such a medium must depend on the bubble size. At large Reynolds numbers $\mathbf{Re}_l$, the indicated effect becomes insignificant considering the background of more intensive transverse turbulent pulsations whose nature is unrelated to the presence of bubbles.

In a well developed, high-velocity turbulent flow ($\mathbf{Re}_l > 10^5$) of a bubbly liquid in a tube, the hydraulic resistance, as in a single-phase flow, is independent of viscosity, and is defined only by the roughness of the interior surface of a tube wall. In this case, the ordinary formulas derived for a single-phase liquid (see L. G. Loitsyanskii, 1973; V. Streeter, 1961) may be used (G. Wallis, 1969); in these formulas, the mixture density must replace density, and the mass-averaged (flow rate) velocity of mixture represent velocity

$$\begin{gathered}\tau_W = {}^1/_2\, C_W \rho v^2, \quad \rho = \overset{\circ}{\rho}_1 \alpha_1 + \overset{\circ}{\rho}_2 \alpha_2, \\ \rho v = \rho_1 v_1 + \rho_2 v_2 = m/S, \quad C_W = [3.48 + 4 \lg R/\Delta]^{-2}\end{gathered} \tag{7.1.7}$$

where Δ is the average height of the surface roughness elements. For steel tubes used in power engineering, $\Delta \approx 50$ micrometers.

Hydrodynamic effects of a dispersed-film flow. The gas-liquid flow in a disperse-annular regime is characterized by a combined motion of two phases in the form of three components of the mixture—gas (vapor), liquid in a form of drops within the core of a flow, and liquid in a form of a film, each of which may have their inherent average velocity and temperature. In this case, mass exchange may occur between the flow core and film, and between liquid and vapor due to vaporization and condensation, and also due to the carrying-away and precipitation of drops. The drop precipitation is accounted for by turbulent pulsations and transverse forces in the gradient field of velocities of the gas phase near the film surface. The transverse flow of vapor during the intensive film vaporization (vapor "ablation") may inhibit such precipitation. The carrying of drops away from film and the thinning of the latter takes place due to gas flow over the film, due to splashing resulting from drop precipitation. In an intensively heated channel, the additional removal of liquid with vapor bubbles in a film during bubbly boiling also becomes possible. The transverse "blow-in" (ablation) of vapor and the carrying-away of drops contribute to the film thinning, but it may vanish completely only because of vaporization.

The presence of liquid film on the channel wall to a significant extent affects the hydraulic resistance during flow of the gas-liquid mixture in a disperse-annular regime, since the structure of the film wave surface (or the "film roughness"), and, consequently, the viscous friction between the flow core and film, depend on its thickness. The so-called crisis of hydraulic resistance may occur when, with the gas-phase velocity growth due to reduction of the "film roughness," the hydraulic resistance decreases rather than grows (see §7.5).

The presence of a liquid film plays a predominant role also for heat exchange, particularly, for heat removal from the heating channel-wall responsible for the film vaporization. Under intensive evaporation, when drops from the flow core have no time to additionally feed the film, the latter can vanish (the flow becomes dispersed) or lose its continuity. In this case, because of the lack of required contact between the heating wall and liquid phase, the heat exchange deteriorates, and the wall is overheated. This phenomenon is referred to as a heat-transfer crisis due to the wall liquid film drying-up (sometimes, it is called a crisis of heat transfer of the second kind (see §7.6)). A crisis of heat transfer during bubbly boiling (crisis of the first kind) also exists. This crisis may occur under heavy heat loadings due to the agglomeration of vapor bubbles generated on the heating wall into a vapor film; this also contributes to a distortion of a contact between a liquid and heating wall which may result in emergency overheating (burnout) of the latter (see §7.8 below). Crises of heat transfer are factors which limit the power of nuclear reactors and steam generators, and which complicate the operation of tube furnaces in technological processes.

The dispersed-film flow, together with film, foamy, drop, and partially inversed disperse-annular and bubbly flows, represent varieties of flows of disperse-annular structure which, in flows of gas-liquid mixtures in channels of various geometrical parameters, are the most widespread in nuclear-power installations, chemical plants used in petroleum refining, industries dealing with processing of raw materials, and also in the transporting of gas-condensate and petroleum products from oil fields. It is noteworthy that in steam generating channels, to whose inlets a saturated or subcooled water is fed, there is at the outlet a vapor-liquid mixture with maximum vapor concentration which can be obtained with no heat-transfer crisis, the disperse-annular regime may occupy 90% of the channel, and the single-phase, bubbly, and slug flows account only for the remaining 10%.

Parameters of thin turbulent wall films which yield to measurements. The described effects, including the heat-transfer crisis and hydraulic resistance, in a disperse-annular regime of flow are determined by the behavior of the wall liquid film. That is why particular importance is attached to its study.

The inherent thicknesses δ of liquid films in dispersed-film flows typical for steam generating plants and nuclear reactors and a number of other technological installations (D = 10–100 mm, p = 0.1–10 MPa, v_g = 5–100 m/s) amount generally to fractions of a millimeter, i.e., the thin films are characteristic for such systems ($\delta \ll D$). Because of difficulties involved in measuring velocity and temperature distribution over a cross section of such films, there are practically no measurements at all. For indicated conditions, and, especially, for vapor-liquid flows of high parameters (p = 1–10 MPa), only methods for measuring the rate of flow (m_3) of liquid in thin turbulent films, their rapid-oscillating thicknesses $\delta'(t)$, and the tracer-concentration variation are presently developed. The tracer (salt) is fed at a relatively small rate m_s ($m_s \ll m_3$) into the film for determining the intensity of moisture exchange between the wall film and the flow core (see §7.4 below).

The liquid flow rate m_3 in the film is measured by means of its suction through the annular slots, or porous segments of the tube in its measurement section (see §7.4 below). This flow rate determines the film Reynolds number which is a major parameter characterizing its state

$$\mathrm{Re}_f \equiv \mathrm{Re}_3 = m_3/(\pi D \mu_l) \qquad (7.1.8)$$

The pattern of distributions of both velocity and temperature in the film depends on the flow regime. At $\mathbf{Re}_3 < 300$–400, a laminar regime is observed, and at $\mathbf{Re}_3 > 400$, a turbulent regime with a generally wavy surface (S. S. Kutateladze and M. A. Styrikovich, 1976) is observed.

Thickness and characteristics of the wave surface of a liquid film. Rapid-oscillating thickness of the wall liquid film of the dispersed-film flow in a tube is measured by three methods: the electrocontact method (P. L. Kirillov et al., 1975; B. I. Nigmatulin et al., 1978), the resistance method (T. Tomida et al., 1974) and the capacitance-resistance method (B. I. Nigmatulin et al., 1981).

The electrocontact method is based on determining the points in time of contact between a gauge-tip and an electrically-conductive liquid at various distances y from the tube wall, which enables us to evaluate the film thickness and amplitude of its wave surface. With small film thicknesses $\delta \leqslant 0.1$ mm, the electrocontact method becomes invalid because of the distortion of the film surface by capillary and dynamic forces; in such situations, the resistance and capacitance-resistance methods are based on measuring the electric conductivity of the film over a segment of length $\Delta z \approx$ 4 mm ($\Delta z \gg \delta$) between two contacts of a sensor installed "flush" with the tube wall. Variation of the film thickness leads to variation of electric conductivity and capacity between these contacts. The corresponding calibration permits the determination of the instantaneous film thickness δ' above the sensor. By means of resistance sensors, the wave profiles $\delta'(t)$ of thin turbulent films were obtained for "cold" air-water dispersed-film flows (T. Tomida et al., 1974). An account of variation of not only conductivity but also of capacitance resulting from variation of the film thickness above the capacitance-resistance sensor increased the accuracy and sensitivity of measurements, and the wave profiles for not only cold flows but also for "hot" steam-water dispersed-film flows (see Fig. 7.1.3) under high pressure (B. I. Nigmatulin et al., 1981) were obtained.

With the known relation $\delta'(t)$, the geometrical average film thickness with respect to time may readily be found

$$\delta = \frac{1}{\Delta t} \int_0^{\Delta t} \delta'(t)\, dt \tag{7.1.9}$$

Here, the averaging procedure should be used for a sufficiently representative period of time Δt when the number of waves on an oscillogram for this time period is large (for instance, more than ten; then, for conditions

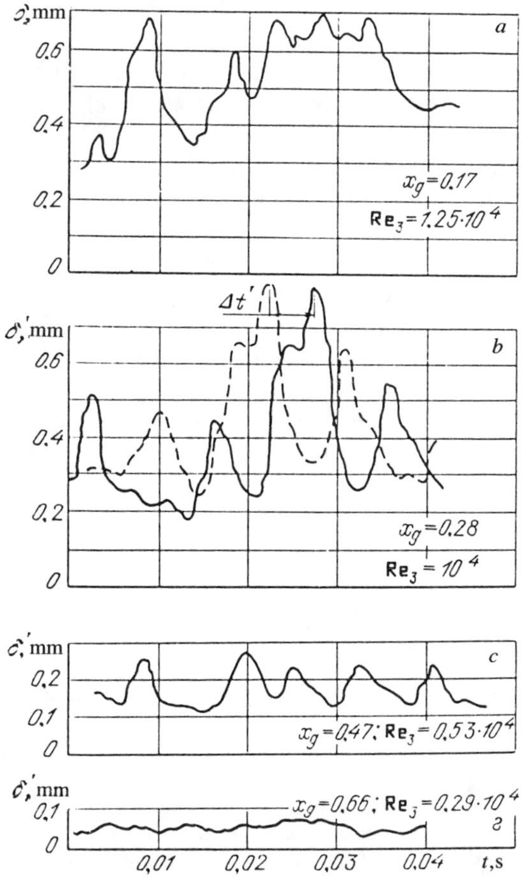

Figure 7.1.3 Oscillogram of the film thickness variation with time $\delta'(t)$, using a capacitance-resistance sensor in a vertical tube (D = 8 mm) with an upward air-water flow (p = 6.9 MPa, m^0 = 1000 kg/(m$^2\cdot$s)) at various mass flow-rate vapor concentrations x_g of mixture, and various Reynolds numbers $\mathbf{Re}_3 \equiv \mathbf{Re}_f$ (B. I. Nigmatulin, A. A. Vinogradov et al., 1982).

depicted in Fig. 7.1.3, it may be assumed that $\Delta t \approx 0.2$ s). Also, the average film crest-thickness δ_{cr}, and the average film throat-thickness δ_{th} for the time period Δt may be found

$$\delta_{cr} = \frac{1}{\Delta N_{cr}} \sum_{j=1}^{\Delta N_{cr}} \delta_{cr}(j), \quad \delta_{th} = \frac{1}{\Delta N_{th}} \sum_{j=1}^{\Delta N_{th}} \delta_{th}(j) \tag{7.1.10}$$

where j is the crest and throat number, ΔN_{cr} and ΔN_{th} are total numbers of crests and throats for the time interval Δt. The values δ_{cr} and δ_{th} represent, respectively, the average maximum and average minimum film thicknesses in its wave regime of flow

$$\delta_{cr} > \delta > \delta_{th} \tag{7.1.11}$$

The more δ_{cr}/δ_{th} exceeds unity, the wavier is the film. Figure 7.1.4 depicts the dependence of average thicknesses of turbulent films under fixed pressure, on the mass flow rate of a steam-water mixture. It was found that δ_{cr}/δ_{th} reaches the value of 5–6, which is indicative of the strong development of the film wave nature.

The simultaneous recording of signals from the two sensors described above located at a distance Δz from each other permits the determination of the time phase displacement $\Delta t'$ (see Fig. 7.1.3b), and, consequently, the phase velocity $C = \Delta z/\Delta t'$ of the motion of both crests and throats of waves (with lengths $L > \Delta z$) on the film surface. By averaging these velocities for the above-indicated representative time interval Δt, the average phase velocities of both crests C_{cr} and throats C_{th} can be omitted. Figure 7.1.5 represents these velocities versus vapor concentrations for the same turbulent regimes with thin films ($\delta \ll R$) shown in Fig. 7.1.3. It is seen that the phase velocities of crests and throats do not differ too much even in strongly

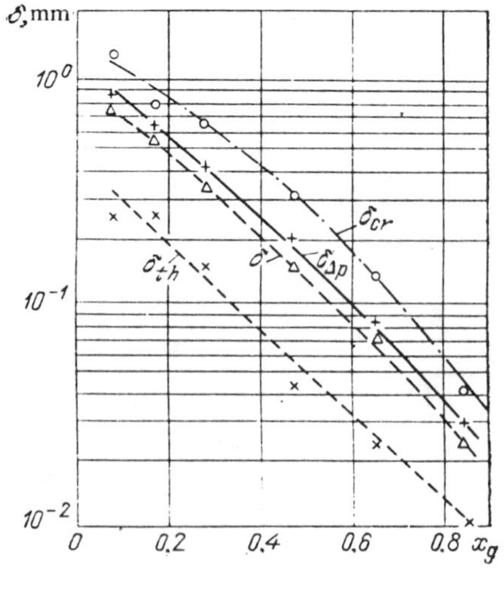

Figure 7.1.4 Effect of the flow-rate vapor concentration on the average thickness of a wall film on crests (δ_{cr}) and throats (δ_{th}), on its geometric average thickness δ, and on the thickness $\delta_{\Delta p}$ reduced in accordance with Blausius (see (7.3.10)) for a steam-water disperse-annular flow. Conditions are the same as in Fig. 7.1.3.

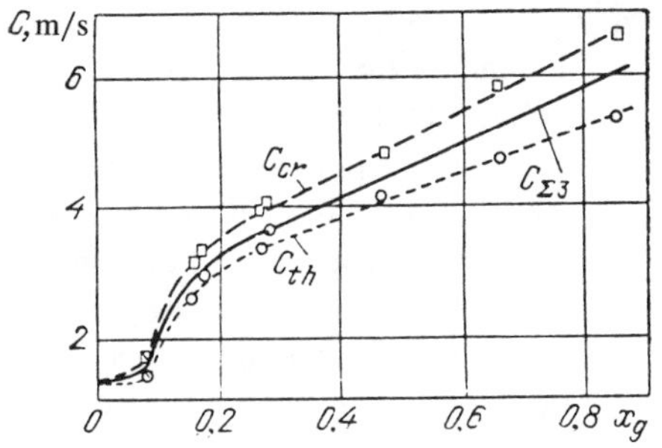

Figure 7.1.5 Effect of the flow-rate vapor concentration on the average phase velocities of crests (C_{cr}), throats (C_{th}), and the average wave velocity ($C_{\Sigma 3}$) on the surface of a wall film in steam-water dispersed-annular ($x_g > 0.08$) flow. The conditions are the same as in Fig. 7.1.3.

turbulent and wavy films ($\delta_{cr}/\delta_{th} \sim 5$). Therefore, it is worthwhile to introduce an average velocity of waves on a surface of a thin film

$$C_{\Sigma 3} = {}^1/_2(C_{cr} + C_{th}) \qquad (7.1.12)$$

Knowing both flow rate m_3 and average thickness δ, the average velocity of liquid in the film may be found as

$$v_3 = m_3/(\pi D\delta) \qquad (7.1.13)$$

As is evident from Fig. 7.1.6, the ratio of the wave average velocity $C_{\Sigma 3}$ on the film surface to the average velocity v_3 of liquid in it, for a well-developed turbulent regime ($\mathbf{Re}_3 > 1000$), with an increase of $\mathbf{Re}_3$ approaches 1.1, i.e., resembles unity ($C_{\Sigma 3} \approx v$). The same result is obtained in the above-mentioned experiments with "cold" air-water flows (T. Tomida et al., 1974). Thus, in turbulent films at $\mathbf{Re}_3 > 1000$, waves do not move relative to liquid, and the liquid in the turbulent thin wall film is transferred mainly in waves.

§7.2 AVERAGED EQUATIONS OF HYDRODYNAMICS OF THE DISPERSED-FILM FLOW

We shall now review the nonstationary axisymmetric flow of gas-liquid mixture in a dispersed-annular regime in a circular channel of radius R, or di-

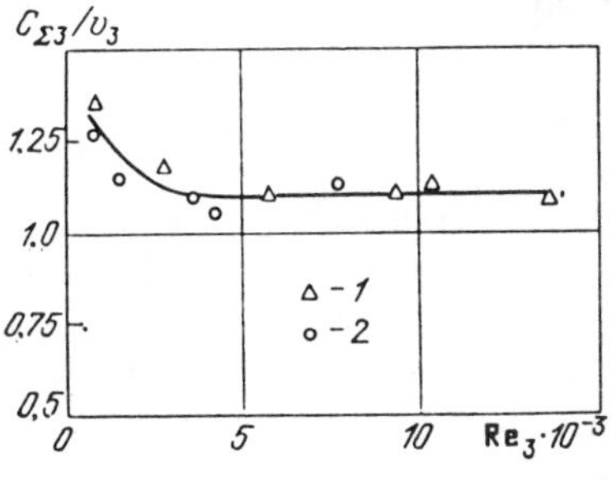

Figure 7.1.6 Ratio between average wave velocity $C_{\Sigma 3}$ on the surface of a wall film and average velocity v_3 in the film (experimental data of the same authors as in Figs. 7.1.3–7.1.5) at various Reynolds numbers $\mathbf{Re}_3$ of the film in dispersed-annular steam-water flows (m^0 = 1000 kg/(m$^2 \cdot$ s), D = 8 mm), at pressures p = 6.9 MPa (points *1*) and p = 10 MPa (points *2*).

ameter D, cross-section area $S = \pi R^2$, in the framework of a quasi-one-dimensional model (and assuming that the channel may undergo only small expansion and small curvature). Since the channel widening is small, a flow may exist in which the velocities of the mixture components are parallel at any point of the cross section. In this case, the velocity components perpendicular to the channel center-line, as well as the transverse components of acceleration, are small compared to components parallel to the channel center-line. Therefore, the difference between velocities and their axial components may be disregarded. The energy of the pulsating motions (including those in the turbulent regime of flow) is also ignored; the transverse pressure gradient is also neglected, and pressure p in any section of the channel is assumed uniform over the entire section, is the same for all phases, and is a function of only the axial coordinate z. The flow core is viewed as a monodisperse gas-suspension consisting of a continuous gas phase, and liquid phase in a form of drops, in the framework of assumptions and equations described in §1.4; the film is considered as a separate phase consisting only of liquid.

We shall discuss here the dispersed-annular flows in a circular tube, although channels of arbitrary cross sections (annular channels, channels with longitudinal bundles of rods, etc.) may also be considered. In this case, films should be individually singled-out on each wetted solid surface.

This approach may also be extended to other regimes of flow of two-phase mixture, in which zones of the channel cross section having significantly different flow characteristics may be singled-out: flow of a boiling liquid in an inversed dispersed-annular regime (vapor film on the wall, and bubbly liquid in the core), bubbly boiling of subcooled liquid, etc.

Parameters related to gas (vapor), drops, and liquid film are hereafter denoted by subscripts 1, 2, and 3, respectively.

Phase-parameter averaging over the channel cross section. Let us assume that the gas-drop core of the flow occupies the cylindrical domain of radius $R - \delta$, and the liquid film occupies the annular domain $R - \delta < r < R$, where δ is geometric average of the film thickness (7.1.9).

Mixture of gas and drops in the core is described as a sum of two continua (see §§1.3 and 1.4) making up the volume of the flow core, and occupying a part of the channel cross section equal to $S_c \equiv S_1 \equiv S_2$. At each point of the axisymmetric flow core with coordinates r, z, where $0 \leqslant r \leqslant R - \delta$, there may be introduced the macroscopic gas velocity $v_1(r)$ and drop velocity $v_2(r)$ (Fig. 7.2.1), gas temperature $T_1(r)$ and drop temperature $T_2(r)$, and their specific internal energies $u_1(r)$ and $u_2(r)$, volumetric concentrations of gas $\alpha_1(r)$ and drops $\alpha_2(r)$, reduced densities of gas $\rho_1(r)$ and drops $\rho_2(r)$, and other averaged parameters. The annular wall film, in which only condensed phase is distributed, occupies part of the channel cross-section area equal to $S_f \equiv S_3$. At each point of this domain with coordinates r, z, where

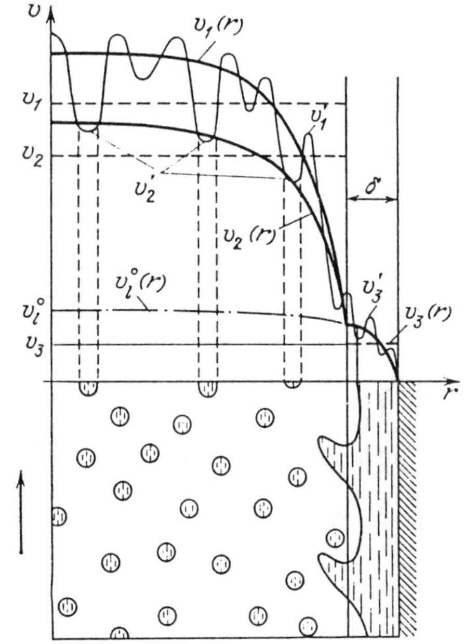

Figure 7.2.1 Schematic of the distribution of various velocity fields in a dispersed-annular flow, where $v_i' = v_i'(t, r', z')$ is a field of local velocities (microvelocities) (thin solid line) of gas ($i = 1$), liquid in drops ($i = 2$), and in film ($i = 3$); $v_i(r)$ fields of global velocities (macrovelocities) (heavy solid lines) when the radial scale Δr of the domain over which the averaging procedure is being performed (the annular section $r < r' < r + \Delta r$, or the corresponding annular cylinder may be used as such domain) are much higher than the drop size, but much smaller than the channel diameter and the film thickness; v_1 is the gas velocity determined over the cross section $S_1 = S_2 = S_c$ of the gas-liquid core of the flow; v_2 is the drop velocity averaged over the cross section S_c of the flow core; $v_3 \equiv v_f$ is the velocity of liquid in the wall film averaged over the film cross section $S_3 \equiv S_f$. The values v_i ($i = 1, 2, 3$) are shown by heavy dashed lines. The dot-dashed line is related to the "reduced" liquid flow, in which the velocity distribution $v_l^0(r)$ extrapolates the field of velocities $v(r)$ of the film liquid in the flow core.

$R - \delta \leqslant r \leqslant R$, the macroscopic average velocity $v_3(r)$, average temperature $T_3(r)$ of liquid, its specific internal energy $u_3(r)$, and other film parameters, may be determined. Because of the insignificant nonhomogeneity of fields of pressure and temperatures over the channel cross section, the variation of the phase-substance density over the channel cross section may be neglected, i.e.

$$\overset{\circ}{\rho}_i(r) = \overset{\circ}{\rho}_i \quad (i = 1, 2, 3), \qquad \overset{\circ}{\rho}_2 = \overset{\circ}{\rho}_3 = \overset{\circ}{\rho}_l$$

Thus

$$\begin{gathered} S_c + S_f = S \qquad (S_1 = S_2 = S_c, \quad S_3 \equiv S_f) \\ \rho_c(r) = \rho_1(r) + \rho_2(r), \quad \rho_i(r) = \overset{\circ}{\rho}_i(r) \cdot \alpha_i(r), \quad i = 1, 2, 3 \\ \alpha_1(r) + \alpha_2(r) = 1, \quad \alpha_i(r) \geqslant 0, \quad \alpha_3(r) \equiv 1 \\ \alpha_2(r) = {}^4/_3 \pi a^3 n(r) \end{gathered} \tag{7.2.1}$$

Here, $\rho_c(r)$ is density of the mixture in the flow core.

In order to give consideration to the nonuniformity of the mixture-component parameters over the channel cross section, each parameter is represented as a sum of an average value S_i of this parameter over the cross-section area and a variable component which reflects the nonuniformity of this parameter distribution over the section S_i, i.e.

$$\psi_i(r) = \psi_i + \Delta\psi_i(r)$$

$$\psi_i = \langle\psi_i(r)\rangle = \frac{1}{S_i}\int_{S_i} \psi_i(r)\, ds, \quad \int_{S_i} \Delta\psi_i(r)\, ds = 0 \qquad (i = 1, 2, 3) \tag{7.2.2}$$

We shall assume that the values ψ_i averaged over the cross-section area are regular functions of z, i.e., they smoothly vary along the channel center line (see (1.2.15))

$$\frac{\partial\psi_i}{\partial z} \sim \frac{\Delta\psi_0}{L}, \tag{7.2.3}$$

where $\Delta\psi_0$ is a characteristic variation of ψ_i in the process under discussion, L is the channel length.

The product of parameters averaged over the cross section is

$$\frac{1}{S_i}\int_{S_i} \psi_i(r)\cdot\varphi_i(r)\, ds = \frac{1}{S_i}\int_{S_i} [\psi_i + \Delta\psi_i(r)]\,[\varphi_i + \Delta\varphi_i(r)]\, ds$$

$$= \psi_i\varphi_i + \langle\Delta\psi_i(r)\cdot\Delta\varphi_i(r)\rangle = k_i^{(\varphi\psi)}\psi_i\varphi_i$$

$$k_i^{(\varphi\psi)} - 1 = \frac{\langle\Delta\psi_i(r)\cdot\Delta\varphi_i(r)\rangle}{\varphi_i\psi_i} \tag{7.2.4}$$

$$\langle\Delta\psi_i(r)\cdot\Delta\varphi_i(r)\rangle = \frac{1}{S_i}\int_{S_i} \Delta\psi_i(r)\cdot\Delta\varphi_i(r)\, ds$$

where $k_i^{(\varphi\psi)}$ is a correlation coefficient which takes into account the nonuniformity of parameters $\psi_i(r)$ and $\varphi_i(r)$ distributions over sections S_i ($i = 1$, 2, 3). Thus, the average values over the sections S_c and S_f are denoted by ψ_i, φ_i, etc. without indication of r. In particular, the average values of void fraction, average velocities, and average internal energies of phases are

$$\alpha_i = \frac{1}{S_i}\int_{S_i} \alpha_i(r)\, ds, \quad v_i = \frac{1}{S_i}\int_{S_i} v_i(r)\, ds, \quad u_i = \frac{1}{S_i}\int_{S_i} u_i(r)\, ds \tag{7.2.5}$$

Then, mass, number of drops, momentum (mass flow), and internal energy of the i-th component of mixture per unit length of the channel, and the corresponding flows of these values through the cross section S_i, become

$$\int_{S_i} \overset{\circ}{\rho}_i\alpha_i(r)\, ds = \overset{\circ}{\rho}_i\alpha_i S_i = \rho_i S_i, \quad \int_{S_c} n(r)\, ds = nS_c = \frac{\alpha_2 S_c}{{}^4\!/_3\,\pi a^3}$$

$$m_i = \int_{S_i} \overset{\circ}{\rho}_i \alpha_i(r) \cdot v_i(r)\, ds$$

$$= \overset{\circ}{\rho}_i \alpha_i v_i S_i \left(1 + \frac{\langle \Delta\alpha_i(r) \cdot \Delta v_i(r)\rangle}{\alpha_i v_i}\right) = \overset{\circ}{\rho}_i \alpha_i v_i S_i k_i^{(\alpha v)}$$

$$\int_{S_i} \overset{\circ}{\rho}_i \alpha_i(r) \cdot u_i(r)\, ds = \overset{\circ}{\rho}_i \alpha_i u_i S_i k_i^{(\alpha u)} \tag{7.2.6}$$

$$\int_{S_3} n(r) \cdot v_2(r)\, ds = n v_2 S_2 \left(1 + \frac{\langle \Delta n(r) \cdot \Delta v_2(r)\rangle}{n v_2}\right) = n v_2 S_2 k_2^{(\alpha v)}$$

$$\int_{S_i} \overset{\circ}{\rho}_i \alpha_i(r) [v_i(r)]^2\, ds = \overset{\circ}{\rho}_i \alpha_i v_i^2 S_i k_i^{(\alpha vv)}$$

$$\int_{S_i} \overset{\circ}{\rho}_i \alpha_i(r)\, v_i(r)\, u_i(r)\, ds = \overset{\circ}{\rho}_i \alpha_i v_i u_i S_i k_i^{(\alpha vu)}$$

Here, $k_i^{(\alpha v)}$, $k_i^{(\alpha u)}$, $k_i^{(\alpha vv)}$, and $k_i^{(\alpha vu)}$ are correlation coefficients which take into consideration the effect of nonuniformity in distributions of void fraction, velocity and internal energy of the i-th component of mixture over the cross section S_i. Note that for a film, because of $\alpha_3(r) \equiv 1$, it follows that $k_3^{(\alpha vv)} = k_3^{(vv)}$, $k_3^{(\alpha vu)} = k_3^{(vu)}$.

Below, the average and flow-rate average phase temperatures are used

$$T_i = \frac{1}{S_i} \int_{S_i} T_i(r)\, ds$$

$$T_i^{(m)} = \frac{1}{\alpha_i S_i v_i} \int_{S_i} \alpha_i(r)\, v_i(r)\, T_i(r)\, ds = k_i^{(\alpha v \mathrm{T})} T_i \tag{7.2.7}$$

Conservation equations. Using (7.2.6), we obtain the following differential equations of conservation of mass, number of drops in the flow core, momentum, and heat influxes for components of mixture

$$\frac{\partial(\rho_1 S_c)}{\partial t} + \frac{\partial\left(\rho_1 v_1 S_c k_1^{(\alpha v)}\right)}{\partial z} = J_{21} + J_{31}, \quad \rho_1 = \overset{\circ}{\rho}_1 \alpha_1$$

$$\frac{\partial(\rho_2 S_c)}{\partial t} + \frac{\partial\left(\rho_2 v_2 S_c k_2^{(\alpha v)}\right)}{\partial z} = -J_{21} - \overset{\circ}{J}_{23} + \overset{\circ}{J}_{32}, \quad \rho_2 = \overset{\circ}{\rho}_2 \alpha_2, \quad \overset{\circ}{\rho}_2 = \overset{\circ}{\rho}_l$$

$$\frac{\partial(\rho_3 S_f)}{\partial t} + \frac{\partial(\rho_3 v_3 S_f)}{\partial z} = -J_{31} + \overset{\circ}{J}_{23} - \overset{\circ}{J}_{32}, \quad \rho_3 = \overset{\circ}{\rho}_3 = \overset{\circ}{\rho}_l$$

$$\frac{\partial(n S_c)}{\partial t} + \frac{\partial\left(n v_2 S_c k_2^{(\alpha v)}\right)}{\partial z} = \overset{\circ}{\psi}_{32} - \overset{\circ}{\psi}_{23}$$

$$\alpha_2 = {}^4/_3\pi a^3 n, \quad S_1 \equiv S_2 \equiv S_c, \quad S_3 \equiv S_f, \quad v_3 \equiv v_f$$

$$\overset{\circ}{J}_{23} = {}^4/_3\, \pi a_{23}^3 \overset{\circ}{\rho}_l \overset{\circ}{\psi}_{23}, \quad \overset{\circ}{J}_{32} = {}^4/_3\, \pi a_{32}^3 \overset{\circ}{\rho}_l \overset{\circ}{\psi}_{32},$$

$$\frac{\partial\left(\rho_1 v_1 S_c k_1^{(\alpha v)}\right)}{\partial t} + \frac{\partial\left(\rho_1 v_1^2 S_c k_1^{(\alpha vv)}\right)}{\partial z}$$

$$= -\alpha_1 S_c \frac{\partial p}{\partial z} - F_{12} - F_{13} + J_{21} v_{21} + J_{31} v_{31} + \rho_1 S_c g^z$$

$$\frac{\partial\left(\rho_2 v_2 S_c k_2^{(\alpha v)}\right)}{\partial t} + \frac{\partial\left(\rho_2 v_2^2 S_c k_2^{(\alpha vv)}\right)}{\partial z}$$

$$= -\alpha_2 S_c \frac{\partial p}{\partial z} + F_{12} - J_{21} v_{12} + \overset{\circ}{J}_{32} v_{32} - \overset{\circ}{J}_{23} v_{23} + \rho_2 S_c g^z \tag{7.2.8}$$

$$\frac{\partial\left(\rho_3 v_3 S_f\right)}{\partial t} + \frac{\partial\left(\rho_3 v_3^2 S_f k_3^{(vv)}\right)}{\partial z}$$

$$= -S_f \frac{\partial p}{\partial z} + F_{13} - F_W - J_{31} v_{13} + \overset{\circ}{J}_{23} v_{23} - \overset{\circ}{J}_{32} v_{32} + \rho_3 S_f g^z$$

$$\frac{\partial\left(\rho_1 u_1 S_c k_1^{(\alpha u)}\right)}{\partial t} + \frac{\partial\left(\rho_1 u_1 v_1 S_c k_1^{(\alpha uv)}\right)}{\partial z}$$

$$= \frac{\alpha_1 p S_c}{\overset{\circ}{\rho}_1} \frac{d\overset{\circ}{\rho}_1}{dt} + J_{21} \frac{(v_{21} - v_1)^2}{2} + J_{31} \frac{(v_{31} - v_1)^2}{2}$$

$$+ J_{31} u_{31} + J_{21} u_{21} + F_{12}(v_1 - v_{12}) + F_{13}(v_1 - v_{13}) - Q_{1(\Sigma 2)} - Q_{1(\Sigma 3)}$$

$$\frac{\partial\left(\rho_2 u_2 S_c k_2^{(\alpha u)}\right)}{\partial t} + \frac{\partial\left(\rho_2 u_2 v_2 S_c k_2^{(\alpha uv)}\right)}{\partial z} = -J_{21} \frac{(v_{12} - v_2)^2}{2} - \overset{\circ}{J}_{23} \frac{(v_{23} - v_2)^2}{2}$$

$$+ \overset{\circ}{J}_{32} \frac{(v_{32} - v_2)^2}{2} - J_{21} u_{12} - \overset{\circ}{J}_{23} u_{23} + \overset{\circ}{J}_{32} u_{32} - Q_{2(\Sigma 2)}$$

$$\frac{\partial\left(\rho_3 u_3 S_f\right)}{\partial t} + \frac{\partial\left(\rho_3 u_3 v_3 S_f k_3^{(uv)}\right)}{\partial z} = -J_{31} \frac{(v_{13} - v_3)^2}{2} - \overset{\circ}{J}_{32} \frac{(v_{32} - v_3)^2}{2}$$

$$+ \overset{\circ}{J}_{23} \frac{(v_{23} - v_3)^2}{2} - J_{31} u_{13} - \overset{\circ}{J}_{32} u_{32} + \overset{\circ}{J}_{23} u_{23} - Q_{3(\Sigma 3)} + Q_W$$

Here, J_{ji} (j, i = 1, 2, 3; $j \neq i$) is the intensity of the mass transition from the j-th to the i-th component of mixture per unit time and unit length of the channel. In this case, J_{31} and J_{21} are intensities of evaporation (or condensation, if J_{21}, $J_{31} < 0$) of vapor from the surfaces of a film and drops, respectively; J_{32}^0 and J_{23}^0 are, respectively, intensities of entrainment of drops from the film surface and their precipitation on it; $J_{23} = -J_{32} = J_{23}^0 - J_{32}^0$; ψ_{32}^0 and ψ_{23}^0 are, respectively, the number of entrained and precipitated drops per unit time and unit length of the channel; a_{32} and a_{23} are, respectively, the average radii of drops, entrained from the film surface and precipitated onto its surface; F_{13}, F_W, and F_{12} are friction forces, respectively, between gas phase and film, film and channel wall, and the force of interaction between gas phase and drops, per unit length of channel; g^z is the intensity of external body forces in the direction of the channel center line; $Q_{1(\Sigma 3)}$, $Q_{3(\Sigma 3)}$, $Q_{1(\Sigma 2)}$, $Q_{2(\Sigma 2)}$, and Q_W are heat fluxes per unit time and unit length of the channel, respectively, from the gas phase to the interface between gas and film (to the Σ_3-phase), from film to its surface (to Σ_3-phase), from the gas phase to the interface between the gas phase and drops (to Σ_2-phase), from the drop interior to the Σ_2-phase, from the wall to film; $\upsilon_{12} = \upsilon_{21}$, and $\upsilon_{13} = \upsilon_{31}$ are, respectively, longitudinal components of the velocities of a substance undergoing the phase transition on surfaces of the Σ_2-phase and Σ_3-phase; υ_{32} and υ_{23} are longitudinal components of drop velocities, with which they are carried away from the film surface and precipitate on its surface, respectively; similarly, u_{32} and u_{23} are specific internal energies of liquid in drops removed from the film surface and precipitated on its surface, respectively; the remaining parameters of type u_{ji} (namely, u_{12}, u_{21}, u_{13}, and u_{31}) are internal energies of the substance of the i-th phase, which undergoes phase transition.

When writing the heat-influx equations, the longitudinal heat conduction in phases was ignored, and liquid was assumed incompressible ($\rho_2^0 = \rho_3^0 = \rho_l^0 =$ constant). Later on, the equations of state for internal energies u of phases are assumed in the approximation of constant heat capacities in the form of linear functions of temperatures (see (1.3.73) and (1.3.72)).

The terms of type $(1/2)J_{ji}(\upsilon_{ji} - \upsilon_i)^2$ in equations of heat influx determine the energy dissipation due to phase transitions in conditions of the phase-velocity nonequilibrium (see §§1.1 and 1.2). These terms are insignificant in the processes under consideration.

The energy conservation equations (or equations for heat influx on the interfaces) should also be given consideration on both the drop surface (Σ_2-phase) and the film surface (Σ_3-phase) where, in addition to vaporization and condensation, the drop entrainment and precipitation take place (cf. the last equation (1.1.56))

$$\begin{gathered}
-Q_{1(\Sigma 2)} - Q_{2(\Sigma 2)} + J_{21}(i_{21} - i_{12}) = 0 \\
-Q_{1(\Sigma 3)} - Q_{3(\Sigma 3)} + J_{31}(i_{31} - i_{13}) = 0 \\
i_{21} = u_{21} + p/\overset{\circ}{\rho}_1, \quad i_{12} = u_{12} + p/\overset{\circ}{\rho}_2 \\
i_{31} = u_{31} + p/\overset{\circ}{\rho}_1, \quad i_{13} = u_{13} + p/\overset{\circ}{\rho}_3
\end{gathered} \tag{7.2.9}$$

To close the obtained system of equations (7.2.8) and (7.2.9), in addition to the equations of state of phases, assumptions for the correlation coefficients $k_i^{(\alpha v)}$, $k_i^{(\alpha v v)}$, $k_i^{(\alpha u)}$, and $k_i^{(\alpha v u)}$, equations of interphase interaction for J_{ji}, F_{ji}, Q_{ji}, a_{ji}, v_{ji}, and u_{ji}, and, moreover, the external effects of the tube wall upon the flow through F_W, and Q_W, must also be used.

From the equations of conservation of mass and number of drops, an equation may be deduced which in an explicit form determines variation of the average size of drops

$$\frac{\partial a^3}{\partial t} + v_2 k_2^{(\alpha v)} \frac{\partial a^3}{\partial z} = \frac{J_{31}}{(4\pi/3)\overset{\circ}{\rho}_l n S_c} - \overset{\circ}{\psi}_{23}\left(a_{23}^3 - a^3\right) + \overset{\circ}{\psi}_{32}\left(a_{32}^3 - a^3\right) \tag{7.2.10}$$

Coefficients of nonuniformity, and interrelation between parameters on the interface between phases with averaged parameters. Analysis of experimental data concerning distributions of concentrations and velocities of the mixture components over the cross section of the flow core (for air-water flows at $p \sim 0.1$ MPa, see G. Hewitt and N. Hall-Taylor (1972); for steam-water flows at $p \sim 7$ MPa, see P. L. Kirillov et al. (1973)) proves that at a turbulent regime of the gas phase in a core, the distributions may be represented in the form of power (exponential) functions

$$\frac{\alpha_i(r) - \alpha_{i(\Sigma 3)}}{\overset{\circ}{\alpha}_i - \alpha_{i(\Sigma 3)}} = \left(1 - \frac{r}{R_c}\right)^{\theta_i}, \quad \frac{v_i(r) - v_{\Sigma 3}}{\overset{\circ}{v}_i - v_{\Sigma 3}} = \left(1 - \frac{r}{R_c}\right)^{\nu_i}$$

$$0 \leqslant r \leqslant R_c, \quad R_c = R - \delta \approx R \tag{7.2.11}$$

The subscript $\Sigma 3$ denotes the parameter average values on the surface of liquid film (Σ_3-phase), and superscript ^{0} denotes parameters on the channel center line.

It may similarly be supposed that distribution of temperatures of the mixture components over the cross section of the flow core also obeys the exponential law

$$\frac{T_i(r) - T_{\Sigma 3}}{T_i^\circ - T_{\Sigma 3}} = \left(1 - \frac{r}{R_c}\right)^{\theta_i}, \quad i = 1, 2 \tag{7.2.12}$$

We shall further use the average shear stresses $\tau(r)$, and the pertinent force $F(r) = 2\pi r \cdot \tau(r)$. In particular, the shear stresses defining the indicated forces F_W and F_{13} may be introduced on the boundaries between film and the wall (W), and between film and the flow core ($\Sigma 3$ or 13)

$$F_W = 2\pi R \tau_W, \quad F_{13} = 2\pi (R - \delta)\tau_{13} \tag{7.2.13}$$

As has been indicated already in §7.1, measurements of the velocity profiles in thin ($\delta \sim 10^{-1}$ mm) but turbulent and wavy films are absent. Therefore, we shall give a rough picture of these profiles in the form of predetermined relations $v_3(r)$ in accordance with the averaged values $m_3 \equiv m_f$, δ, and $v_3 \equiv v_f$, because of which the velocity profile $v_3(r)$ must satisfy the normalization conditions following from Eqs. (7.2.5) and (7.2.6) for $i = 3$ when $\alpha_l = 1$

$$\frac{1}{S_3}\int\limits_{S_3} v_3(r)\,ds = v_3, \qquad \int\limits_{S_3} \overset{\circ}{\rho}_l v_3(r)\,ds = m_3$$

These conditions, given that the annular domain S_3 is thin ($\delta \ll R$), assume the form

$$\frac{1}{\delta}\int\limits_{R-\delta}^{R} v_3(r)\,dr = v_3, \quad 2\pi R\delta\overset{\circ}{\rho}_l v_3 = m_3 \tag{7.2.14}$$

It should be remembered that the liquid flow rate m_3 in a film and its average thickness are measurable quantities (see §7.1), m_3 defining the Reynolds number introduced in (7.1.8)

$$\mathrm{Re}_3 = \frac{\overset{\circ}{\rho}_3 v_3 \delta}{\mu_3} = \frac{m_3}{2\pi R\mu_l} \tag{7.2.15}$$

Let us now isolate an annular layer of liquid in a film between radii $R_c = R - \delta$ and $r < R$. Then, the momentum equation for this layer becomes

$$\left[\rho\,(g^z - \tilde{a}) - \frac{dp}{dz}\right]\pi\,(r^2 - R_c^2) + [\tau_{13}\cdot R_c - r\cdot\tau(r)]\,2\pi = 0 \tag{7.2.16}$$

where $\tilde{a}$ is an average acceleration of liquid particles in the annular layer under consideration.

Two limiting cases are evolved in the analysis of film flows:

1) flow with constant shear stress affected by the gas core

$$\tau(r) = \tau_{13} = \tau_W = \text{const} \tag{7.2.17a}$$

2) free flow of a film affected only by gravity forces, when the impacts from the gas core and inertia forces are small, and there is no pressure variation along the flow. Then, using Eq. (7.2.16) for $r = R$, we obtain

$$\tau_W = \overset{\circ}{\rho}_l g^z \delta \qquad (\tau_{13} = 0, \quad \tilde{a} = 0, \quad dp/dz = 0) \tag{7.2.17b}$$

Let us first examine a laminar regime of thin films ($\delta/R \ll 1$), which is realized at $\mathbf{Re}_3 < 300$–400. For the indicated two limiting cases, the following distribution can be, respectively, obtained in terms of the relative distance from the channel wall $\bar{y} = (R - r)/\delta$ from the equations for viscous liquid, when $\tau = \mu_l(\partial v(r)/\partial r)$, which are defined by the average velocity v_3 and the average film thickness δ

$$1)\ v_3(r) = 2v_3\bar{y}, \quad \tau(\bar{y}) = \tau_{13} = \tau_W = 2v_3\mu_l/\delta \tag{7.2.18a}$$

$$2)\ v_3(r) = 3v_3\left(\bar{y} - {}^1/_2\bar{y}^2\right)$$

$$\bar{y} = 1:\ \tau = \tau_{13} = 0; \quad \bar{y} = 0:\ \tau = \tau_W = \overset{\circ}{\rho}_l g^z\delta = 3v_3\mu_l/\delta \tag{7.2.18b}$$

Note that at laminar flow with a fixed shear (7.2.18a), the nonuniformity of the liquid velocity distribution in the film is maximum, and, in this case, we have $v_{\Sigma 3} = 2v_3$ and $k_3^{(vv)} = 1.33$.

If the heat fluxes along the channel are much lower compared to the transverse fluxes, the heat-conduction equation in the laminar film together with conditions determined by heat fluxes on its boundaries are

$$v_3(r)\frac{\partial T_3(r, z)}{\partial z} = \nu_l^{(T)}\frac{\partial^2 T_3}{\partial r^2}$$

$$r = R: \quad \frac{\partial T_3}{\partial r} = -\frac{Q_W}{2\pi R\lambda_l}, \qquad T = T_W \tag{7.2.19}$$

$$r = R - \delta: \quad \frac{\partial T_3}{\partial r} = \frac{Q_{3(\Sigma 3)}}{2\pi (R - \delta)\lambda_l}$$

where T_W is the liquid temperature on the channel wall. Hence, a formula for the temperature profile in a thin film may be readily obtained as

$$T_3(\bar{y}) - T_W = A\left[\frac{1 - \Omega}{v_3}\int_0^{\bar{y}}\int_0^{\bar{y}'} v_3(\bar{y}')\, d\bar{y}''dy' - \bar{y}\right] \tag{7.2.20}$$

$$A = \frac{Q_W\delta}{2\pi R\lambda_l}, \qquad \Omega = \frac{Q_{3(\Sigma 3)}}{Q_W}$$

which for flow with a fixed shear (7.2.18a) and free flow (7.2.18b), respectively, yields

$$T_3(\bar{y}) - T_W = A\left[\bar{y}^3(1 - \Omega)/3 - \bar{y}\right] \qquad (\tau_{13} = \tau_W) \tag{7.2.21a}$$

$$T_3(\bar{y}) - T_W = A\left[({}^1\!/\!_2\bar{y}^3 - {}^1\!/\!_2\bar{y}^4)(1 - \Omega) - \bar{y}\right] \qquad (\tau_{13} = 0) \tag{7.2.21b}$$

If both mass-averaged temperature T_3 and flow-rate-averaged temperature $T_3^{(m)}$ of liquid in a film (see (7.2.7)) are employed, then, e.g., for a flow with a fixed shear (7.2.18a), and for two limiting values of thermal factor $\Omega = 0$ and $\Omega = 1$, the distribution (7.2.21a) yields the following relations

$$\Omega = 0, \quad T_{\Sigma 3} - T_W = {}^8\!/\!_5(T_3 - T_W), \quad T_3^{(m)} - T_W = {}^{32}\!/\!_{25}(T_3 - T_W)$$
$$\tag{7.2.22}$$
$$\Omega = 1, \quad T_{\Sigma 3} - T_W = 2(T_3 - T_W), \quad T_3^{(m)} - T_W = {}^4\!/\!_3(T_3 - T_W)$$

The difference of the numerical coefficients appearing in these relations from unity characterizes, in the case under consideration, the degree of temperature nonuniformity across the laminar film.

The turbulent regime in the film, which is realized at $\mathbf{Re}_3 > 400$ and at which the major friction resistance and the primary resistance to heat transfer are defined by the laminar sublayer and a buffer domain localized in a very narrow zone of thickness δ_* near the wall, is of the greatest importance for practical applications. The small thickness of this zone ($\delta_* \ll \delta$) provides the basis for assuming that the distinctions and peculiarities in turbulent pulsations, which are realized outside the indicated zone depending on whether the same flow rate m_3 of liquid passes through the fixed annular domain $R - \delta \leqslant r \leqslant R$ together with the gas-drop core, or together with a single-phase flow[2] of the same liquid, do not affect the nature of flow in both laminar sublayer and buffer zone $R - \delta_* \leqslant r \leqslant R$.

We shall assume here (A. A. Armand, 1946; G. Wallis, 1969; G. Hewitt and N. Hall-Taylor, 1972), in connection with the above discussion, that the distribution profiles for liquid velocity and temperature over the cross-section area of this zone are similar to those which exist in flow of a single-phase liquid throughout the channel. Their extension for the entire thickness of the film does not lead to significant errors since, in addition to above-indicated data, in the main volume of the turbulent film beyond the viscous sublayer and buffer zone, the differences in both velocity and temperature are fairly small and are not significantly sensitive to methods of their approximations.

Moreover, for both laminar and turbulent regimes of flow in the film, we shall adopt the exponential approximation for both the velocity profile and the temperature profile

$$v_3(r) = v_{\Sigma 3}\bar{y}^{\bar{\nu}_2}, \quad T_3(r) - T_W = (T_{\Sigma 3} - T_W)\bar{y}^{\bar{\theta}_3}$$
$$(\bar{y} = (R - r)/\delta, \quad 0 \leqslant \bar{y} \leqslant 1) \tag{7.2.23}$$

Given $\delta/R \ll 1$, it becomes possible to interrelate velocities and temperatures on the boundary film-core (Σ_3-phase) for these distributions with their mass-averaged and flow-rate-averaged values in the film

$$\frac{v_{\Sigma 3}}{v_3} = 1 + \nu_3, \quad \frac{T_{\Sigma 3} - T_W}{T_3 - T_W} = 1 + \theta_3, \quad \frac{T_3^{(m)} - T_W}{T_3 - T_W} = \frac{(1+\nu_3)(1+\theta_3)}{1+\nu_3+\theta_3} \tag{7.2.24}$$

Similarly, for exponential distributions (7.2.11) and (7.2.12) in the flow core, relations which interrelate the phase parameters on the boundary with a film ($v_{i(\Sigma 3)}$, $T_{i(\Sigma 3)}$, and $\alpha_{i(\Sigma 3)}$), and on the channel center line (v_i^0, T_i^0, α_i^0), with corresponding average values (v_i, T_i, α_i), are available

$$\frac{v_i - v_{i(\Sigma 3)}}{v_i^0 - v_{i(\Sigma 3)}} = \frac{1}{(1+\nu_i)(1+{}^1/_2\nu_i)}, \quad \frac{T_i - T_{i(\Sigma 3)}}{T_i^0 - T_{i(\Sigma 3)}} = \frac{1}{(1+\theta_i)(1+{}^1/_2\theta_i)} \tag{7.2.25}$$

For distributions (7.2.11), (7.2.12), and (7.2.23), using (7.2.4), the correlation coefficients can be easily calculated. In particular, for $k_i^{(vT)}$ we have

[2]By ordinary single-phase flow, we mean "modified" in an indicated manner with respect to flow rate.

$$k_i^{(vT)} = 1 + \left(\frac{5}{4}\,\nu_i\theta_i + \frac{3}{4}\,\nu_i^2\theta_i + \frac{3}{4}\,\nu_i\theta_i^2 + \frac{1}{4}\,\nu_i^2\theta_i^2\right)\frac{v_i - v_{i(\Sigma3)}}{v_i}\cdot\frac{T_i - T_{i(\Sigma3)}}{T_i}$$

$$(i = 1, 2) \qquad (7.2.26)$$

$$k_3^{(vT)} = 1 + \frac{\nu_3\theta_3}{1+\nu_3+\theta_3}\,\frac{T_3 - T_W}{T_3}$$

As is proven by the above-described experiments with turbulent flows of gas in a core and liquid in a film, the exponents in the assumed approximations (7.2.11), (7.2.12), and (7.2.23) of the parameter distribution over the flow cross section are generally less than 1/5. It should be remembered that the higher the exponent, the more elongated the distribution profile, and the stronger the impact of nonuniformity. From formulas for $k_i^{(vT)}$, it is evident that for the exponent values $\nu_i \approx \theta_i \leqslant 1/5$ (i = 1, 2, 3) (characteristic for turbulent flows), the correlation coefficients $k_i^{(vT)}$ are close to unity even when the parameters on the channel center line, film surface, and the channel wall significantly differ from each other. Therefore, for turbulent flows of gas in the core and liquid in a film, the issue of determining the coefficients of nonuniformity is not acute, since we may assume

$$k_i^{(\alpha v)} \approx k_i^{(\alpha vv)} \approx k_i^{(\alpha u)} \approx k_i^{(\alpha vu)} \approx 1 \qquad (7.2.27)$$

Note that, in conformity with measurements of wave velocities $C_{\Sigma3}$ in a turbulent film indicated in §7.1, the liquid velocity on its surface may be assumed as $v_{\Sigma3} \approx C_{\Sigma3} \approx 1.1\ v_3$, which resembles the value $v_{\Sigma3} = 1.14\ v_3$ obtained from (7.2.11) at $\nu_3 = 1/7$.

The drops precipitating with an intensity J_{23}^0 onto the film have a longitudinal velocity v_{23} resembling the average velocity of drops in the flow core ($v_{23} \approx v_2$). This is accounted for by the fact that the drop retardation in the layer with sharp variation of lonogitudinal velocity near the film is small, since this layer is too thin for the precipitating drops (which have many times higher inertia compared to gas) to have sufficient time to slow down.

The precipitating drops knock out of the film secondary drops (splashes) which return back into the core. The intensity of this removal of splashes, denoted by $J_{32}^{(s)}$, is a component of the intensity J_{32}^0 of removal of drops (see §7.4 below). Evaluations (B. I. Nigmatulin, V. E. Nikolaev, and S. I. Ivandaev) show that drop splashes leave the film with a velocity $v_2^{(s)}$ that is close to the velocity of falling drops ($v_{32}^{(s)} \approx v_{23} \approx v_2$), and that the remaining drops carried away by gas with an intensity $J_{32}^0 - J_{32}^{(s)}$ into the core from the film surface, as well as evaporating, or condensing with an intensity J_{31} vapor, have a longitudinal velocity ($v_{\Sigma3}$) of the substance on the interface. Thus, we assume

$$v_{13} = v_{31} = v_{\Sigma 3}, \quad v_{21} = v_{12} = v_2$$

$$v_{23} = v_2, \quad \overset{\circ}{J}{}'_{32} v_{32} = J_{32}^{(s)} v_2 + \left(\overset{\circ}{J}_{32} - J_{32}^{(s)}\right) v_{\Sigma 3} \tag{7.2.28}$$

In determining u_{ij}, the following assumptions are the most appropriate (cf. (1.1.55))

$$u_{21} = u_1(T_{\Sigma 2}), \quad u_{12} = u_2(T_{\Sigma 2}), \quad u_{13} = u_3(T_{\Sigma 3}), \quad u_{31} = u_1(T_{\Sigma 3})$$

$$u_{23} = u_2(T_2), \quad u_{32} = u_3(T_{\Sigma 3}) \tag{7.2.29}$$

The relations discussed in §1.4 may be used as closing relations for interphase interaction in the flow core between gas and drops (J_{21}, F_{21}, $Q_{1(\Sigma 2)}$, and $Q_{3(\Sigma 3)}$). Therefore, later, in §§7.3 and 7.4, only force and thermal interactions between the channel wall and film (F_W, Q_W) are discussed together with force, thermal, and mass interactions between the flow core and film (F_{13}, $Q_{1(\Sigma 3)}$, $Q_{3(\Sigma 3)}$, J^0_{32}, J^0_{23}, and $J^{(s)}_{32}$).

Equations of phase heat influx in conditions of thermodynamic phase equilibrium and velocity equilibrium in the flow core. In a wide range of processes, the phase temperature difference may be ignored, and in "hot" single-component flows, the departure of these temperatures from the saturation temperature may also be neglected, i.e., we assume

$$T_1 = T_2 = T_3 = T_{\Sigma 2} = T_{\Sigma 3} = T_S(p), \quad u_1 = u_{gS}(p), \quad u_2 = u_3 = u_{lS}(p) \tag{7.2.30}$$

Moreover, it may often be suggested that velocities of drops and gas in the flow core coincide

$$v_1 = v_2 = v_c. \tag{7.2.31}$$

Then, for this "single-temperature" and "single-velocity" core, a total drop-and-gas momentum equation and a total heat-influx equation should be considered. The equations for momentum and film, assuming stationary regimes and using Eqs. (7.2.27) and (7.2.28), become

$$\frac{d}{dz}\left(\rho_c v_c^2 S_c\right) = -S_c \frac{dp}{dz} - F_{13} - P_{cf} + \rho_c S_c g^z$$

$$\frac{d}{dz}\left(\overset{\circ}{\rho}_l v_3^2 S_3\right) = -S_3 \frac{dp}{dz} + F_{13} + P_{cf} - F_W + \rho_3 S_3 g^z$$

$$\rho_c = \overset{\circ}{\rho}_g \alpha_1 + \overset{\circ}{\rho}_l \alpha_2 \qquad (\alpha_1 + \alpha_2 = 1) \tag{7.2.32}$$

$$S_c \equiv S_1 = \pi (R - \delta)^2, \qquad S_3 \equiv S_f = 2\pi R \delta$$

$$P_{cf} = \left(\overset{\circ}{J}_{23} - J_{32}^{(s)}\right) v_c - \left(J_{31} + \overset{\circ}{J}_{32} - J_{32}^{(s)}\right) v_{\Sigma 3}$$

The mass-conservation equations for the species are written in the form

$$\frac{dm_1}{dz} = J_{21} + J_{31}, \quad \frac{dm_2}{dz} = -J_{21} + \overset{\circ}{J}_{32} - \overset{\circ}{J}_{23}, \quad \frac{dm_3}{dz} - J_{31} += \overset{\circ}{J}_{23} - \overset{\circ}{J}_{32}$$

$$\left(m_1 = \overset{\circ}{\rho}_g \alpha_1 v_1 S_c, \quad m_2 = \overset{\circ}{\rho}_l \alpha_2 v_2 S_c, \quad m_3 = \overset{\circ}{\rho}_l v_3 S_3\right) \tag{7.2.33}$$

The phase heat-influx equations (the last three equations (7.2.8)) and the heat-influx equations on interfaces—given that the kinetic energy dissipation is insignificant, $u_{32} = u_{23} = u_2 = u_3 = u_{lS}(p)$, and $i_{21} - i_{12} = i_{31} - i_{13} = i_{gS} - i_{lS} = l(p)$—are reduced to

$$\rho_1 \frac{di_g}{dz} + \rho_2 \frac{di_l}{dz} - \frac{dp}{dz} = -\frac{Q_{1(\Sigma 3)}}{S_c v_c} - \frac{J_{21} l}{S_c v_c}$$

$$\overset{\circ}{\rho}_l \frac{di_l}{dz} - \frac{dp}{dz} = \frac{Q_W - Q_{3(\Sigma 3)}}{S_f v_f} \tag{7.2.34}$$

$$Q_{3(\Sigma 3)} + Q_{1(\Sigma 3)} = J_{31} l, \quad Q_{1(\Sigma 2)} + Q_{2(\Sigma 2)} = J_{21} l.$$

Giving consideration to phase equilibrium, i.e., $i_g = i_{gS}(p)$, $i_l = i_{lS}(p)$, from these equations we find intensity of phase transitions on both film and drops in terms of pressure gradients, when $Q_W \neq 0$

$$J_{31} l = Q_W - \overset{\circ}{\rho}_l v_3 S_3 \left(\frac{di_{lS}}{dp} - \frac{1}{\overset{\circ}{\rho}_{lS}}\right) \frac{dp}{dz}$$

$$\left(|Q_{1(\Sigma 3)}| \ll |Q_{3(\Sigma 3)}|, \ |J_{21} l|\right) \tag{7.2.35}$$

$$J_{21} l = -\left\{\overset{\circ}{\rho}_g \alpha_1 \left(\frac{di_{gS}}{dp} - \frac{1}{\overset{\circ}{\rho}_g}\right) + \overset{\circ}{\rho}_l \alpha_2 \left(\frac{di_{lS}}{dp} - \frac{1}{\overset{\circ}{\rho}_l}\right)\right\} S_c v_c \frac{dp}{dz}$$

In cases when the phase velocities are much smaller than the equilibrium speed of sound in the core, and the pressure drops due to friction and the phase acceleration over the section under consideration are small ($\Delta p \ll p$), the phase enthalpy variation may be ignored. Then, it may be assumed that

$$J_{31} l = Q_W, \quad J_{21} \approx 0 \tag{7.2.36}$$

i.e., the total heat influx is consumed for the film vaporization (until it is completely dry) and the drops do not evaporate.

§7.3 INTERPHASE FRICTION AND HEAT EXCHANGE IN A DISPERSED-FILM FLOW

Based on the formulated supposition, or the principle of analogy which states that averaged flow and transfer processes in a film are similar to these processes in a wall zone of an equivalent, or "reduced" single-phase steady flow of liquid in the entire channel, the relations may be deduced for both resistance coefficients and heat exchange between film and the channel wall as functions of average film parameters. It is obvious that this supposition holds true for laminar wave-free films. For both wave and turbulent films, the result of such an approach, based on the above-formulated principle (or hypothesis) of analogy, must be experimentally verified.

Interaction between film and the channel wall. The wall film of liquid with a flow rate $m_3 = 2\pi R\delta\rho_3^0 \upsilon_3$ (average thickness of the film is δ, and temperature on its boundaries are $T_{\Sigma 3}$ and T_W) is set to match with an equivalent, or "reduced," single-phase flow of the same liquid occupying the entire cross section of the channel, and having in a near-wall annular layer of thickness $\delta (R - \delta < r < R)$ the same flow rate m_3, and the same temperature difference from T_W to $T_{\Sigma 3}$. We shall, in this case, assume that distributions of velocities and temperatures in the equivalent flow extrapolate the exponential distributions (7.2.23) of both velocities and temperatures in the film upon the entire cross section of the channel, namely

$$v_l(r) = \overset{\circ}{v_l}(1 - r/R)^{\nu_3}, \quad T_l(r) - T_W = (\overset{\circ}{T_l} - T_W)(1 - r/R)^{\theta_3} \tag{7.3.1}$$

where υ_l^0 and T_l^0 are, respectively, the values of velocity and temperature of the equivalent single-phase flow on the channel center line whose parameters are denoted by a subscript l. From conditions stipulated by both flow-rate and temperature-difference in the near-wall layer of thickness δ, expressions may be obtained that interrelate the average velocities υ_l and υ_3, and temperature differences $T_l^{(m)} - T_W$ and $T_{\Sigma 3} - T_W$. These conditions, using $\nu_3 = \theta_3$, are of the form

$$v_l = \frac{1}{\pi R^2}\cdot \int_0^R 2\pi r \cdot v_l(r)\, dr$$

$$= \frac{2}{(\nu_3+1)(\nu_3+2)}\left(\frac{\delta}{R}\right)^{-\nu_3} v_{\Sigma 3} = \frac{2}{\nu_3+2}\left(\frac{\delta}{R}\right)^{-\nu_3} v_3$$

$$\overset{\circ}{v_l} = v_3(\delta/R)^{-\nu_3}(1+\nu_3)^{-1} \tag{7.3.2}$$

$$T_l^{(m)} = \frac{1}{\pi R^2 v_l}\int_0^R 2\pi r T_l(r)\, v_l(r)\, dr = T_W + \frac{1 + {}^1\!/_2\nu_3}{1 + 2\nu_3}(T_{\Sigma 3} - T_W)\left(\frac{\delta}{R}\right)^{-\nu_3}$$

The friction force between the film and channel wall is represented as

$$F_W = 2\pi R \tau_W, \qquad \tau_W = C_W \frac{\overset{\circ}{\rho_l} v_3^2}{2} \tag{7.3.3}$$

where C_W is a coefficient of friction between the film and channel wall.

In a laminar axisymmetrical film, the shear stress equals $\tau(r) = \mu_l(\partial \upsilon_3/\partial r)$. Then, using the velocity-distribution law (7.2.18a), we obtain

$$C_W = 4/\mathrm{Re}_3, \quad \mathrm{Re}_3 < 300 - 400 \tag{7.3.4}$$

For a turbulent film, in conformity with the assumed analogy between friction in film and in an equivalent single-phase flow, shear stress on the surface of a smooth tube is determined by the Blausius formula (see L. G. Loitsyanskii, 1973) based on an exponential law (7.2.23) of the velocity distribution with an exponent $\nu_3 = 1/7$

$$\tau_W = C_{W(l)} \frac{\overset{\circ}{\rho}_l v_l^2}{2}, \quad 4C_{W(l)} = \frac{0.316}{\mathrm{Re}_l^{0.25}} \quad \left(\mathrm{Re}_l = \frac{\overset{\circ}{\rho}_l v_l D}{\mu_l}\right) \tag{7.3.5}$$

Now, from both (7.3.2) and (7.3.3), we have

$$v_l = \frac{14}{15}\left(\frac{\delta}{R}\right)^{-1/7} v_3, \quad \mathrm{Re}_l = 2\frac{R}{\delta}\frac{v_l}{v_3}\mathrm{Re}_3, \quad C_W = \frac{v_l^2}{v_3^2} C_{W(l)} \tag{7.3.6}$$

As a result, we obtain the "modified" Blausius law for a turbulent film

$$C_W = 0.0589/\mathrm{Re}_3^{0.25}, \qquad \mathrm{Re}_3 = m_3/(\pi D \mu_l) > 300 - 400 \tag{7.3.7}$$

The relations (7.3.4) and (7.3.7) are in satisfactory agreement with values C_W (Fig. 7.3.1) obtained as a result of measurements of pressure drop Δp on an experimental segment of the channel of length ΔL, liquid flow rate m_3 in the film, and also from measurements of the geometric average thickness δ of the film, which was determined based on averaging of local thicknesses of the film in conformity with (7.1.9), measured using the method of electrical conductivity (see §7.1) in vertical, upward, hydrodynamically stabilized gas-liquid flows (C. Shearer et al., 1965; P. Whalley, 1973) and high-pressure steam-water flows (B. I. Nigmatulin et al., 1982) in unheated channels. Analysis of these experiments and calculations of C_W and $\mathbf{Re}_3$ for each regime have been performed on the basis of the following considerations. Under stabilized regimes ($J_{23} = -J_{32} = 0$, $J_{23}^0 = J_{32}^0$), provided $\Delta p/p \ll 1$, both evaporation and condensation (even if they occur) are virtually insignificant over the segment of the channel under consideration ($J_{21} \approx J_{31} \approx 0$), and the flow parameters along the channel length are constant. Therefore, the inertia forces due to phase acceleration associated with either expansion or compression of the gas phase may be neglected. Then, the momentum conservation equation for the mixture may be written in the following form representing the tube wall friction forces

$$F_W = -S\frac{dp}{dz} - G, \qquad G = g^z \sum_{i=1}^{3} \rho_i S_i = g^z \rho S \quad (\rho = \overset{\circ}{\rho}_l \alpha_l + \overset{\circ}{\rho}_g \alpha_g) \tag{7.3.8}$$

Weight G of a column of liquid of unit length, which is to be taken into account for vertical flows ($g^z = 9.81$ m/s^2), is determined based on the void fraction $\varphi \equiv \alpha_g$, measured by the cutoff method, or by γ-radiation (see §7.1). The value of G for the dispersed-annular regimes ($\alpha_g \geqslant 0.8$) is gen-

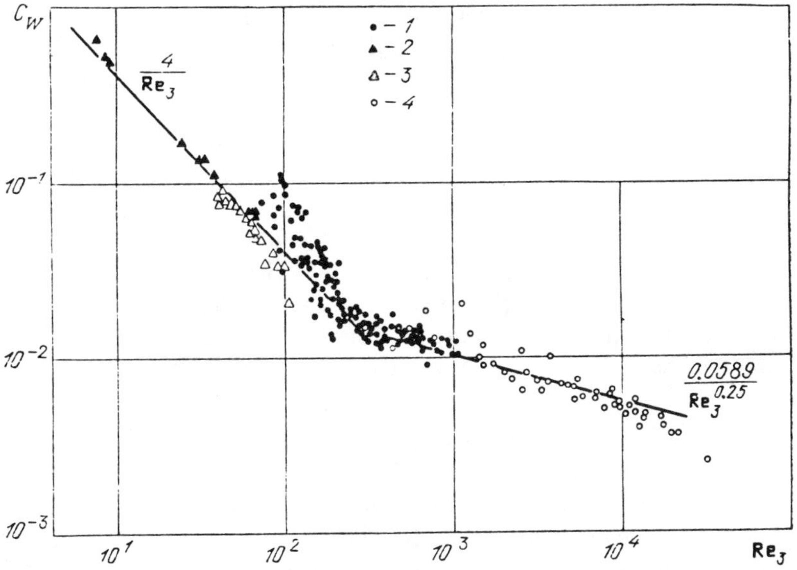

Figure 7.3.1 The coefficient C_W of friction between the film and channel wall as a function of Reynolds number $\mathbf{Re}_f \equiv \mathbf{Re}_3$, obtained from the following experimental studies: *1)* $p = 0.1$ MPa, $D = 31.8$ mm, air-water (P. Whalley et al., 1976); *2)* $p = 0.1$ MPa, $D = 15.9$ mm, air-water (C. Shearer et al., 1965); *3)* $p = 0.1$ MPa, $D = 31.8$ mm, air-trichlorethane (P. Whalley et al., 1973); *4)* $p = 3$–10 MPa, $D = 8$ mm, steam-water (B. I. Nigmatulin et al., 1982).

erally within 15% of the pressure gradient dp/dz. Therefore, there is no special need for great accuracy in determining $G(\alpha_g)$, and for α_g, there may be used empirical formulas of type (7.1.4) obtained from the processing of data of other experiments conducted in similar conditions. Note that, because of the film presence, $\alpha_g < \alpha_1$.

Having calculated F_W, and with m_3 and δ available, we obtain values of C_W and $\mathbf{Re}_3$, represented in Fig. 7.3.1

$$C_W = \frac{2F_W}{\pi D \overset{\circ}{\rho}_l v_3^2} \quad \left(v_3 = \frac{m_3}{\pi D \delta}\right), \quad \mathrm{Re}_3 = \frac{m_3}{\pi D \mu} \tag{7.3.9}$$

Suitability of the "reduced"-flow model in conformity with the "one-seventh" law and Blausius law for predicting the friction force F_W between turbulent films and the channel wall may be illustrated also in a different way. Knowing the measured values m_3 and Δp, the quantities F_W and $\mathbf{Re}_3$ may be calculated as before. And, using C_W, calculated based on the "reduced" Blausius law (7.3.7), the "modified" film thickness $\delta_{\Delta p}$ (reduced to the measured pressure drop and the Blausius law) can now be calculated

$$\delta_{\Delta p} = \frac{m_3}{\pi D v_3 \overset{\circ}{\rho}_l} \quad \left(v_3 = \sqrt{\frac{2F_W}{\pi D \overset{\circ}{\rho}_l C_W}}\right) \tag{7.3.10}$$

The values $\delta_{\Delta p}$ coincide with the measured average thicknesses of turbulent films (see Fig. 7.3.4), in spite of well-defined wave oscillations δ' which lead to a situation when the height of crests exceeds the average film thickness many times.

A more detailed experimental investigation of the relation $C_W(\mathbf{Re}_3)$ in the range of Reynolds numbers about $\mathbf{Re}_3 \sim 100$, where laminar flow changes to a turbulent regime, reveals a departure of experimental points from relations (7.3.4) and (7.3.7), and, also, the extent of this departure depending on the flow core parameters (M. E. Deich and G. A. Filippov, 1981).

The intensity of heat exchange Q_W between a film and channel wall shall be related to the total temperature gradient $T_W - T_{\Sigma 3}$ in the film by introducing the heat-transfer coefficient, or the Nusselt number

$$Q_W = \pi D q_W, \qquad q_W = \lambda_l \, \mathsf{Nu}_W \frac{T_W - T_{\Sigma 3}}{\delta} \tag{7.3.11}$$

In a laminar film both $q(r) = \lambda_l(\partial T/\partial r)$ and the heat-transfer coefficient significantly depend on hydrodynamic conditions and thermal factor $\Omega = Q_{3(\Sigma 3)}/\theta_W$, which determines the portion of the heat-influx to the film which is released from film towards its boundary with the flow core. In particular, for a flow with fixed shear (7.2.18a) and (7.2.21a), when $\tau_W = \tau_{13}$, and for a free flow (7.2.18b) and (7.2.21b), when $\tau_{13} = 0$, the related formulas can be readily obtained

$$\mathsf{Nu}_W = 3/(2+\Omega), \quad \mathsf{Nu}_W = 8/(5+3\Omega) \tag{7.3.12}$$

With $\Omega = 1$, both formulas yield $\mathbf{Nu}_W = 1$, and with $\Omega = 0$, they produce, respectively, 1.5 and 1.6. Thus, the difference in hydrodynamic conditions at a laminar regime of flow affects the heat exchange between film and wall to a slight extend. In this case, there is an equation for the thermal factor $0 < \Omega < 1$, which follows from the heat-influx equation (7.2.9) on the film surface, provided that $T_{\Sigma 3} = T_S(p)$

$$\Omega = \frac{J_{31} l - Q_{1(\Sigma 3)}}{Q_W} \tag{7.3.13}$$

It should be remembered that during evaporation and condensation, $\Omega \to 1$, and during heating or cooling with no phase transitions, $\Omega \to 0$.

Under turbulent regime of flow in a film, the major thermal resistance of the film is concentrated in a thin near-wall layer; therefore, the effect of conditions on the boundary with the flow core, characterized by the parameter Ω, upon the temperature profile in it, and, also, upon the heat exchange parameter $\mathbf{Nu}_W$, is insignificant. Then, as well as for friction (7.3.5), the heat exchange in a turbulent film may be described using the analogy of heat exchange in a reduced single-phase flow by means of the following well-known semiempirical formula

$$q_W = \lambda_l \, \mathrm{Nu}_{W(l)} (T_W - T_l^{(m)})/D$$
$$\mathrm{Nu}_{W(l)} = 0.023 \, \mathrm{Re}_l^{0.8} \, \mathrm{Pr}_l^{0.4} \qquad (\mathrm{Pr}_l = \mu_l c_l / \lambda_l) \tag{7.3.14}$$

For an exponential distribution of both velocity and temperatures in an equivalent flow with exponents $\nu_3 = \theta_3 = 1/7$, Eq. (7.3.6) holds, and we have

$$T_l^{(m)} - T_W = \frac{5}{6}\left(\frac{\delta}{R}\right)^{-1/7} (T_{\Sigma 3} - T_W), \qquad \mathrm{Nu}_W = \mathrm{Nu}_{W(l)} \frac{2R}{\delta} \frac{T_W - T_l^{(m)}}{T_W - T_{\Sigma 3}} \tag{7.3.15}$$

As a consequence, we arrive at

$$\mathrm{Nu}_W = 0.016 \, \mathrm{Re}_3^{0.8} \, \mathrm{Pr}_l^{0.4} \, (\delta/R)^{-0.057} \tag{7.3.16}$$

Figure 7.3.2 represents a comparison of the obtained relation (curves *1* and *1'*) with experimental data of various authors based on measurements of Q_W, T_W, $T_{\Sigma 3}$, and m_3, observed in film flows of various liquids along a flat wall, over interior and exterior surfaces of tubes of various diameters and lengths, in both horizontal and vertical descending directions, with concurrent gas flow and without it, with evaporation and condensation, heating and cooling. These data have been plotted as a function of the dimensionless groups $\mathbf{Nu}_W$ and $\mathbf{Re}_3$ (see B. I. Nigmatulin et al., 1981). In cases when $\mathbf{Pr}_l$ differed from 1.75, the experimental values $\mathbf{Nu}_W$ were "reduced" to $\mathbf{Pr}_l = 1.75$ using an additional multiplication by $(1.75/\mathbf{Pr}_l)^{0.5}$, in order to single-out the experimental dependence of $\mathbf{Nu}_W$ only on $\mathbf{Re}_3$. This procedure is quite justified because all relations in the case of turbulent flows reveal the effect of $\mathbf{Pr}_l$ in the form of a multiplier $\mathbf{Pr}_l^{\varkappa}$ in a rather unambiguous manner ($\varkappa = 0.4$–0.5 (see (7.3.16) and (7.3.17) below)). From Fig. 7.3.2, it is evident that the indicated treatment of data enables us to reveal the general-purpose nature of the relationship $\mathbf{Nu}_W(\mathbf{Re}_2, \mathbf{Pr}_l)$, and clearly indicates the absence of significant impact of other independent parameters in such a variety of conditions. Now, formula (7.3.16), to which curves *1* and *1'* correspond, quite properly describes the nature of the experimental relation for turbulent films ($\mathbf{Re}_3 > 300$) being discussed, but overpredicts the values $\mathbf{Nu}_W$ (on an average by a factor of 1.5). This is indicative of the fact that the thermal resistance of a turbulent film, as distinct from the friction resistance, is somewhat higher than the thermal resistance of the same layer of liquid in an equivalent single-phase flow. This situation may be given consideration by using for the equivalent single-phase flow (instead of (7.3.14)) a formula that has been obtained from analysis of heat exchange of a film freely flowing due to gravity force ($g^z = 9.81$ m/s^2) (D. A. Labunskii, 1957), which, using our nomenclature, may be written in the form

$$\frac{q_W \delta^{(g)}}{\lambda_l (T_W - T_{\Sigma 3})} = 0.0325 \, \mathrm{Re}_3^{0.25} \, \mathrm{Pr}_l^{0.5}, \qquad \delta^{(g)} = \sqrt[3]{\left(\frac{\mu_l}{\rho_2^\circ}\right)^2 \frac{1}{g}} \tag{7.3.17}$$

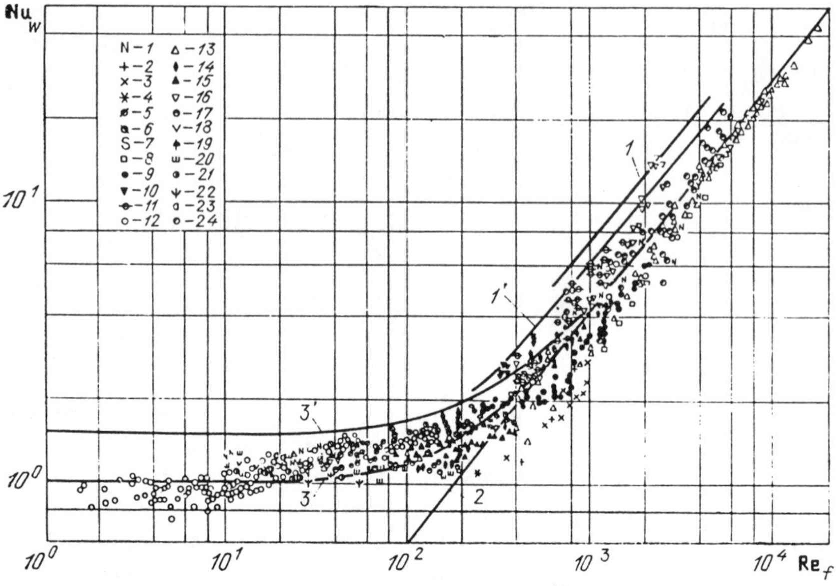

Figure 7.3.2 Relationship between $\mathbf{Nu}_W$ for heat transfer between the channel wall and film on Reynolds number $\mathbf{Re}_f$; which has been "reduced" to $\mathbf{Pr}_l = 1.75$ (see the discussion following Eq. (7.3.16)). There are represented experimental data (about 500 points) of 24 authors (see, for more detail, B. I. Nigmatulin et al., 1981) obtained during 1934–1978 in both horizontal and vertical downward film flows of various liquids (water, freon, liquid oxygen, nitrogen, ammonia, and diphenyl at $\mathbf{Pr}_l = 1.0$–8.4, but the major data are obtained for water at 100°C, when $\mathbf{Pr}_l = 1.75$) along a flat wall, over interior and exterior surfaces of tubes of various diameters ($D = 14$–61 mm) and lengths ($L = 0.2$–3.6 m) with a concurrent gas flow and without it, at evaporation and condensation, with heating and cooling. Lines *1* and *1'* relate to formula (7.3.16) at values $\delta/R = 10^{-3}$ and 10^{-1}. Line *2* relates to (7.3.20); lines *3* and *3'* relate to (7.3.21) at $\Omega = 1$ and 0.

From the equilibrium condition (7.2.17b) of a freely-flowing film, using (7.3.7), we have

$$\tau_W = \delta \overset{\circ}{\rho}_l g = \frac{0.0589}{\mathrm{Re}_3^{0.25}} \frac{\overset{\circ}{\rho}_l v_3^2}{2}, \qquad v_3 = \frac{\mu_l}{\overset{\circ}{\rho}_l \delta} \mathrm{Re}_3 \tag{7.3.18}$$

whence

$$\delta/\delta^{(g)} = 0.309\,\mathrm{Re}_3^{0.58} \tag{7.3.19}$$

Then, from (7.3.17), we have

$$\mathrm{Nu}_W = 0.010\,\mathrm{Re}_3^{0.83}\,\mathrm{Pr}_l^{0.5} \tag{7.3.20}$$

which describes the experimental data on heat exchange in turbulent films ($\mathbf{Re}_3 > 300$) in a satisfactory manner.

Matching the relations (7.3.12) and (7.3.20), we obtain a formula which expresses heat exchange of both laminar and turbulent films with the channel wall

$$\mathrm{Nu}_W = \sqrt{\left(\frac{3}{2+\Omega}\right)^2 + (0.010\,\mathrm{Re}_3^{0.83}\,\mathrm{Pr}_l^{0.5})^2} \tag{7.3.21}$$

This formula generalized the experimental data in the above-indicated range with a relative error of + 19% (see curves *3* and *3′* in Fig. 7.3.2).

Interaction between gas-drop core and a wall liquid film. Friction between a core and film is directly associated with regimes of flow of a wavy film surface defined by phase velocities and film thickness. Three types of film-surface regimes may be pointed out: a wavy regime with large-scale waves, a wavy regime with ripplemarks, and a smooth-film regime. A thorough experimental investigation of three regimes and the transition modes between them is still imperative.

A film swelling, a change of its wavy surface characteristics, and, accordingly, a change of interphase friction force arise in the process of a bubbly boiling in a heated channel. The transverse flow of evaporating or condensating vapor on the film surface must also affect the friction force.

The force of friction between the gas core and film, similarly to (7.3.3), is represented in the form

$$F_{13} = 2\pi R_c \tau_{13}, \quad \tau_{13} = {}^1\!/_2 C_{13} \overset{\circ}{\rho}_g (v_1 - v_{\Sigma 3})^2, \quad C_{13} = C_{13}(\delta/R, \mathrm{Re}_1, \bar{J}_{31}) \tag{7.3.22}$$

$$\bar{J}_{31} = \frac{w_1}{v_1}, \quad w_1 = \frac{J_{31}}{2\pi R_c \overset{\circ}{\rho}_g}, \quad \mathrm{Re}_1 = \frac{2(v_1 - v_{\Sigma 3}) R_c \overset{\circ}{\rho}_g}{\mu_1}, \quad R_c = R - \delta$$

where C_{13} is a coefficient of friction between gas core and film, $\bar{J}_{31}$ is the permeability parameter (S. S. Kutateladze and A. I. Leont'ev, 1972) characterizing the effect of the transverse velocity w_1 of gas (vapor) on the film surface, upon the friction force.

During a turbulent flow in the flow core, the coefficient of friction C_{13} must mostly depend on the mode of retardation over the waves as if it occurs under a well-developed turbulent flow of a single-phase liquid in rough tubes, since the flow core moves in a channel with liquid walls. A supposition that processes arising during a gas flow past individual waves on the film surface are analogous to those occurring near protrusions of a rough surface is reported in a well-known work by P. L. Kapitsa (1948). The "liquid-wall roughness" strongly varies in a rather wide range depending on the flow regime in both film and core. Systematic experimental studies of the impact of "roughness" of the liquid-film surface on the value of C_{13} have by now been conducted by a number of researchers (C. Shearer and R. Nedderman, 1965; G. Hewitt and N. Hall-Taylor, 1972; B. I. Nigmatulin, 1982 and 1983). We shall now discuss the major results of these investigations. The quantity C_{13} was determined based on measurements of: a total pressure drop Δp over a relatively short segment (Δz = 0.15–0.2 m) of the channel, flow rates of gas (m_g) and liquid (m_l) phases at both the entrance to and exit from this segment in a stabilized stationary gas-liquid and vapor-liquid "cold" air-water and "hot" steam-water flows. Experiments were conducted in both unheated and heated channels under conditions of small pressure drops ($\Delta p \ll p$) and thermodynamic equilibrium (phase temperatures are equal and equal the saturation temperature T_S), when gas and drop velocities may be regarded as almost equal to each other ($v_1 \approx v_2 \approx v_c$).

Equations (7.2.36) hold good in the indicated regimes, from which it follows that in unheated channels ($Q_W = 0$) there are no phase transitions

$$m_1 = \overset{\circ}{\rho}_g \alpha_1 v_1 S_c = \text{const}, \quad m_2 = \overset{\circ}{\rho}_l \alpha_2 v_1 S_c, \quad m_3 = \overset{\circ}{\rho}_l v_3 S_f \tag{7.2.23}$$

The momentum equations for the core and entire mixture in a stabilized flow ($J^0_{23} = J^0_{32} = J^0$) in accordance with (7.2.32) assume the form

$$F_{13} = -S_c \frac{dp}{dz} - \left(J^\circ - J^{(s)}_{32}\right)(v_1 - v_{\Sigma 3}) + G_c, \quad F_W = -S\frac{dp}{dz} + G$$

$$\left(G_c = \rho_c S_c g^z, \quad G_f = \overset{\circ}{\rho}_l S_f g^z, \quad G = G_c + C_f\right) \tag{7.2.24}$$

In experiments with unheated channels, the quantities dp/dz, m_1, m_3, and m_l are measured. Using these quantities, and given that $m_2 = m_l - m_3$, and using also the phase densities ρ^0_g and ρ^0_l, the volumetric concentrations of both liquid and gas in the flow core may be determined as

$$\alpha_1 = \frac{m_1/\overset{\circ}{\rho}_g}{m_1/\overset{\circ}{\rho}_g + m_2/\overset{\circ}{\rho}_l}, \quad \alpha_2 = 1 - \alpha_1 \tag{7.3.25}$$

and the weight G_C (assuming that $S_c \approx S$) of a unit column of the flow core may now be found. As in the case of (7.3.8), using the approximation for φ, the weight of G of a unit column of the entire flow can also be calculated, and, then, the quantities F_W (from the second equation (7.3.24)), $\mathbf{Re}_3$, C_W, v_3 may be successively found; finally, the film thickness $\delta_{\Delta p}$ which, as was shown above, coincides with the geometric average value of film thickness δ, may be determined similar to (7.3.10). After the above-indicated calculations, a refinement may be accomplished based on $S_c = S - 2\pi R\delta$. Now, both velocity in the flow core $v_1 = m_1/(\rho^0_g \alpha_1 S_c)$ and velocity $v_{\Sigma 3} \approx 1.14\ v_3$ are calculated. Having obtained v_1, $v_{\Sigma 3}$, $\rho^0_g \rho^0_l$, and other physical properties of phases, the values J^0 and $J^{(s)}_{32}$ may be calculated using the empirical formulas for the mass exchange intensity between film and the flow core (see §7.4 below); then, the values F_{13} and $C_{13} = F_{13}/\pi R_c \rho^0_g (v_1 - v_{\Sigma 3})^2)$ may be singled-out of the first equation (7.3.24).

Note that errors resulting from using empirical formulas for φ, J^0, and $J^{(s)}_{32}$, which define both terms G and $(J^0 - J^{(s)}_{32})(v_1 - v_{\Sigma 3})$, are not substantial since the indicated terms in turbulent flows do not represent major constituents of both F_W and F_{13}.

Figure 7.3.3 depicts the results of the above-discussed treatment of experiments of various researchers for steam-water and air-water flows in the

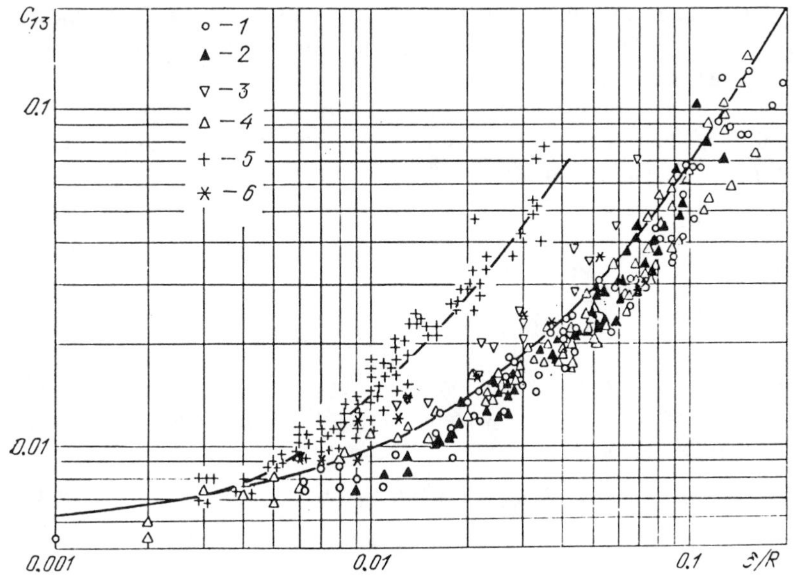

Figure 7.3.3 Coefficient C_{13} of friction between the gas-drop turbulent core ($\mathbf{Re}_{13} \geqslant 10$) and liquid film as a function of the relative thickness δ/R of the film obtained from the following experimental conditions: *1)* p = 5–10 MPa, D = 8 mm, steam-water (N. V. Tarasova et al., 1965; G. Gaspari et al., 1964); *2, 3)* p = 3–9 MPa, D = 10 mm (*2*) and D = 20 mm (*3*), steam-water (J. Wurtz, 1978); *4)* p = 1–10 MPa, D = 13.3 mm, steam-water (B. I. Nigmatulin et al., 1978, 1983); *5)* p = 0.28 MPa, D = 31.8 mm, air-water (P. Whaley et al., 1975); *6)* p = 0.18–0.45 MPa, D = 31.5 mm, air-water (S. V. Netunaev, 1982). The lower solid line passing through points *1–4* and *6* is plotted in accordance with relation (7.3.26).

form of a relationship between C_{13} and relative film thickness δ/R for well-developed turbulent flows. Experimental points, characterized by a mean-root-square error of + 20%, are generalized by a formula

$$C_{13} = 0{,}005 + 0{,}84(\delta/R)^{1,1} \approx 0{,}005 + 0{,}6\delta/R \qquad (7.3.26)$$

Note that the "air-water" points *6* relate to an annular regime (with no drops, and, consequently, with no splash removal) and are in a good agreement with "steam-water points" and the relation (7.3.26). At the same time, the "air-water points" *5* relate to the dispersed-annular regime with the presence of splash removal, and they depart from the remaining points to a considerable extent. This may be explained by the fact that in air-water flows, the assumption (7.2.28), in a condition when drops are knocked off the film surface by the precipitating drops, is not satisfied, and $v_{(32)}^{(s)} < v_2$.

The fact that C_{13} depends only on δ/R substantiates the hypothesis on the similarity between friction of a turbulent gas-drop core over a near-wall liquid film and friction of a well-developed turbulent flow of a single-phase liquid ($\mathbf{Re} \geqslant 10^5$) in a rough tube when the coefficient of friction is independent of the Reynolds number, depending only on the tube roughness. In this case, the effective "roughness" of the film is unequivocally defined by its average thickness.

To describe the exchange of momentum between the core and film, a simplified model may be employed in which the loss of pressure for accelerating the drops broken off the film surface is not identified individually, namely

$$\Phi_{13} = -S_c \frac{dp}{dz} + G_c = C^*_{13} \frac{\overset{\circ}{\rho}_g (v_1 - v_{\Sigma 3})^2}{2} 2\pi R_c \qquad (7.3.27)$$

$$\left(\Phi_{13} = F_{13} + (J^\circ - J^{(s)}_{(32)})(v_c - v_{\Sigma 3}), \quad S_c = \pi R_c^2, \quad R_c \approx {}^1/_2 D\right)$$

Here, Φ_{13} is determined based on measurements of the pressure drop and evaluations of the flow core weight G_c (see (7.1.4)). Such treatment leads to an approximation

$$C^*_{13} = 0.005 + 0.6\,(\delta/R) + 3.3\cdot 10^4\,(\delta/R)^{5.5} \qquad (7.3.28)$$

§7.4 DROPWISE MOISTURE EXCHANGE BETWEEN A CORE AND WALL LIQUID FILM IN A TURBULENT DISPERSED-FILM FLOW

An experimental approach is presently a major method of acquisition of quantitive information on the dropwise moisture exchange. We will now briefly discuss the results of related studies described by B. I. Nigmatulin and his co-workers. Experiments were conducted with stationary upward flows of thermodynamically equilibrium steam-water and air-water mixtures in tubes with inner diameters D = 8 and 13 mm (for steam-water mixtures), and D = 13 and 31.5 mm (for air-water mixtures) with parameters varying in the following ranges: for steam-water mixtures p = 1–12 MPa, T = 450–600 K, v_1 = 4–120 m/s, α_2 = 0.005–0.1; for air-water mixtures p = 0.18, 0.3, and 0.45 MPa, T = 293 K, v_1 = 8–60 m/s, α_2 = (0.9–5) × 10^{-3}. The experimental segments consisted of two sections: the flow stabilization section of a length of about 200 D, and the measuring section. At the inlet to the measuring section, the mixture species flow rates m_1, $m_l = m_2 + m_3$, pressure p, and mixture temperature T were measured. At several points along the measuring section, the liquid flow rate m_3 in the film was measured by the suction method, and the pressure drop Δp was also measured at several points. Using the pressure drop along the measuring section, the shear stress τ_{13} and τ_W may be determined. The salt-tracer concentration in liquid samples, obtained by suction out of the film, was measured at various distances from the inlet (the salt-tracer was introduced into the wall film at the inlet).

The film suction method. The determination of the drop removal (breaking away) intensity and their deposition onto the film are based on measurements of the liquid flow-rate in the film in the absence of vaporization and condensation when the flow-rate equation in the film is

$$dm_3/dz = \overset{\circ}{J}_{23} - \overset{\circ}{J}_{32} \tag{7.4.1}$$

In general, both drop breaking-away and deposition occur simultaneously and over the segment after stabilization balance each other ($J^0_{32} = J^0_{23} = J^0$). Therefore, in order to find them, conditions must be realized when one of these processes does not happen. For example, to find the deposition intensity J^0_{23} at predetermined flow core parameters at the inlet to the measuring section, (at $z = 0$), the entire film is sucked out in such a manner that nothing can be broken away at the inception portion of the measuring section (it may be regarded that $J^0_{32} = 0$), and flow rate in the film at the inception of the measuring section grows linearly (with respect to z from zero) due to drop deposition with intensity J^0_{23}. And, only downstream (where because of the drop deposition, a film of a sufficient thickness is produced to ensure the drop removal) the breaking-away process comes into effect, and the growth $m_3(z)$ will slow down (see points *1* and *2* in Fig. 7.4.1).

Similarly, in order to determine the breaking-away intensity J^0_{32} at prescribed parameters of both film and gas, the entire liquid at the inlet to the measuring section is fed (through a porous insertion) in the form of a film so that, at the inception of this section nothing can deposit, and, therefore, it may be considered that $J^0_{23} = 0$. And, only downstream (where because of the breaking away) a sufficient amount of drops is produced in such a manner that their deposition becomes significant, and the reduction process slows down (see points *3* and *4* in Fig. 7.4.1). In fact, at these two extreme conditions, J^0_{23} and J^0_{32} are determined by the slope of linear portion of the relation $m_3(z)$ at $z = 0$ plotted based on the measurement results.

Salt-tracer method. It is the difference between intensities of deposition on the "dry" wall and on the film, and also, the difference between inten-

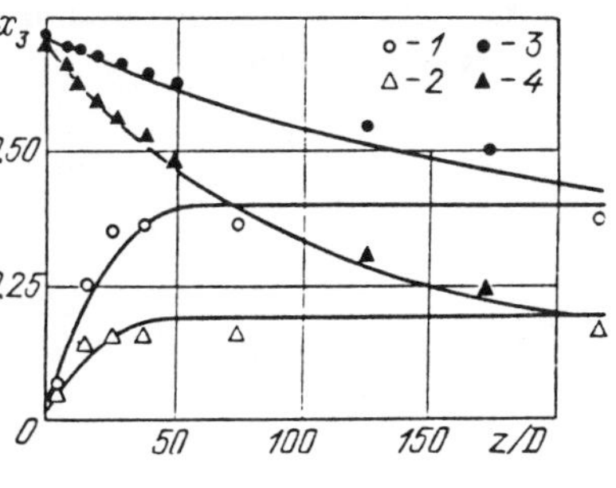

Figure 7.4.1 Variation of the relative discharge $x_3 = m_3/m$ of liquid in the film with distance from the channel inlet where the entire liquid was fed either to the flow core (points *1* and *2*) or to film (points *3* and *4*) at fixed pressure $p = 6.9$ MPa and channel diameter $D = 13.3$ mm conveying a steam-water flow (B. I. Nigmatulin et al., 1978, 1981). Points *1* and *3* relate to $m^0 = 1000$ kg/(m$^2 \cdot$ s), $x = 0.27$; points *2* and *4* relate to $m^0 = 1600$ kg/(m$^2 \cdot$ s), $x_1 = 0.30$; lines correspond to an analysis using formulas (7.4.8), (7.4.19), and (7.4.23).

sities of moisture removal from the film by gas flow and its removal from the same film but by the gas-drop flow that must be scrutinized. The film effects may be accounted for by variation of the gas-liquid core boundary layer which is responsible for the transverse motion of drops, which, in turn, defines their deposition. The effect of drops in the flow core upon the moisture removal from the film may occur due to splashing when a drop strikes the film. The indicated verification has been performed by B. I. Nigmatulin, I. V. Dolinin et al. (1978) by means of a salt-tracer method of measuring the moisture exchange intensity in a hydrodynamically developed dispersed-film steam-water flow where $J^0_{23} = J^0_{32} = J^0$, and J^0 is the measured moisture exchange intensity. This technique was earlier employed in measurements of the moisture exchange intensity in air-water flows under pressures resembling atmospheric pressure (E. Quandt, 1965; L. Cousins and G. Hewitt, 1968; A. Jagota et al., 1973). The substance of the technique is as follows. A salt solution (brine) is fed at a flow rate $m_s \ll m_3$ to the film at the inlet of the measuring section with a fully-developed flow ($J^0_{23} = J^0_{32} = J^0$). Then, at several sections along the flow, the salt concentration in the film is measured (this concentration reduces downstream because of moisture exchange between core and film). The indicated concentrations are measured in the samples of liquid sucked out of film at corresponding sections. In order to determine the relation between the moisture exchange intensity J^0 and variation of the salt concentration c_3 in the film along the length of a channel, we use the salt conservation equation for both film and drops

$$m_3 dc_3/dz = J^\circ (c_{23} - c_{32}), \quad m_2 dc_2/dz = J^\circ (c_{32} - c_{23})$$
$$\left(m_2,\ m_3 = \text{const},\ m_j c_j = \int_{S_j} \alpha_j(r)\, \overset{\circ}{\rho}_j v_j(r)\, c_j(r)\, ds,\ j = 1, 2, 3 \right) \tag{7.4.2}$$

where c_j is the flow-rate average salt concentration in the j-th mixture component ($c_1 = 0$). It is generally envisaged that drops with average salt concentration c_{23} undergo deposition. The evaluations prove that the salt diffusion in a turbulent film is so strong that the salt concentration profile in it resembles a uniform distribution as early as at distances $z \geqslant 10\delta$ to 0.01 m from the salt feeding section. Then, it may be said that $c_{32} = c_3$. Diffusion of salted drops over the core cross section is also sufficiently intensive to ensure that the salt-concentration profile in the flow core is uniform at distances $z > 10D$ (at $D \sim 10$ mm, this distance is of the order of 0.1 m), and $c_{23} = c_2$. The boundary conditions for Eq. (7.4.2) are

$$\begin{aligned} z = 0&: \ c_3 = c_{30}, \quad c_2 = 0 \\ z \to \infty&: \ c_2 = c_3 = c_\infty \end{aligned} \tag{7.4.3}$$

The integral of the system under consideration may be written as

$$m_2 c_2 + m_3 c_3 = m_3 c_{30} = (m_2 + m_3)\, c_\infty \tag{7.4.4}$$

Upon integrating, we obtain

$$\ln\left[\frac{c_3(z)}{c_\infty}-1\right]=A_0-\frac{z}{L_J}$$
$$\left(A_0=\ln\frac{m_2}{m_3}\equiv\ln\left(\frac{c_{30}}{c_\infty}-1\right),\quad L_J=\frac{m_2m_3}{(m_2+m_3)\,J^\circ}\right) \tag{7.4.5}$$

The obtained solution is represented by a straight line with an inclination determined by the relaxation length L_j, and intercepting the ordinate (z = 0) at a point defined by A_0 in a semilogarithmic system of coordinates. Hence, using the measured salt concentrations $c_3(z)$ in liquid samples obtained from the wall film at various distances z from the points of salt injection (including sufficiently large distances z, where $c_3 = c_\infty$, i.e., c_3 is unvarying, since it is more convenient to measure c_∞ than c_{30}), both values L_j and A_0 are readily found. These values, together with the measured liquid flow rate $m_2 + m_3$, enable us to simultaneously determine J^0, m_2, and m_3. Thus, the salt-tracer method provides simultaneous determination of both moisture exchange intensity J^0 and liquid distribution between the core (m_2) and film (m_3) in a fully-developed flow without disturbing the latter.

Figure 7.4.2 represents the dependence of ln $[c_3(z)/c_\infty - 1]$ on z for various regimes. All experimental points group (with insignificant scatter) around straight lines which correspond to Eq. (7.4.4). It should be noted that the location of experimental points is unaffected by the flow rate m_1 = (0.02–0.04) m_3 of brine. Hence, it follows that the conditions assumed for the derivation of Eq. (7.4.2) are ensured in the process of this experiment, and the measurements themselves are sufficiently accurate.

It should be remembered that in undeveloped flows, when $J^0_{23} \neq J^0_{32}$, the

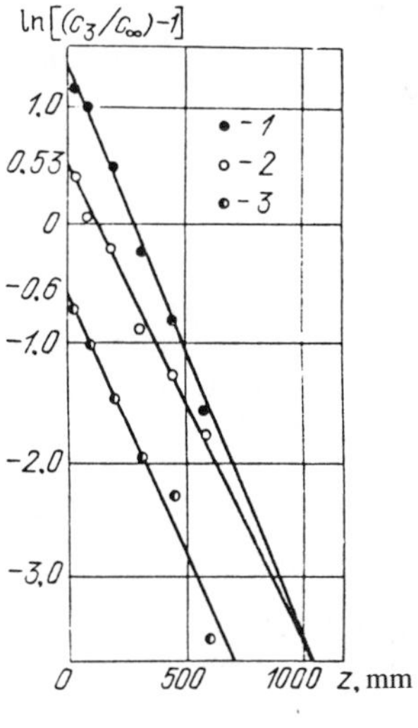

Figure 7.4.2 Results of measurements of salt-tracer concentration along an unheated measuring section (D = 13.1 mm) with an upward steam-water flow (p = 2.94 MPa, m^0 = 1000 kg/m$^2\cdot$s)) of various flow-rate concentrations of vapor (steam): x_1 = 0.6 (*1*); 0.4 (*2*); 0.2 (*3*).

salt-tracer technique alone (without measuring $m_3(z)$) does not identify the intensity of mass exchange between the core and film, i.e., the intensities of both deposition J^0_{23} and removal (breaking away) J^0_{32}. The variation of parameters along the flow and with time in such flows depends also on the imbalance between J^0_{23} and J^0_{32}.

Comparison of experimental results on the relative flow rate x_3 of liquid in film and the intensity of moisture exchange $J^0 = J^0_{32} = J^0_{23}$ obtained by the salt-tracer technique and by the method of film suction through a porous insertion showed that at $x_3 > 0.25$, the values x_3, obtained by two different methods, are in an agreement within $\pm 20\%$. The scatter of results is slightly higher at lower values $x_3 < 0.25$, which is associated with a deterioration of the accuracy of measurements of small flow rates in the film by both methods.

Intensity of drop deposition. A theoretical study of drop deposition on the film and the determination of its quantitive characteristic, intensity J^0_{23} of deposition, is impeded by two factors: first, the picture of interaction between drops and turbulent flow of gas phase is extremely complicated, and second, the function of drop distribution with respect to their size in the core is unknown for most practical cases, whereas the drop size is the basic parameter needed for development of the process theory. The most complete review of information concerning particle precipitation on the tube wall in the absence of the wall film as well as their analysis may be found in a number of works (R. Boothroad, 1971; A. A. Shraiber et al., 1980; E. P. Mednikov, 1981). A review of experimental investigations of drop precipitation on a film is presented in an article by D. McCoy and T. Hanratty (1977).

The value J^0_{23} may be conveniently represented through a dimensionless quantity $\bar{J}^0_{23}$ which equals the ratio of the specific transverse drop flow $J^0_{23}/(2\pi R_c) = \rho^0_l \alpha_2 w_2$ (where w_2 is the average transverse velocity of depositing drops) to specific longitudinal drop flow $m_2/(\pi R_c^2) = \rho^0_l \alpha_2 v_2$. Then, given that $2R_c \approx D_1$, we have

$$J^\circ_{23} = \frac{4m_2}{D} \bar{J}^\circ_{23}, \qquad \bar{J}^\circ_{23} = \frac{w_2}{v_2} \tag{7.4.6}$$

Analysis of drop motion affected by random turbulent pulsations permitted the identification of a dimensionless parameter (A. E. Kroshilin, V. N. Kukharenko et al., 1985) defining the deposition intensity

$$\Pi = \frac{t^{(\mu)} v_*}{D} \frac{v_*}{v_1} \qquad \left(t^{(\mu)} = \frac{2a^2 \rho^\circ_l}{9\mu_g}, \quad v_* = \sqrt{\frac{\tau_W}{\rho^\circ_g}} \right) \tag{7.4.7}$$

Here, $t^{(\mu)}$ is the Stokes' time of particle velocity relaxation (1.4.33), v_* is the dynamic velocity of gas in turbulent pulsations. In the regimes discussed below, the Reynolds numbers $\mathbf{Re}_* = 2av_*/v_g^{(v)}$, characterizing the motion of drops in turbulent pulsations equal 0.1–10 (by order of magnitude); it is, therefore, the Stokes time $t^{(\mu)}$ that in these regimes characterizes the velocity relaxation in the process of turbulent deposition.

Analysis of available experimental data concerning deposition in unheated (the respective values J_{23}^0 are denoted by J_{23}^{00}) air-water and steam-water flows showed that these data are described by the following linear relation with an average relative error within ±50% in the range $\Pi = 2 \times 10^{-3}$–2×10^{-1}

$$\bar{J}_{23}^{\circ} = \bar{J}_{23}^{\circ\circ} = 0.3\,(1 - 7.5\alpha_2)\,\Pi \tag{7.4.8}$$

In this case, the drop size in the flow core, which appears in the parameter Π, was determined using an empirical formula (P. Whaley, 1978) obtained based on measurements in vertical steam-water flows; this formula determines the drop radius as a function of regime parameters (p, v_1, m^0, x_g, D)

$$\frac{a}{D} = 1.9 \cdot \mathrm{Re}_1^{0.1} \cdot \mathrm{We}_1^{-0.6} \left(\frac{\overset{\circ}{\rho}_g}{\overset{\circ}{\rho}_l}\right)^{0.6} \qquad \left(\mathrm{Re}_1 = \frac{\overset{\circ}{\rho}_g v_1 D}{\mu_g}, \quad \mathrm{We}_1 = \frac{\overset{\circ}{\rho}_g v_1^2 D}{\Sigma}\right) \tag{7.4.9}$$

Accordingly, the dynamic velocity v_* for processing the experimental results was calculated using the well-known formula for single-phase flows (see L. G. Loitsyanskii, 1983), which is consistent with the Blausius formula (7.3.5)

$$v_*/v_1 = 0.2\,\mathrm{Re}_1^{-0.125} \tag{7.4.10}$$

The multiplier (1–7.5 α_2) takes into account the reduction of the turbulent pulsation intensity in gas due to the presence of drops, and also, the effects of drop multiplicity on the deposition process.

The relation (7.4.8) is in disagreement with experimental results on solid-particle deposition. These data indicate a rather ineffectual impact of Π on $\bar{J}_{23}^{00}$. In particular, experiments conducted by B. Liu and G. Agarwal (1974) yield $\bar{J}_{23}^{00} \approx 6 \times 10^{-3}$ in the range Π = 0.001–0.01. The relation (7.4.8) is also in disagreement with data on precipitation of drops of water which are produced by the injection of moisture through a nozzle. In this case, a much narrower spectrum of drop sizes is produced compared to ordinary dispersed-film flows, and the measured values of $\bar{J}_{23}^{00}$ (E. Ganic and K. Mastanaiah, 1981) vary from 6×10^{-3} to 3×10^{-3} with an increase of Π from 3×10^{-3} to 7×10^{-2}.

The further refinement and generalization (including gas with solid particles) of the relation for J_{23}^{00} is associated with the exact specification of the drop size, and, moreover, the specification of the drop-size spectrum in the flow core and in the annulus of depositing drops.

A transverse flow of vapor (steam) occurs in a heated channel due to film vaporization; this flow may hinder the drop deposition process on the film surface. The effects of this transverse flow on drop deposition from the gas-drop core of the flow has been studied (B. I. Nigmatulin et al., 1982) using an air-water flow through a horizontal segment of a channel in which the transverse gas flow was generated by injecting air through a porous bronze insertion sealed flush with the lower channel wall. The drop flow was produced by means of a nozzle installed before the operational segment of the channel. Experiments were conducted in the range of velocities v_1 = 20–50 m/s at a pressure p = 0.1 MPa. Velocity of the transverse (injected) air flow varied in the range of w_1 = 0–0.83 m/s that is adequate to w_1/v_1 = 0–0.042. Drops in the near-wall zone were photographed from one side with a special lighting provided. The photography technique (V. V. Guguchkin et al., 1978) enabled the experimentators to determine the number N of drops in the frame, to measure their size and velocity vector, and also, to evaluate the drop quantity distribution $f = N^{-1}(\Delta N/\Delta\beta_2)$ with respect to angles β_2 of their velocity deviation (trajectory deviation) from the longitudinal direction (Fig. 7.4.3). In the absence of the transverse injection, the average angle β_{2m} of the drop trajectory deviation from the longitudinal direction of the wall is zero. In the presence of the transverse injection, the drops, on the average, deviate from the longitudinal direction ($\beta_{2m} > 0$), tan β_{2m} linearly growing with the growth of w_1/v_1; for conditions related to Fig. 7.4.3, tan $\beta_{2m} \approx 3(w_1/v_1)$ takes place. The number of drops departed from the wall is determined by the number of drops whose trajectories make a positive angle β_2.

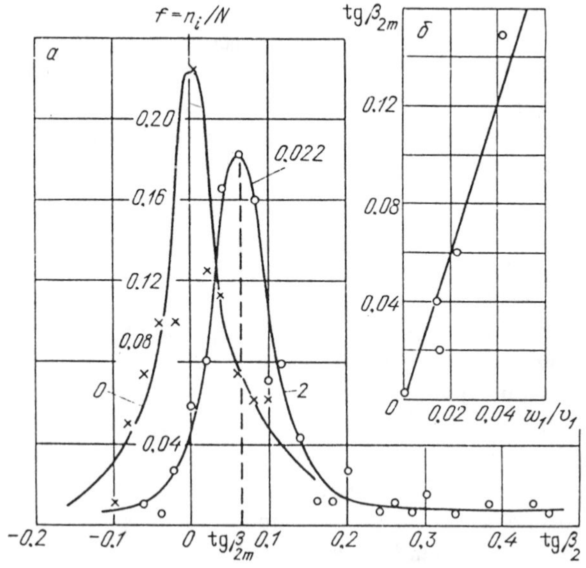

Figure 7.4.3 Distribution f of drops with respect to angles β_2 of deviation of their velocities from the longitudinal direction (a), and average values β_{2m} of these angles (b) during a transverse gas injection in an air-water drop flow (p = 0.1 MPa; v_1 = 35 m/s). Numerical labels 0 and 0.022 on curves $f(\beta_2)$ correspond to values w_1/v_1.

The intensity of drop deposition decreased due to the transverse flow of evaporating steam may be represented in the form

$$\overset{\circ}{J}_{23} = \begin{cases} \overset{\circ\circ}{J}_{23} - \Delta J_{23}^{(q)}, \quad \Delta J_{23}^{(q)} = k^{(q)} \rho_2 J_{31} / \overset{\circ}{\rho}_1, \quad \text{if} \quad \Delta J_{23}^{(q)} \leqslant \overset{\circ\circ}{J}_{23} \\ 0, \quad \text{if} \quad \Delta J_{23}^{(q)} > \overset{\circ\circ}{J}_{23} \quad (J_{31} = Q_W / l) \end{cases} \tag{7.4.11}$$

Here, $k^{(q)}$ is the blowing coefficient which was evaluated (B. I. Nigmatulin, 1973) based on experiments with heat flows when particle deposition in steam-water flows under the above-indicated regimes comes completely to an end. This qualitative evaluation yielded $k^{(q)} \sim 2$.

Intensity of the drop breaking away (removal) from the film surface. A portion of liquid in a dispersed-annular regime of gas-liquid flow in rectilinear channels is generally broken away from crests of large-scale waves and transferred to the flow core. This process is referred to as a dynamic (wave) break-away, and its intensity is denoted by $J_{32}^{(\alpha)}$. In the presence of drops in the flow core, an additional removal (as was indicated in §7.2) of droplets of moisture knocked out of a film in the form of secondary drops (splashes) is likely; these splashes are caused by a shock produced by drops of the flow core striking the film. The process is referred to as a shock splash break-away, and its intensity is denoted by $J_{32}^{(s)}$. In an intensively heated channel, when the film undergoes a bubbly boiling, the removal of moisture out of film in the form of splashes is also possible (splashes occur when vapor (steam) bubbles appear on the film surface). This process is referred to as bubbly splash break-away (removal), and its intensity is denoted by $J_{32}^{(b)}$. Both shock and bubbly splash removals are said to be by-products which take place during deposition of drops from the flow core and bubbly flashing of the film. Each of the three indicated break-away (removal) processes is characterized by the conditions at which the particular type of removal initiates and also by its intensity. In general, the intensity of the overall break-away of drops from the film surface is

$$\overset{\circ}{J}_{32} = J_{32}^{(d)} + J_{32}^{(s)} + J_{32}^{(b)} \tag{7.4.12}$$

The process of dynamic drop removal ($J_{32}^{(\alpha)}$) is characterized by interaction of turbulent pulsations, forces due to surface tension, viscosity, and inertial forces. Direct observations, photography, and holography of the film surface in a wake air flow, and those of an aqueous film under pressure close to atmospheric (G. Hewitt and N. Hall-Taylor, 1972; V. I. Bystrov and M. E. Lavrent'ev, 1976) showed that drop break-away occurs only from crests of large-scale waves; both wave break-down and drop formation take place in a manner similar to the breaking-up of liquid streams in gas; they depend significantly on liquid viscosity. At relatively low gas velocities (below 25–30 m/s), the wave break-down follows the deformation of the wave as a whole: a wave falls apart with its crests ejecting streamers which break up into individual drops. At larger gas velocities, the scale of perturbations is smaller by one order of magnitude than the size of the wave itself, and its complete break-down resembles atomization or stripping.

Only such issues as wave formation in laminar films blown around by gas lend themselves to theoretical analysis. The phenomena of drop break-away, all the more from turbulent films, are based solely on qualitative considerations.

Conditions for initiation of the dynamic drop break-away from a film surface by gas flow. The conditions for the film fragmentation, or the dynamic drop break-away by gas from the film surface, are defined by the mechanism leading to the Kelvin-Helmholtz instability. This mechanism is characterized by the Weber number $\mathbf{We}_{13}$, equal to the ratio of dynamic effect to capillary forces. A measure of the dynamic effect of gas-liquid flow on the film wavy surface is shear stress τ_{13} (for turbulent flow $\tau_{13} \sim \rho_g^0(v_1 - v_2)^2$). Supposing that, in similarity with drop fragmentation (see §2.2), the waves with lengths smaller than film thickness δ, and with amplitudes of order of δ, are "dangerous" for films, we arrive at the conclusion that a quantity Σ/δ is a measure of capillary forces. Then, the condition for the initiation of the dynamic break-away may be written in the form

$$\mathrm{We}_{13} = \tau_{13}\delta/\Sigma > \mathrm{We}_{13*} \tag{7.4.13}$$

The effect of viscous forces in phases, as well as in the process of drop fragmentation, is characterized by Laplace numbers

$$\mathrm{Lp}_g = \frac{\rho_g^\circ \Sigma\delta}{\mu_g^2}, \quad \mathrm{Lp}_l = \frac{\rho_l^\circ \Sigma\delta}{\mu_l^2} \tag{7.4.14}$$

with the viscosity increase (reduction of $\mathbf{Lp}_i$; $i = g, l$) enhancing the film stability, i.e. increasing the values $\mathbf{W}_{13*}$. Furthermore, the turbulent viscosity $\mu^{(t)} \sim \rho_3 v_3 y$ (where y is the distance to the channel wall) may become a stabilizing factor hampering the liquid breaking-away in the process of a turbulent flow. On the film surface ($y = \delta$), we have $\mu^{(t)} \sim \mu_l \mathbf{Re}_3$.

In addition to capillary, dynamic, and viscous forces, the liquid is affected by gravity forces which exert a significant effect on the conditions of initiation of the dynamic drop breaking-away from the film. In particular, this is exhibited by the fact that the experimental values $\mathbf{We}_{13*}$ are different for upward, downward, and horizontal flows. The ratio between these forces and capillary forces are determined by

$$\bar{g} = g\rho_l^\circ \delta^2/\Sigma \tag{7.4.15}$$

If the break-away process must depend on the phase densities ρ_g^0 and ρ_l^0, the dependence of the critical Weber number, which specifies the beginning of the dynamic break-away of drops, on the process parameters may be represented as

$$\mathrm{We}_{13*} = f\left(\mathrm{Lp}_l, \mathrm{Lp}_g, \rho_g^\circ/\rho_l^\circ, \bar{g}, \mathrm{Re}_3\right) \tag{7.4.16}$$

Experimental investigations of the initiation of the dynamic drop break-away from the film surface are mainly devoted to determining the flow parameters (flow rate in the film and critical gas velocity) at which the break-away starts (N. A. Mozharov, 1959; A. Ya. Zhivaikin, 1961; V. A. Chernukhin, 1963; G. Wallis, 1969; G. Hewitt and N. Hall-Taylor, 1972; M. Ishii and M. Grolmes, 1975; S. S. Kutateladze and M. A. Styrikovich, 1976; A. A. Andreevskii, 1978). The pressure drop due to friction, whose magnitude is required for determining the average shear stress τ_{13} on the film surface, was not measured in most of the works indicated above. Therefore, shear stress τ_{13} was not used in these works as a major defining parameter in relations specifying criteria for determining the break-away initiation; consequently, the related formulas have very limited significance, and are sometimes of controversial nature.

Extensive experimental studies of the initiation and intensity of dynamic break-away together with measurements of pressure drop due to friction during an upward flow of steam-water and air-water mixtures have been carried out under conditions described in the beginning of this section. Experimental data of the break-away initiation were processed in the form of criterion relations (7.4.16), where τ_{13} was determined using Eqs. (7.3.22)–(7.3.24) based on measurements of pressure drop due to friction and of the weight G_c of a unit column of the flow core. It was found that experimental results are described (B. I. Nigmatulin and V. E. Nikolaev) by the following relation (Fig. 7.4.4)

$$\mathsf{We}_{13*} = \begin{cases} 8.5\,\bar{\mu}^{(\Sigma g)} & (\mathsf{Re}_3 \leqslant 290) \\ 4.4\cdot 10^{-3}\bar{\mu}^{(\Sigma g)}\,\mathsf{Re}_3^{4/3} & (290 \leqslant \mathsf{Re}_3 \leqslant 3\cdot 10^3,\ g^z v_1^z < 0) \\ 0.20\,\bar{\mu}^{(\Sigma g)}\,\mathsf{Re}_3^{2/3} & (290 \leqslant \mathsf{Re}_3 \leqslant 3\cdot 10^3,\ g^z v_1^z > 0) \\ 5.5\cdot 10^{-4}\,(\bar{\mu}^{(\Sigma g)})^{-0,75} & (\mathsf{Re}_3 \geqslant 3\cdot 10^3,\ g^z v_1^z < 0) \end{cases}$$

$$\left(\bar{\mu}^{(\Sigma g)} \equiv \frac{\bar{g}^{1/4}}{\mathsf{Lp}_l^{1/2}} \equiv \mu_l \sqrt[4]{\frac{g}{\overset{\circ}{\rho}_l \Sigma^3}}\right) \tag{7.4.17}$$

where the first approximation relates to laminar ($\mathbf{Re}_3 < 290$) films, the second and fourth to concurrent ($v_1 v_3 > 0$) turbulent ($\mathbf{Re}_3 > 290$) films in vertical ($g^z = \pm g$) upward ($g^z v_1 < 0$) flows, the third to concurrent turbulent films in vertical downward ($g^z v_1 > 0$) flows. The Weber number $\mathbf{We}_{13*}$ for horizontal ($g^z = 0$) turbulent flows assumes intermediate values between the second and third approximations. The fourth approximation yields assymptotic values $\mathbf{We}_{13*}$ for highly turbulent films (at large values of $\mathbf{Re}_3$); it was obtained by the extrapolation of data on drop break-away intensity (see below).

The represented relations generalize the experimental data with an accuracy of $\pm 20\%$ under different conditions by many authors in concurrent air-water and steam-water flows in tubes at p = 0.1–10 MPa, D = 10–80 mm, $\mathbf{Re}_3$ = 30–3000 corresponding to $\mathbf{We}_{13*}$ = 0.014–0.175; $\bar{\mu}^{\Sigma g}$ = (0.5–

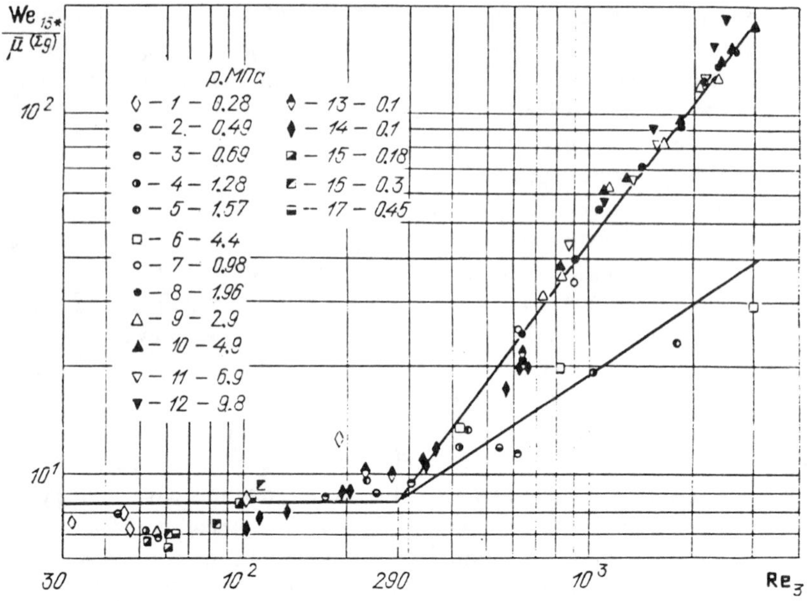

Figure 7.4.4 Experimental results concerning the initiation of dynamic break-away: *1)* p = 0.28 MPa, D = 9.5 mm, air-water, upward flow (L. Cousins et al., 1965); *2–6)* p = 0.49–4.41 MPa, D = 25 mm, steam-water, downward flow (N. A. Mozharov, 1959); *7–12)* p = 0.98–10 MPa, D = 13 mm, steam-water, upward flow (B. I. Nigmatulin et al., 1982); *13)* p = 0.1 MPa, D = 26 mm, air-water, horizontal flow (A. A. Armand, 1946); *14)* p = 0.1 MPa, D = 79 mm, air-water, horizontal flow (A. I. Shevskii, 1975); *15–17)* p = 0.1–0.45 MPa, D = 13 and 31.5 mm, air-water, upward flow (B. I. Nigmatulin et al., 1982).

0.3) $\times 10^{-3}$. Although the gas density[3] does not appear in equations which determine the conditions for the initiation of the dynamic drop break-away, the gas density effect is exhibited through the Weber number $\mathbf{We}_{13}$ proportional to $\rho_g^0(\upsilon_1 - \upsilon_{\Sigma 3})^2$. Therefore, the greater ρ_g^0, the lower velocities at which the break-away initiation conditions are achieved.

Intensity of the dynamic drop removal (break-away). The value $J_{32}^{(\alpha)}$ is more conveniently represented in the form of a dimensionless quantity $\bar{J}_{32}^{(\alpha)}$, which is a ratio of $J_{32}^{(p)}$ to the mass flux $m_3/(\pi D)$ arriving at the channel wall

$$\bar{J}_{32}^{(d)} = J_{32}^{(d)}\pi D/m_3 \tag{7.4.18}$$

Analysis of experimental data regarding $m_3(z)$ for both upward steam-water (D = 13.3 mm) and upward air-water (D = 13.3 and 31 mm) flows in the absence of drops in the flow core permitted their generalization of

[3]Gas density varied in these experiments within a range of $\rho_g^0/\rho_l^0 = 10^{-3}$–$10^{-1}$ by means of pressure variation.

using the following formula (B. I. Nigmatulin and V. E. Nikolaev), with a maximum relative error ±20%

$$\bar{J}_{32}^{(d)} = 0.55\,(\overset{\circ}{\rho}_l/\overset{\circ}{\rho}_g)^{0.5}\,\bar{\mu}^{(\Sigma g)}\,\mathrm{Re}_3^{-1.0}\,(\mathrm{We}_{13} - \mathrm{We}_{13*})^{0.85} \tag{7.4.19}$$

where $\overset{\circ}{\rho}_l/\overset{\circ}{\rho}_g = 10 - 500$, $\bar{\mu}^{(\Sigma g)} = (0.5 - 3.0)\cdot 10^{-3}$, $\mathrm{Re}_3 = 700 - 33000$

$\mathrm{We}_{13} - \mathrm{We}_{13*} = 0 - 3$

Note that if $\mathbf{Re}_3/\delta$ grows at fixed gas parameters (v_1 and ρ_g^0), i.e., v_3 increases, then $J_{32}^{(\alpha)}$ decreases because of growth of turbulent viscosity in the film. Increase of gas-phase density ρ_g^0 leads to an increase of $\mathbf{We}_{13}$ proportional to ρ_g, and also to the growth of $J_{32}^{(\alpha)}$ proportional to $(\rho_g^0)^{0.5}$.

We shall mention that the parameter $\bar{\mu}^{(\Sigma g)}$ characterizing the effect of the gravity force appears in formula (7.4.19), which defines the dynamic break-away. Therefore, this formula is applicable only for upward flows. The effect of forces due to gravity becomes insignificant only if

$$\frac{\overset{\circ}{\rho}_g (v_1 - v_3)^2}{\overset{\circ}{\rho}_l g D} \sim \frac{\overset{\circ}{\rho}_g}{\overset{\circ}{\rho}_l}\,\mathrm{Fr} \gg 1 \tag{7.4.20}$$

The radius a_{32} of drops broken-away from films may be evaluated using a formula similar to (2.2.17), which follows from the results reported by M. Adelberg (1968)

$$\mathrm{We}_{32} = \frac{k_{32}^{3/2}}{\mathrm{Lp}_{32}^{1/2}} \tag{7.4.21}$$

$$\left(k_{32} = 100 - 300, \quad \mathrm{We}_{32} = \frac{2a_{32}\overset{\circ}{\rho}_g (v_1 - v_{\Sigma 3})^2}{\Sigma}, \quad \mathrm{Lp}_{32} = \frac{2a_{32}\overset{\circ}{\rho}_l \Sigma}{\mu_l^2}\right)$$

Intensity of the shock break-away of splashes. The effect of drops in the flow core upon the break-away intensity may be evaluated based on comparison of the data on the moisture exchange intensity J^0 obtained by a salt-tracer method, with the relation (7.4.19) obtained from the data on break-away in flows with a drop-free core. It appeared that the break-away intensity (governed by formula (7.4.19)) in a purely annular regime of film flow (with no drops in the flow core) is sometimes 3–5 times lower than in a hydrodynamically fully-developed dispersed-annular flow where the break-away intensity was found based on the salt-tracer method. In these cases, the shock break-away of splashes (i.e., the break-away of splashes produced from drops depositing on the film surface) is of significant importance. The intensity $J_{32}^{(s)}$ of this process may be determined from the condition

$$\overset{\circ}{J}_{32} = J_{32}^{(d)} + J_{32}^{(s)} \tag{7.4.22}$$

The latter is obtained from data on the liquid distribution between the core and film in a developed flow in unheated channel when the bubbly break-away ($J_{32}^{(b)} = 0$) does not exist. In this case, $J_{32}^{0} = J_{23}^{0}$, and this value is found from formula (7.4.8), and $J_{32}^{(\alpha)}$ is determined from formula (7.4.19), if m_3 is measured.

Based on such treatment of experimental results (B. I. Nigmatulin et al., 1976) obtained for upward flow of steam-water mixture in vertical tubes with D = 8–20 mm in the range of pressure p = 1–10 MPa, it was found (B. I. Nigmatulin and V. E. Nikolaev, 1983) that

$$J_{32}^{(s)} = k^{(s)} J_{23}^{\circ}, \qquad k^{(s)} = 1 - 0.01\alpha_2^{-0.17}\left(\frac{\overset{\circ}{\rho}_g}{\overset{\circ}{\rho}_l}\right)^{0.67} \qquad (k^{(s)} \geqslant 0) \qquad (7.4.23)$$

where $k^{(s)}$ is the film splashing coefficient.

Intensity of bubbly break-away. The effect of film boiling on the drop break-away from film was studied (B. I. Nigmatulin et al., 1983) based on measurements of the total removal of moisture from the film during an upward flow of steam-water mixture in a short heated vertical tube (D = 13.3 mm, heating length $L^{(q)}$ = 0.15 m) with regime parameters ranging as follows: p = 5–10 MPa, m^0 = 1000–2000 kg/(m$^2\cdot$s), v_1 = 5–30 m/s, q_W = 1–4 MW/m^2. The total removal (break-away) was determined by means of measuring the liquid flow rates m_{30} and m_{3e} in film at the inlet and outlet of the heated segment, respectively. Then, from the liquid flow rate equation for the film over the heated segment, we obtain

$$\frac{m_{30} - m_{3e}}{L^{(q)}} = -J_{31} - J_{32}, \qquad J_{31} = \frac{q_W \pi D}{l} \qquad (7.4.24)$$

Now, the value J_{32} is easily calculated. Recall that $J_{32} = J_{32}^{(\alpha)} + J_{32}^{(s)} + J_{32}^{(b)} - J_{23}^{0}$. In accordance with evaluations based on using (7.4.11) for indicated regimes with high q_W, there is no drop deposition used by their removal by vaporizing steam ($J_{23}^{0} = 0$). Nor is the removal (break-away) of splashes ($J_{32}^{(s)} = 0$) feasible. As a result, we obtain $J_{32}^{(b)} = J_{32} - J_{32}^{(\alpha)}$, where $J_{32}^{(\alpha)}$ is calculated using (7.4.19). The bubbly break-away intensity data processed in this manner may be generalized (B. I. Nigmatulin and V. E. Nikolaev) by the following formula

$$J_{32}^{(b)} = k^{(b)} J_{31}$$
$$k^{(b)} = 2.0\cdot \mathrm{Re}_3 \left(\frac{\overset{\circ}{\rho}_g}{\overset{\circ}{\rho}_l}\right)^{1/5}\left(\frac{w_1 \mu_l}{\Sigma}\right)^{2/3}, \qquad w_1 = \frac{J_{31}}{\pi D \overset{\circ}{\rho}_g} = \frac{q_W}{\overset{\circ}{\rho}_g l} \qquad (7.4.25)$$

where $k^{(b)}$ is the bubbly break-away coefficient. The value $J_{32}^{(b)}$ may be evaluated based on the heat-transfer crisis data (see (7.6.16), R. I. Nigmatulin, 1977; S. I. Ivandaev, 1982).

§7.5 HYDRODYNAMICS OF STATIONARY DISPERSED-FILM VAPOR-LIQUID FLOW IN AN UNHEATED TUBE

The relations obtained in §§7.3 and 7.4 for force, thermal, and mass interactions between mixture components in a dispersed-film flow enable us to close the system of conservation equations (7.2.8) and (7.2.9) and to analyze, by means of numerical experiment, the effect of regime parameters (pressure p, specific discharge m^0 of mixture, flow-rate mass concentration x_1 of vapor, diameter and length of the tube, heating intensity and its distribution along the channel) upon the principal hydrodynamic characteristics of the flow (pressure drop Δp, liquid distribution x_2 and x_3 between core and film, respectively, film thickness δ, phase slip v_1/v_3, etc.).

This section presents turbulent, stationary, adiabatic ($Q_W = 0$) flows of gas-liquid mixtures in a tube under a dispersed-annular regime, when the mixture may be considered in a state of a thermodynamic equilibrium (phase temperatures are equal to each other, $T_1 = T_2 = T_3 = T$, and, in the case of a one-component mixture, equal to the saturation temperature T_S), and velocity equilibrium ($v_1 = v_2 = v_c$) exists in the flow core. These conditions are ensured if the residence time of both gas and liquid in a channel is many times greater than the inherent times of the temperature profile flattening between gas, film, and drops, and than the inherent time of the velocity profile flattening between gas and drops. Moreover, we confine ourselves to regimes when the differences in pressure and temperature along the channel are small ($\Delta p \ll p$, $\Delta T \ll T$), and when velocities of gas and drops are many times smaller than the equilibrium speed of sound in the channel core. Then, density variation of both liquid and gas may be neglected

$$\overset{\circ}{\rho}_l = \text{const}, \quad \overset{\circ}{\rho}_g = \text{const} \tag{7.5.1}$$

With indicated limitations, the phase transitions (evaporation and condensation) may always be ignored, i.e., $J_{21} = J_{31} = 0$. Then, denoting the parameters at the channel inlet ($z = 0$) by subscript 0, the following relations may be used as mass conservation equations

$$m_1 = m_{g0}, \quad m_2 + m_3 = m_{l0}, \quad dm_3/dx = \overset{\circ}{J}_{23} - J_{32}^{(d)} - J_{32}^{(s)} \tag{7.5.2}$$

and Eqs. (7.2.32) may be employed as the momentum equations. These equations are closed by equations of interphase interactions for F_W, F_{13}, J_{23}^0, $J_{32}^{(\alpha)}$, and $J_{32}^{(s)}$ represented in §§7.3 and 7.4. Of all physical characteristics, only densities and viscosities of gas and liquid and surface tension appear in these equations

$$\overset{\circ}{\rho}_g, \overset{\circ}{\rho}_l, \mu_g, \mu_l, \Sigma \tag{7.5.3}$$

For numerical integration of the represented system of three ordinary differential equations, i.e., for the Cauchy problem, it is necessary to prescribe at the channel inlet ($z = 0$) the flow rates and velocities of species (m_{10}, m_{20}, m_{30}, υ_{10} and υ_{30}), and also pressure p_0, which, together with temperature T_0, define the gas density ρ_g^0.

With the channel geometry (diameter and length of the tube) known, the following regime parameters are usually kept constant throughout experiments: specific flow rate m^0, mass flow-rate gas concentration x_1 (flow quality) of mixture, and pressure p_0 at the channel inlet. Then, the flow rates of vapor and liquid at the inlet are found from

$$\overset{\circ}{m}_{10} = m^\circ x_1, \quad \overset{\circ}{m}_{l0} = \overset{\circ}{m}_{20} + \overset{\circ}{m}_{30} = m^\circ (1 - x_1) \tag{7.5.4}$$

To determine all channel-inlet parameters ($z = 0$), the values $m_{30}^0 = k_{30}^{(m)} m_{l0}^0$ and $S_{30} = k_{30}^{(S)} S$ ($k_{30}^{(m)} \leqslant 1$, $k_{30}^{(S)} \ll 1$) should be predetermined, in terms of which the remaining parameters are expressed as

$$m_{20} = \left(1 - k_{30}^{(m)}\right) m_{l0}, \quad S_{10} = \left(1 - k_{30}^{(S)}\right) S$$

$$\upsilon_{10} = \frac{m_{10}}{\overset{\circ}{\rho}_g \alpha_{10} S_{10}} = \upsilon_{20} = \frac{m_{20}}{\overset{\circ}{\rho}_l \alpha_{20} S_{10}} \qquad (\alpha_{10} + \alpha_{20} = 1) \tag{7.5.5}$$

$$\upsilon_{30} = \frac{k_{30}^{(m)} m_{l0}}{\overset{\circ}{\rho}_l k_{30}^{(S)} S}, \quad \delta_0 = \frac{k_{30}^{(S)} S}{\pi D}$$

Note that the values α_{10} and α_{20} are readily found from the condition $\upsilon_{10} = \upsilon_{20}$.

Stabilization of a stationary dispersed-film flow. The system of equations (7.5.2) and (7.2.32) was numerically integrated with various coefficients $k_{30}^{(m)}$ and $k_{30}^{(S)}$ (which define the distribution of both liquid flow rate m and the channel section S between core and film at the channel inlet) for each regime given by the prescribed values D, p, m^0, and x_1. Varying the coefficients $k_{30}^{(m)}$ and $k_{30}^{(S)}$ permits the investigation of their impact on the stabilization segment length; at the end of this segment, a state of flow is established when velocities and mixture components distribution along the channel do not change. The analysis was continued up to a section of the channel where such stabilized flow was reached.

Figure 7.4.1 represents a comparison between analytical and experimental data concerning the film liquid flow rate x_3 variation along the channel when at the inlet segment one of two limiting initial distributions of liquid in a flow was established: either the entire liquid flows in the form of drops in the flow core ($k_{30}^{(m)} = 0$, $k_{30}^{(S)} = 0$), or the entire liquid flows in the film ($k_{30}^{(m)} = 1$, $k_{30}^{(S)} \ll 1$). In case a film is present at the channel inlet, the value of $k_{30}^{(S)}$ may first be prescribed arbitrarily, e.g., $k_{30}^{(S)} = 0.1$, since as early as $z/D \leqslant 1$, the liquid velocity υ_3 in film, and also, its thickness δ, obtain their steady values (complying with relations for τ_W, τ_{13},

J^0_{23}, and J^0_{32}), which are independent of $k^{(S)}_{30}$ and $k^{(m)}_{30}$. From Fig. 7.4.1, it is seen that analytical and experimental data with respect to x_3 (z) are in quite satisfactory agreement between each other; the fully-developed relative flow rate x^0_3 of liquid in a film is independent of the initial distribution of liquid and is determined by such integral flow characteristics as flow-rate of liquid and vapor in the mixture and pressure in the channel. As to the length of the stabilization segment, it depends significantly on the initial distribution. So, when the whole liquid is present in a film, the length of the stabilization segment amounts to 2–3 meters (L/D = 150–200), and, is, accordingly, 4–6 times larger than in the case when the whole liquid is distributed at the inlet to the channel in the flow core in a form of drops.

Figure 7.5.1 represents an example of variation of characteristics of a dispersed-film flow as a function of both distance from the channel entrance and initial liquid distribution in the flow (the example under consideration corresponds to conditions shown by points *1* and *3* in Fig. 7.4.1). It is evident, as indicated above, that velocity of liquid in a film assumes its developed magnitude over a very short distance.

The effect of regime parameters upon thickness and flow rate of liquid in a film in a stabilized stationary flow. Figure 7.5.2 compares the analytical and experimental data of relative flow rates x^0_3 of liquid in a film as a function of x_1 obtained for stabilized flow with various specific mass flow

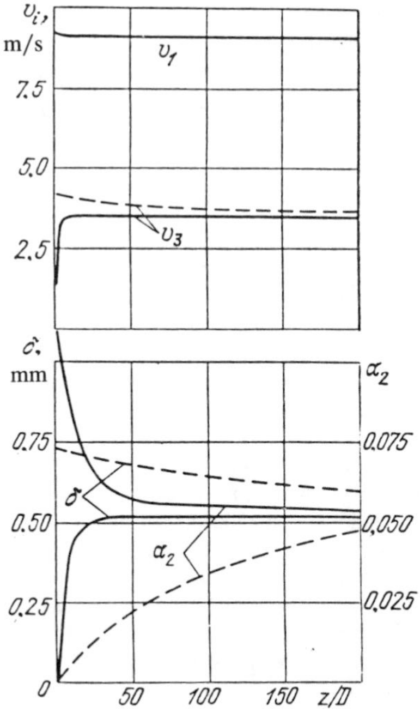

Figure 7.5.1 Variation of characteristics of an upward steam-water dispersed-annular flow (υ_1, υ_3, δ, α_2) along the tube (analytical data) for conditions specified by experimental points *1* and *3* in Fig. 7.4.1 (p = 6.9 MPa, m^0 = 1000 kg/(m^2 × s), x_1 = 0.27, D = 13.3 mm). Solid lines represent the whole liquid at the inlet (z = 0) present in the core ($k^{(m)}_{30}$ = 0); dashed lines represent the whole liquid at the inlet present in a film ($k^{(m)}_{30}$ = 1).

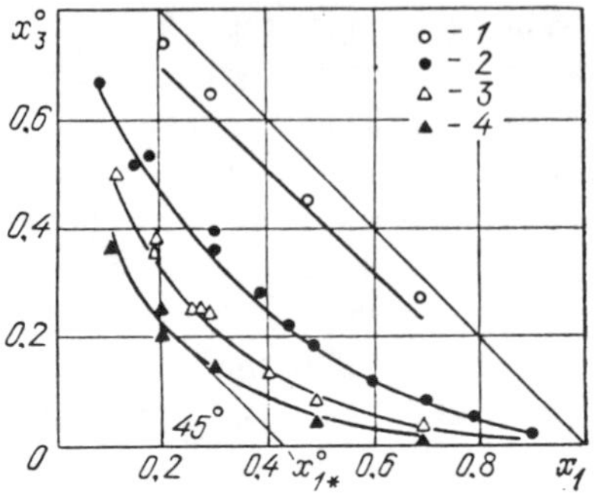

Figure 7.5.2 Dependence of relative mass flow rate of liquid in a film ($x_3^0 = m_3/m$) on mass flow-rate vapor concentration x_1 in a steam-water, upward, stationary and stabilized dispersed-annular flow (p = 4.9 MPa, D = 13.1 mm). Dots represent the experiment (B. I. Nigmatulin et al., 1978), solid lines represent analytical results: *1*) m^0 = 500 kg/(m^2 × s); *2*) 1000; *3*) 1500; *4*) 2000.

rates m^0 of the mixture. It is evident that analytical and experimental data with respect to x_3^0 are in good agreement. Continued comparison of analytical values of x_3^0 with experimental data (G. Hewitt and N. Hall-Taylor, 1972; B. I. Nigmatulin et al., 1976 and 1983; J. Wurtz, 1978) in a wide range of experimental parameters, D = 8–20 mm, p = 1.0–10 MPa, m^0 = 500–3000 kg/(m^2 × s), x_1 = 0.1–0.9, showed that the discrepancy does not exceed 20%.

Hydraulic resistance and its crisis in a dispersed-annular flow. Figure 7.5.3 depicts analytical and experimental results concerning hydraulic resistance in stationary, stabilized, upward steam-water flows in a vertical tube of a fixed diameter, as a function of flow-rate vapor concentration x_1, pressure p, and specific flow-rate m^0. The ordinate indicates the relative loss of pressure Π due to friction, which equals the ratio between pressure loss due to friction in a two-phase flow (calculated based on measurements of dp/dz along the developed flow with an account of gravity force) and pressure

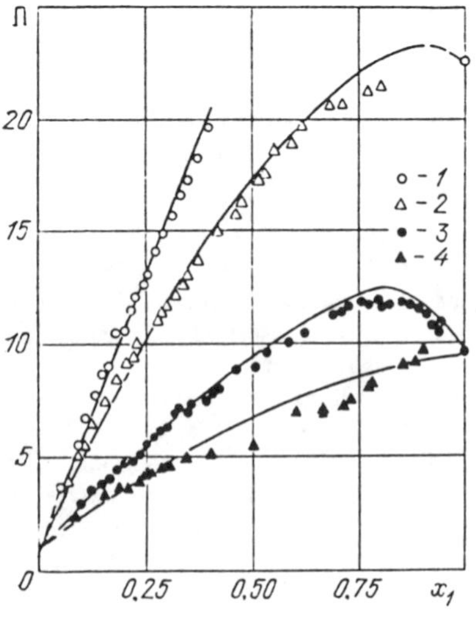

Figure 7.5.3 Dependence of relative pressure drop Π due to friction on x_1 in an upward, steam-water, stationary, stabilized flow in a tube (D = 8 mm). Dots represent experimental data (N. V. Tarasova and A. I. Leont'ev, 1965), solid lines represent analytical data. Light dots *1* and *2*) p = 4.9 MPa, black dots *3* and *4*) p = 9.8 MPa; dots *1* and *3*) m^0 = 1000 kg/(m2 × s), 2 and *4*) m = 2000 kg/(m^2 × s).

loss due to friction in flow of water with the same mass flow-rate m^0 and at the same temperature and pressure (see (7.3.24))

$$\Pi = \frac{(dp/dz)_f}{(dp/dz)_0},$$

$$\left(\frac{dp}{dz}\right)_f = \frac{dp}{dz} - \frac{G}{S} = -\frac{F_W}{S} = -\frac{F_{13}}{S_c} - \frac{J^\circ - J_{32}^{(s)}}{S_c}(v_1 - v_{\Sigma 3}) + \frac{G_c}{S_c} - \frac{G}{S}$$

(7.5.6)

$$\left(\frac{dp}{dz}\right)_0 = -C_{Wl}\frac{\overset{\circ}{\rho}_l v_{l0}^2}{2} \quad \left(v_{l0} = \frac{m^\circ}{\overset{\circ}{\rho}_l}, \quad C_{Wl} = \frac{0.316}{\mathrm{Re}_{l0}^{0.25}}, \quad \mathrm{Re}_{l0} = \frac{\overset{\circ}{\rho}_l v_{l0} D}{\mu_l}\right)$$

The magnitude of Π at $x_1 = 1$ corresponds to a hydraulic resistance of "pure" gas with the prescribed flow rate $m^0 = \overset{0}{\rho}_g v_{g0}$

$$\Pi(1) = \frac{C_{Wg}}{C_{Wl}} \cdot \frac{v_{g0}}{v_{l0}} = \left(\frac{\mu_g}{\mu_l}\right)^{0.25} \frac{\overset{\circ}{\rho}_l}{\overset{\circ}{\rho}_g} \quad \left(C_{Wg} = \frac{0.316}{\mathrm{Re}_{g0}^{0.25}}, \quad \mathrm{Re}_{g0} = \frac{\overset{\circ}{\rho}_g v_{g0} D}{\mu_g}\right)$$

The velocity nonequilibrium due to relative motion of the core and film in a dispersed-film flow leads to the dependence of hydraulic resistance Π on the specific flow rate m^0, and under fairly low both pressure p and specific flow rates m^0, an interesting phenomenon of a "crisis" of hydraulic-resistance is clearly observed. This effect consists of a reduction of hydraulic resistance at fixed p and m^0, but at an increase of the part of flow rate x_1 related to the gas phase. Although with an increase of x_1 the phase velocities and, especially, the gas velocity v_1 grow (that in itself contributes to an increase of pressure drop due to friction proportional to $\overset{0}{\rho}_g (v_1 - v_{\Sigma 3})$) nevertheless, at $x_1 > x_*(p, m^0)$, a reduction of pressure drop due to friction may be observed. The latter is accounted for by a thinning of the liquid film with an increase of gas velocity v_1, which leads to the diminishing of its effective roughness, i.e., to the decrease of the coefficient C_{13} defining F_{13} (see (7.3.26)).

The continuing comparison of analytical data with available experimental results (G. Gaspari et al., 1964; N. V. Tarasova and A. I. Leont'ev, 1965; P. L. Kirillov et al., 1973; B. I. Nigmatulin et al., 1978 and 1983; J. Wurtz, 1978) on pressure drop due to friction in steam-water and vapor-liquid helium flows in a vertical tube (B. S. Petukhov et al., 1980) revealed satisfactory agreement, with deviations within 15%.

§7.6 CRISIS OF HEAT TRANSFER IN A DISPERSED-FILM VAPOR-LIQUID FLOW

In a flow of vapor-liquid mixture through a heated channel, a deterioration, or crisis, of heat transfer, which is characterized by a sharp temperature rise of the heating surface, may occur under certain conditions; such a crisis is associated with the disruption of the contact between liquid and this surface, which has already been mentioned in §7.1. The occurrence of the heat transfer crisis significantly depends on both the vapor-liquid flow structure and presence of an exterior specific (per unit area) heat flux $q_W = Q_W/(\pi D)$.

With a flow of liquid subcooled relative to the boiling point[4] (vapor-liquid mixture in a bubbly regime of flow), the heat transfer crisis arises as a result of the transition of a bubbly boiling on the heating surface to a film boiling (boiling with a vapor film) with formation of vapor film due to bubble agglomeration (bubbly boiling crisis).

At lower-level specific heat fluxes, gradual growth of volumetric vapor concentration occurs. It is accounted for by bubbly boiling and evaporation from the interface surface; such growth is, finally, responsible for the transition of the bubbly flow to a slug-flow regime followed by a dispersed-annular regime of flow of mixture. As a consequence of both bubbly and dynamic moisture removal from the film, and also of evaporation or boiling, the liquid flow rate in a film and film thickness decrease. A situation may occur when film-thickness reduces to an extent when its continuity and contact between liquid and heating surface become disrupted, resulting in the formation of dry "spots." With the formation of dry spots on the heating surface, heat transfer deteriorates, which, with an intensive heating, causes uneven rise of temperature of the tube wall (heat-transfer crisis due to the wall liquid film drying-up).

Analysis of experimental data on the heat transfer crisis. The experimental data concerning the heat transfer crisis obtained at both fixed pressure and specific mass flow rates of mixture are generally represented in coordinates $x_{1*}q_*$, where q_* and x_* are, respectively, specific heat flux q_W and mass flow-rate vapor concentration x_1 at the location of the heat transfer crisis. Figure 7.6.1*a* depicts, by way of example, the relations $q_*(x_*)$ for a steam-water flow at both fixed pressure and tube diameter. It is seen that on each of the represented curves, four regions may be specified that are shown in Fig. 7.6.1*b*.

Domains *II* (*DB*), *III* (*BC*), and *IV* (*CE*) correspond to dispersed-annular and dispersed regimes of flow of vapor-liquid mixture $x_1 \geqslant x_1(D) = x_{1d}$. Hereafter, x_{1d} denotes the vapor concentration in a fully-developed vapor-liquid flow (at predetermined p, m^0, and D in the direction of the flow relative to forces due to gravity), above which a dispersed-annular regime of flow is realized. The heat fluxes in domain *II* are strong enough to support an intensive bubbly boiling in the film, which may be responsible for a bubbly break-away of liquid from film into the flow core. With a decrease

[4]Formally, a "negative quality" $x_g \equiv x_1 = (i - i_{lS})/l < 0$, corresponds to a single-phase liquid subcooled ($T < T_S(p)$) relative to the boiling point.

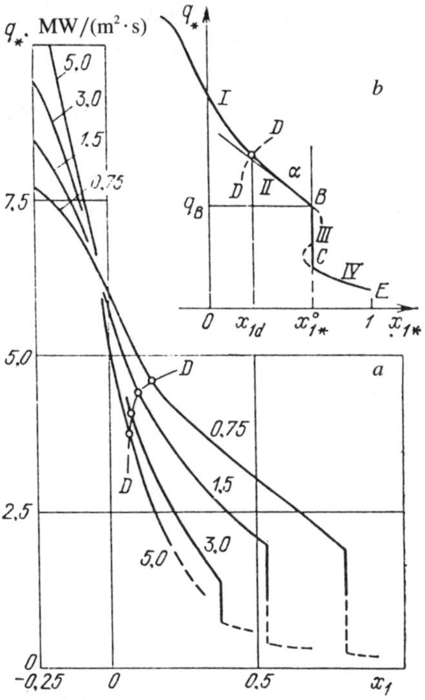

Figure 7.6.1 Dependence of specific heat flux q_* on both mass flow-rate concentration x_{1*} of vapor at the location of heat transfer crisis and specific flow-rate m^0 in the process of boiling of water in a circular tube of diameter $D = 8$ mm at a pressure $p = 7$ MPa (*a*). Numerical labels on curves indicate the value m^0 in 1000 kg/(m^2 × s); thin line *DD* separates the domain of the dispersed-film regime; *b* depicts the schematic of a characteristic relation $q_*(x_{1*})$.

of q_W, the contribution of the bubbly break-away to the intensity of the drop removal from the surface diminishes significantly (see §7.4). Therefore, disappearance of a film (heat transfer crisis) with reduction of q_W may occur at a higher value of x_1. With q_B reaching a certain magnitude, the further decrease of specific heat flux down to q_C leads to a very slight change of x_{1*}. This is associated with the fact that at $q < q_B$, redistribution and mutual compensation of processes of bubbly, dynamic, and drop break-away and deposition over the heated length of the channel take place in such a manner that the value x_{1*} begins to depend on the specific heat flux to an insignificant extent. In particular, the drop deposition may be absent because of its removal by vaporizing vapor. This situation is reflected by domain *III* (or vertical line *BC*) with an abscissa x^0_{1*}. In this case, the relation $q_*(x_{1*})$ in domain *II* (*DB*) corresponds to a straight line passing through point *B* ($q_* = q_B$, $x_{1*} = x^0_{1*}$), whose inclination is determined by a dimensionless parameter

$$b = -q_B \left(\frac{dq_*}{dx_{1*}}\right)^{-1}_B = \operatorname{tg}\alpha \tag{7.6.1}$$

Under high pressure (at $p \geqslant 16$ MPa for water), when the thermophysical properties of vapor and liquid approach each other, the vertical portion (*BC*) degenerates and becomes insignificant. Increase of specific flow rate m^0 (see relation $q_*(x_{1*})$ for $p = 7$ MPa and $m^0 = 5000$ kg/(m^2 × s) in Fig. 7.6.1*a*) also contributes to degeneration of the indicated vertical portion.

With the further reduction of specific heat flux (domain *IV* or *CE*), the drop removal by vapor diminishes, and drop deposition on the film or wall begins to play a significant role, thereby enhancing the contact between liquid phase and wall, which intensifies heat transfer from the wall. Therefore, with a decrease of q_W, the magnitude of x_{1*} grows, i.e., the lower the intensity of heating, the higher the vapor concentration of vapor-liquid flow heated in a crisis-free regime. The domain $q_W < q_C$ is referred to as the domain of a post-crisis heat exchange, and is discussed in §7.7.

Investigation of the effects of various regime and geometrical parameters upon both relation $q_*(x_{1*})$ and the value x_{1*} is reported in a number of published works and reviews (L. Tong, 1972; A. A. Andreevskii et al., 1974; P. L. Kirillov, 1983; B. S. Petukhov et al., 1974; D. Butterworth and G. Hewitt, 1977; A. M. Kutepov et al., 1977; J. Wurtz, 1978; G. Hewitt, 1978; S. S. Kutateladze, 1978; B. I. Nigmatulin, 1979; V. N. Smolin et al., 1979; V. I. Tolubinskii, 1980; Recommendations . . ., 1980; M. A. Styrikovich et al., 1982; V. E. Doroshchuk, 1983). They present empirical relations for calculating $\overset{0}{x}_{1*}$ and q_* for mainly uniform (over the length) heat release.

Table 7.6.1 represents values of parameters x_{1d}, $\overset{0}{x}_{1*}$, q_B, and b defining the generalized experimental relations $q_*(x_{1*})$ (see Recommendations . . ., 1980) for the heat transfer crisis in dispersed-annular regimes, which correspond to sections *II* and *III* (or *DBC*) in Fig. 7.6.1*b*; the values presented in this Table relate to steam-water flows in tubes of fixed diameter (D = 8 mm) for various pressures (p = 3–14 MPa) and specific flow rates (m^0 = 750–5000 kg/(m^2 × s).

Recalculation of these parameters for tubes of different diameters is based on that in accordance with experimental data (Recommendations . . ., 1980), at other conditions (p, m^0, x_1) fixed, the following takes place in the domain D = 4–20 mm

$$\frac{x_{1d}(D)}{x_{1d}(D_0)} = \left(\frac{D}{D_0}\right)^0 = 1, \quad \frac{q_*(D)}{q_*(D_0)} = \left(\frac{D}{D_0}\right)^{-0.5}, \quad \frac{\overset{\circ}{x}_{1*}(D)}{\overset{\circ}{x}_{1*}(D_0)} = \left(\frac{D}{D_0}\right)^{-0.2} \tag{7.6.2}$$

Then, taking into consideration that according to Fig. 7.6.1 and formula (7.6.1)

$$q_*(D_0, x_1) = q_B(D_0)\left[1 + \frac{\overset{\circ}{x}_{1*}(D_0) - x_1}{b(D_0)}\right] \qquad (x_1 < \overset{\circ}{x}_{1*}(D_0)) \tag{7.6.3}$$

we obtain

$$q_*(D, x_1) = q_*(D_0, x_1)\left(\frac{D}{D_0}\right)^{-0.5} = q_B(D_0)\left(\frac{D}{D_0}\right)^{-0.5}\left[1 + \frac{\overset{\circ}{x}_{1*}(D_0) - x_1}{b(D_0)}\right]$$

$$b(D) = q_B(D)\left(-\frac{dq_*(D, x_1)}{dx_1}\right)^{-1} = \frac{q_B(D)}{q_B(D_0)}\left(\frac{D}{D_0}\right)^{0.5} b(D_0)$$

Table 7.6.1

p, MPa	m°, kg/(m^2·s)	x_{1d}	$\overset{\circ}{x}_{1*}$ (±0.04)	q_B, MW/m^2	$b = q_B\left(\frac{dx_{1*}}{dq_{1*}}\right)_B$
3.0	750	0.09	0.75	3.25	0.51
	1000	0.08	0.67	3.25	0.51
	1500	0.07	0.54	2.60	0.33
	2000	0.06	0.47	2.30	0.25
	3000	0.05	0.38	2.25	0.24
	4000	0.04	0.33	2.30	0.24
	5000	0.04	0.29	2.45	0.25
5.0	750	0.12	0.80	2.80	0.69
	1000	0.10	0.70	2.80	0.61
	1500	0.08	0.55	2.80	0.57
	2000	0.07	0.46	2.60	0.49
	3000	0.06	0.37	2.30	0.33
	4000	0.06	0.30	2.05	0.19
	5000	0.06	0.26	2.15	0.19
7.0	750	0.15	0.80	1.80	0.47
	1000	0.12	0.68	1.80	0.37
	1500	0.09	0.53	1.80	0.37
	2000	0.08	0.43	1.80	0.32
	3000	0.07	0.37	1.30	0.16
10	750	0.19	0.66	1.40	0.41
	1000	0.15	0.56	1.40	0.34
	1500	0.10	0.42	1.40	0.27
	2000	0.09	0.36	1.30	0.22
	3000	0.09	0.30	1.20	0.20
14	750	0.25	0.50	1.00	0.50
	1000	0.17	0.43	0.90	0.31
	1500	0.15	0.33	0.82	0.17
	2000	0.15	0.33	0.62	0.11

Recall that $q_B(D) = q_*(D, x^0_{1*}(D))$. Then, based on data $x_{1d}(D)$, $x^0_{1*}(D_0)$, $q_B(D_0)$, and $b(D_0)$ outlined in Table 7.6.1 for $D_0 = 8$ mm, these parameters can be readily determined for tubes of any diameter D

$$x_{1d}(D) = x_{1d}(D_0), \quad \frac{\overset{\circ}{x}_{1*}(D)}{\overset{\circ}{x}_{1*}(D_0)} = \left(\frac{D}{D_0}\right)^{-0.2}, \quad \frac{b(D)}{b(D_0)} = \frac{q_B(D)}{q_B(D_0)}\left(\frac{D}{D_0}\right)^{0.5}$$

$$\frac{q_B(D)}{q_B(D_0)} = \left(\frac{D}{D_0}\right)^{-0.5}\left\{1 + \frac{\overset{\circ}{x}_{1*}(D_0)}{b(D_0)}\left[1 - \left(\frac{D}{D_0}\right)^{-0.2}\right]\right\} \tag{7.6.4}$$

Parameters represented in Table 7.6.1 and relations (7.6.4) approximate the experimental data for conditions specifying the occurrence of the heat-transfer crisis in uniformly heated, fully-developed, stationary, dispersed-annular $x_1d \leqslant x^0_{1*}$ steam-water flows (p = 3–14 MPa, m^0 = 500–5000 kg/(m^2 × s) in tubes (D = 4–20 mm).

The prediction of conditions of the heat-transfer crisis onset in a circular tube based on a hydrodynamic model of a dispersed-annular flow, with the determination of the location of the wall-film disappearance ($m_{3*} = 0$), was first made by B. I. Nigmatulin (1973), who used approximate relations for intensity of moisture exchange processes which were not by then substantiated by direct experimental data. A similar approach was developed by P. Whalley (1974) and S. I. Ivandaev (1982).

Experimental study of the heat-transfer crisis and liquid flow rate in films in dispersed-film steam-water flows. Results of film flow rate m_3 measurements near the location of occurrence of the heat-transfer crisis in a dispersed-annular regime of a mixture flow (E. Moeck, 1970; G. Hewitt and N. Hall-Taylor, 1972; D. Butterworth and G. Hewitt, 1977; S. P. Kaznovskii et al., 1978; J. Wurtz, 1978; B. I. Nigmatulin, 1979) show that the heat-transfer deterioration crisis occurs as a consequence of the liquid flow rate in a film continuously approaching zero at the location of occurrence of the crisis. However (S. P. Kaznovskii et al., 1978), at great specific heat fluxes ($q_W \approx 4$ MW/m^2), relatively low pressures ($p \lesssim 1.5$ MPa), and fairly small mass flow-rate vapor concentrations ($x_1 < 0.4$), a considerable residual flow rate (m_{3*} = 9–35 g/s) of liquid in a film was observed at the location of crisis (where the tube wall temperature rose), which corresponds to a relative flow rate of liquid in a film x_{3*} = 0.18–0.35 (the lower the vapor concentration x_{1*} at the location of the heat-transfer crisis, the larger the residual flow rate of liquid in a film). Note that all measurements by S. P. Kaznovskii et al., (1978) were related to the domain *II* (Fig. 7.6.1). Apparently, at such strong specific heat fluxes, the "dry" spots are produced in sufficiently thick films between the wave crests; therefore, the residual flow rate is so large that it virtually resembles elimination of crises of the first and second kind. New experimental results have recently confirmed an analogous situation for $q_W > 4$ MW/m^2 and at higher pressure.

B. I. Nigmatulin reported (1979) results of continuous experimental study of a liquid flow rate in a film in uniformly heated upward steam-water dispersed-annular flow approaching the heat-transfer crisis at q_W = 0.6–2.0 MW/m^2, m^0 = 500–2000 kg/(m^2 × s), and at relatively high pressures p = 1–10 MPa. Experiments were conducted in a vertical tube (D = 13.3 mm, L = 4.0 m) consisting of segments of hydrodynamic development of a steam-water flow, each of length $L_s \approx 2.6$ m ($L_s/D \approx 200$), and of a heated segment made of a thin-wall tube of length 1.5 m. Heating was effected by transmitting an electric current through the tube. Displacement of the lower current-conducting flange permitted the variation of the length of a heated segment. The heated length L_q in most experiments was 0.13 and 0.64 m. Ten thermocouples were installed at the end of the heated segment: four thermocouples spaced at 10 mm along two opposite generatrix for determining the location of the heat-transfer crisis occurrence. A device for the film suction through a porous insert of length 45 mm was installed next to the heated segment for measuring the liquid flow rate^{m_3} in it. The pressure drop over the heated segment was also measured. At every fixed regime (p, m^0, x_{10}), the electrical power delivered to the experimental segment was gradually increased up to the occurrence of the heat-transfer crisis, which was identified by a sharp jump of temperature of the outer surface of the heated segment.

The heat-transfer crisis during a uniform heating over the channel length occurred always at the outlet of the heated segment. Both mass flow-rate concentration x_{1e} of vapor at the outlet of the experimental segment and the liquid flow rate m_{2e} in the flow core were calculated based on the measured flow rate of liquid in a film, and, also, on thermal and material balance in the process of experiment. The mass conservation equations for each species of the mixture in a dispersed-annular stationary flow in a heated channel may be modified to the form (see (7.2.33) and (7.2.36))

$$\frac{dx_1}{dz} = \frac{J_{31}}{m}, \quad \frac{dx_2}{dx_1} = \frac{\overset{\circ}{J}_{32} - \overset{\circ}{J}_{23}}{J_{31}}, \quad \frac{dx_3}{dx_1} = -1 + \frac{\overset{\circ}{J}_{23} - \overset{\circ}{J}_{32}}{J_{31}}$$

$$\left(x_j = \frac{m_j}{m} \quad (j = 1, 2, 3), \quad x_1 + x_2 + x_3 = 1, \quad J_{31} = \frac{\pi D q_W}{l}\right) \tag{7.6.5}$$

The relationship between J^0_{23} and J^0_{32} is more convenient to analyze based on variation of x_3 as a function of x_1 at fixed q_W and regime parameters of the mixture.

Figure 7.6.2 depicts the relation between the liquid flow rate m_3 in a film at the end of the heated segment and specific heat flux q_W. It is evident that the liquid flow rate in the film decreases with an increase of heating power, the heat transfer crisis (points K) arising at the liquid flow rate in a film approaching zero. The dot-dashed line indicates the anticipated reduction of m_3 due only to vaporization of liquid in a film, i.e., if no moisture exchange between a core and film or their mutual compensation are as-

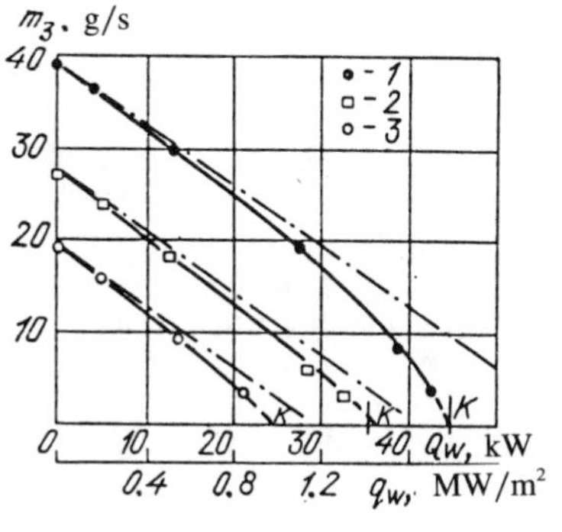

Figure 7.6.2 Variation of the liquid flow rate in a film at the end of the heated segment of length $L_q = 0.64$ m ($D = 13.3$ mm, $p = 6.9$ MPa, $m^0 = 1000$ kg/(m^2 × s)) with change of both input heating power $Q_w = 2\pi D g_W$ and vapor concentration x_{10} at the inlet: *1*) for $x_{10} = 0.40$, *2*) for $x_{10} = 0.49$, *3*) for $x_{10} = 0.62$. (B. I. Nigmatulin's data).

sumed. It can be seen that the measured flow rate in a film reduces more rapidly than if it was due only to pure vaporization; the larger the liquid flow rate in a film, the larger the departure from the line of pure vaporization.

Figure 7.6.3 depicts the dependence of relative flow rate x_3 and x_2 of liquid in a film and in the flow core, respectively, on the mass flow rate concentration x_1 of vapor in the mixture at various heat fluxes. The liquid flow rates x_3^0 in a film measured in an unheated channel are obtained under conditions of a hydrodynamically developed flow. The same figure represents experimental results (A. Bennet et al., 1969) obtained in a vertical tube with ID = 12.6 mm, length 3.66 m, when water at inlet to the channel was fed at a temperature lower than the saturation temperature. The liquid flow rate in a film was measured at a location next to the heated segment. The heat transfer crisis also occurred here at flow rates of liquid in a film approaching zero. The intensity of moisture break-away from the film surface is predominant compared to drop deposition on the film surface for all regimes.

Vaporization of vapor (steam) (i.e., its arrival from a film) leads to its acceleration, which may result in the alteration of the intensities of drop deposition and break-away. In a steady fully-developed film, the relative flow rate should not exceed the value of x_3^0 corresponding to relative flow rate in a film in a developed unheated flow (see Figs. 7.5.2 and 7.6.2) and depending on regime parameters p, m^0, x_{10}, and D. If $\partial x_3^0/\partial x_1 > -1$, the increase of x_1 does not result in an additional thinning of the film, other than vaporization, caused by intensification of break-away. Moreover, if deposition is not impeded by break-away of vapor from film, the latter will be additionally fed in the process of its calming from flow core due to the intensification of drop deposition. As small vapor concentrations, when

$$\partial x_3^\circ/\partial x_1 < -1 \tag{7.6.6}$$

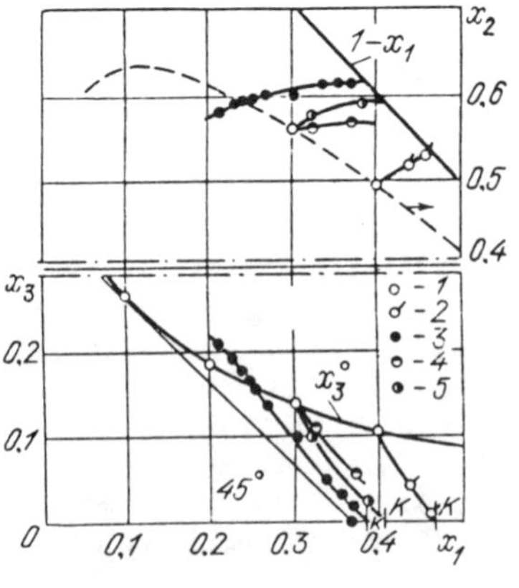

Figure 7.6.3 Relative flow rates of liquid in a film (x_3) and in the flow core (x_3) (p = 6.9 MPa, m^0 = 2000 kg/(m^2 × s), D = 12.6–13.3 mm) at various values of heat flux q_W (MW/m^2, uniformly distributed over the channel length and equal to 0 (*1*), 0.5 (*2*), 0.9 (*3*), 1.1 (*4*), 1.4 (*5*). Points *1* and *4* represent the data of A. Bennet et al., (1969), *2*, *3*, and *5* represent the data of B. I. Nigmatulin (1979).

the increase of x_1 due to vaporization leads to intensification of break-away and its additional thinning.

In the absence of both drop deposition (because of drop removal by vaporizing liquid) and bubbly break-away

$$J^{\circ}_{23} = J^{(s)}_{32} = J^{(b)}_{32} = 0 \qquad (7.6.7)$$

the vapor concentration x_{1*} at the location of film disappearance, if $x_{10} < x_{1c}$ (Fig. 7.6.4) at the inlet to the heated segment, may be found using the relations $x_3^0(x_1)$ for an unheated channel ($q_W = 0$) by means of the tangent fcK to line $x_3^0(x_1)$ inclined at an angle of 45° to axis x_1 (see also Figs. 7.5.2 and 7.6.2). If at the inlet to the heated segment $x_{10} > x_{1c}$, the vapor concentration at $x_3 = 0$ equals $x_{10} + x_{30}$, which is reflected by line $c'h'$ in Fig. 7.6.4.

The extent of the influence of various constituents on the intensities of mass exchange between core and film is different at various combinations of regime parameters.

With an increase of the heat flux q_W, the amount of bubbles in a film increases; the increased amount of bubbles enhances the bubbly break-away $J_{32}^{(b)}$, thereby increasing the flow of vaporizing liquid from the liquid film surface, which hampers drop J_{23}^0 deposition. These processes, with growth of q_W, reduce the film moisture consumed by vaporization.

In so far as the dynamic break-away $J_{32}^{(d)}$ is concerned, it does not directly depend on q_W, but q_W may affect $J_{32}^{(d)}$ through other parameters, in particular, at small vapor concentrations when (7.6.6) takes place, and when neither deposition nor bubbly break-away (7.6.7) is present. In this case, if the vapor acceleration due to its vaporization (removal from the film) proportional to q_W is quite small, the break-away has enough time to stabilize the film adjusting it to $x_1(z)$, so that the liquid distribution between film and core is the same as in a fully-developed unheated flow. Then, the amount of moisture broken-away (due to $J_{32}^{(d)}$) over a segment between z_0 and z (where the condition (7.6.6) is satisfied) is independent of q_W and is determined by the relation $x_3^0(x_1)$ and values $x_1(z_0)$ and $x_1(z)$

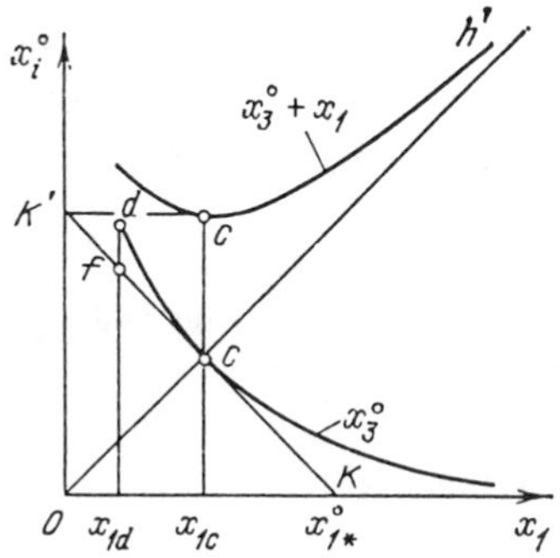

Figure 7.6.4 Schematic for the determination of x_{1*} provided that neither precipitation nor bubbly break-away (7.6.3) exists in accordance with relation $x_3^0(x_1)$ in a fully-developed flow at prescribed p, m^0, and D.

$$\frac{1}{m}\int_{z_0}^{z} J_{32}^{(d)} dz = x_2(z) - x_2(z_0) = \overset{\circ}{x}_3(z_0) - \overset{\circ}{x}_3(z) - [x_1(z) - x_1(z_0)] \tag{7.6.8}$$

In this case, with both conditions (7.6.7) holding good, and the change to higher vapor concentrations along the flow (when $\partial x_3^0/\partial x_1 > -1$, i.e., $x_1 > x_{1c}$), the break-away ceases.

If the gas acceleration is high (because of relatively strong heat flux q_W), the intensity of a dynamic break-away $J_{32}^{(d)}$ may become insufficient to adjust x_3 to rapidly growing $x_1(z)$, i.e., we have

$$x_3 > \overset{\circ}{x}_3(x_1) \tag{7.6.9}$$

This situation increases (with growth of q_W) the film moisture consumed by vaporization, and leads to a loop-like configuration of the relation $q_*(x_{1*})$ shown in Fig. 7.6.1*b*.

Elementary theory of the heat-transfer crisis at a nonuniform heat flux along the channel. Distribution of heat release along the operational channel of active zones of nuclear reactors is not uniform, rather, it resembles a cosine curve with a maximum in the middle of the channel. Therefore, the account of the impact of the heat-release distribution nonuniformity upon conditions of the heat-transfer crisis onset is of important practical significance. A review of experimental investigations concerned with the heat-transfer crisis with a nonuniform (with respect to length) heat release is outlined in "Recommendations for prediction of heat-transfer crisis . . ." (1980).

A continuing experimental study of the effects of heat-release nonuniformity over the channel length on the conditions of occurrence of the heat-transfer crisis was conducted by Rs. I. Nigmatulin (1975, 1977). Two experimental segments (D = 8 mm, L_q = 1800 mm) were used with various degrees of nonuniformity of heat release along the channel length. The cosine-curve law of heat release along the channel was produced by an appropriate reduction of the wall thickness towards the middle of a tube; consequently, the electrical resistance along the tube adequately decreased. The coefficients of nonuniformity $\varepsilon = q^{max}/q^{min}$ were $\varepsilon = 3$ and $\varepsilon = 11$.

A number of experiments on the second segment ($\varepsilon = 11$) were conducted with heated lengths 1500 and 1200 mm by means of appropriate displacement of the upper outlet of the current-conducting flange downward by 300 and 600 mm.

The experiments were performed at p = 2.9–16.7 MPa, m^0 = 750–3000 kg/(m^2 × s) and $x_{10} = (i_0 - i_{lS})/l > -0.1$ where i_0 is the enthalpy of the medium at the inlet to the heated segment. The heat-transfer crisis was monitored at the point in time of the occurrence of the temperature jump over the experimental segment by means of thermocouples whose readings were recorded on the loop oscillograph. First, the heat-transfer crisis occurred in the intermediate section between the middle and the outlet section of the channel, closer to the outlet, with further propagation upstream. Figure 7.6.5 shows the experimental data processed in the form q_* (x_{1*}). It is seen that, in domain *II* (see Fig. 7.6.1), the values of critical heat fluxes q_* at the location of crisis obtained at heat-release governed by a cosine-curve law along the channel is significantly lower than that at uniform heat release, the degree of the difference depending on the regime parameters (p, m^0, ε). This difference, with growth of both pressure p and specific mass flow rate m^0 of mixture, reduces. Growth of the degree of nonuniformity ε leads to an increase in the above-indicated difference. It should be noted that, with reduction of x_{1*}, the differences in q_* also reduce, and become insignificant in domain *I* shown in Fig. 7.6.1.

Experimental points in domain *III* (see Figs. 7.6.1 and 7.6.5) clearly concentrate near a vertical line with an abscissa x^0_{1*} (p, m^0, D).

From Fig. 7.6.5, it is also evident that, with displacement of the upper current-conducting flange on the segment with ε = 11 upstream the flow ($L^{(q)}$ = 1500 and 1200 mm), i.e., with a decrease of the length corresponding to the descending part of the cosinusoid, the length of the "vertical" portion (on the $q_*(x_{1*})$ diagram) increases approaching the experimental results obtained on uniformly heated tubes.

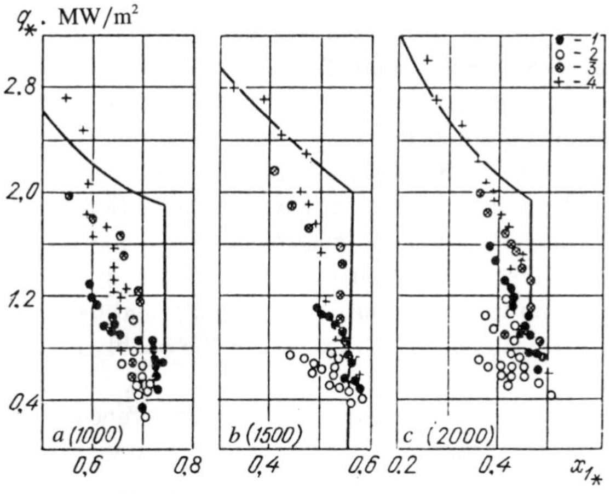

Figure 7.6.5 Dependence of specific heat flux q_* on mass concentration x_{1*} of vapor at the location of the crisis at nonuniform (cosine-curve law) heat $q_W(z)$ input along the channel length (Rs. I. Nigmatulin, 1975 and 1977) in an upward steam-water flow (p = 6.9 MPa, D = 8 mm) at various specific flow rates m^0, kg/(m² × s) (equal to 1000 (*a*), 1500 (*b*), 2000 (*c*)), heating lengths L_q, m (equal to 1.8 (*1* and 2), 1.5 (*3*), 1.2 (*4*), and nonuniformities of heat flux ε (equal to 3 (*1*) and 11 (*2*, *3*, and *4*)). Solid lines represent the approximation of experimental data on the crisis for corresponding conditions (p, m^0, and D) at uniform heat q_W input along the channel length (see Table 7.6.1).

As was proven by experimental results, the nonuniformity of heat release along the channel length has an insignificant impact on the heat exchange crisis in the region of the boiling of water not heated to the saturation temperature, and for small concentrations of vapor, i.e., in the region of a bubbly regime of flow. Nonuniformity of heat release does not effect the value of x^0_{1*}. In the region of intermediate vapor concentrations ($x_{1d} < x_{1*} < x^0_{1*}$, where x_{1d} is the concentration of vapor at which a dispersed-annular regime of flow initiates (see fig. 7.6.1)); the influence of nonuniformity, becomes, however, very significant. This effect of the heat-release nonuniformity upon the heat-exchange crisis in the intermediate range of vapor concentrations is associated with the fact that the mechanism of moisture exchange between the flow core and liquid film in this region fundamentally changes.

It has been found that departure of experimental points from the "vertical" line $x_{1*} = x^0_{1*}$ (p, m^0, D) begins when the maximum specific heat flux $q^{\max}$ on the experimental segment in the middle of a tube exceeds q_B, which is the maximum magnitude of the specific heat flux on the "vertical" line obtained in experiments with uniformly heated tubes (see Fig. 7.6.1). All dots located on the left-hand side of the vertical line relate to a case when the maximum heat flux $q^{\max}$ in experiments exceeded q_B.

Thus, the following pattern of the heat transfer crisis in a dispersed-annular regime of flow may be assumed. With a definite combination of regime parameters, the heat transfer crisis arising as a consequence of the wall liquid film break is independent of the heat flux intensity. This fact is reflected by the fact that all experimental points for these regime parameters are described by a vertical line. As soon as specific heat flux q_W exceeds q_B, the intensity of moisture break-away from the film surface significantly increases, thereby leading to a reduction of the critical concentration of vapor, since liquid that should have vaporized is carried away to the flow core. Consequently, the crisis in this region arises also by virtue of the disappearance of the annular film; but, in addition to vaporization and the dynamic break-away of liquid from the film surface, thinning, and even an additional breaking of the film, take place on account of the bubbly break-away intensification.

As is evident from Fig. 7.6.5, the heat-transfer crisis as a function of the heat-release profile along the length of flow occurs in the region of a dispersed-annular regime at various q_*, but at values x_1 of vapor concentration which resemble x^0_{1*}. As has already been indicated, with a reduction of flow rate m^0, the discrepancy between critical heat fluxes for a uniform and cosinusoidal heat release grows. This is said to be natural, since with a reduction of m^0, the film thickness increases, and, as a result, the bubbly break-away also increases, thereby leading to the q_* discrepancy augmentation.

Based on these facts, a technique was proposed (Rs. I. Nigmatulin,

1977) for predicting the critical concentration x_{1*} of vapor, and the localization z_* of the occurrence of crisis in a dispersed-annular regime of flow. The analysis is carried out based on the following assumptions:

1) the heat exchange crisis in a dispersed-annular regime of flow occurs when the liquid film breaks, i.e., when the flow rate of liquid in a film is zero: $x_{3*} = 0$;

2) there is no drop precipitation from the flow core on the film, and, consequently, there is no removal of splashes due to their break-away by liquid vaporizing from film: $J_{23}^0 = J_{32}^{(s)} = 0$;

3) the film, in addition to the dynamic break-away of moisture, also undergoes thinning due to bubbly break-away whose intensity strongly increases with the growth of q_W at $q_W > q_B$;

4) in the absence of an indicated strong increase of bubbly break-away ($q_W < q_B$), the heat exchange crisis occurs when the limiting vapor concentration x_{1*}^0 is reached.

From the above-indicated assumptions, it follows that the critical concentration of vapor reduces due to bubbly break-away intensification characterized by ΔJ_{32}, and its magnitude is determined by the expression

$$x_{1*} = \overset{\circ}{x}_{1*} - \frac{1}{m}\int_{z_d}^{z_*} \Delta J_{32} dz \tag{7.6.10}$$

where the lower limit of integration z_d is a coordinate of the section where either a film flow or dispersed-film flow originates

$$x_1(z_d) = x_{1d}(p, \overset{\circ}{m}, D) \tag{7.6.11}$$

The relation $\Delta J_{32}(q_W)$ is sought, based on the experimental results related to the heat exchange crisis with uniform heat release.

The relation $x_{1*}(q_W)$ at conditions of uniform heat release may be represented in the following form (see Fig. 7.6.1; Eqs. (7.6.1) and (7.6.3))

$$x_{1*} = \begin{cases} \overset{\circ}{x}_{1*} - b\,[(q_W/q_B) - 1], & \text{if} \quad q_W \geqslant q_B \\ \overset{\circ}{x}_{1*}, \quad \text{if} \quad q_W \leqslant q_B \end{cases} \tag{7.6.12}$$

Since

$$\pi D q_W\, dz = ml\, dx_1 \tag{7.6.13}$$

expression (7.6.10) may be rewritten as

$$x_{1*} = \overset{\circ}{x}_{1*} - l\int_{x_{1d}}^{x_{1*}} \frac{\Delta j_{32}^{(b)}}{q_W}\, dx_1 \quad \left(\Delta j_{32}^{(b)} = \frac{\Delta J_{32}}{\pi D}\right) \tag{7.6.14}$$

For a uniform heat release, we obtain

$$x_{1*} = \overset{\circ}{x}_{1*} - (x_{1*} - x_{1d})\, \Delta j_{32}^{(b)} l/q_W \tag{7.6.15}$$

Comparing formulas (7.6.12) and (7.6.15), we find an expression that defines $\Delta j_{32}^{(b)}$

$$\Delta j_{32}^{(b)} = \begin{cases} \dfrac{B(\bar{q}_W - 1)}{\Delta - b(\bar{q}_W - 1)}, & \text{if} \quad \bar{q}_W \equiv \dfrac{q_W}{q_B} \geqslant 1 \\ 0, & \text{if} \quad \bar{q}_W \leqslant 1 \end{cases} \tag{7.6.16}$$

$$(\Delta = \overset{\circ}{x}_{1*} - x_{1d},\ B = bq_B/l)$$

Here, the values b, q_B, $\overset{0}{x}_{1*}$, and x_{1d} are functions of regime parameters p, m^0, and D. These functions are determined from experiments on the crisis under conditions of a uniform heat input. Using (7.6.13), we have

$$\overset{\circ}{x}_{1*} - \frac{\pi D}{m} \int_{z_{1d}}^{z_*} \Delta j_{32}^{(b)}(z)\, dz = x_{1d} + \frac{\pi D}{ml} \int_{z_{1d}}^{z_*} q_W(z)\, dz \tag{7.6.17}$$

Substituting the obtained expression for $\Delta j_{32}^{(b)}$, we obtain the equation for the coordinate z_* of the section where the crisis occurs ($m_3(z_*) = 0$)

$$\frac{4q_B}{Dm^{\circ}} \int_{z_{1d}}^{z_*} \frac{\bar{q}_W(z)\, dz}{\Delta - b(\bar{q}_W(z) - 1)} = 1 \tag{7.6.18}$$

Unless the solution of this equation is realized on the segment under consideration, where there is heat release ($q_W > 0$), there is no crisis at a predetermined $q_W(z)$. Substituting expression (7.6.16) into (7.6.10), we find the critical concentration of vapor

$$x_{1*} = \overset{\circ}{x}_{1*} - \frac{4bq_B}{Dm^{\circ}l} \int_{z_{1d}}^{z_*} \frac{\bar{q}_W(\bar{q}_W - 1)}{\Delta - b(\bar{q}_W - 1)}\, dz \tag{7.6.19}$$

Thus, having the distribution of a specific heat flux $q_W(z)$ along the experimental segment, we find from (7.6.18) the coordinate z_* of the section where the heat-transfer crisis is realized, following which $q_* = q(z_*)$ may be determined, and from Eq. (7.6.19), we obtain $x_{1*} = x_1(z_*)$.

Analysis of experimental data shows that detection of the location of the heat-transfer crisis occurrence presents the greatest complication in a study of the heat-transfer crisis under conditions of cosinusoidal heat input. The ambiguity in detecting this location very strongly affects the results of investigations at high gradients of the heat-flux distribution along the channel. In this context, it is advisable to use, when processing the experimental data, a regime characteristic such as a principal parameter, which is not interrelated with the flow parameters in the zone of occurrence of the crisis. The maximum heat flux $q^{\max}$ in the middle of a tube may be used as such a characteristic; at predetermined laws of heat release and coefficient ε of nonuniformity, this heat flux is proportional to the input thermal power N.

Then, the analysis is reduced to determination of $q_*^{\max}$, or N_*, at which the heat-transfer crisis occurs under given conditions at the inlet (p, m^0, and x_{10}) and the channel geometry (D and L_q).

The values of N_* predicted in conformity with the above-described technique are in agreement with experimental results with an accuracy of $\pm 10\%$ in a range of regime parameters: $p = 2.9$–16.7 MPa, $m^0 = 750$–3000 kg/(m^2 $\times$ s).

Thus, the heat-transfer crisis at nonuniform distribution of heat flux along a channel in the region of dispersed-annular structure of the flow may be predicted based on heat-transfer-crisis experimental data obtained at uniform heat release.

The heat-transfer crisis due to the liquid wall film drying at nonstationary conditions. We shall now consider a stationary one-dimensional flow of vapor-liquid mixture in a heated vertical tube. Let a step increase of heat release in a tube wall at time $t = 0$, or a smooth reduction of the mixture flow rate at the tube inlet, take place. It is necessary to find the conditions and point in time of occurrence of the heat-transfer crisis caused by the liquid film drying.

A brief review of available experimental studies of the heat-transfer crisis in nonstationary conditions was reported by B. I. Nigmatulin et al. (1980). The investigations may be divided into three major groups based on the nature of an exterior disturbance: variation of the heat-release power (intensity), reduction of the heat-transfer agent flow rate, and change of pressure at either inlet or outlet of the channel. It is obvious that, unlike stationary conditions, the feasibility of experimental investigation of the heat-transfer crisis in nonstationary conditions is limited because of a great number of ways to produce disturbance, and, also, because of the arbitrariness of the development of the introduced disturbance with time. Therefore, great importance is attached to the generation of closed hydrodynamic models, and to the development of suitable numerical algorithms.

Before we discuss a more complicated nonequilibrium model of the process based on the equations outlined in §§7.2–7.4, it is worthwhile to mention the simplest, single-velocity, quasi-stationary model, with no consideration given to either time or kinetic effects, which is often used for predicting the nonstationary heat-transfer crisis (D. Moxon and P. Edwards, 1967; G. Gaspari et al., 1973; O. K. Smirnov et al., 1973). In the framework of this model, it is assumed that under nonstationary conditions, the heat-transfer crisis is realized when the heat-carrying agent flow rate, specific heat flux, and mass flow-rate concentration of vapor at certain section of the channel are equal to corresponding values obtained for the heat-transfer crisis under stationary conditions. Comparison of the results of analysis with experimental results shows that the values of the major parameter of the process (time interval between the onset of a nonstationary process and a point in time of occurrence of the heat-transfer crisis) may differ by a factor of two and more. The maximum discrepancy is observed at relatively rapid nonstationary processes, when the inherent time of the process development t_* = 0.1–1 second. The fact is that in nonstationary processes, due to finite velocities of heat exchange, the same pattern of liquid distribution between core and film realized in stationary flows may not have enough time to be established.

The nonequilibrium effects of mass exchange must be given consideration for analysis of these processes. We shall discuss this issue in the framework of an equilibrium single-velocity and single-temperature pattern of a flow core in a state of a thermodynamic (but not mechanical) equilibrium with film. Then, the mass equations for components (vapor, film, and drops) and momentum equations for both core and film become generalizations of Eqs. (7.5.2) and (7.2.32) for nonstationary flows. These generalizations may be readily obtained using the general form of Eqs. (7.2.8).

The distributions of volumetric concentrations of the mixture species are transferred along the channel length $L \sim 1$ m with velocities $v_1 \sim 10$ m/s and v_3, ($v_1 > v_3$). The disturbances of both velocity and pressure introduced into flow propagate with the speed of sound $C > C_e$, where C_e is an equilibrium speed of sound in vapor-drop core of the flow (see §§4.1 and 4.2). If $v_1 \ll C_e$, $t_* \gg L/v_j$, the distributions of both velocity and pressure are of a quasi-stationary nature

$$\frac{\partial v_j}{\partial t} \sim \frac{v_j}{t_*} \ll v_j \frac{\partial v_j}{\partial z} \sim \frac{v_j^2}{L}, \quad \frac{\partial p}{\partial t} \sim \frac{\Delta p}{t_*} \ll v_j \frac{\partial p}{\partial z} \sim v_j \frac{\Delta p}{L} \qquad (j = 1, 2, 3) \tag{7.6.20}$$

Supposition of a quasi-stationary distribution of velocities and pressure of the mixture species means that the time-dependent variation of these parameters is determined by the nonstationarity of boundary conditions for both flow rates and concentrations of the mixture species, and by the nonstationarity of the external thermal loading. Note that this supposition cannot be used at step pressure variation at the channel inlet (outlet) with an inherent time of pressure variation within 0.001–0.01 s when the wave effects are predominant. With a smoother variation of pressure, flow rates, and heat fluxes with an inherent time of 0.1 s and more, the outlined assumption holds good. Using (7.6.20), the system of equations presented in §7.2 for such flow may be reduced to the following form

$$\frac{\partial \rho_1 S_c}{\partial t} + v_c \frac{\partial \rho_1 S_c}{\partial z} = J_{21} + J_{31} - \rho_1 S_c \frac{\partial v_c}{\partial z}$$

$$\frac{\partial \rho_2 S_c}{\partial t} + v_c \frac{\partial \rho_2 S_c}{\partial z} = \overset{\circ}{J}_{32} - J_{21} - \overset{\circ}{J}_{23} - \rho_2 S_c \frac{\partial v_c}{\partial z}$$

$$\overset{\circ}{\rho}_l \frac{\partial S_f}{\partial t} + \overset{\circ}{\rho}_l v_f \frac{\partial S_f}{\partial z} = \overset{\circ}{J}_{23} - J_{31} - \overset{\circ}{J}_{32} - \overset{\circ}{\rho}_l S_f \frac{\partial v_f}{\partial z}$$

$$\rho_c S_c v_c \frac{\partial v_c}{\partial z} + S_c \frac{\partial p}{\partial z} = -F_{13} - \rho_c S_c g + \left(J_{31} + \overset{\circ}{J}_{32} - J_{32}^{(s)}\right)(v_{\Sigma 3} - v_c)$$

$$\overset{\circ}{\rho}_l S_f v_f \frac{\partial v_f}{\partial z} + S_f \frac{\partial p}{\partial z}$$

$$= -F_W + F_{13} - \overset{\circ}{\rho}_l S_f g + \left(\overset{\circ}{J}_{23} - J_{32}^{(s)}\right)(v_c - v_f) - \left(\overset{\circ}{J}_{32} - J_{32}^{(s)}\right)(v_{\Sigma 3} - v_f) \tag{7.6.21}$$

$$S_c \frac{\partial v_c}{\partial z} + S_f \frac{\partial v_f}{\partial z} + \frac{\alpha_1 S_c}{\overset{\circ}{\rho}_g}\left(\frac{\partial \overset{\circ}{\rho}_g}{\partial p}\right)_S v_c \frac{\partial p}{\partial z}$$

$$= (J_{31} + J_{21})\left(\frac{1}{\overset{\circ}{\rho}_g} - \frac{1}{\overset{\circ}{\rho}_l}\right) + (v_c - v_f)\frac{\partial S_f}{\partial z}$$

$$\rho_c = \rho_1 + \rho_2, \quad S_c + S_f = S, \quad v_1 \equiv v_2 \equiv v_c, \quad v_3 \equiv v_f$$

$$J_{21} = -\frac{v_c}{l}\left[\overset{\circ}{\rho}_g \alpha_1 S_c \left(\frac{\partial u_g}{\partial p}\right)_S + \overset{\circ}{\rho}_l \alpha_2 S_c \left(\frac{\partial u_l}{\partial p}\right)_S - \frac{\alpha_1 S_c p}{\overset{\circ}{\rho}_g}\left(\frac{\partial \overset{\circ}{\rho}_g}{\partial p}\right)_S\right]\frac{\partial p}{\partial z}$$

$$J_{31} = \frac{1}{l}\left[Q_W - \overset{\circ}{\rho}_l S_f v_f \left(\frac{\partial u_l}{\partial p}\right)_S \frac{\partial p}{\partial z}\right]$$

The derivatives of ρ_g^0, u_g, u_l, and p are taken along the saturation line.

To close this system of equations, there must be predetermined relations for the break-away intensity J_{32}^0, drop deposition intensity J_{23}^0 from the film surface, force interaction between the mixture species and channel wall F_{13} and F_W, and, also, for the external heat flux Q_W.

Analysis of results of experimental studies of processes of drop break-away from the film surface, and drop deposition on its surface (§7.4) in steam-water flows, shows that the inherent time t_j of establishment of stationary values J_{32}^0 and J_{23}^0 is about 0.01 s. In most cases, the inherent time t_* of occurrence of the heat-transfer crisis associated with the wall liquid film drying-out is of the order 0.1–1 s and more, i.e., $t_j \ll t_*$. Therefore, when investigating such processes, a hypothesis of a quasi-stationary nature of intensity of interaction between a film and core of the flow may be adopted, i.e., the relations may be used for drop break-away and deposition obtained for stationary conditions.

In a similar manner, the force interaction between the flow core and film, between the film and the channel wall, is also described by relations obtained for stationary conditions since the inherent time for establishment of the mixture species velocities is about 0.001 s. This is substantiated by the results of analysis of stationary regimes. For the variant represented in Fig. 7.5.1, the velocity υ_f of liquid in a film assumes its developed value over the length $\Delta z/D \sim 1$ at velocity $\upsilon_f \sim 3$ m/s. Hence, the developed values of velocity of liquid in a film are obtained in a time period of $\Delta t \lesssim \Delta z/\upsilon_f \sim 0.001$ s.

Note that, with variation of external thermal loading with time, it is imperative to take into consideration the thermal inertia of the channel wall. Generally, the thickness of experimental segments is $h \ll D$; then, a parabolic profile of temperature variation over the cross section of a channel in stationary conditions takes place. The temperature of the interior surface of the channel may be considered equal to the liquid saturation temperature $T_W = T_j(p)$ during boiling of liquid on this surface. The exterior surface of experimental segments is generally heat-insulated; therefore, thermal losses into the environment may be ignored. From the heat-influx equation for a wall with stepped variation of the thermal loading, and in supposition of similarity of the temperature profile to a stationary profile, we obtain

$$N(t) = N^{(e)} - (N^{(e)} - N^{(0)})\exp(-t/t_N)$$
$$t_N = \frac{h^2}{3\nu_W^{(T)}}, \quad Q_W = \frac{N}{L_q}, \quad q_W = \frac{Q_W}{\pi D} \tag{7.6.22}$$

where $N^{(0)}$ and $N^{(e)}$ are, respectively, initial and final magnitudes of external thermal loading; L_q is the length of heated segment; $\nu_W^{(T)}$ is the thermal diffusivity of the wall material.

Analogously, the law of the liquid flow rate reduction at the channel inlet was predetermined

$$m(t) = m^{(e)} + (m^{(0)} - m^{(e)})\exp(-t/t_m) \tag{7.6.23}$$

As a rule, in experiments on the heat-transfer crisis in a steam-water dispersed-annular flow, subcooled liquid is fed at the channel inlet. Thus, over the initial segment of a channel, first, subcooled liquid flows followed by a flow of steam-water mixture in a bubbly regime, which transforms into a dispersed-annular regime at the volumetric concentration of steam $\alpha_g \equiv \varphi \equiv 0.7$–$0.8$. In order to compare the predicted and experimental data related to the heat-transfer crisis, the analysis of nonstationary flow must be performed for the initial segment of the channel.

The heat-influx equation for flow of underheated liquid is written in the form

$$\frac{\partial T_l}{\partial t} + \upsilon_l \frac{\partial T_l}{\partial z} = \frac{Q_W}{\overset{\circ}{\rho}_l c_l S} \tag{7.6.24}$$

Index l denotes parameters of a subcooled liquid. From the assumption of the liquid incompressibility, it follows that υ_l = constant. Solution of Eq. (7.6.24) enables us to find the coordinate of the channel section, where T_l

$= T_S$ and the flow of underheated liquid is transformed into bubbly regime. This regime of flow will be described in the framework of the equilibrium flow model. Since both pressure and temperature of the mixture vary insignificantly over the initial segment of the channel, density of vapor (steam) is said to be constant (ρ_g^0 = constant). Then, from conservation equations for mass and internal energy, we obtain for an equilibrium two-phase flow

$$\frac{\partial v}{\partial z} = \frac{Q_W}{Sl}\left(\frac{1}{\overset{\circ}{\rho}_g} - \frac{1}{\overset{\circ}{\rho}_l}\right) \qquad (Q_W = \pi D q_W)$$
$$\frac{\partial \alpha_g}{\partial t} + v\frac{\partial \alpha_g}{\partial z} = \frac{Q_W}{Sl}\left(\frac{1-\alpha_g}{\overset{\circ}{\rho}_g} - \frac{\alpha_g}{\overset{\circ}{\rho}_l}\right) \tag{7.6.25}$$

Solution of this system permits the determination of the channel section where the volumetric concentration of vapor in a mixture, $\alpha_g = \varphi_d$ (φ_d is the limiting magnitude), at which a transition takes place from a bubbly regime to a disperse-annular regime of flow of mixture. It may be assumed, in this case, that $\varphi_d \approx 0.75$.

Unless the two-velocity mode of flow of liquid is given consideration (i.e., its separation between a high-velocity core and "slow" film), the variation of vapor concentration over the length of the tube with a two-phase flow ($z_b \leqslant z \leqslant L$) is governed by Eq. (7.6.25). Then, a simplified single-velocity model may be considered for determining the time of occurrence of the heat-transfer crisis using the relationship $x_{1*}(q_W)$ for a fully-developed flow (Fig. 7.6.1), having assumed that the crisis is realized as soon as (in conformity with relations (7.6.25)) vapor concentration $x_g = \rho_g^0 d_g/(\rho_g^0 d_g + \rho_l^0(1 - d_g)$ at the outlet ($z = L$) becomes equal $x_{1*}(q_W)$. This pattern takes into consideration a nonstationarity due only to the finite rate of redistribution of the vapor-concentration profile at a disturbance $q_W(z)$ or $m(0, t)$, but gives no consideration to the nonstationarity effect due to the finite rate of the mass exchange between film and the flow core due to drop breakaway and deposition.

In order to solve the system of equations (7.6.21), (7.6.24), and (7.6.25), both boundary and initial conditions must be predetermined. The initial conditions define the liquid temperature distribution over a segment of flow of subcooled liquid, the volumetric concentration of vapor over the segment of the bubbly regime of flow, and also, the volumes of liquid film and drops over the segment of a dispersed-annular regime of flow. These initial distributions for the variants being discussed below were stationary, and, therefore, determined from a solution of a related stationary problem (see §§7.5 and 7.6).

The boundary conditions required for both stationary and nonstationary problems define the velocity v_0, temperature T_0 of a subcooled, and pressure p_0 at the channel inlet. In variants of analytical methods and experimental regimes discussed below, change of the heat-carrying agent flow-rate $m(0,$

t) and of the input heating power $Q_W(t)$ was carried out at a fixed inlet pressure (p_0 = constant) that was ensured by experimental conditions. The liquid velocity at the inlet to the segment with a bubbly regime of flow ($z = z_b$) equals its velocity at the channel inlet. The volumetric concentration of vapor at the inlet to the segment of bubbly regime of boiling is said to equal zero. Finally, there must be predetermined: the velocity of the vapor-drop core of both flow and liquid film, pressure, and volumetric concentrations of drops and liquid film in the section where the bubbly regime undergoes transition to a dispersed-annular regime of flow of mixture. Because of the smallness of the pressure drop at the initial segment, it may be assumed that pressure at the inlet to a segment with a dispersed-annular regime of flow equals pressure at the channel inlet. Boundary conditions at the inlet to the segment with a dispersed-annular regime of flow ($z = z_d$, $\alpha_g = \varphi_d \approx 0.75$, $\upsilon = \upsilon_d$) follow from the mass conservation equations for vapor and liquid, and momentum conservation equations for the entire flow, having assumed that the flows of these values are continuous

$$\begin{aligned}
&\overset{\circ}{\rho}_g (1-\alpha_2)\, S_c v_c = \overset{\circ}{\rho}_g \varphi_d S v_d \qquad \left([\overset{\circ}{\rho}_g \varphi_d + \overset{\circ}{\rho}_l (1-\varphi_d)]\, v_d = \overset{\circ}{\rho}_l v_0\right)\\
&\overset{\circ}{\rho}_l \alpha_2 S_c v_c + \overset{\circ}{\rho}_l S_f v_f = \overset{\circ}{\rho}_l (1-\varphi_d)\, S v_d \qquad (S_f + S_c = S)\\
&\overset{\circ}{\rho}_g (1-\alpha_2)\, S_c v_c^2 + \overset{\circ}{\rho}_l \alpha_2 S_c v_c^2 + \overset{\circ}{\rho}_l S_f v_f^2 = \left(\overset{\circ}{\rho}_g \varphi_d + \overset{\circ}{\rho}_l (1-\varphi_d)\right) S v_d^2
\end{aligned} \tag{7.6.26}$$

To determine α_2, S_c, υ_1, and υ_f based on known ρ_g^0, ρ_l^0, φ_d and υ_0 in a section $z = z_d$, there has been prescribed a liquid separation between core and film by a relation $S_f/(\alpha_2 S_c) \approx 0.95$, i.e., almost the entire volume of liquid in a section where the dispersed-annular structure of flow originates is contained in a film.

When performing the analysis, the heat-transfer crisis was considered to occur at a relative flow rate of liquid in a film $x_{3*} = 0.01$.

To verify the adequacy of the above-described model, a comparison was made between predicted and experimental data on the heat-transfer crisis in stationary conditions. Figure 7.6.6 depicts typical distributions of both relative flow rate $x_f \equiv x_3$ of liquid in a film and mass flow rate concentration $x_g = x_1$ of vapor along the heated tubes at various initial subcoolings and thermal loadings in developed conditions. The predicted curves are in agreement with experiments where the heat-transfer crisis occurred in an outlet section of a channel. It can be seen that analytical coordinates of occurrence of the heat-transfer crisis differ from those obtained experimentally by not more than 10%. The analytical values of x_{1*}, which are independent of q_W in the range under consideration, are in good agreement with above-discussed experimental values presented in Recommendations . . . (1980). The relative flow rate x_f of liquid in a film in a heated channel under stationary conditions is determined by pressure, specific flow rate of heat carrying agent, relative flow rate of vapor x_g, and is virtually independent of the heat flux (the latter is illustrated in Fig. 7.6.6*b*).

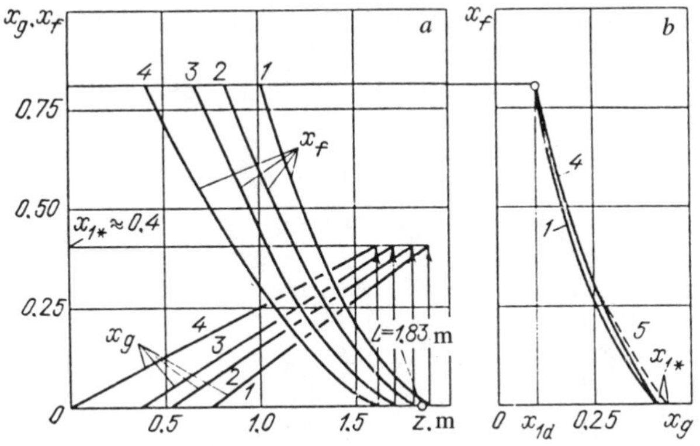

Figure 7.6.6 Analytically predicted distributions of x_g and x_f along (*a*) the tube ($D = 9$ mm, $L = 1.83$ m), and relationship $x_f(x_g)$ (*b*) in stationary conditions ($p = 6.9$ MPa, $m^0 = 2700$ kg/(m^2 × s)), when at the channel inlet ($z = 0$) a subcooled water is fed (at a temperature lower by ΔT than saturation temperature), and the channel is uniformly heated at an intensity q_W along its length: this regime of heating caused the heat-transfer crisis at the outlet ($z = 1.83$ m) in an experiment reported by D. Moxon et al., 1967. Curves *1–4* correspond to various ΔT, K, and q_W, MW/m^2: *1*) (83.4; 2.45); *2*) (55.6; 2.23); *3*) (27.8; 2.01); *4*) (0; 1.79). Vertical lines with arrows indicate the predicted coordinates z_*, and horizontal line ($x_{1*} \approx 0.4$) specifies the predicted concentration x_{1*} of vapor at the location of crisis ($x_f = 0$). Dashed line *5* relates to $D = 10.8$ mm, $L = 3.66$ m, $p = 6.9$ MPa, $m^0 = 2020$ (kg/(m^2 × s), subcooled temperature $\Delta T = 0–56$ K at the inlet, and to critical heat fluxes $q_W = 0.97–1.32$ MW/m^2, which cause the heat-transfer crisis at the outlet ($z = 3.66$ m).

Figure 7.6.7 depicts analytical results of nonstationary flow with an uneven variation of the input heating power N at both fixed flow rate and liquid subcooling at the tube inlet. These conditions were maintained throughout the experiment. In accordance with analytical determinations, increase of thermal loading leads to the intensification of the film vaporization, and to the increase of both the flow rate and velocity of vapor. The latter also results in the acceleration of the liquid. As a consequence, at the outlet and neighboring sections of the tube, flow $m(z, t)$ of the mass of mixture at fixed $m(0, t)$ rapidly grows. Further on, the flow rate of vapor-liquid mixture stabilizes in all sections of the tube, and drops to the magnitude of flow rate at the inlet section. Increase of velocity v_f in the film at initial time leads to some growth of the liquid flow rate in it. Later, velocity in the film drops because of decrease of its thickness and corresponding reduction of the force of its interaction with the core.

Figure 7.6.8 shows the predicted results obtained during a smooth reduction of the underheated water flow rate at the inlet to a heated channel. A significant reduction of x_f is realized beginning at a middle portion of a channel, which corresponds to the initial zone of dispersed-annular flow, with the rate of reduction of x_f slowed down toward the end portion of a

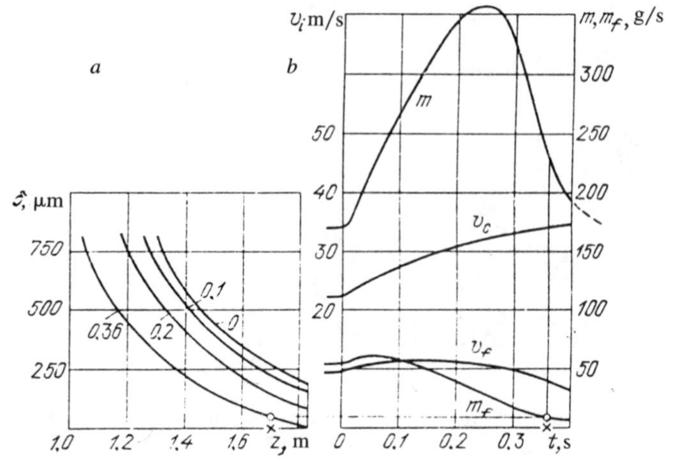

Figure 7.6.7 Distribution of the film thickness δ along the length of the tube (*a*) ($D = 9$ mm, $L = 1.83$ m, $h = 2.4$ mm) with steam-water flow at various times ($t = 0$; 0.1; 0.2; 0.36 s are indicated by numerical labels on curves); and variation of the flow parameters (v_c, v_f, m_f, and m) with time (*b*) in a fixed section $z = 1.7$ m (indicated by a cross on axis z) following an increase of thermal loading at $t = 0$ from a stationary value $N^{(0)} = 100.3$ kW ($q_W = 1.94$ MW/m^2) to $N^{(e)} = 142.5$ kW ($q_W = 2.69$ MW/m^2) in accordance with the law (7.6.22) with $t_N = 0.29$ s. There were maintained fixed At the tube inlet, liquid subcooling ($T_0 - T_S(p_0)$ $= -62.8$ K), and its flow rate ($m = 1/4D^2m^0 = 0.172$ kg/s) were maintained fixed. The represented analysis yields the heat-transfer crisis (film drying-out at the inlet: $z = L = 1.83$ m) at $t = 0.36$ microsecond (indicated by a cross on axis t); the experiment (D. Moxon et al., 1967) shows $t = 0.44$ s, and the single-velocity model yields $t = 0.29$ s. The initial points on curves $\delta(z)$ correspond to the transition of the bubbly regime of flow to a dispersed-annular regime.

channel. This is accounted for by the fact that under a nonstationary regime, with the liquid velocity reduction, the initiation of boiling is displaced closer to the inlet, and both vapor concentration and core velocity v_c at fixed sections of the channel increase, the film-liquid flow rate decreasing with a certain delay compared to the reduction of the liquid flow rate at the inlet section.

The delay is determined by the difference in the velocities of the vapor-drop core v_c and film v_f, where $v_f \ll v_c$. As a result, at the initial segment of the dispersed-film flow, the liquid break-away from the "slow" film is intensified, and the removed liquid with a "high" velocity v_c is carried in the flow core towards the outlet portion of the channel, where deposition onto the film is thereby intensified compared to the original stationary state. Therefore, at initial points in time, the film thickness δ and liquid flow rate m_f in a film grow somewhat over the outlet portion of the channel, and only later on begin to reduce.

Figure 7.6.9 illustrates the predicted displacement of segments of a

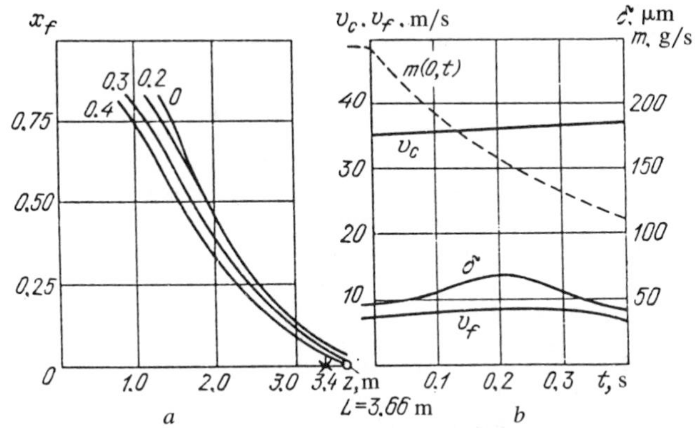

Figure 7.6.8 Distribution of the portion of flow rate x_f related to liquid in film along (*a*) the heated tube (D = 10.8 mm, L = 3.66 m) with a steam-water flow at various points in time (t = 0; 0.2; 0.3; 0.4 seconds, indicated by labels on curves), and variation of parameters of a dispersed-annular flow (v_c, v_f, and δ) with time in a fixed section z = 3.4 m (indicated by a cross on axis z) following the reduction (at t = 0) of the flow rate $m(0, t)$ of subcooled liquid at the channel inlet (z = 0) in accordance with the law (7.6.23) with t_m = 0.275 s (see dashed line in Fig. *b*) from a stationary value $m(0, 0)$ = 247 g/s (m^0 = 2700 kg/(m^2 × s)) to $m(0, \infty)$ = 71 g/s (m^0 = 774 kg/(m^2 × s)). Both pressure (p_0 = 6.9 MPa) and the liquid subcooling ($\Delta T = T_0 - T_S(p) = -13.3$ K) at the tube inlet (z = 0) were maintained constant. The heating intensity was not changed: N = 144.7 kW (q_W = 1.17 MW/m^2). This analysis yields the heat-transfer crisis (film drying-out at the outlet; $z = L$ = 3.66 m) at t = 0.4 s, compared to experiment (D. Moxon et al., 1967) with t = 0.41 s, and a single-velocity model with t = 0.2 s.

single-phase, bubbly, and dispersed-film flow in an upstream direction, when feeding of the subcooled liquid to a channel uniformly heated with a constant intensity is reduced.

The above-indicated examples of analytical prediction and the results of comparison of points in time of occurrence of the heat-transfer crisis prove that, in the framework of the presented discussion, this time is determined quite satisfactorily. The described model is significantly more accurate and physically more substantiated than the single-velocity, quasi-stationary approach.

§7.7 POST-CRISIS HEAT EXCHANGE DURING FLOW OF A DISPERSED (DROP) FLOW IN VAPOR-GENERATING CHANNEL

A film in a heated channel may vaporize, and dispersed-annular regime transforms into a purely dispersed regime—flow of a mixture of vapor and drops. In this case, since there is no appropriate contact between an inten-

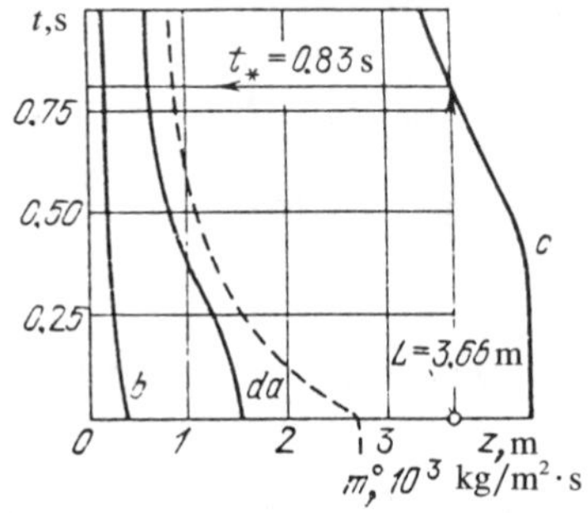

Figure 7.6.9 Predicted displacements of boundaries along the tube with a heated steam-water flow as a function of time (the indicated boundaries specify initiation of bubbly regime) (*b*), dispersed-annular (*da*) and dispersed (*c*—the heat-transfer crisis due to film drying-out) regimes of flow following the beginning (at $t = 0$) of reduction of flow rate $m(0, t)$ of underheated liquid at the inlet ($z = 0$) indicated by a dashed line (the same law as specified for Fig. 7.6.8). Both pressure ($p_0 = 6.9$ MPa) and water subcooling ($\Delta T = T - T_S(p_0) = -12.8$ K) at the tube inlet were maintained constant. The intensity of a uniform heating along the tube was not changed: $N = 119.3$ kW ($q_W = 0.961$ MW/m^2). The represented analysis yields the heat-transfer crisis (film drying) at the tube outlet ($z = L = 3.66$ m) at $t = 0.83$ s; experimental data (D. Moxon et al., 1967), at $t = 0.89$ s, and the single-velocity quasi-stationary model, at $t = 0.66$ s.

sively heated wall and liquid phase, the wall may become overheated, and the downstream heating surface has a post-crisis region of deteriorated heat exchange. The heat exchange in this region is characterized by significantly lower heat-transfer coefficients than in the pre-crisis area, which may result in significant overheating of the heating surfaces compared to saturation temperatures. We shall discuss below a particular case of heat exchange of this kind, viz., heat exchange in a post-crisis area in a smooth vertical tube.

Investigation of the intensity of heat exchange in a post-crisis area is of considerable interest from the standpoint of analysis of emergency regimes of operation of nuclear reactors, and also in designing single-tube steam generators where the area of the post-crisis heat exchange may occupy up to 40% of the total steam-generating surface and more. The heat exchange in the post-crisis area defines the tube-wall maximum temperature which affects both the life and reliability of the steam-generating channel operation to a significant extent.

The single-velocity, single-temperature ("homogeneous") model is the simplest model for investigating the heat exchange in a post-crisis area; according to this model, the two-phase mixture is viewed as some homogeneous liquid with its own velocity and temperature, and the heat-exchange experimental data are processed using empirical formulas (D. Groenveld, 1972; M. A. Styrikovich et al., 1982) of type

$$\mathrm{Nu}_W = A\,\mathrm{Re}^b \mathrm{Pr}_g^c \qquad (\mathrm{Re} = m^\circ D/\mu_l, \quad \mathrm{Pr}_g = \nu_g^{(v)}/\nu_g^{(T)}) \tag{7.7.1}$$

where A, b, and c are empirical parameters. In such relations, slipping between phases and the absence of thermal equilibrium is taken into consideration by parameters A, b, and c, and also by true mass concentration of vapor in the mixture, which affects the Reynolds number **Re**. In this case, two limiting models are often used: the first is based on an assumption of a complete thermodynamic equilibrium of the mixture, and the second is founded on an assumption of the absence of phase transitions past the section where heat-transfer crisis occurs.

According to the first model, the mass flow-rate concentration of vapor in the mixture is said to be equal its equilibrium value. This holds true at high specific mass flow rates of the mixture, when an intensive heat exchange exists between wall and flow, and also near the section where the heat-transfer crisis occurs when the vapor phase has not had enough time to be overheated. The relations obtained based on the hypothesis of a complete thermodynamic equilibrium of the mixture yield the lower limiting temperature of the wall over the segment of a post-crisis heat exchange.

In accordance with the second model, there is no phase transition during a flow of vapor-drop mixture in a heated channel in the area of a post-crisis heat exchange, and the mass concentration of vapor along the channel is constant. The total input heat is assumed to be used for superheating vapor relative to the saturation temperature. The relations obtained based on this model yield substantially higher temperatures of the channel wall, and may be used for evaluating their upper boundary over the segment of post-crisis heat exchange.

Figure 7.7.1 shows an example of comparison between experimental data on the wall temperature and relations obtained based on indicated limiting schemes.

Failure to take into consideration the real characteristics of heat transfer from wall to vapor and drops, as well as the characteristics of heat exchange between phases, limits applications of relations of type (7.7.1). Therefore, the one-dimensional models of vapor-drop flow became favorable for predicting heat transfer over a segment of post-crisis heat exchange (A. Bennet et al., 1967; R. Forslung and W. Rohsenow, 1968; V. A. Vorob'ev, P. L. Kirillov et al., 1972; V. K. Koshkin, E. K. Kalinin et al., 1973); generally, a two-stage mechanism for heat transfer is used in such flows: from the wall to overheated vapor phase, and from this phase to drops of liquids. In conformity with this model, with known initial conditions at the location of

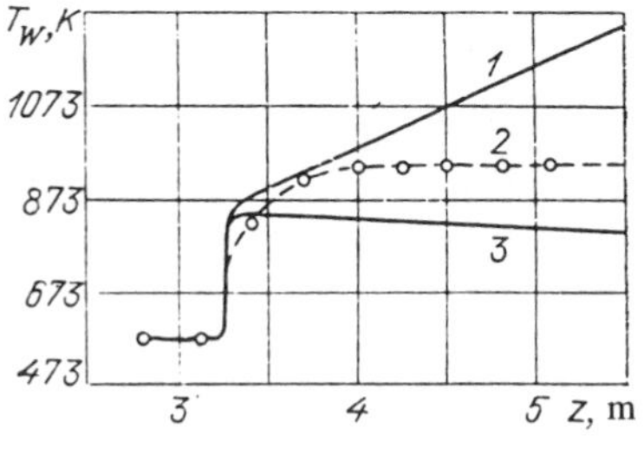

Figure 7.7.1 The effect of thermal nonequilibrium of a dispersed flow on the channel-wall temperature in a post-crisis area of heat exchange (M. I. Marinov, 1977) in steam-water drop flow (p = 6.8 MPa, x_{1*} = 0.585 (z_* = 3.25 m), m^0 = 1000 kg/(m$^2 \cdot$ s), q_w = 0.92 MW/m^2). Line *1*) prediction in accordance with "frozen" phase transitions, *2*) experimental results, *3*) prediction based on a complete thermodynamic equilibrium.

crisis ($z = z_*$ is the beginning of the post-crisis heat exchange zone), the channel-wall temperature may be found. The failure to take into account the direct thermal interaction between drops and heated surface, and, as a consequence, its inapplicability for predicting heat transfer over the channel segment where precipitating drops moisten the heating surface, is a major shortcoming of this model.

A hydrodynamic model of a dispersed vapor-drop flow, in which all major processes of interaction between phases and the channel wall are given consideration, is discussed below.

A system of differential equations for a stationary vapor-drop flow in a tube. Equations for a drop (dispersed) flow follow from a similar system written for a dispersed-annular flow (§7.2), if the liquid-film equations are dropped, and the terms giving consideration to effects of liquid film are omitted in equations for vapor and drops. In addition to the assumptions of §7.2, the following must be taken into consideration: the relative velocity of vapor and drops is small, i.e., much smaller than the velocity of vapor (drops) ($|v_1 - v_2| \ll v_1$); therefore, slip is taken into account only when determining the heat flux $Q_{1(\Sigma 2)}$ from vapor to the drop surface Σ_2, where even a slight slip may be significant. Slip is determined without any account of inertial forces of drops in their motion relative to vapor, which is small compared to friction force between vapor and drops. The vapor-drop flow may be thermodynamically nonequilibrium to a significant extent—vapor is overheated relative to the saturation temperature, and liquid in drops is on the saturation line. The system of conservation equations for a stationary vapor-drop flow in a tube (S = constant) is written in the form

$$\frac{d\overset{\circ}{m}_1}{dz} = \frac{J_{21}}{S}, \quad \frac{d\overset{\circ}{m}_2}{dz} = -\frac{J_{21}}{S}, \quad \overset{\circ}{m}_2 \frac{da}{dz} = -\frac{J_{21}a}{3S} \qquad (nv = \text{const})$$

$$\overset{\circ}{m} \frac{dv}{dz} = -\frac{dp}{dz} - \frac{F_W}{S} \qquad (v_1 \approx v_2 \approx v)$$

$$\overset{\circ}{m}_1 \frac{du_1}{dz} = \frac{\overset{\circ}{m}_1 p}{(\overset{\circ}{\rho}_1)^2} \frac{d\overset{\circ}{\rho}_1}{dz} + \frac{Q_{W1} - Q_{1(\Sigma 2)} + J_{21}(i_{1S} - i_1)}{S}$$

$$\overset{\circ}{m}_2 \frac{du_2}{dz} = \frac{Q_{W2} - Q_{2(\Sigma 2)}}{S} \qquad (u_2 = u_{2S} \ \text{or} \ T_2 = T_S(p)) \tag{7.7.2}$$

$$(\overset{\circ}{m}_1 \equiv \overset{\circ}{m}_g = \overset{\circ}{\rho}_1 \alpha_1 v, \quad \overset{\circ}{m}_2 \equiv \overset{\circ}{m}_l = \overset{\circ}{\rho}_2 \alpha_2 v, \quad \overset{\circ}{m} = \overset{\circ}{m}_1 + \overset{\circ}{m}_2, \quad \alpha_1 + \alpha_2 = 1)$$

$$Q_{1(\Sigma 2)} + Q_{2(\Sigma 2)} = J_{21} l, \quad Q_{W1} + Q_{W2} = Q_W$$

where Q_{W1} and Q_{W2} are external heat fluxes, respectively, from wall to vapor, and from vapor to drops.

Here, in the equation for internal energy of vapor, the term pertinent to work of the wall friction force is omitted. This work is negligible compared to exterior, relative to vapor, heat fluxes.

Force and thermal interactions between phases and channel wall. The friction force between vapor and wall in variants being discussed below is determined by the Blausius formula (cf. (7.3.5)) for a single-phase flow of vapor in smooth tubes (with no consideration given to the effects of drops on this force)

$$F_W = \pi D \tau_W, \qquad \tau_W = {}^1\!/_2 C_{W1} \overset{\circ}{\rho}_g v^2$$
$$C_{W1} = 0.316/\mathrm{Re}_1^{0.25}, \quad \mathrm{Re}_1 = \overset{\circ}{\rho}_g v D/\mu_g \tag{7.7.3}$$

To predict the heat transfer between a wall and vapor-drop flow in the post-critical area of heat exchange, the processes of thermal interaction between phases, and also between phases and heating surface, play a predominant role.

The heat flux from the tube wall to vapor may be represented in the form (cf. (7.3.14))

$$Q_{W1} = \pi D q_{W1}, \qquad q_{W1} = \lambda_g \mathrm{Nu}_{W1} (T_W - T_1)$$
$$\mathrm{Nu}_{W1} = 0.023\, \mathrm{Re}_1^{0.8} \mathrm{Pr}_g^{0.4} \qquad (\mathrm{Pr}_g = \nu_g^{(v)}/\nu_g^{(T)} \equiv \mu_g c_g/\lambda_g) \tag{7.7.4}$$

The heat flux between vapor and vapor-drop interface may be represented in the form (1.4.11). We shall now evaluate the effect of the slip velocity w_{12} of gas relative to drops upon heat exchange between vapor and drops under conditions specific for vapor-generating channel: $w_{12} = 1$ m/s, $2a =$ 10 micron, and $p = 7.0$ MPa ($\rho_g^0 = 36$ kg/m^3, $\mu_g = 1.9 \times 10^{-5}$ kg/(m · s)). For these parameters $\mathbf{Re}_{12} \approx 20$, and, according to (1.4.11), $\mathbf{Nu}_{1(\Sigma 2)} = 6$, i.e., the phase slipping in this example increases the number $\mathbf{Nu}_{1(\Sigma 2)}$ by a factor of three.

As a result, we obtain for $Q_{1(\Sigma 2)}$

$$Q_{1(\Sigma 2)}/S = k^{(1)} (1 + k^{(2)} |w_{12}|^{0.5})$$
$$k^{(1)} = 3\alpha_2 \lambda_g (T_1 - T_S)/a^2, \quad k^{(2)} = 0.6\, (2a\overset{\circ}{\rho}_g/\mu_g)^{0.5} \mathrm{Pr}_g^{0.4} \tag{7.7.5}$$

Under conditions of a vapor-drop flow in the vapor-generating channel (the characteristic times of parameter change along the flow $t \gtrsim 1$ second), the major force of interaction between vapor and drop is friction force. The remaining forces are small compared to it. Then the drop-motion equation becomes

$$\frac{8}{3} a \overset{\circ}{\rho}_l v \frac{dv}{dz} = C_\mu \overset{\circ}{\rho}_g w_{12}^2 \tag{7.7.6}$$

where the relation (1.4.9) may be used for determining the coefficient of friction C_{μ}.

Expression for the flow acceleration dv/dz is obtained as a result of solving the system (7.7.2) for the derivatives. Further, having substituted dv/dz into (7.7.6), we obtain from (7.7.5) the equation for the slip velocity w_{12}

$$a\overset{\circ}{\rho}_l v\left[k - b^{(5)}k^{(1)}\left(1 + k^{(2)}|w_{12}|^{0,5}\right)\right] = {}^3/_8 C_{\mu}\overset{\circ}{\rho}_g w_{12}^2$$
$$\left(k = b^{(1)}Q_{W1} + b^{(2)}Q_{W2} + b^{(3)}Q_{2(\Sigma 2)} + b^{(4)}F_W\right) \qquad (7.7.7)$$

where $b^{(i)}$ ($i = 1, \ldots, 5$) are complexes obtained as a result of solving the system (7.7.2) for derivatives.

From a solution of a transcendental equation (7.7.7), the value w_{12} is obtained, and upon its substitution into (7.7.5), the magnitude of the heat flux $Q_{1(\Sigma 2)}$ from vapor to drops may be readily calculated.

Heat exchange between drops and heating surface. The direct thermal interaction of drops with the heating surface may be essential in the overall heat removal by a dispersed vapor-drop flow from the channel wall. The review of related experimental works may be found in M. A. Styrikovich et al., (1982).

Investigation of vaporization of an individual drop on a hot surface showed that time t_d of a complete vaporization of a drop depends, in a rather complicated manner, on the heating surface temperature T_W (Fig. 7.7.2). The entire range of variation of T_W may be divided into four regions. Region *I*, where the heating-surface temperature is below the saturation temperature of the vaporizing substance, the drop of a liquid spreads over the surface and slowly evaporates. With growth of T_W, time t_d reduces. With T_W exceeding the saturation temperature T_S, a bubbly boiling initiates in the spreading liquid (region *II*). With growth of T_W, intensity of bubbly boiling grows, the heat flux q_{W2} to the drop increases, and, consequently, time of complete vaporization of the drop reduces.

With further growth of T_W, a transient boiling of the drop ensues (region *III*), both frequency and area of contact between liquid and wall reduce, the magnitude of q_{W2} reduces, and, accordingly, time t_d grows. The boundary between regions *II* and *III* is temperature T_c (time corresponding to this temperature is $t_d = t_{min}$), and the heat flux q_{W2} from wall to drop reaches its maximum value, i.e., an analogue of the bubbly-boiling crisis in bulk volume takes place in this situation.

With further growth of T_W, the transient boiling of a drop transforms to a film boiling (region *IV*), drop vaporization begins when it is in its spheroidal state, the drop is separated from wall by its vapor. The boundary between regions *III* and *IV* is temperature $T_W = T_L$, the Leidenfrost temperature (time corresponding to this temperature is $t_d = t_{max}$). Later on, with growth of T_W, time of complete vaporization of drop slowly reduces.

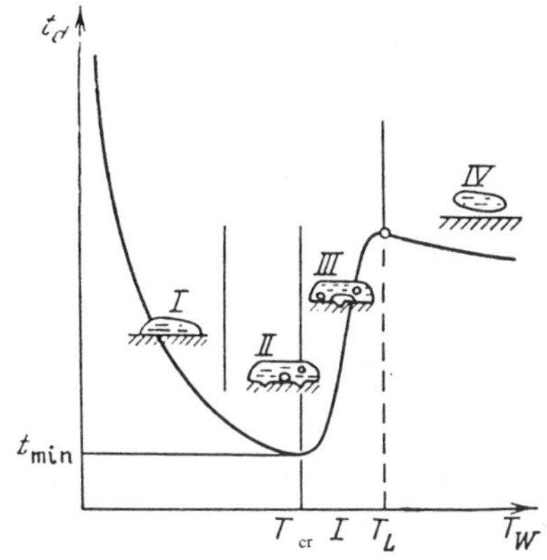

Figure 7.7.2 Schematic relationship between drop-vaporization time t_d and wall temperature T_W.

Experimental investigations of interaction between drops and heated surface showed that the Leidenfrost temperature greatly depends on: the liquid critical temperature, thermal properties of the surface material, the roughness and impurity (contamination) of the surface, the velocity of falling of drops on the wall, and the extent of a drop flow on the hot surface. In the presence of roughness on the heated surface, the Leidenfrost temperature is considerably higher than in the presence of roughness on a smooth surface.

For determining the amount of heat q_{W2} transmitted from the heating surface to a single drop per unit time, M. Cumo's model may be employed (1969); this model implies that during the drop contact t_W with the heating surface, heat is transmitted to a drop through a thin layer δ_* of vapor generated during a time period t_*, which is significantly shorter than $t_W (t_* \ll t_W)$. Then, the magnitude $\overset{0}{q}_{W2}$ is

$$\overset{\circ}{q}_{W2} = ct_W \pi a^2 \lambda_g (T_W - T_S)/\delta_* \tag{7.7.8}$$

where C is the coefficient of proportionality. It is assumed here that the contact area equals by order the mid-section of a drop. The thickness of the vapor layer δ_* is of the order of a roughness protrusion of a metal tube wall, and for cases under consideration, is assumed as $\delta_* = 50$ micron.

M. A. Styrikovich et al., (1982) reported results of measurements of contact time t_W at sufficiently high temperature of the heating surface ($T_W > 450$–500 K at $p = 0.1$ MPa). Analysis of these data shows that t is of the order resembling the period of free oscillations of a drop due to capillary forces

$$t_W \approx t_\Sigma = \pi \sqrt{\overset{\circ}{\rho}_l a^3/(2\Sigma)} \tag{7.7.9}$$

The amount of drops N_{23} entering the near-wall zone from the flow core per unit time and per unit length of a channel is

$$N_{23} = \overset{\circ}{J}_{23}/\left(\tfrac{4}{3}\pi a^3 \overset{\circ}{\rho}_l\right) \tag{7.7.10}$$

Then, heat flux from wall to drops becomes

$$Q_{W2} = \overset{\circ}{q}_{W2} N_{23} = \frac{3c\overset{\circ}{J}_{23}\lambda_g t_W (T_W - T_S)}{4\overset{\circ}{\rho}_l a \delta_*} \tag{7.7.11}$$

The approximate nature of evaluations of magnitudes of τ_W, δ_*, and also, the presence of the drop-removal effect by vapor from the wall (vapor is produced in the process of drop vaporization), is taken into consideration by introducing a coefficient of proportionality c into appropriate formulas.

The above-described laws of phase interaction permit the system of equations (7.7.2) to be solved numerically.

The boundary conditions for this solution are prescribed in the section $z = 0$, which is taken as the beginning of a post-crisis area due to film drying-out ($\delta = 0$). In this case, the precipitation intensity J_{23}^0 is predetermined by formula (7.4.8), and the original radius a of drops is given by the empirical formula (7.4.9).

Selection of the model parameter c was based (B. I. Nigmatulin and V. S. Kukharenko) on the condition of best agreement between predicted and experimental distributions of temperature of an interior surface of the heated tube in the post-crisis zone of heat exchange. Experimental data (Z. D. Miropol'skii, 1963; A Bennet et al., 1967; V. I. Subbotin et al., 1973; M. I. Marinov and L. P. Kabanov, 1977; S. Nijawan et al., 1980; G. Barzoni and R. Martini, 1982; P. Kirillov et al., 1982; G. V. Tsiklauri et al., 1982; O. V. Remizov et al., 1983) were used for this purpose, obtained in upward steam-water flows in heated tubes (D = 8 to 12.7 mm, m^0 = 35–5600 kg/(m$^2\cdot$s), q_W = 0.035–1.8 MW/m^2, p = 0.3–18.5 MPa, x_g = 0.3–1.0). As a result, the following formula was obtained

$$c = 1.3\cdot 10^{-4}\mathrm{Re}_1^{0.75}\left(\frac{\overset{\circ}{\rho}_g}{\overset{\circ}{\rho}_l}\right)^{0.4}\exp\left[1-\left(\frac{T_W}{T_S}\right)^2\right] \qquad (\delta_* = 50\ \mu) \tag{7.7.12}$$

Parameter c proved to depend significantly on $\mathbf{Re}_1$. This, apparently, is associated with the increase of kinetic energy of the turbulent motion of the drop that may increase its heat-removal capacity when the drop occurs on a hot wall. The exponential multiplier takes into consideration the drop-deposition reduction due to their removal by their own vapor in a gradient field of temperatures near an overheated wall. The account of the T_W effect in this form was proposed by E. Ganic and W. Rohsenow (1976).

An extensive comparison of analytical and experimental results concerning the wall temperature T_W distribution along the channel has been carried out using formula (7.7.12). Examples of such comparison are shown in Fig. 7.7.3.

It is evident that the relation $T_W(z)$ depends on the specific flow rate m^0 of mixture. At small m^0, the temperature T_W, after a sharp jump at location of the heat-transfer crisis, continues to grow rapidly. At relatively large m^0,

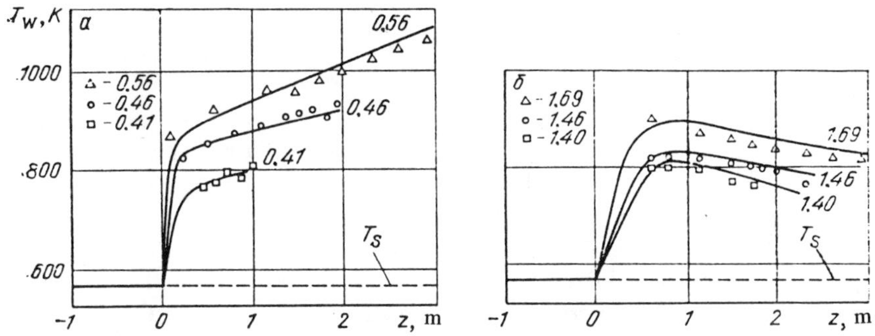

Figure 7.7.3 Variation of the wall temperature T_W along a flow in a heated tube (D = 12.7 mm) with a specific heat flux q_W, MW/m^2 (indicated by numerical labels on curves and experimental dots); upward steam-water drop flow (p = 6.9 MPa) with a specific flow-rate m^0 = 394 kg/(m$^2 \cdot$ s) (*a*) and 3850 kg/(m$^2 \cdot$ s) (*b*). Distance z is measured from location of the heat-transfer crisis ($z = z_* = 0$) due to wall-film drying out, followed by realization of a dispersed flow. Lines represent analytical prediction, dots represent experiment (A. Bennet et al., 1967).

the wall temperature T_W, after a sharp increase due to film drying out, gradually reduces along the flow.

§7.8 BUBBLY BOILING, AND ITS CRISIS ON A HORIZONTAL SURFACE UNDER CONDITIONS OF A FREE CONVECTION[5]

The thermal balance on the boundary W (Fig. 7.8.1), through which heat q_W is an input to a vapor-liquid medium, may be represented in the form

$$q_W = q_l + q_g \tag{7.8.1}$$

where q_l is heat removed by liquid, and q_g is heat removed by vapor. With sufficient contact with liquid (absence of crisis), given that $\lambda_g \ll \lambda_l$, $q_g \ll q_l$ takes place. Later on, during boiling, part of the heat flux q_l, which is denoted by q_{lg}, is consumed for vaporization whose intensity per unit area of the boundary W is denoted by ξ_g, and the part q_l' is removed by liquid that is accounted for by heat conduction and convection, or mixing intensified by detached and rising vapor bubbles

$$q_l = q_{lg} + q_l', \qquad q_{lg} = \xi_g l \tag{7.8.2}$$

Figure 7.8.2 represents dependence on the heat flux q_W of both tem-

[5]This section was written by Rs. I. Nigmatulin.

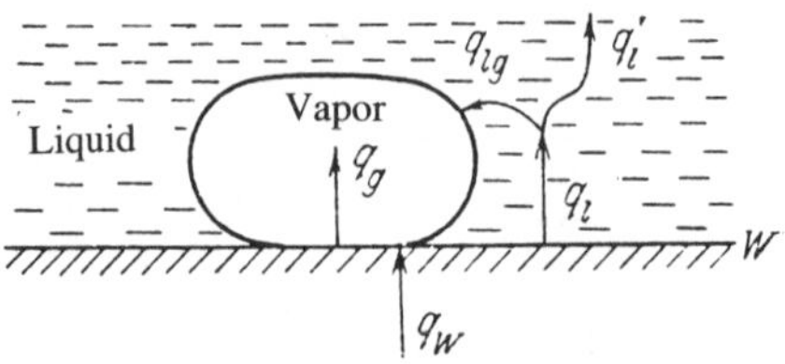

Figure 7.8.1 Heat balance on the surface W through which heat input to a vapor-liquid mixture is realized.

perature difference $\Delta T = T_W - T_l$ (where T_W is the heating wall temperature, and T_l is the liquid temperature at some distance from the wall W where this temperature is sufficiently uniform) and heat-transfer coefficient β during boiling of saturated liquid ($T_l = T_S$) on a horizontal face-up surface in the field of gravity, and in the absence of a forced flow, or flow past heating surface. It is seen that with sufficiently small heat fluxes (segment AB), when bubbly boiling is weak, heat flux q_W is proportional to ΔT^n ($n > 1$), and the relations $q_W(\Delta T)$ and $\beta(\Delta T)$ are analogous to a single-phase liquid under conditions of free convection. A highly developed bubbly boiling is realized over the segment BC, when both formation and break-away of bubbles from the heating surface contribute to the intensification of heat exchange on account of an increase in q_l', because liquid undergoes mixing by broken-away bubbles. The further increase of heat flux leads to the growth of vapor concentration α_g in a near-wall layer, and at $\alpha_g \approx 0.8$, the bubbly structure is virtually broken due to bubble agglomeration, and a foamy, or vapor film is produced on the wall, which leads to the film-boiling regime. The contact between the heating surface W and liquid phase is disrupted in this regime, thereby reducing sharply the heat-transfer coefficient β, and with forced input of heat flux q_W, as it takes place in heat-removing elements of nuclear reactors and in electrically heated walls, temperature T_W of the heating wall sharply rises (transition $C \to G$). As was indicated in §7.1, the described phenomenon is referred to as a *bubbly boiling crisis*, or a heat-transfer crisis of the first kind. Segment DE relates to a boiling with vapor, or more specifically, a foamy film on the heating surface when this film realizes a substantial thermal resistance and small heat-transfer coefficient.

Transition from a well-developed bubbly boiling to film boiling may, under certain conditions, be achieved in the region of high heat fluxes (towards point C'). In this case, the process is said to be an *extended near-critical* regime.

It is noteworthy that with reducing q_W in a regime of film boiling over the segment ED, the reversed transition to a bubbly-boiling regime ($D \to H$) takes place with a hysteresis at sufficiently lower heat fluxes than $q_W(C)$. Point D determines the heat flux $q_W(D)$, and generally we have

$$q_W(D) \approx 0.2\, q_W(C) \tag{7.8.3}$$

It should be remembered that the heat-transfer intensity with a well-devel-

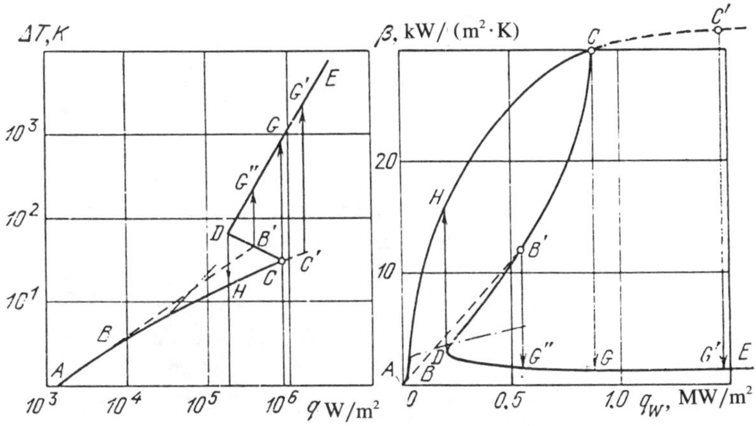

Figure 7.8.2 Dependence of both thermal head (driving force) $\Delta T = T_W - T_S$ and heat-transfer coefficient $\beta = q_W/\Delta T$ on the heat flux during boiling of saturated liquid (water at $p = 0.1$ MPa) on a horizontal face-up surface under conditions of a free convection. Segment *AB* represents heat exchange with a free convection of a single-phase liquid; *BC* represents well-developed bubbly boiling; *CD* represents transitional boiling; *DE* represents stable film (with a vapor film) boiling (crisis); *BB'* represents heat exchange from a surface "leaned" by centers of vapor (steam) formation accounted for by the appropriate selection of both material and treatment of the heating surface. The dot-dashed line corresponds to bubbling of water by gas with a reduced velocity $W = q_W/(l\rho_g^0)$.

oped bubbly boiling depends considerably on the heating-surface material, its treatment, and duration of the operation process. Thus, on a heating surface with fewer centers of vapor formation on account of appropriate selection of material and its treatment, film boiling may occur after the transition $B' \to G''$ directly from the regime of a free single-phase convection (which is related to line ABB') avoiding the bubbly regime.

Analysis of liquid superheating is presented in N. Afgan (1976).

The S. S. Kutateladze analogy for bubbly boiling and bubbling. Hydrodynamics of a bubbly boiling is similar to hydrodynamics of the process of bubbling a liquid with gas forced through a semipermeable surface (with void fraction $\varphi_0 = \alpha_{g0}$) with the same volumetric flow rate, or reduced gas velocity $W = \xi_g/\rho_g^0 = v_{g0}\varphi_0$, which take place during boiling when $W = q_{lg}/(\rho_g^0 l)$. Based on this consideration, S. S. Kutateladze proposed to simulate bubbly boiling by a process of bubbling, i.e., to investigate boiling under the conditions of a "cold" experiment. This idea has been realized for investigating boiling and bubbling on a horizontal surface under the conditions of a free convection; these conditions are reflected on Fig. 7.8.2 (see S. S. Kutateladze and M. A. Styrikovich, 1976).

Experimental technique for investigating both bubbling and boiling. Figure 7.8.3 depicts a schematic for performing the experiment. The height of the liquid column h, when pressure drop Δp along the depth of a layer was negligible ($\Delta p \ll p$), is about 5 cm. Pressure p in a system was maintained using a layer of gas above liquid. If the diameter D of the porous surface (or the boiling surface) and the height of the column of liquid are sufficiently high (h, $D \gg a$, where a is a characteristic radius of bubbles), the effect of both h and D on hydrodynamics and heat exchange in a near-wall layer during boiling and bubbling does not manifest itself. Note also that D is much smaller compared to the diameter of a vessel.

The bubbly boiling crisis at point C in Fig. 7.8.2 is related to a critical regime of bubbling when, with a sufficiently high reduced velocity $W > W_*$ of gas, the bubbly structure of the near-wall layer is destroyed, and liquid is virtually displaced from the porous surface through which gas is forced. In this situation, to prevent a jetting regime of gas input, and to ensure the bubble size the same as in boiling, the pore-channel diameters and distances between them must be sufficiently small. In particular, in experiments conducted by S. S. Kutateladze and I. G. Malenkov (1976), porous plates of porosity $m = \varphi_0 \approx 0.2$ were used, and the number of openings $N \approx 2{,}000$ cm^{-2}, which corresponds to the opening diameter $\delta_0 \sim 0.1$ mm and distances between them $b_0 \sim 0.2$ mm. The size of bubbles generated under these conditions is much higher than the opening diameter δ_0 and distances between them b_0, i.e., each broken-away bubble is fed by gas out of several pores at a time. The further increase of the number of openings N and decrease of their diameter δ_0 does not affect the size of the bubbles produced ($a \sim 1$ mm).

The critical regime (point in time of actual displacement of liquid during bubbling) is determined by three methods: 1) by change of gas flow rate W as a function of pressure drop $p_p - p$, where p_p is the gas pressure before its entrance into the porous plate; 2) by change of dielectric constant of the

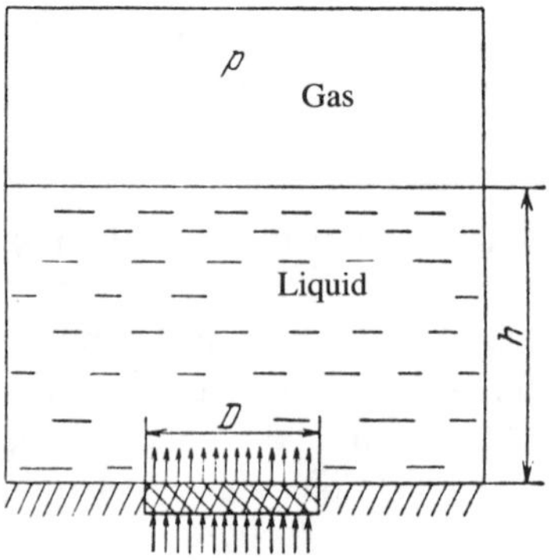

Figure 7.8.3 Schematic for an experiment on bubbling and boiling on a horizontal surface.

wall layer, which depends on void fraction α_g of this layer; 3) by change of electric conductivity which also depends on α_g.

The last of these methods is the most convenient. Its essence consists of the fact that if two electrodes are placed in the wall layer (the porous plate may be one of them), which are fed by a fixed voltage, the current I in a circuit will reduce with an increase of gas concentration, and with liquid displacement (crisis), I sharply drops. This method permits the determination of the void fraction of the wall layer using a formula

$$I = I_0(1 - \alpha_g)$$

where I_0 is electric current at prescribed voltage in pure liquid ($\alpha_g = 0$).

In an investigation of heat transfer, the porous plate was heated by an electric current whose power was monitored. This power, related to the plate surface, determines the heat influx q_W. In addition, the temperatures of both porous plate T_W and wall gas-liquid layer T were also measured. Finally, the heat-transfer coefficient $\beta_W = q_W/\Delta T$, where $\Delta T = T_W - T$, is found.

For determining, in accordance with formula $W = q_{lg}/(\rho_g^0 l$, the reduced bubble-rise velocity during boiling, it is imperative to separate that part of heat q_{lg} of the total heat flux q_W which is consumed for vaporization of the wall layer. For this purpose, mass of vapor which vaporized precisely in the wall layer should be measured, separating this from vapor generated on the free surface of liquid.

Analysis of experimental data concerning critical velocity of injection which displaces liquid. The gas mass conservation law under a stationary regime, if both variation of gas density with height of layer and gas spreading across the layer are ignored, interrelates volumetric concentration of gas and its mass velocity

$$\alpha_g v_g = \alpha_{g0} v_{g0} = W = \text{const} \qquad (\alpha_{g0} = \varphi_0) \tag{7.8.4}$$

Distortion of bubbly structure in a wall layer occurs at a sufficiently high volumetric concentration of gas ($\alpha_{go} \gtrsim 0.8$), which grows with gas velocity dropping (as it happens, e.g., in a wall layer when gas flows out of the porous channel into liquid). The indicated retardation of gas is accounted for by pressure forces, liquid viscosity (defined by the coefficient of viscosity μ_l for liquid), and surface tension (defined by the coefficient of surface tension Σ) during the bubble-formation process. Because of gas retardation, its volumetric concentration increases compared to φ_0, which facilitates the destruction of the bubbly structure in a wall layer, resulting in a sharp reduction of contact between liquid and solid wall. Bubbles above the wall layer accelerate, and gas concentration reduces with height of the layer, and, consequently, a bubbly structure may be realized above the wall layer.

The critical gas-injection velocity W_*, at which the stationary bubbly structure of the wall layer is destroyed, depends on the removal of bubbles from this wall layer (in addition to dependence on the conditions defining both formation and detachment of bubbles). The removal of bubbles out of wall layer depends on the size of the bubbles being removed; this size is determined by the frequency of break-away, which, in turn, depends on surface tension Σ, density, and viscosity of liquid (ρ_l^0 and μ_l), and also, on acceleration due to gravity. Moreover, the break-away frequency may depend on pressure p in liquid, since the bubble detachment is virtually an oscillating process which depends on gas compressibility determined by pressure (cf. natural frequency of radial oscillations of a bubble in accordance with Minnaert formula (1.6.22), where this frequency ω_r is proportional to $p^{1/2}$).

Given the above-indicated considerations (including the impact of geometrical parameters of the experimental installation), and also, the possible effects of both density and viscosity of gas, or vapor (which are small compared to those of liquid ($\rho_g^0 \ll \rho_l^0$, $\mu_g \ll \mu_l$), the critical reduced (modified) gas injection velocity W_* depends on the following parameters

$$W_* = f\left(\overset{\circ}{\rho}_g, \overset{\circ}{\rho}_l, \mu_g, \mu_l, \Sigma, p, g\right) \tag{7.8.5}$$

It can be readily proven that the above-indicated relationship yields five, and only five, independent parameters which may, in particular, be represented as

$$\overline{W}_* = \frac{W_*}{C^{(\Sigma g)}}, \quad P = \frac{p}{p^{(\Sigma g)}}, \quad \overline{\mu}^{(\Sigma g)} = \frac{\mu_l}{\overset{\circ}{\rho}_l C^{(\Sigma g)} \delta^{(\Sigma g)}} = \mu_l \sqrt[4]{\frac{g}{\overset{\circ}{\rho}_l \Sigma^3}}$$

$$\widetilde{\rho}^{\circ} = \overset{\circ}{\rho}_g / \overset{\circ}{\rho}_l, \quad \widetilde{\mu} = \mu_g / \mu_l \tag{7.8.6}$$

$$C^{(\Sigma g)} = \sqrt[4]{\frac{\Sigma g}{\overset{\circ}{\rho}_l}}, \quad \delta^{(\Sigma g)} = \sqrt{\frac{\Sigma}{\overset{\circ}{\rho}_l g}}, \quad p^{(\Sigma g)} = \overset{\circ}{\rho}_l \left(C^{(\Sigma g)}\right)^2 = \sqrt{\Sigma g \overset{\circ}{\rho}_l}$$

Here, $C^{(\Sigma g)}$, $\delta^{(\Sigma g)}$, and $p^{(\Sigma g)}$ determine characteristic velocities, linear dimensions, and pressure drops in capillary-gravitational waves; parameter $\bar{\mu}^{(\Sigma g)} = \mathbf{Ar}^{-1/2}$, where **Ar** is the Archimedes number characterizing the liquid viscosity effect in capillary-gravitational waves. Any dimensionless combinations of parameters (7.8.5) other than these are expressed through dimensionless parameters (7.8.6). Thus, the sought relation (7.8.5) may be written in the following dimensionless form

$$\overline{W}_* = f\left(P, \overline{\mu}^{(\Sigma g)}, \widetilde{\rho}^{\circ}, \widetilde{\mu}\right) \tag{7.8.7}$$

In the above-cited works by S. S. Kutateladze, an isothermic speed $C_g^{(T)}$ of sound in gas and dimensionless parameters k_* and M are used

$$k_* = \frac{W_* \sqrt{\overset{\circ}{\rho}_g}}{\sqrt[4]{\Sigma g \overset{\circ}{\rho}_l}}, \quad M = \frac{C^{(\Sigma g)}}{C_g^{(T)}} = \sqrt{\frac{\overset{\circ}{\rho}_g}{p}} \sqrt[4]{\frac{\Sigma g}{\overset{\circ}{\rho}_l}} \quad \left(C_g^{(T)} = \sqrt{p/\overset{\circ}{\rho}_g}\right) \tag{7.8.8}$$

These may be easily expressed in terms of dimensionless parameters (7.8.6)

$$k_* = \overline{W}_* \sqrt{\tilde{\rho}^{\circ}}, \quad M = \sqrt{\tilde{\rho}^{\circ}/P} \tag{7.8.9}$$

It is these particular parameters that are used for processing the experimental results obtained by S. S. Kutateladze and I. G. Malenkov. And, if both k_* and M are employed instead of $\bar{W}_*$ and $\tilde{\rho}^0$, the relation (7.8.7) may be rewritten in the form

$$k_* = f\left(P, M, \overline{\mu}^{(\Sigma g)}, \tilde{\mu}\right) \tag{7.8.10}$$

The stability criterion k_* is first introduced by S. S. Kutateladze. We shall now set forth a schematic explaining the physical meaning of this criterion. If liquid is displaced by gas, then on the interface between liquid and gas, due to the effects of a mechanism leading to a Taylor instability (heavy phase is above; light phase below), waves whose lengths b satisfy the condition (see the discussion of (2.2.6))

$$b \gtrsim b^{(\Sigma g)} = 2\pi\delta^{(\Sigma g)} \tag{7.8.11}$$

destabilize the indicated interface ($I(b) > 0$ and the wave amplitude grows with time). These capillary-gravitational waves result in occurrence of streamers or liquid protrusions in a direction opposite to gas; the characteristic transverse dimension of streamers is about $\delta > 1/2b^{(\Sigma g)}$. These streamers tend to penetrate gas and destroy the gas or vapor film. If the aerodynamic effects $F^{(v)}$ of gas flowing with velocity $v_g \sim W$ past this protrusions (or liquid streamers) balance the gravity force $F^{(g)}$ of the streamer, such a regime with a gas film, or a pad, displacing liquid, is relatively stable. Thus, the stability condition becomes

$$\begin{gathered} F^{(v)} \gtrsim F^{(g)}, \qquad F^{(g)} \sim \overset{\circ}{\rho}_l g \delta^3 \\ F^{(v)} = C_\mu \frac{\overset{\circ}{\rho}_g W^2}{2} \delta^2 \qquad \left(C_\mu = f(\mathrm{Re}_g), \quad \mathrm{Re}_g = \frac{2\delta \overset{\circ}{\rho}_g W}{\mu_g} \right) \end{gathered} \tag{7.8.12}$$

If we assume that $C_\mu \sim 0.5$, i.e., C_μ is independent of the phase viscosity (which may be anticipated at $\mathbf{Re}_g \gtrsim 1000$), the stability condition for the regime with gas film, or a pad, displacing liquid from the wall is reduced to the form

$$k = \frac{W\sqrt{\overset{\circ}{\rho}_g}}{\sqrt[4]{\overset{\circ}{\rho}_l \Sigma g}} > k_* \sim O(1) \tag{7.8.13}$$

Experiments on earth ($g = 9.81$ m/s^2) at normal or slightly lowered temperature $T = 277$–293 K with liquids like water ($C^{(\Sigma g)} = 0.165$ m/s), aqueous-glycerine solutions of various viscosities, and ethanol subjected to bubbling by various gases (hydrogen, helium, nitrogen, argon, xenon, etc., which permitted varying the isothermal speed of sound $C_g^{(T)}$ in a range 130–1050 m/s) showed that at fairly high pressures $p = 0.1$–20 MPa ($P \gtrsim 1000$), the value k_*—at which the liquid displacement is realized—at fixed gas and temperature is independent of pressure (Fig. 7.8.4), and with a change of kind of gas, it varies as M^n (n = 0.66–1.0) (the value M being $M = 10^{-4}$–10^{-3}), and with all other conditions fixed, M is proportional to $\sqrt{\mu_m}$, where μ_m is the molecular weight of gas.

Under indicated conditions, at fixed temperature of gas and liquid, and with variation of P, values of $\bar{\mu}^{(\Sigma g)}$ and $\tilde{\mu}$, as well as M, do not change, and small relative density of gas $\tilde{\rho}^0$ varies proportional to P. Thus, the experiment shows that under the above-specified conditions, the value k_* is proportional to M^n (n = 0.66–1) or $\mu_m^{n/2}$, and is independent of pressure P. Then, recalling that in bubbly flows the gas viscosity, small compared to liquid viscosity ($\tilde{\mu} \ll 1$), does not exhibit itself, we obtain a simple form of relation (7.8.10)

$$k_* = K_*\left(\overline{\mu}^{(\Sigma g)}\right) M \qquad \left(P \leqslant 10^5 - 10^6,\quad \tilde{\rho}^\circ = 10^{-3} - 10^{-1}\right) \qquad (7.8.14)$$

Note that it is assumed here[6] that n = 1. Based on experimental data, Fig. 7.8.5 represents the following values

$$K_* = k_*/M \equiv W_* \sqrt{p/(\Sigma g)}$$

as a function of dimensionless viscosity $\bar{\mu}^{(\Sigma g)}$. Good correlation between K_* and $\bar{\mu}^{(\Sigma g)}$ for bubbling and boiling of various liquids is evident. Experimental points for boiling of alkaline earth metals (sodium, potassium, rubidium, and caesium), related to small values of viscosity $\bar{\mu}^{\Sigma g)}$, were processed based

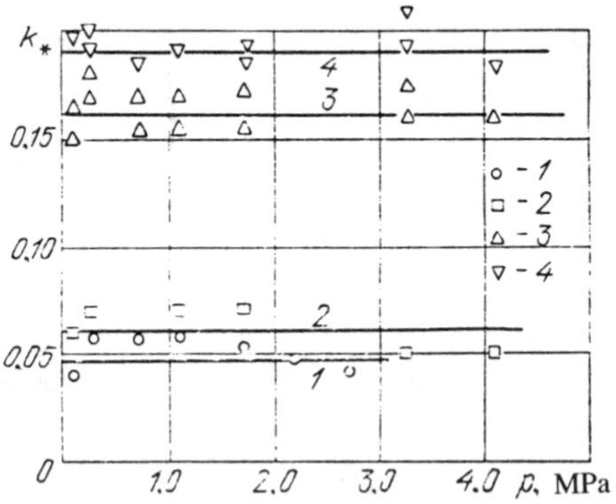

Figure 7.8.4 The effect of pressure and kind of gas used in experiment on the stability criterion k_* (S. S. Kutateladze) during bubbling of water at temperature T_0 = 280 K by hydrogen (*1*), helium (2), nitrogen (*3*), and argon (*4*).

[6]The experimental data on $k_*(M)$ are approximated by relation $k_* = A(\bar{\mu}^{(\Sigma g)})\, M^{2/3}$ (n = 2/3 (S. S. Kutateladze and I. G. Malenkov (1976)). This relation, as well as (7.8.14), describes the field of experimental points in an almost similar manner (if the scatter of experimental points is taken into consideration). In the same work, the authors indicate that the effect of pressure was observed at small subatmospheric pressures. The corresponding relation (7.8.10) is written as

$$k_* = 5.8\, P^{1/3} M^{5/6}, \text{ or } W_* = 5.8(\tilde{\rho}^0 P)^{-1/2}$$

on total heat flux (i.e., it was assumed that $W = q_W/(\rho_g^0 l)$), since a part of heat flux q_{lg} consumed for vapor formation in the wall layer did not vary in these experiments. Because of high heat conductivity of alkaline earth metals, part of the heat input equal $q_l' = q_W - q_{lg}$ and not being consumed for vapor formation in the wall layer may be very substantial. The remaining points related to both bubbling and boiling (and associated with separating of q_{lg}) indicate that at $\bar{\mu}^{(\Sigma g)} \leqslant 10^{-2}$, viscosity does not affect K_*, and at $\bar{\mu}^{(\Sigma g)} > 10^{-2}$, the value K_* reduces with growth of $\bar{\mu}^{(\Sigma g)}$, i.e., with growth of viscosity μ_l, the liquid displacement takes place at lower velocities of gas injection. The corresponding approximation may be written in the form

$$K_*\left(\bar{\mu}^{(\Sigma g)}\right) = \begin{cases} 310, & \bar{\mu}^{(\Sigma g)} < 10^{-2}, \\ 110/\left(\bar{\mu}^{\Sigma g}\right)^{0.2}, & \bar{\mu}^{(\Sigma g)} > 10^{-2} \end{cases} \tag{7.8.15}$$

The relation (7.8.14), using (7.8.9), may be transformed to (7.8.7)

$$\overline{W}_* = \frac{K_*\left(\bar{\mu}^{(\Sigma g)}\right)}{\sqrt{\bar{p}}}, \quad \text{or} \quad W_* = K_*\left(\bar{\mu}^{(\Sigma g)}\right)\sqrt{\frac{\Sigma g}{p}} \tag{7.8.16}$$

The following conclusion may finally be made: unless the dimensionless parameter k_* depends on p and is proportional to M (which is realized in experiments on bubbling and boiling), the critical velocity of injection is independent of gas properties (in particular, of its density and speed of sound in it), inversely proportional to $p^{1/2}$, proportional to $g^{1/2}$, $\Sigma^{1/2}$, and slowly decreases with growth of the liquid viscosity μ_l.

The conclusion on a paradoxical effect of acoustic properties of gas and small Mach number M on the process of displacement of liquid by gas is a consequence only of treatment of the experimental results in the form of (7.8.10). This conclusion is related only to parameter k_*, since it involves the multiplier $\sqrt{\rho_g^0}$, though the gas density ρ_g^0 does not affect the critical injection velocity W_* and corresponding dimensionless parameters $\bar{W}_*$ and K_* (or may affect them insignificantly (as $(\rho_g^0)^{1/6}$, if $k_* = A\,(\bar{\mu}^{(\Sigma g)})\,M^{2/3}$); this conclusion in no way reflects the very essence of the issue.

In order to qualitatively comprehend the implication of formula (7.8.16), we shall evaluate the characteristic values of bubble sizes equal by order the lengths $\delta^{(\Sigma g)}$ of capillary-gravitational waves, velocities $C^{(\Sigma g)}$ of these waves, and corresponding Reynolds numbers $\mathbf{Re}_l = 1/\bar{\mu}^{(\Sigma g)}$, $\mathbf{Re}_g$, Weber numbers $\mathbf{We}_l$ and $\mathbf{We}_g$, using water bubbled by nitrogen, as an example

$$\begin{gathered} \delta^{(\Sigma g)} \sim 3 \cdot 10^{-3}\ \text{m}, \qquad C^{(\Sigma g)} \sim 0.2\ \text{m/s} \\ \mathrm{Re}_l \sim 3 \cdot 10^2, \qquad \mathrm{Re}_g \sim 3 \cdot 10 \\ \mathrm{We}_l = \overset{\circ}{\rho}_l \left(C^{(\Sigma g)}\right)^2 \delta^{(\Sigma g)}/\Sigma \sim 1, \quad \mathrm{We}_g = \overset{\circ}{\rho}_g \left(C^{(\Sigma g)}\right)^2 \delta^{(\Sigma g)}/\Sigma \sim 10^{-3} \end{gathered} \tag{7.8.17}$$

An increase of Σ leads to an increase of the bubble size, which together

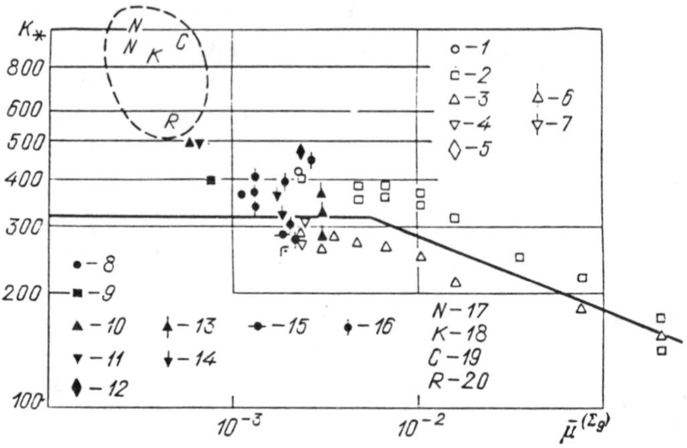

Figure 7.8.5 The effect of liquid viscosity upon parameter K_*, which defines occurrence of crisis (liquid displacement) during bubbling and boiling. The light points *1–7* relate to bubbling at $p = 0.1$–4.1 MPa, $T = 280$ K; points *1–5* correspond to water and aqueous-glycerine solutions of various viscosities subjected to bubbling by various gases: *1*) hydrogen, *2*) helium, *3*) nitrogen, *4*) argon, *5*) xenon; points *6* and *7* relate to ethanol subjected to bubbling by nitrogen (*6*) and argon (*7*). Dark points *8–16* relate to boiling of various liquids at various pressures p (MPa); points *8–12*) for boiling of water (*8*) at 0.02 MPa, *9*) 0.1 MPa, *10*) 4.5 MPa, *11*) 5.4 MPa, *12*) 18.6 MPa), points *13* and *14*) for boiling of ethanol (*13*) at 0.1 MPa, *14*) at 1.0 MPa); *15*) for boiling of benzene at 0.1 MPa, *16*) for boiling methanol at 0.1 MPa. Points *1–16* reflect experimental data obtained by S. S. Kutateladze and I. G. Malenkov (1976), and I. G. Malenkov (1978). Points *17–20* relate to boiling of sodium, potassium, cesium, and rubidium, for which velocity W was determined based on total heat flux (data obtained by V. I. Subbotin et al., 1968, 1969).

with an increase of g, leads to the growth of the bubble rise velocity at $\mathbf{Re}_l \leqslant 500$, impeding destruction of the bubbly structure.[7]

The fact that gas density virtually does not affect the critical injection velocity is associated with the fact that the force interaction of gas bubbles with liquid under these conditions (characterized by parameters (7.8.17)), because of relatively small Reynolds numbers $\mathbf{Re}_l \leqslant 100$, is governed by equations characteristic for bubble rise ($C_\mu \sim 24/\mathbf{Re}_l$) when the Archimedes force $F^{(g)}$ and drag force $F^{(v)}$ (which is defined by the liquid viscosity (see §2.2)) are balanced

$$F^{(g)} = F^{(v)}, \quad F^{(g)} \sim \overset{\circ}{\rho}_l g \delta^3, \quad F^{(v)} \sim \mu_l W \delta \tag{7.8.18}$$

(rather than by expressions (7.8.12)). Hence, instead of k_*, we obtain criterion $\bar{k}_*$

[7]At $\mathbf{Re}_l > 500$, an increase of δ or volume δ^3 of a bubble does not contribute to the increase of the bubble rise velocity, with all other conditions the same (g, ρ_l^0, μ_l) (see S. S. Kutateladze and M. A. Styrikovich, 1976).

$$\bar{k}_* = \frac{\mu_l W_*}{\rho_l^{\circ} \delta^2 g}, \qquad \frac{\delta}{\delta^{(\Sigma g)}} = f(p, \mu_l) \tag{7.8.19}$$

An increase of viscosity, which slows down the rise of bubbles, contributes to the increase of gas concentration of the wall layer, thereby facilitating its destruction (Fig. 7.8.6). The viscosity effect is significantly weakened by the fact that the growth of μ_l leads to an increase of the detached bubble size $\delta \sim 2a$, which partially compensates for the indicated reduction of the bubble rise velocity.

The pressure rise, as shown by the cine photography, leads to a significant reduction in the size of detached bubbles, and, correspondingly, to an increase of their break-away frequency similar to the effect of pressure rise on growth of frequency of natural volumetric oscillations of bubbles (see (1.6.22)).

Analysis of the heat-transfer experimental data. Intensification of heat transfer when gas is injected during bubbling and bubbly boiling is accounted for by two causes. The first is the intensification of convective heat exchange due to mixing of liquid by bubbles in the wall layer, i.e., due to increase of q_l' (see Fig. 7.8.1). The second is the formation of heat sinks q_{lg} on bubble surfaces, which are consumed for liquid evaporation.

The bubbling experiments, in addition to determining the critical injection velocity, enables us to single-out the convective component of the heat-exchange intensification by reducing evaporation to zero on account of small thermal heads (driving forces) or heat fluxes ($q_W \sim$ 20–40 kW/m^2), low temperature of liquid (the temperatures used for water were 5–8° C), and preliminary saturation of injected gas by vapor of liquid subjected to bubbling.

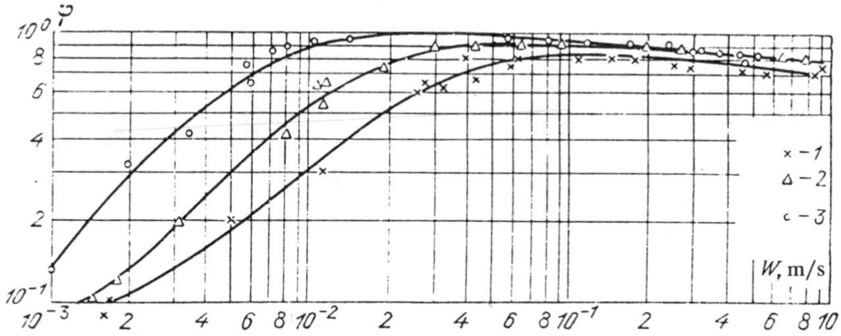

Figure 7.8.6 Relationship between concentration φ of gas in a wall layer and velocity W of the gas (nitrogen) injected in the process of bubbling: water (*1*), 20% solution (*2*) and 50% solution (*3*) of glycerine in water (p = 0.1 MPa, T = 280 K). Experimental data obtained by S. S. Kutateladze and I. G. Malenkov (1976).

Figure 7.8.7 depicts values of the heat-transfer coefficient $\beta = q_W/(T_W - T)$ between the heating wall with temperature T_W and liquid with temperature T in the process of bubbling of a "cold" water by nitrogen, and during boiling of saturated water. Three major areas are observed in the process of bubbling: the first—at small velocities W of bubbling when β is unaffected by W; second—area of a well-developed bubbling when coefficient β is proportional to W^n ($n \approx 2/3$); third—area of saturation of β when intensification of both mixing of liquid and convective transfer due to gas injection increase is compensated by reduction of liquid content in the wall layer. The extent of the second area with respect to W increases with liquid-temperature growth and with the injection of dry gas because of vaporization in the wall layer. The second area is the most extensive during boiling (see solid line in Fig. 7.8.7), when its interruption is caused by crisis when the condition (7.8.16) is satisfied.

Thus, in the second area, or the area of a well-developed bubbling, an analogy between bubbling and boiling was observed for both heat-transfer coefficient β and value K_*.

When analyzing heat transfer processes, in addition to parameters (7.8.5), the coefficients of thermal conductivity and the thermal capacity of phases must be added

$$\lambda_l, \lambda_g, c_l, c_g \tag{7.8.20}$$

Similar to density and viscosity, small values (compared to liquid) of thermal conductivity and volumetric thermal capacity of gas ($\lambda_g \ll \lambda_l$, $\rho_g^0 c_g \ll \rho_l^0 c_l$) should not affect heat transfer which, in this case, may be predeter-

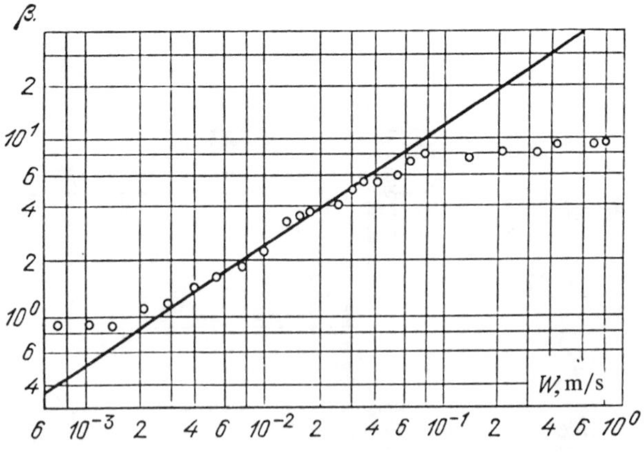

Figure 7.8.7 Relationship between the heat-transfer coefficient β, $kW/(m^2 \cdot K)$, and the injection velocity W during bubbling of water by nitrogen (light circles) and during boiling of water (solid line) at pressure $p = 0.1$ MPa. Data were obtained by S. S. Kutateladze and I. G. Malenkov (1976).

mined by a dimensionless relation whose parameters have been already defined in (7.8.6)

$$\mathrm{Nu} = f\left(\overline{W}, P, \overline{\mu}^{(\Sigma g)}, \overline{\lambda}^{(\Sigma g)}\right)$$

$$\mathrm{Nu} = \frac{\beta\delta^{(\Sigma g)}}{\lambda_l}, \quad \overline{W} = \frac{W}{C^{(\Sigma g)}}, \quad \overline{\lambda}^{(\Sigma g)} = \frac{\lambda_l}{\overset{\circ}{\rho}_l c_l C^{(\Sigma g)}\delta^{(\Sigma g)}} = \frac{\lambda_l}{c_l}\sqrt[4]{\frac{g}{\overset{\circ}{\rho}_l \Sigma^3}} \tag{7.8.21}$$

Experiment proved that with pressure p rise, the heat-transfer coefficient due to convection (but not on account of evaporation) increases by displacement of the second region (where coefficient β is proportional to $W^{2/3}$) towards smaller velocities. In this case, points corresponding to various pressures are described by a relation of type $\beta = \beta$ (pW), with a well-developed bubbly regime (the second region), β being proportional to $(\mathrm{pW})^{2/3}$. Growth of β with rise of p is associated with the increase of the bubble-detachment frequency accompanied by reduction of their size at both bubbling and boiling, rather than by increase of gas density (multiple variation of gas density because of the use of various gases does not affect coefficient β).

Increase of liquid viscosity μ_l lowers that fraction of the heat-transfer coefficient β (but nonuniformly in various ranges of $\bar{\mu}^{(\Sigma g)}$ and W), which is due to convection, leaving the fraction of β which is accounted for by evaporation unvaried. As a result, the indicated reduction of β is realized by displacement of the second region towards higher velocities. In this case, points corresponding to various viscosities are described by a single (common) function of $\bar{W}\bar{\mu}^{(\Sigma g)}$.

S. S. Kutateladze and I. G. Malenkov (1976) generalized experimental data in the form (Fig. 7.8.8)

$$\mathrm{Nu} = N_* \cdot \left(\frac{p\overline{W}}{\overline{\lambda}^{(\Sigma g)}}\right)^{2/3}, \quad N_* = N_*\left(\overline{W}\overline{\mu}^{(\Sigma g)}\right)$$

$$\left(\frac{p\overline{W}}{\overline{\lambda}^{(\Sigma g)}} = \frac{pWc_l}{\lambda_l g}, \quad \overline{W}\overline{\mu}^{(\Sigma g)} = \frac{W\mu_l}{\Sigma}\right) \tag{7.8.22}$$

where $\bar{W}$ is determined during bubbling (injection) by a measurable reduced injection velocity W in accordance with (7.8.21), and it was determined based on a total heat flux in a case of boiling

$$\overline{W} = \frac{W_q}{\sqrt[4]{\Sigma g/\overset{\circ}{\rho}_l}}, \qquad W_q = \frac{q_W}{\overset{\circ}{\rho}_g l} \tag{7.8.23}$$

Since formation of vapor consumes only a part of q_W equal $q_{lg} < q_W$, $W < W_q$.

Treatment of heat-transfer experimental results in the form of (7.8.22) showed that for boiling $N_* = N_*^{(\mathrm{boil})} = (1.5 \pm 0.4) \cdot 10^{-3}$, and for bubbling (blow-in)

$$=\begin{cases} N_*^{(1)}=1.5\cdot10^{-3} & \text{at } \overline{W}\overline{\mu}^{(\Sigma g)}<10^{-3} \\ A/(\overline{W}\overline{\mu}^{(\Sigma g)}) & \text{at } 10^{-3}<\overline{W}\overline{\mu}^{(\Sigma g)}<10^{-2} \quad (A=0.2\cdot10^{-5}) \\ N_*^{(3)}=0.3\cdot10^{-3} & \text{at } \overline{W}\overline{\mu}^{(\Sigma g)}>10^{-2} \end{cases} \tag{7.8.24}$$

The heat-transfer coefficient may, in conformity with this relation, be represented in the form

$$\beta = N_*\cdot\left(\frac{\overset{\circ}{\rho}_l}{\Sigma}\right)^{1/2}\left(\frac{\lambda_l c_l p^2 W^2}{\sqrt{g}}\right)^{1/3} \tag{7.8.25}$$

Hence, the effect of various parameters on heat transfer during bubbling and bubbly boiling on a horizontal surface in the absence of a forced motion (flow past a plate) can be observed.

The "cold" heat-transfer experiments during bubbling simulate (from the standpoint of boiling analysis) a heat flux q_l' carried away by liquid:

q_l' is proportional to $N^{(\text{inj.})}W^{2/3}$
W is proportional to $q_{lg} = q_W - q_l'$

On the other hand, in conformity with treatment based on (7.8.22) and (7.8.23)

q_W is proportional to $N_*^{(\text{boil})}W^{2/3}$
W is proportional to $\overset{*}{q}_W$

Then, for evaluating the part of heat $\bar{q}_l'$ carried away by liquid during boiling, we have an algebraic equation

$$\bar{q}_l' = \overline{N}(1-\bar{q}_l')^{2/3}$$
$$(\overline{N} = N^{(\text{inj})}/N^{(\text{boil})}, \quad \bar{q}_l' = q_l'/q_W, \quad \bar{q}_{lg} = q_{lg}/q_W) \tag{7.8.26}$$

For the case of $\bar{W}\bar{\mu}^{(\Sigma g)} < 10^{-3}$, when $\bar{N} = 1$, we obtain $q_l' \approx 0.6$, and $q_{lg}' \approx 0.4$. For the case of large flows of vapor ($\bar{W}\bar{\mu}^{(\Sigma g)} > 10^{-2}$) characteristic for occurrence of a crisis, when $\bar{N} \approx 0.2$, $\bar{q}_l' \approx 0.2$ and $\bar{q}_{lg} \approx 0.8$ may be calculated from (7.8.26).

On the other hand, in "cold" experiments on critical injection in the process of bubbling, a heat flux q_{lg} is simulated, which is consumed for formation of vapor in the wall layer. In accordance with (7.8.15) (see Fig. 7.8.5), this heat flux is determined, at small viscosities ($\bar{\mu}^{(\Sigma g)} < 10^{-2}$) and near-critical states, by the value $K_* \approx 310$. For alkaline earth metals (for which K_* was calculated based on a total heat flux), $K_* \approx 900$. Hence, it follows that for alkaline earth metals under near-critical regimes, $\bar{q}_{lg} \approx 310/900 \approx 0.3$. This value of $\bar{q}_{lg}$ strongly differs from the value $\bar{q}_{lg} \approx 0.8$ obtained above from the heat-transfer data.

To resolve the indicated discrepancy for q_{lg}/q_W in two types of "cold" experiments, it is imperative to carry out experiments on a critical injection during bubbling under conditions $\bar{\mu}^{(\Sigma g)} < 10^{-3}$, for instance, in bubbling of

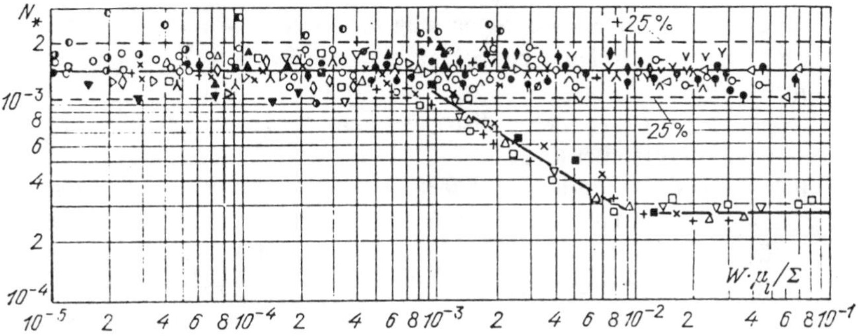

Figure 7.8.8 General relation of the heat transfer coefficient β to the physical properties of liquid and velocity of injection during bubbling (water, hydroglycerin solution) and boiling (water, sodium, potassium, cesium, ethanol, benzol, liquid nitrogen, liquid helium, and freon) using the parameter $N_* = (\lambda_g g/(ph/c_l))^{2/3}$ Nu from parameter $\bar{W} \cdot \bar{\mu}^{(\Sigma g)} = W \cdot \mu_l/\Sigma$. Pressure $p = 0.025$–10 MPa.

mercury by air or other gases in order to finally prove that $K_* \approx 310$, and is independent of $\bar{\mu}^{(\Sigma g)}$. Then the heat-transfer experiments during bubbling of mercury and other liquid metals must show that parameter $N_*^{(3)}$ defining q_l'/q_W grows with increase of the liquid heat conduction, characterized by parameter $\bar{\lambda}^{(\Sigma g)}$, since the value $N_*^{(3)} = 0.3 \cdot 10^{-3}$ is obtained from experiments on bubbling of water, alcohols, etc., whose parameter $\bar{\lambda}^{(\Sigma g)}$ is the same, namely $\bar{\lambda}^{(\Sigma g)} = (0.3\text{–}0.5)10^{-3}$. The value $N_*^{(\text{boil})} = (1.5 \pm 0.4) \cdot 10^{-3}$ is obtained in the boiling experiments for a much broader class of substances (including alkaline earth metals) covering a many times wider range of $\bar{\lambda}^{(\Sigma g)} = 0.3 \cdot 10^{-3}$–0.24. In order to find the relationship $N_*^{(3)}(\bar{\lambda}^{(\Sigma g)}$, "cold" experiments on heat transfer during bubbling when $\bar{\lambda}_{(\Sigma g)} \gg 10^{-3}$, e.g., in the process of bubbling mercury ($\bar{\lambda}^{(\Sigma g)} \approx 0.22$), must be conducted.

§7.9 HYDRAULICS AND HEAT EXCHANGE OF A MULTICOMPONENT GAS-LIQUID MIXTURE WITH CHEMICAL REACTIONS AND PHASE TRANSITIONS IN TUBE FURNACES

A combined motion of gas and liquid in channels is widely used in chemical technology, petroleum refining, and other related industries. In particular, many technological processes are associated with an intensive heating of large masses of multicomponent liquid accompanied by phase transitions and chemical reactions. Tube furnaces with *ID* about 0.1 m are often used in these processes; these tubes convey gas-liquid multicomponent flows with heat input through tube walls by burning fuels outside the tubes. The level of heat fluxes in petroleum processing reaches 50 kW/m^2. And, with the volumetric concentrations of gas $\alpha_g \equiv \varphi \geqslant 0.8$ and velocities $W \geqslant 10$ m/s (**Fr** > 100), the axisymmetric dispersed-annular regimes of flows are realized even in horizontal tubes.

It is the main purpose of this section to show, by way of an example of the so-called "installation for retarded coking" (IRC) in the process of production of petroleum coke[8] (Z. I. Syunyaev, 1981), the generalization of equations and representations of §§7.2–7.5 for the account of the multicomponent nature of phases and chemical reactions in phases. This generalization was accomplished by R. I. Nigmatulin, R. G. Shagiev et al. (1977). It should be remembered that large pressure drops are realized in intensive processes, and the process of a chemical reaction depends not only on heating intensity but also on purely hydrodynamic effects defining, in particular, pressure variation and, interrelated with it, intensity of vaporization or condensation. It is these particular situations that are specific for a number of modern intensive and energy-consuming processes whose analysis and design require a combined solution of a complete system of equations of mass, momentum, phase energy, and kinetics of both interphase and intraphrase processes.

Each component of a mixture—gas, drops, and film—is a homogeneous mixture of a number of species. We shall now give consideration to situations when chemical reactions proceed only in a liquid phase (drops and film). The mass exchange processes between components: gas vaporization from liquid, break-away and deposition of drops are also possible. The average velocities of gas and drops, and the temperatures of all three components of the flow—gas, drops, and film—are assumed equal ($v_1 = v_2$, $T_1 = T_2 = T_3$), since in processes discussed below, the inherent times $t_{12}^{(v)}$ and $t^{(T)}$ of the equalization of, respectively, velocities of gas and liquid, temperatures of three components of mixture are much smaller than inherent time $t_{i0} = L/v_{i0}$ of residence of a fixed mass of any phase in a channel, and, accordingly, the characteristic lengths $t_{12}^{(v)}v_i$ and $t^{(T)}v_i$, over which the above-indicated parameters equalize, are much smaller than the channel length L. The conservation equations for mass of gas, drops, and film, momentum equations for both core and film, and also, the heat influx equation for the entire mixture, are written in the form (7.2.8).

In tube furnaces for heating petroleum raw materials, the gas phase consists of a light component (L) with the molecular weight $\mu_m \approx 100$, heavy component (O) (vapor of oil with $\mu_m \approx 400$), and inert component (W) (vapor of water with $\mu_m = 18$). The latter does not participate in phase transitions and chemical reactions. Each component of the liquid phase—

[8] Focus on this process was suggested by A. S. Eigenson.

film and drops—consists of four species: 0—oil ($\mu_m \approx 400$), R—resin ($\mu_m \approx 700$), A—asphaltene ($\mu_m \approx 900$), C—carboids ($\mu_m \approx 1000$), which undergo chemical transformations as follows

$$\begin{array}{ccccccc} 0 & \rightleftarrows & R & \rightleftarrows & A & \rightleftarrows & C \quad \text{(liquid phase)} \\ \downarrow & & \downarrow & & \downarrow & & \downarrow \\ L & & L & & L & + & W \quad \text{(gas phase)} \end{array} \tag{7.9.1}$$

Parameters related to species O, R, A, and C in drops ($j = 2$) and film ($j = 3$) are denoted by indices $j(2)$, $j(3)$, $j(4)$, and $j(5)$, respectively, and parameters related to inert gas W and species L and O in gas phase are denoted, respectively, by indices 1(0), 1(1), and 1(2), i.e., the first index identifies the number of phase, and the second (in parenthesis) identifies the number of component (species). Then, the mass equations for components become

$$\frac{dm_{1(0)}}{dz} = 0, \quad \frac{dm_{1(1)}}{dz} = J_{21(1)} + J_{31(1)}, \quad \frac{dm_{1(2)}}{dz} = J_{21(2)} + J_{31(2)}$$

$$\frac{dm_{2(k)}}{dz} = -J_{21(k)} - \overset{\circ}{J}_{23(k)} + \overset{\circ}{J}_{32(k)} + S_c \sum_{q=2, q\neq k}^{5} (K_{3(qk)} - K_{3(kq)})$$

$$\frac{dm_{3(k)}}{dz} = -J_{31(k)} - \overset{\circ}{J}_{32(k)} + \overset{\circ}{J}_{23(k)} + S_f \sum_{q=2, q\neq k}^{5} (K_{3(qk)} - K_{3(kq)}) \tag{7.9.2}$$

$$m_{j(k)} = k_{j(k)} m_j, \quad \overset{\circ}{J}_{ji,k} \approx k_{j(k)} \overset{\circ}{J}_{ji}$$

$$(i, j = 1, 2, 3; \quad k, q = 0, 1, 2, 3, 4, 5)$$

Here $m_{j(k)}$ and $k_{j(k)}$ are, respectively, flux and mass concentration of the k-th component in the j-th phase; $J_{ji(k)}$ is the intensity of mass transfer of the k-th component during the phase transition $j \to i$. When prescribing $J^0_{ji(k)}$ by the last expression of (7.9.2), it was assumed that composition of mass undergoing the transition $j \to i$ is identical to composition of species of j-th component. And, $K_{j(kq)}$ is the rate of formation of q-th component from the k-th component in liquid phase ($j = 2$ in film, $j = 3$ in drops) due to chemical reaction.

We shall consider, in the interest of simplicity, that gas phase is a mixture of thermally perfect gases with gas constants $R_{g(k)}$, heat capacities $c_{g(k)}$ at constant pressure ($k = 0, 1, 2$), and specific heat of vaporization $l_{(1)}$ and $l_{(2)}$ of components O and L ($k = 1, 2$)

$$p = \overset{\circ}{\rho}_1 R_1 T_1, \quad R_1 = \sum_{k=0}^{2} k_{1(k)} R_{g(k)}, \quad i_1 = \sum_{k=0}^{2} k_{1(k)} i_{1(k)}$$

$$i_{1(0)} = c_{g(0)} (T - T_0), \quad i_{1(1)} = c_{g(1)} (T - T_0) + l_{(1)} \tag{7.9.3}$$

$$i_{1(2)} = c_{g(2)} (T - T_0) + l_{(2)}, \quad i_2 = i_3 = c_2 (T - T_0) + (p - p_0)/\overset{\circ}{\rho}_2$$

The thermophysical data of hydrocarbon raw materials were taken from a manual (Methods . . ., 1974).

Expressions for F_{13} and F_W and for intensities of break-away J^0_{32} and deposition J^0_{23} were used in accordance with §§7.3 and 7.4. The minor losses due to tube elbows were taken into consideration by introducing an additional correction factor $\chi > 1$ in force F_W.

The entire component L generated in liquid phase was assumed instantaneously vaporizing ($k_{2(1)} = k_{3(1)} = 0$). Intensity of vaporization of the heavy (second) component O was determined based on an equilibrium approach

$$p_{1(2)} = p_{S(2)}(T) \tag{7.9.4}$$

where $p_{S(2)}(T)$ and $p_{1(2)}$ are, respectively, saturation pressure and partial pressure of the component O. The pressure relationship for the oil-vapor (O) saturation was approximated in the form (M. L. Kreimer and R. I. Ilembitova et al., 1974)

$$p_{S(2)} = \exp\left\{0.0794 + \frac{8.262\,[f(T) - f(T^{\circ}_{S(2)})]}{0.8541 - f(T^{\circ}_{S(2)})}\right\}$$
$$f(T) = (T - 273)/T \tag{7.9.5}$$

where $T^0_{S(2)}$ is the saturation temperature (boiling point) at normal pressure $p_0 = 0.1$ MPa.

The chemical-reaction rate was prescribed by the Arrhenius law

$$K_{j(qk)} = K^{\circ}_{j(qk)} \exp(-T_{(qk)}/T)$$
$$K^{\circ}_{j(qk)} = K^{\circ} k_{j(q)} \qquad (j = 2, 3; \quad (qk) = (23), (34), (45)) \tag{7.9.6}$$

where the values of kinetic constants $K^0_{j(qk)}$ and $T_{(qk)}$ for liquid-phase reactions were used in conformity with the data of the Institute of Catalysis (Siberian Department of Academy of Sciences of USSR), which are outlined in an article by G. G. Valyavin et al., (1975).

Parameters used in the following discussion are set forth in the Appendix (Table A.3).

The above-indicated equation, together with equations outlined in §§7.3 and 7.4, forms a closed system for which (with the inlet parameters predetermined) the Cauchy problem for ordinary differential equations may be solved. Figure 7.9.1 (where $x_j = m_j/m$) represents analytical results together with experimental data obtained on a functioning reactor battery of an installation for retarded coking (IRC) with the tube $ID = 0.1$ m, total length $L = 850$ m and U-turns spaced at 15 m. We shall discuss an operation regime with the mixture flow rate $m = 10$ kg/s, gas concentration at the inlet $x_{10} = m_1(0)/m = 0.03$, inlet pressure $p_0 = 2, 5$ MPa, and the heat-input distribution $Q_w(z)$ depicted in Fig. 7.9.1. The above-mentioned coefficient of minor losses due to elbows was assumed $\chi_W = 1.54$ in the magnitude of force F_W based on coincidence of predicted and experimental relation $p(z)$ in the same regime of operation of the system under consideration. The observed agreement between theoretical and experimental results for the remaining parameters should be viewed as major evidence of the model adequacy. Analysis showed satisfactory agreement between theoretical and experimental data also on all other regimes (other m, x_{10}, p_0, and $Q_W(z)$). Thus, it has been proved that adequate numerical experiment based on the proposed model for studying the process on IRC is possible and relevant.

Analysis predicts the presence of a "crisis of the second kind," which is characterized by reduction of the flow rate x_3 and liquid film thickness at section $z = z_*$ approaching zero because of vaporization; these must lead to the deterioration of heat transfer from the wall and the rise of its temperature. The latter is a main cause of the observed overburning of tubes and a gradual plugging of the open section by coke ("sclerosis") as a consequence of drop deposition on a hot tube-wall unprotected by film. This is confirmed by data which show that overburning and plugging by coke are specific mostly for the tube segment $z > z_*$, where the flow, according to predictions, is totally dispersed and film-free; it is noteworthy that the film vanishes in spite of the presence of a great amount of liquid phase at section z_* ($x_2 \gtrsim 0.7$; see Fig. 7.9.1, where $z_* \approx 650$ m is marked by a cross). The

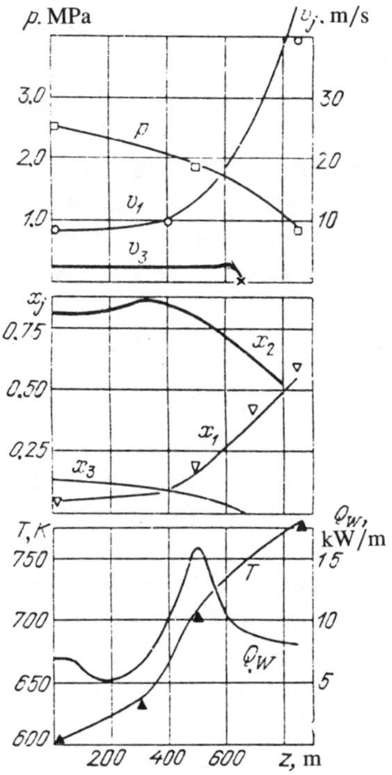

Figure 7.9.1 Predicted and experimental data on parameter distribution (pressure, temperature, phase velocities, and flow rates) along the reactor battery of installation for retarded coking (IRC) with *ID* = 0.1 m, total length L = 850 m, and U-turns spaced at every 15 m. The mixture flow rate m = 10 kg/s (m^0 = 1273 kg/(m$^2 \cdot$ s)), gas concentration at the inlet (z = 0) in the form of vapor of water, $x_{1(0)} = m_{1(0)}/m = 0.03$, pressure at the inlet p = 2.5 MPa. Distribution of heat input is given by curve $Q_W(z)$.

fact is that a predominant part of liquid flows in the form of drops in a core, and its velocity is the same as the gas velocity ($v_1 \approx v_2 = 2$–50 m/s). This fact has been found from theoretical analysis, and is substantiated by coincidence of residence time of both liquid phase and gas phase in a tube.

By means of numerical experiment, the optimum magnitudes of structural parameters D, L, and controlling parameters p_0, m, x_{10}, and $Q_W(z)$ may be found for a variety of original raw materials.

The method for prediction of the gas-liquid flows of multicomponent mixtures with chemical reactions and heat-transfer processes in a reactor battery of IRC discussed in this section by way of example is applicable to processes of chemical cracking, primary fractionation of petroleum, pyrolysis of oil products, etc.

§7.10 GAS-DYNAMIC CRISIS (CRITICAL FLOW) OF DISPERSED AND DISPERSED-FILM VAPOR-LIQUID FLOWS

When studying a stationary flow of a two-phase mixture discharged out of a large-capacity vessel through channels of various geometrical configurations, the critical (maximum) flow rate of the mixture is an important characteristic of flow. By definition, the flow of mixture is said to be critical if, at fixed rest-parameters and phase flow-rate ratio at the channel inlet, the further increase of the mixture flow rate by the pressure reduction at the channel outlet becomes impossible. Accordingly, the flow rate corresponding to these conditions is referred to as critical, or maximum. Knowledge of characteristics of critical outflow of gas-liquid mixtures and flows of flashing (cavitating) liquids is of great importance for evaluating the consequences of emergency loss of sealing of high-pressure vessels, and for estimating the maximum discharge through channels and nozzles in which two-phase liquid undergoes acceleration.

Review of experimental data on the critical flow rate of a gas-liquid mixture under various conditions of outflow (through long and short tubes, nozzles and orifices, in a bubbly, drop, dispersed-annular, and other regimes of flow) can be found in works by G. Wallis (1969), G. V. Tsiklauri et al., (1973), V. A. Zysin (1976), V. V. Fisenko (1978), K. Ardon (1978), and V. G. Selivanov (1976). It should be kept in mind that experiments on high-velocity discharge through channels of large diameters ($D = 0.1$–1 m), which are of particular interest from the standpoint of analysis of emergency situations, are very expensive. Therefore, development of theoretical models enabling us to describe the outflow process in a wide range of regime parameters, including large-diameter channels, are of considerable importance.

We shall discuss below the major principles of the theory of critical stationary outflow (discharge) of two-phase liquids.

First, we shall confine ourselves to a case of a dispersed gas-liquid flow (either bubbly or drop) governed by equations of quasi-one-dimensional flow (§7.2), but in the absence of film. Parameters related to continuous (carrying) and dispersed phases are denoted by subscripts 1 and 2, respectively. And, we assume a quasi-equilibrium state of the interface (Σ-phase), i.e., $T_\Sigma = T_S(p)$, and also $v_{12} = v_2$. As a result, the conservation equations for mass, momentum, and heat influx for each phase in a state of stationary flow become (see also (1.4.8))

$$\frac{dm_1}{dz} = J_{21}, \quad \frac{dm_2}{dz} = -J_{21} \qquad (m_1 = \rho_1 v_1 S, \quad m_2 = \rho_2 v_2 S$$

$$\rho_1 = \overset{\circ}{\rho}_1 \alpha_1, \quad \rho_2 = \overset{\circ}{\rho}_2 \alpha_2, \quad \alpha_1 + \alpha_2 = 1)$$

$$m_1 \frac{dv_1}{dz} = -\alpha_1 S \frac{dp}{dz} + F_{21} - F_{W1} + J_{21}(v_2 - v_1)$$

$$m_2 \frac{dv_2}{dz} = -\alpha_2 S \frac{dp}{dz} - F_{21} - F_{W2}$$

$$m_1 \frac{di_1}{dz} = \alpha_1 v_1 S \frac{dp}{dz} + J_{21}\left[\frac{(v_2 - v_1)^2}{2} + i_{1S} - i_1\right] \qquad (7.10.1)$$

$$+ F_{21}(v_2 - v_1) - Q_{1\Sigma} - F_{W1} v_1 + Q_{W1}$$

$$m_2 \frac{di_2}{dz} = \alpha_2 v_2 S \frac{dp}{dz} - J_{21}(i_{1S} - i_2) - F_{W2} v_2 - Q_{2\Sigma} + Q_{W2}$$

$$Q_{1\Sigma} + Q_{2\Sigma} = J_{21}(i_{1S} - i_{2S}), \ p = p_1 \quad (Q_{i\Sigma} \equiv Q_{i(\Sigma 2)}, \ i = 1, 2)$$

These equations must be supplemented by thermodynamic equations of state of phases

$$\overset{\circ}{\rho}_j = \overset{\circ}{\rho}_j(p_j, T_j), \quad i_j = i_j(p_j, T_j) \qquad (j = 1, 2) \qquad (7.10.2)$$

equations for drop or bubble sizes (e.g., Eq. (1.3.9) for the size of incompressible, unbreakable drops, or the Rayleigh-Lamb equation (1.3.15) for bubble size), and equations defining the force F_{21} interaction of phases and interphase heat exchange ($Q_{1\Sigma}$, $Q_{2\Sigma}$). Then, if the external effects upon flow are predetermined (i.e., the channel geometry ($S(z)$), force (F_{W1} and F_{W2}) and thermal (Q_{W1} and Q_{W2}) effects from the channel walls are prescribed), we obtain a closed system of equations of two-velocity, two-temperature flow.

With the predetermined parameters at the inlet ($z = z_b = 0$) (which are denoted by a subscript b)

$$z = 0: \ p = p_b, \quad v_1 = v_{1b}, \quad v_2 = v_{2b}$$

$$T_1 = T_{1b}, \quad T_2 = T_{2b}, \quad \alpha_1 = \alpha_{1b}, \quad a = a_b \qquad (7.10.3)$$

the parameter distribution along the channel of length L ($0 \leqslant z \leqslant L$) is found by means of solving the Cauchy problem (7.10.1)–(7.10.3).

Adding both phase-mass equations, we obtain an integral of mass or flow rate of mixture

$$m_1 + m_2 = m = m_0 = \text{const} \tag{7.10.4}$$

For adiabatic flows with no heat exchange with walls ($Q_{W1} = Q_{W2} = 0$), an integral of the mixture energy takes place

$$m_1\left(i_1 + \frac{v_1^2}{2}\right) + m_2\left(i_2 + \frac{v_2^2}{2}\right) = m_0 i_0 = \text{const} \tag{7.10.5}$$

where i_0 is the rest enthalpy.

Formulation of a problem on a stationary discharge of a two-phase liquid out of a large-capacity vessel through a channel. Critical regime. Let a two-phase liquid flow out of a large-capacity vessel through a channel whose cross section is characterized by dimensions $S^{1/2}$, which are much smaller than the inherent size of the vessel. The rest parameters denoted by subscripts 0 are realized within the vessel

$$\begin{gathered} \rho_{10},\ \rho_{20},\ v_{10} = v_{20} = 0, \quad T_{10} = T_{20} = T_0,\ p_0 \\ (\rho_0 = \rho_{10} + \rho_{20}, \quad \rho_0 i_0 = \rho_{10} i_1(p_0, T_0) + \rho_{20} i_2(p_0, T_0)) \end{gathered} \tag{7.10.6}$$

The rest parameters are assumed time-independent, which is ensured by large dimensions of the vessel compared to $S^{1/6}$, and which is a necessary condition for a stationary outflow. At fixed rest parameters, different flow rates m_0 are realized depending on pressure (back pressure) in the space into which the two-phase liquid discharges. We shall confine ourselves to adaiabatic outflow processes

$$Q_{W1} = Q_{W2} = 0 \tag{7.10.7}$$

It is necessary to determine the maximum possible flow rate m_0. The flow regime, pressure at the channel outlet $p(L) \geqslant p_\infty$, and the maximum flow rate itself, which correspond to this flow rate, are referred to as *critical*. Both critical outlet pressure and flow rate are denoted by p_* and m_*, respectively.

If the discharging medium is a perfect compressible liquid for which, in addition to (7.10.1), there is no friction between liquid and the channel wall ($F_W = 0$), then, as is a well-known fact, the critical flow is realized when the flow velocity at the minimum cross section of a channel, or over the entire length of a constant-cross-section channel ($S' = 0$), equals the speed of sound ($v = C$). Interphase nonequilibrium and friction between liquid and the channel wall alter this canonical statement.

Equilibrium ideal model of outflow. During an equilibrium flow of vapor-liquid mixture when

$$v_1 = v_2 = v, \quad T_1 = T_2 = T_S(p) \tag{7.10.8}$$

in the absence of interaction with a wall, when in addition to adiabaticity (7.10.7), friction between liquid and the wall is also absent.

$$F_{W1} = F_{W2} = 0 \tag{7.10.9}$$

Eqs. (7.10.1) are reduced to equations of isoentropic ($s = s_0 =$ constant) flow of a perfect compressible liquid whose state is predetermined by a barotropic equation of state

$$p = p(\rho, s_0), \qquad s_0 = s(p_0, x_{10}) \tag{7.10.10}$$

The flow velocity is determined from the Bernoulli integral

$$\frac{v^2}{2} = -\int_{p_0}^{p} \frac{dp}{\rho} \quad \text{or} \quad \frac{v^2}{2} = i_0 - i(p)$$
$$(i = x_1 i_{1S}(p) + x_2 i_{2S}(p), \quad \rho = \rho_1 + \rho_2, \quad x_j = \rho_j/\rho) \tag{7.10.11}$$

where i_{jS} is the enthalpy of phases in state of saturation ($j = 1,2$), x_j is the mass concentration of phases. Condition of the entropy constancy is written in the form

$$x_1 s_{1S}(p) + x_2 s_{2S}(p) = s_0 = \text{const} \tag{7.10.12}$$

The above-indicated condition of a critical regime of flow of perfect liquid is

$$v(p) = C(p) \qquad (C = \sqrt{(\partial p/\partial \rho)_s}) \tag{7.10.13}$$

From the relation

$$\frac{1}{\rho} = \frac{x_1}{\overset{\circ}{\rho}_1} + \frac{x_2}{\overset{\circ}{\rho}_{2S}}$$

and using both (7.10.12) and the condition of phase saturation ($\rho_j^0 = \rho_{jS}^0(p)$), we obtain the expression for the speed of sound written in derivatives of ρ_j^0 and s_j along the saturation line

$$\frac{1}{C^2} = \rho^2 \sum_{j=1}^{2} \left[\frac{x_j}{(\overset{\circ}{\rho}_{jS})^2} \frac{d\overset{\circ}{\rho}_{jS}}{dp} - \frac{1}{\overset{\circ}{\rho}_{jS}} \left(\frac{dx_j}{dp}\right)_S \right]$$
$$\left(\left(\frac{dx_1}{dp}\right)_S = -\left(\frac{dx_2}{dp}\right)_S = \frac{1}{s_{2S} - s_{1S}} \left[x_1 \frac{ds_{1S}}{dp} + x_2 \frac{ds_{2S}}{dp} \right] \right. \tag{7.10.14}$$
$$\left. \frac{ds_{jS}}{dp} = \frac{1}{T_S} \left(\frac{di_{jS}}{dp} + \frac{1}{\overset{\circ}{\rho}_{jS}} \right) \right)$$

As a consequence, the prediction of a critical regime for an equilibrium outflow is reduced to determining the critical pressure p_*, for which the condition (7.10.13) is satisfied with prescribed p_0, T_0, and x_{10}. Because of the transcendental nature of relations $\overset{0}{\rho}_{jS}(p)$, $i_{jS}(p)$, and $s_{jS}(p)$, an analytical-tabular technique is employed for such predictions based on the use of $i-s$ diagrams of state of substance in a two-phase domain; a method of successive approximations for solving the algebraic equation (7.10.13) may also be employed.

Models of inlet segments of channels. A quasi-one-dimensional flow described by Eqs. (7.10.1) is realized only in channels of $0 \leqslant z < L$. In this case, the channel-inlet parameters (7.10.3) differ from the rest parameters (7.10.6), since during outflow ($m_0 \neq 0$) we have $v_{10} \cdot v_{20} \neq 0$. In this connection, the process at the inlet segment must be prescribed so that inlet parameters (7.10.3) can be found based on the rest parameters (7.10.6) and flow rate m_0. We shall now indicate some feasible analytical schemes of inlet segments. If a vessel, from which a discharge through a given channel takes place, contains a single-phase (subcooled) liquid (the first phase)

$$\rho_0 = \overset{\circ}{\rho}_{10} = \rho_{10}, \quad \rho_{20} = 0 \tag{7.10.15}$$

then, the density variation and phase transitions at the inlet may be ignored. This paves the way for using the Bernoulli integral for an incompressible liquid

$$T_b = T_0, \quad \rho_b = \rho_0, \quad \rho_{2b} = 0, \quad p_b = p_0 - {}^1/{}_2\rho_b v_b^2, \quad v_b = m_0/\rho_b S_b \tag{7.10.16}$$

This scheme may be generalized also for a case when the vessel contains a two-phase medium ($\rho_{10} \cdot \rho_{20} \neq 0$), but pressure drop at the inlet section is small ($p_0 - p_b \ll p_0$). Then, the following may be assumed

$$\begin{gathered} T_b = T_0, \quad \rho_{1b} = \rho_{10}, \quad \rho_{2b} = \rho_{20} \quad (\rho_b = \rho_0) \\ p_b = p_0 - {}^1/{}_2\rho_b v_b^2, \quad v_b = v_{1b} = v_{2b} = m_0/(\rho_b S_b) \end{gathered} \tag{7.10.17}$$

To evaluate the maximum effect of phase transitions at the inlet segment, Eqs. (7.10.10)–(7.10.12) may be used for an equilibrium isoentropic flow of a two-phase liquid

$$\begin{gathered} T_b = T_S(p_b), \quad v_b = m_0/((\rho_{1b} + \rho_{2b})\, S_b) \\ \rho_{1b} = (1 - \alpha_{2b})\, \overset{\circ}{\rho}_{1S}(p_b), \quad \rho_{2b} = \alpha_{2b}\overset{\circ}{\rho}_{2S}(p) \\ x_{1b} i_{1S}(p_b) + x_{2b} i_{2S}(p_b) + {}^1/{}_2 v_b^2 = i_0 \\ x_{1b} s_{1S}(p_b) + x_{2b} s_{2S}(p_b) = s_0 \\ (\rho_0 s_0 = \rho_{10} s_1(p_0, T_0) + \rho_{20} s_2(p_0, T_0) \\ x_{jb} = \rho_{jb}/(\rho_{1b} + \rho_{2b}), \quad x_{1b} + x_{2b} = 1) \end{gathered} \tag{7.10.18}$$

These equations enable us to calculate p_b, T_b, α_{2b}, v_b, and other magnitudes based on predetermined values of p_0, T_0, and α_{20}.

If the entrance to the channel is smooth, the configuration of the channel may be extrapolated to domain $z_0 < z < 0$, occupied by the liquid-containing vessel, using a law of type $S = S_b + Az^2 + Bz$ (assuming that the corresponding surface of an "extrapolated channel" is the flow surface). In this case, in order to determine the boundary conditions at point z_0', one of the above-indicated schemes may be used, and the assumed equations of a quasi-one-dimensional flow may be employed in the domain $z > z_0'$.

In case the entrance into the channel is not smooth (i.e., has either rectangular or acute-angle edges), flow lines (surfaces) within the vessel ($z < 0$) (whose contribution to a flow rate m_0 is significant) cannot be obtained by extrapolation of channel walls (because the flow lines at the entrance segment are not parallel to channel walls, and the jet has a vena-contracta (Fig. 7.10.1)). In this case, the beginning $z = 0$ of integration of quasi-one-dimensional equations must be shifted a distance (0.5–1.0)D (where D is diameter of the inlet) into the channel.

If the vessel contains a single-phase (subcooled) liquid, phase transitions do not have enough time to occur within the vena-contracta. Then, we have the following equations for analysis of a sharp-edge entrance portion

$$p_b = p_0 - \frac{\rho_b v_b^2}{2}, \quad v_b = \frac{m_0}{\rho_b S_b^0}, \quad S_b^\circ = \eta S_b, \tag{7.10.19}$$

where η is a coefficient of contraction at the entrance segment. For an entrance with a rectangular edge, $\eta \approx 0.61$, which is characteristic for a flow of a single-phase liquid and is close to 0.595 obtained in experiments on outflow of flashing water conducted by A. K. Tikhonov et al. (1978).

Condition for realization of critical flow. The system of equations (7.10.1) may be solved for derivatives. In this expression for the pressure gradient in an adiabatic flow of vapor-drop mixture with incompressible condensed phase (ρ_2^0 = constant), Eq. (7.10.7) is written in the form

$$\frac{dp}{dz} = \frac{\overset{\circ}{\rho}_1 v_1^2 C_1^2}{\alpha_1 C_f^2} \cdot \frac{S'/S + \zeta}{1 - v_1^2/C_f^2}$$

$$S' = \frac{dS}{dz}, \quad C_f^2 = C_1^2\left(1 + \frac{\alpha_2 \overset{\circ}{\rho}_1 v_1^2}{\alpha_1 \overset{\circ}{\rho}_2 v_2^2}\right) \tag{7.10.20}$$

$$S\zeta = -\frac{1}{(\overset{\circ}{\rho}_1)^2 v_1}\left(\frac{\partial \overset{\circ}{\rho}_1}{\partial i_1}\right)_p (Q_{1\Sigma} - F_{21}(v_2 - v_1)) + F_{21}\left(\frac{1}{\overset{\circ}{\rho}_1 v_1^2} - \frac{1}{\overset{\circ}{\rho}_2 v_2^2}\right)$$

$$-\frac{F_{W1}}{\overset{\circ}{\rho}_1 v_1^2} - \frac{F_{W2}}{\overset{\circ}{\rho}_2 v_2^2} - J_{21}\left(\frac{1}{\overset{\circ}{\rho}_1 v_1} - \frac{1}{\overset{\circ}{\rho}_2 v_2}\right) + \frac{J_{21}(v_2 - v_1)}{\overset{\circ}{\rho}_1 v_1^2}$$

where C_1 is the speed of sound in the carrying (continuous) phase (vapor), and for its expression in terms of functions $\rho_1^0(p,T)$ and $i_1(p,T)$ see (6.11.12);

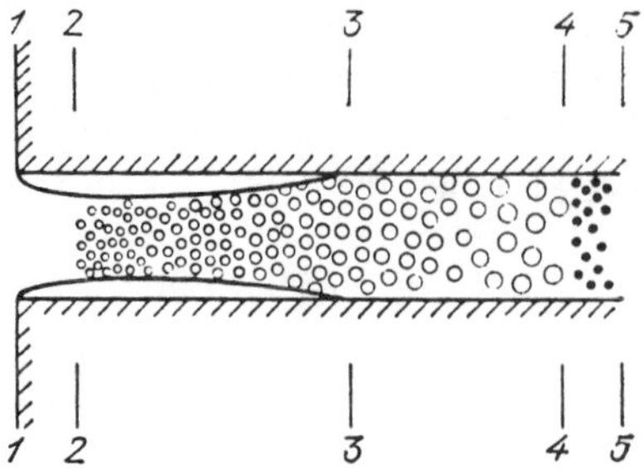

Figure 7.10.1 Structure of a flashing-liquid flow in a tube with a sharp-edge entrance.

then, C_f is the "frozen," or characteristic speed of sound in mixture; ζ is a function which is determined by values J_{21}, F_{12}, $Q_{1\Sigma}$, and $Q_{2\Sigma}$ (interphase interaction in a flow), and also by values F_{W1} and F_{W2} (interaction between a flow and channel wall).

At critical, or near-critical states, the flow accelerates ($dv_1/dz > 0$), and pressure along the channel drops ($dp/dz < 0$). A continuous transition through the characteristic speed of sound C_f, when $v_1 = C_f$, may occur only at a section where $S'/S + \zeta = 0$. For an accelerating flow of perfect gas in the absence of heat exchange and wall friction $\zeta = 0$, and transition through the speed of sound happens, as was already mentioned, within the throat of a nozzle ($S' = 0$, $S = S_{\min}$). Interphase nonequilibrium of a two-phase flow and the presence of the wall friction lead to $\zeta \neq 0$, and transition through the speed of sound does not occur inside the throat; the location of this transition depends on ζ.

Identification of the behavior of the function ζ is generally hampered by a great amount of possible magnitudes of parameters appearing in this function. However, analysis of important flows enables us to indicate the sign of ζ based on following simple considerations.

In accelerating flows at $v_1 < C_f$ in tubes (channels where $S' = 0$), pressure decreases ($dp/dz < 0$). Then, from the first equation (7.10.20), it follows that in these regimes $\zeta < 0$. In a channel with reducing cross-section area ($S' < 0$), the flow undergoes an additional acceleration caused by contraction. This particular situation (when the flow at the channel entrance is in near-equilibrium state) enhances (compared to flow in a tube) interphase nonequilibrium which must augment the absolute value of ζ. This circumstance indicates that at $v_1 < C_f$, and during an accelerating regime ($dv_1/dz > 0$, $dp/dz < 0$), in narrowing segments of a channel $\zeta < 0$ takes place. Since the value ζ does not directly depend on S'/S, nor on the ratio between v_1 and C_f, a conclusion can be made that for accelerating segments of a channel (i.e., where $dv_1/dz > 0$ and $dp/dz < 0$), the condition $\zeta < 0$ is a characteristic feature of the process. Hence, an essential conclusion follows: in accelerating flows ($dv_1/dz > 0$) when liquid at its initial state is near equilibrium, a continuous transition through the characteristic speed of sound

C_f ($S'/S + \zeta = 0$) occurs in the expanded portion of the channel ($S' > 0$). The corresponding point, denoted by t, whose longitudinal coordinate is denoted by z_*, is a singular point in the plane zp, which, as in the perfect-gas case, is a saddle-type point (Fig. 7.10.2). As distinct from perfect gas, when z_* was known beforehand and was related to a minimum cross section (throat) of a channel, determining z_* and corresponding flow rate m_* for a nonequilibrium flow are possible only in the process of the integration of the system of differential equations of motion of the medium; such integration permits the coordinate z_* of a critical section to be found, and successfully takes into consideration the effects of flow "prehistory" (over the segment $0 < z < z_*$) on magnitudes of parameters at the critical section.

Returning back to the above-formulated problem on determining a stationary critical regime of outflow with maximum discharge m_* for both prescribed rest parameters and channel, we shall specify two possible cases. The first case: if in the process of integrating a system of differential equations, the singular point t is located within the channel under consideration (see Fig. 7.10.2a), i.e.

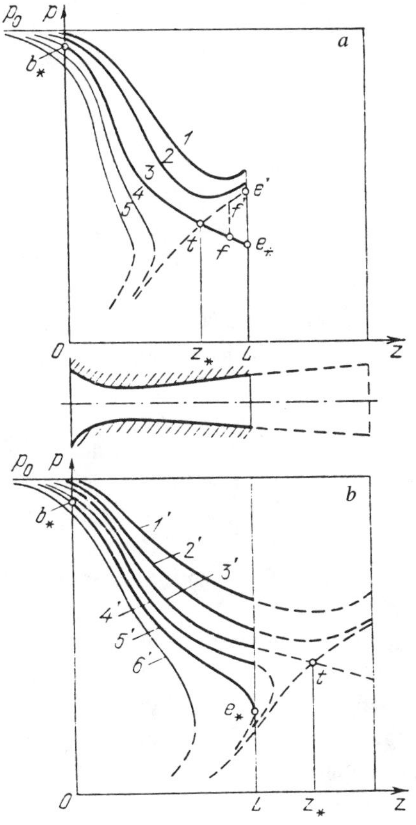

Figure 7.10.2 Integral curves in a plane zp corresponding to a stationary discharge through a Laval nozzle: a) when the singular point t (where velocity v_1 of the continuous phase reaches the characteristic speed C_f of sound) with a coordinate $z = z_*$ is located within a nozzle; b) when the singular point t is outside the nozzle ($z_* > L$).

$$z = z_* \qquad (0 < z_* < L)$$

$$\frac{1}{S}\frac{dS'}{dz} + \zeta = 0 \tag{7.10.21}$$

$$v_1 = C_f, \quad dp/dz = 0/0$$

then, separatrix *3* passing through points 0 ($z = 0$, $p = p_0$) and $t(z = z_*)$, which corresponds to the flow rate m_*, reflects the critical regime being sought. In this case, the curves lying above the separatrix (curves *1* and *2*) pertain to regimes with flow rates smaller than m_* (such regimes are referred to as subcritical), and curves below separatrix (lines *4* and *5*, for which $m > m_*$) correspond to physically irrelevant flows of type of "overturned" waves (see §§4.4 and 6.4) for a given channel with prescribed rest parameters.

The second case: if within a nozzle $S'/S + \zeta < 0$, i.e., the singular point ($0 < z < L$) is not realized within a channel, or the singular point t is located formally outside the channel under consideration ($z_* > L$, see Fig. 7.10.2*b*), then, line *5'* corresponds to the sought critical regime when characteristic velocity ($\upsilon_1 = C_f$) is reached by continuous (carrying) phase at the outlet of the channel ($z = L$). In this case, the outlet pressure gradients and velocities of the continuous phase tend to infinity (see (7.10.20))

$$z = L\colon\ \frac{1}{S}\frac{dS}{dz} + \zeta < 0, \quad v_1 = C_f, \quad \frac{dp}{dz} = -\infty \tag{7.10.22}$$

Curves lying above the indicated line *5'* (including the separatrix *3'*) correspond to subcritical regimes with flow rates smaller than those related to line *5'*. Curves lying below the line *5'* ("overturned" waves) are physically irrelevant.

It has been shown above that for accelerated flows (and critical flows are such), $\zeta < 0$ takes place. Therefore, for tube (channels with constant cross section: $S' \equiv 0$) and channels with reducing cross sections ($S' < 0$), both critical regime and maximum flow rate are governed only by the second condition (namely, condition (7.10.22)) for the carrying phase velocity to reach the characteristic speed C_f at the outlet section ($z = L$). The values of pressure p_* and other parameters at the critical (outlet) section, whose gradients along axis z in this section approach infinity, must be determined in phase planes of type $\upsilon_1 p$ by extrapolating $p(\upsilon_1)$ at the "section" $\upsilon_1 = C_f$.

Critical stationary outflow of flashing liquid through tubes and nozzles. We will now analyze a problem concerned with stationary quasi-one-dimensional outflow of flashing liquid, which is described by a system of equations (7.10.1)–(7.10.3) but in a single-velocity approximation ($\upsilon_1 = \upsilon_2 = \upsilon$, and instead of two phase momentum equations, only one equation of the mixture momentum, which is a sum of these two equations, should be used); also, an approximation of the vapor saturation ($T_2 = T_S(p_2)$), and Eqs. (2.6.48) (where $p_\infty = p$), (1.3.56), and (1.6.20) defining $\dot{a}$ and $q_{\Sigma 1} = -Q_{1\Sigma}/n$ (see §6.11) for thermal bubble growth are used in solving the above-formulated problem. Then, instead of Eq. (7.10.20), we have the following differential equation for p

$$\frac{dp}{dz} = \rho v^2 \frac{S'/S + \zeta}{1 - v^2/C_f^2}, \quad \frac{dv}{dz} = -\frac{S}{m}\frac{dp}{dz} - F_W$$

$$C_f^2 = \frac{1}{\alpha_g \rho}\left\{\frac{1}{\overset{\circ}{\rho}_g}\left(\frac{\partial \overset{\circ}{\rho}_g}{\partial p}\right)_S + \frac{1}{l}\left[\overset{\circ}{\rho}_g\left(\frac{\partial i_g}{\partial p}\right)_S - 1\right]\left(\frac{1}{\overset{\circ}{\rho}_g} - \frac{1-\vartheta}{\overset{\circ}{\rho}_l}\right)\right\}^{-1} \approx \frac{\overset{\circ}{\rho}_g}{\alpha_g \rho}\left(\frac{\partial p}{\partial \overset{\circ}{\rho}_g}\right)_S \tag{7.10.23}$$

$$\zeta = \frac{Q_{1\Sigma}}{vS}\left[\frac{1}{l}\left(\frac{1}{\overset{\circ}{\rho}_l} - \frac{1}{\overset{\circ}{\rho}_g}\right) - \frac{1-\vartheta'}{(\overset{\circ}{\rho}_l)^2}\left(\frac{\partial \overset{\circ}{\rho}_l}{\partial i}\right)_S\right] - \frac{F_W}{\rho v^2 S} \approx \frac{Q_{1\Sigma}}{vSl\overset{\circ}{\rho}_g} - \frac{F_W}{\rho v^2 S}$$

$$\left(\vartheta = \frac{i_{lS} - i_l}{\overset{\circ}{\rho}_l}\left(\frac{\partial \overset{\circ}{\rho}_l}{\partial i}\right)_S, \quad \vartheta' = \frac{i_{lS} - i_l}{l}\right)$$

where C_f is the characteristic speed of sound for a given model of a two-phase medium; this speed of sound is a "frozen" speed with respect to phase transitions and heat exchange in liquid phase, but is equilibrium with respect to phase velocities. The friction force $F_W = \pi D \tau_W$ between the gas-liquid flow and channel wall is prescribed by Eq. (7.1.5) with coefficients $A = 1$, $n = 1.75$. The inlet segment is described by Eqs. (7.10.16) and (7.10.19). The schematic view of the flow is shown in Fig. 7.10.1. During outflow of hot water at pressures $p > 0.5$ MPa through a tube with rectangular-edge entrance, there are no phase transitions over the initial segment up to section *2*. This has been experimentally proved by L. K. Tikhonov et al., (1978). An indicated local minimum of pressure at section *2*, where the cross section of a liquid jet is minimum, proves that maximum superheating of liquid is reached here, and, consequently, the most favorable conditions occur for generating vapor phase in the bulk volume of liquid. Growth of vapor bubbles in the volume of a jet of superheated liquid leads to the jet expansion and its "adherence" to walls of a channel (section *3*). Motion of a separated jet over the segment *2–3* may be assumed approximately isobaric. With growth of α_g above a value limiting the existence of a bubbly structure in a flow, its restructuring from a bubbly to a vapor-drop structure (segment *4–5*) is possible.

The initial stage of flashing in a jet separated from a channel wall is defined by a heterogeneous nuclei formation in the volume of superheated liquid. The model of this process is discussed in §1.7 and utilized in §6.11. Since an inherent diameter of a viable nucleus of the vapor phase depends on thermophysical parameters of liquid and its superheating, various numbers of impurities become the virtual centers of vapor formation in identical specimens of liquids at various degrees of superheating. The spectrum of admixed particles $N(a)$ may be approximately found by solving the "inverse" problems of stationary outflow of flashing liquid (B. I. Nigmatulin, K. I. Soplenkov, and V. N. Blinkov, 1982) using related experiments.

To plot the relationship $n(a)$ representing a number of admixed particles of radius larger than a, i.e., post-critical, or viable particles per unit volume at $p < p_S - 2\Sigma/a$, it is advisable to use experimental results concerning the critical stationary outflow of saturated water through short tubes ($1 < L/D \leqslant 10$, $L \leqslant 0.3$ m) with a rectangular-edge entrance. The lower boundary of values L/D is limited by the validity of a quasi-one-dimensional model of flow, and the upper by the condition that superheated water has no contacts with the channel surface over the major part of the tube, and also that flashing on the channel walls is insignificant. The upper boundary L is determined by the fact that kinetics, or more precisely, flashing delay in long tubes (because of long liquid residence-time in the channel) exhibits itself to an insignificant extent, and the flow resembles an equilibrium regime.

When solving the "inverse" problems, all conditions were predetermined by an experiment, and such n was chosen at which the predicted pressure distribution and critical length of tube were in agreement with experimental results. The critical regime of flow was said to be realized if condition (7.10.22) was satisfied with a prescribed accuracy.

Figure 7.10.3 represents a result of the described treatment of experimental data for water used on thermal electric power plants. It should be remembered that for water-outflow regimes at higher pressures ($p > 16$ MPa), such superheatings (corresponding to Gibbs numbers **Gi** = 20–30) are realized in the minimum cross section of the liquid jet (section *2* in Fig. 7.10.1), when there occurs a homogeneous (spontaneous) nuclei-formation on whose background the impact of admixed particles becomes insignificant. In this situation, critical flow rates turn out to resemble flow rates found in accordance with an equilibrium model.

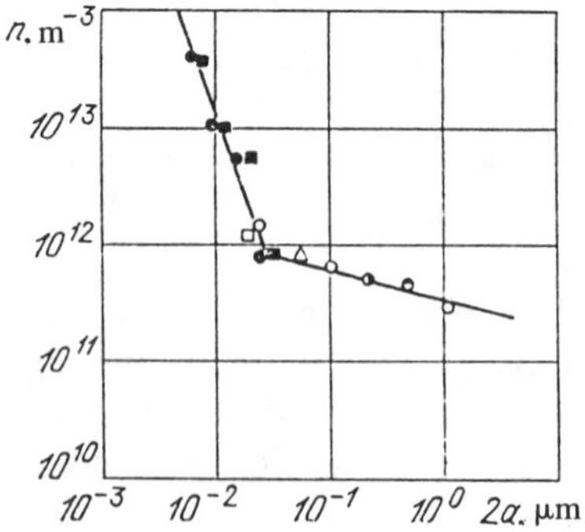

Figure 7.10.3 Number of admixed particles (nucleation sites) whose diameter exceeds $2a$ in treated water used on thermal electric-power plants. The shown points were obtained based on processing experimental data (D. A. Khlestkin et al., 1977, L. R. Kevorkov et al., 1977, L. K. Tikhonov et al., 1979) on outflow of flashing water through cylindrical nozzles of diameter D = 3.6–25 mm and L/D = 4–10 at pressures p_0 = 1.0–14 MPa. The extreme left points ($2a \approx 7 \cdot 10^{-3}$ micron correspond to pressure p_0 = 14 MPa when the minimum Gibbs number during the efflux (see Eq. (1.7.13)) was equal to **Gi** ≈ 60. By virtue of this (see (1.7.14)), the homogeneous nuclei formation was insignificant.

and R. I. Nigmatulin, 1972; B. I. Nigmatulin et al., 1977; A. I. Ivandaev et al., 1977) is generally based on a three-velocity three-temperature model (§7.2). If corresponding conservation equations for mass, momentum, heat influxes, and states of phases are solved with respect to derivatives, we obtain

$$\frac{dX^{(k)}}{dz} = \Psi^{(k)}, \quad X^{(k)} = (m_1, m_2, m_3, T_2, T_3, a), \quad k = 1, 2, \ldots, 6$$

$$\frac{dY^{(j)}}{dz} = \frac{\Delta^{(j)}}{\Delta}, \quad Y^{(j)} = (v_1, v_2, v_3, p, T_1), \quad j = 1, 2, \ldots, 5$$

$$\Delta \sim 1 - v_1/C_f \tag{7.10.24}$$

In this case, parameters of a dispersed-annular flow are divided into two types. The first type—six parameters $X^{(k)}$, differential equations for which have a characteristic form without singularities. The second type—five parameters $Y^{(j)}$. The initial differential equations of conservation and state for these parameters contain derivatives of two and more variables of the total number of $Y^{(j)}$. Upon solving these equations with respect to derivatives, equations with singularities at points where $\Delta = 0$ are obtained. The last condition defines a set of parameters of mixture in the space $Y^{(j)}$ or a hypersurface, at points of which the gradients of $Y^{(j)}$ along axis z equal infinity and alter their sign, with the exception of singular points of this hypersurface where in addition to $\Delta = 0$, $\Delta^{(j)} = 0$ ($j = 1, 2, \ldots, 5$) takes place. For a three-velocity, three-temperature model of a flow, this hypersurface is determined by equation $v_1 = C_f$, where

$$\frac{C_f^2}{C_1^2} = \left[1 + \frac{\alpha_2 \overset{\circ}{\rho}_g v_1^2}{\alpha_1 \overset{\circ}{\rho}_l v_2^2} + \frac{\alpha_3 \overset{\circ}{\rho}_g v_1^2}{\alpha_1 \overset{\circ}{\rho}_l v_3^2}\right] \equiv \left[1 + \varepsilon^2 \frac{x_2 K_2^3 + x_3 K_3^3}{x_1}\right]$$

$$\left(K_i = v_1/v_i \quad (i = 2, 3), \quad \varepsilon = \overset{\circ}{\rho}_g/\overset{\circ}{\rho}_l\right) \tag{7.10.25}$$

Here, K_i are so-called parameters of the slip of mixture species. The magnitude of a slip parameter averaged over the entire liquid is also of particular interest

$$K_{gl} = \frac{v_g}{v_l} \quad \left(v_g = v_1, \quad v_l = \frac{\varkappa_2 v_2 + \varkappa_3 v_3}{\varkappa_2 + \varkappa_3}, \quad \varkappa_i = \rho_i S_i \Big/ \left(\sum_{j=1}^{3} \rho_j S_j\right)\right.$$

$$\left. S_1 \equiv S_2 \equiv S_c, \quad S_3 \equiv S_f \right) \tag{7.10.26}$$

where $\varkappa_i$ is the mass concentration of i-th component (species) of mixture, which is different from its flow-rate concentration x_i.

The expression for specific critical discharge m_*^0, corresponding to (7.10.25), is written in the form

Comparison of predicted data with experimental results on maximum specific flow rates $m_*^0 = m_*/S(L)$ and pressure distribution along the channel during the critical outflow proves the adequacy of the above-described model. In order to obtain the relation $n(a)$ depicted in Fig. 7.10.3, this comparison was carried out also for outflow of saturated and subcooled water in a wider range of the channel parameters compared to the used range. In addition to experimental data for tubes with rectangular-edge outlets ($3.6 < D < 38$ mm, $4 < L/D < 20$, $1 < p_0 < 16$ MPa), experimental data obtained by H. Fauske (1965) were also used, as were data obtained by L. K. Tikhonenko et al., (1980) for outflow through Laval nozzles. Comparison of theoretical and experimental results on specific critical discharges in the range of $m_*^0 = (8\text{–}60) \cdot 10^3$ kg/(m$^2 \cdot$ s) showed that discrepancies were within 10%.

In the represented model, two irreversible processes are taken into consideration. The first is the interphase heat and mass exchange ($T_1 > T_2 = T_S(p)$), whose nonequilibrium is responsible for the superheating of liquid and for the delay of its transition to vapor. The second is friction between liquid and the wall (hydraulic resistance), which is responsible for the irreversible drop of pressure associated with the dissipation of kinetic energy of liquid. Intensity of the first process is determined by: concentration and distribution $n(a)$ of nuclei particles, thermophysical properties of the medium (λ_l, c_l, ρ_l^0, ρ_g^0, and l) appearing in Eq. (2.6.48) for the bubble-growth velocity, pressure difference $p_0 - p_\infty$, and the jet contraction coefficient η, affecting the depth of penetration of state of discharged liquid into the metastable area, i.e., the value of pressure p_b in the minimum section of the jet; this pressure defines the maximum superheating of liquid $\Delta T = T_0 - T_S(p)$ in the process under consideration. Intensity of the second process is defined by the friction force $F_W = \pi D \tau_W$ between liquid and the channel wall.

The flashing delay leads to the reduction of vapor concentration $\alpha_g \equiv \alpha_2$, and, consequently, to an increase of specific critical flow rate m_*^0 compared to the equilibrium scheme; this effect is stronger, the higher the degree of liquid purity (regarding the admixed particles) and the shorter the residence time of liquid in a channel, i.e., the shorter the tube length L. Irreversible loss of pressure due to friction (as well as reversible pressure drop due to both jet contraction at the inlet and liquid acceleration) contributes to the speeding up of the flashing process and to the decrease of m_*^0. In connection with the indicated opposite effects of two irreversible processes on the critical flow rate, the latter may be both smaller and larger than the corresponding magnitude of m_*^0, determined in accordance with the perfect equilibrium scheme, which is observed in the experiment depicted in Fig. 7.10.4. In the range of parameters represented in Fig. 7.10.4, hydraulic losses of pressure due to channel-wall friction F_W are small compared to the reversible pressure drop associated with the flow acceleration. Since the channel diameter D affects the process only through the force F_W, change of D at both fixed p_0 and L (the latter defines the flashing delay at fixed p_0) affects the critical flow rate insignificantly.

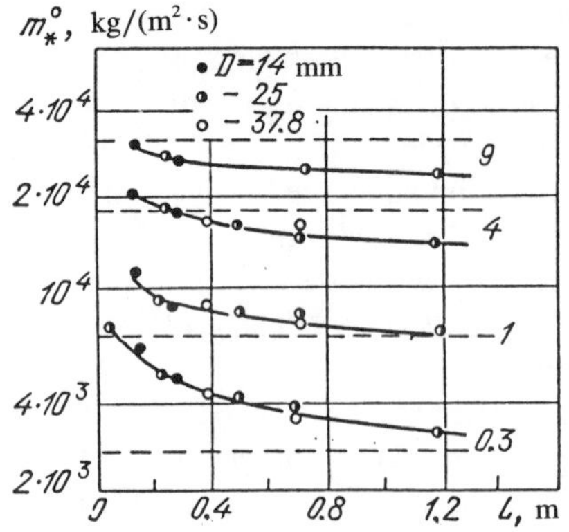

Figure 7.10.4 Effects of pressure p_0, length L, and diameter D of a tube upon the critical flow rate m_*^0 in the process of stationary outflow of initially saturated (in state o of retardation) water through a rectilinear tube with a rectangular-edge entrance (experimental data obtained by L. R. Kevorkov et al. (1978)). Numerical labels 0.3; 1; 4; 9 on curves indicate pressure p_0, MPa, dashed lines represent predicted data based on an equilibrium model of perfect liquid.

Figure 7.10.5 represents distribution of parameters during critical efflux of flashing liquid out of a high-pressure vessel (p_0 = 8.5 MPa) through two tubes of equal relative length (L/D = 4), but differing in diameters D and lengths L by a factor of 20. In both cases, the pressure losses due to friction F_W, which depend on the channel diameter, are small. And, the tube residence-time of liquid flowing out of the tube with D = 25 mm, L = 100 mm (see Fig. 7.10.5*a*) equals $t_L \approx 1$ microsecond, and in a tube of D = 500 mm, L = 2000 mm (Fig. 7.10.5*b*) is $t_L \approx 22$ microsecond; accordingly, the thermal, or phase, nonequilibrium, and flashing delay in a large tube are much smaller than in a small tube. Therefore, the specific critical discharge through a large tube is considerably smaller than in a small tube. If the tube entrance is rounded such that $\eta \approx 1$, then the critical flow rate for a large tube increases, according to analytical prediction, by 13% because of lower pressure drop at the inlet, which reduces the overheating and intensity of flashing. The flashing delay is also slightly increased, accounted for by an increase of the outflow velocity, which reduces t_L.

Note that effects of near-wall formation of vapor in superheated liquid on elements of microroughness of the channel surface are insignificant for all variants represented here. This process becomes significant for water at fairly low pressure ($p_0 \leqslant 1$ MPa) and temperatures when the realized superheatings are minor, and the admixtures in liquid cannot become viable centers of flashing in flows through longer and smaller-diameter channels ($D \leqslant 10^{-2}$ m, $L \geqslant 0.5$ m) and also in nozzle configurations which ensure both separation-free external flow and larger contact surface with liquid. The fact of a well-developed near-wall formation of vapor during flow of flashing water in Laval nozzles has been experimentally observed by a number of authors (G. A. Mukhachev et al., 1973; V. N. Blinkov et al., 1981) and the hydrodynamic nonequilibrium models based on hypotheses of a near-wall nuclei-formation permit the satisfactory description of such flows (A. A. Avdeev et al., 1977; V. N. Blinkov and S. D. Frolov, 1982).

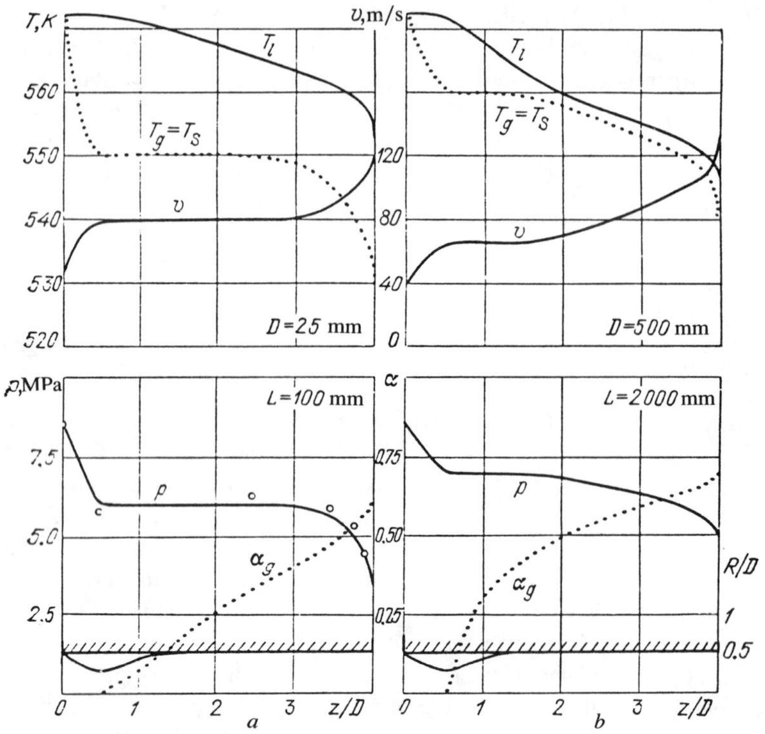

Figure 7.10.5 Distribution of parameters of stationary critical flow along a channel in the process of outflow of flashing water from a large-capacity vessel (where water is in state of saturation: p_0 = 8.51 MPa, T_0 = 572 K) through tubes (L/D = 4) with rectangular-edge entrance; *a*) for D = 25 mm, L = 100 mm, m_*^0 = 35,700 kg/(m²·s); *b*) for D = 500 mm, L = 2000 mm, m_*^0 = 27,770 kg/(m²·s). Lines represent analytical prediction; dots represent experimental measurements of pressure (L. K. Tikhonov et al., 1978).

Flows in the expanded portion of the Laval nozzle may be complicated by strong nonuniformity of velocities and vapor concentrations caused by separation of the liquid jet from the channel walls (M. E. Deich and G. A. Filippov, 1981).

The impact of nonequilibrium and hydraulic resistance (dissipation) upon critical discharge of liquid in a bubbly regime was theoretically studied by Yu. V. Mironov (1975), B. I. Nigmatulin and K. I. Soplenkov (1978), K. Ardron (1978), J. Boure (1972).

Critical flow in a dispersed-annual regime. In the process of outflow of flashing liquid through long channels, concentrations of vapor may become high enough that in a predominant portion of a channel, a dispersed-annular regime is realized ($\alpha_g \equiv \varphi \geqslant 0.8$). Analysis of such flows (A. I. Ivandaev

$$\overset{\circ}{m}_* = \overset{\circ}{\rho}_g C_g \left[\frac{1}{x_1} \left(\sum_{j=1}^{3} x_j \varepsilon_j^2 K_j^3 \right) \Big/ \left(\sum_{j=1}^{3} x_j \varepsilon_j K_j \right) \right]$$

$$\left(\varepsilon_2 = \varepsilon_3 = \overset{\circ}{\rho}_g / \overset{\circ}{\rho}_l, \quad \varepsilon_1 = K_1 = 1\right) \tag{7.10.27}$$

Determination of the values ρ_1^0, C_1, x_2, x_3, K_2, and K_3 appearing in this equation becomes possible only upon complete solution of the critical-flow problem at both prescribed p_0 and flow-rate vapor-concentration x_{10} at the channel inlet. Mathematically, this problem is reduced to a solution of a system of conservation equations (§7.2) together with closing relations (§§7.3 and 7.4). A critical outflow through a tube of a given diameter corresponds to an outlet ($z = L$) condition $\Delta = 0$. The solution can be found by selection of such magnitude of the mixture flow rate m_* (at fixed p_0 and x_{10}) which realizes $\Delta(L) = 0$.

At the inlet to an experimental segment ($z = 0$), two parameters are usually known directly from experiments: mass flow-rate concentrations $x_{10} = m_{10}/m_0$ of vapor and pressure p_0. To carry out calculations, i.e., to solve the Cauchy problem for a system of ordinary differential equations, a number of additional parameters must be prescribed: temperatures T_{i0} of the mixture components ($i = 1,2,3$), their velocities v_i, determined by the slip coefficients K_{20} and K_{30}, relative flow rate x_{30} of liquid in a film, and average radius a of drops in the flow core.

With sufficiently long tubes, the flow parameters at the inlet to the segment under investigation may not be considered as arbitrary values. They are developed resulting from dynamic interaction between components of mixture when passing through the section of stabilization. The mode of their further variation is monotonic. Based on the described premise, the following method for prescriping the unknown flow parameters at the channel inlet may be proposed. First, they are prescribed arbitrarily: a corresponding numerical solution is prepared, from which the flow parameters are found, which are constituted over the section of stabilization after parameter profiles enter the zone of smooth monotonic variation. The sought values of inlet parameters may be obtained, as the first approximation, by means of linear extrapolation of corresponding monotonic profiles to the inlet-section coordinate, $z = 0$.

From Fig. 7.10.6, it is evident that the mass-averaged parameter K_{gl} of slip between vapor and liquid over a major portion of the tube found in conformity with (7.10.26) is practically constant, and equal, approximately, to 1.25; it grows only in the vicinity of the channel outlet up to $K_{gl} \approx 2$. Results of these and other calculations showed that critical slip is generally significantly lower than values recommended by well-known estimating formulas (H. Fauske, 1965) obtained from heuristic considerations

$$K_{gl*} = (\overset{\circ}{\rho}_l / \overset{\circ}{\rho}_g)^{1/2} \quad \text{or} \quad K_{gl*} = (\overset{\circ}{\rho}_g / \overset{\circ}{\rho}_l)^{1/3} \tag{7.10.28}$$

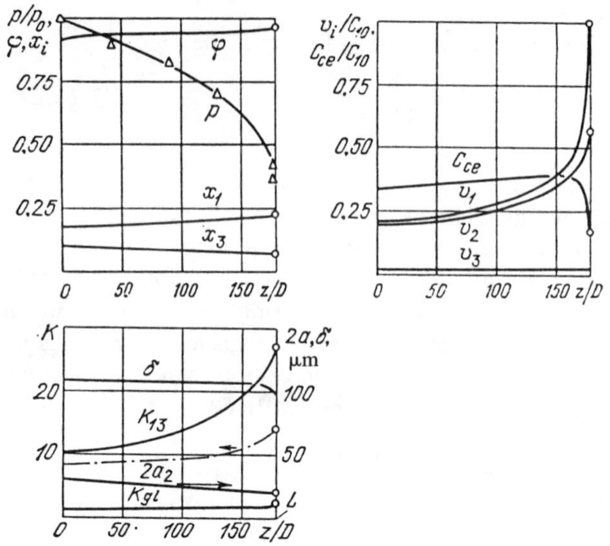

Figure 7.10.6 Distribution of parameters of a dispersed-film steam-water flow (pressure p, steam velocities υ_1, drop velocities υ_2, film velocities υ_3, volumetric concentration φ of steam, mass flow-rate concentrations of steam x_1 and film x_3, film thickness δ, drop diameter $2a$, and also steam slip K_{13} relative to film and average slip K_{gl} of steam relative to liquid along the tube (D = 6.8 mm, L/D = 179)) under conditions of critical outflow ($m^0_* = 6100$ kg/(m$^2 \cdot$ s); pressure and flow-rate concentration of steam at the inlet are p_0 = 2.45 MPa, and x_{10} = 0.177, respectively). Speed C_1 of sound in steam along the entire length of a tube is constant and equal, approximately, to 500 m/s. Dot-dashed line corresponds to $(\rho_l^0/\rho_g^0)^{1/2}$ and is shown here for comparison with K_{13} and K_{gl}. Experimental points for p are obtained by H. Fauske (1965).

Under conditions of critical flow near the tube outlet, a zone of high gradients of parameters $Y^{(j)}$ exists, namely: velocities, phase temperatures, and pressures. Length $C_1 t_0$ (where t_0 is the inherent time of relaxation) of this zone for the variant depicted in Fig. 7.10.6 equals, approximately, 10 mm. The two-phase mixture in this zone becomes nonequilibrium: liquid is superheated, vapor is supercooled, and the temperature difference reaches several tens of degrees. Thus, tubes whose length is of order of $C_1 t_0$ should be considered as short. An equilibrium flow of mixture over a major portion of tube under conditions of a gas-dynamic crisis may occur only in a sufficiently long tube: $L \gg C_1 t_0$. The longitudinal gradients over a predominant portion of such tubes cannot cause noticeable differences in velocities and temperatures of drops and gas in the flow core ($\upsilon_1 \approx \upsilon_2 \approx \upsilon_c$, $T_1 = T_2 = T_c$). Then, the equilibrium speed $C_e = C_c$ (see (4.2.10)) of sound in the core determined by mass concentration of phases in the flow core may be used as a characteristic velocity of the flow core causing choking of the latter.

The example shown in Fig. 7.10.6 relates to a flow in a long tube where the condition $\upsilon_c \approx C_c$ is realized near the tube outlet, and the equilibrium speed of sound is exceeded ($\upsilon_c > C_c$) over a small portion of the tube.

The only way to verify applicability of a chosen hydrodynamic model to analysis of critical flow rates of two-phase mixtures under different conditions is through a broad comparison of results of numerical analysis with experimental data not only on flow rates but also on the flow-parameter profiles along a channel. This comparison may generally be accomplished only with respect to the pressure profile along the channel (example of such comparison is depicted in Fig. 7.10.6), since measurement of profiles of other flow-parameters along the channel are not available. Note that a general picture of flow for long tubes ($L \gg C_1 t_0$) are unaffected by variations of initial temperatures and slip in their actual range. Formation of critical conditions at the outlet section of a tube may be affected to a considerable extent by initial (at the inlet: $z = 0$) relative flow rate x_{30} of liquid in a film and average radius a_0 of drops. These parameters are characterized by significantly slower relaxation towards their stabilized values than K_2, K_3, T_2, and T_3. As a consequence, the rates of flow-parameter variation along the channel with change of both x_{30} and a_0 may be different.

Values of x_{30} may be determined by means of the above-mentioned method of adjustment of boundary conditions at the inlet. However, this method is inapplicable for prescribing the effective mass-averaged radius of drops in the flow core at the inlet to the tube since the inherent time of variation of the radius is long. Evaluations and analysis show that the mass-averaged radius a of drops in a core at an arbitrary section of a sufficiently long channel is proportional to diameter a_{32} of drops removed from film at the given section

$$a = k^{(a)} a_{32} \qquad (k^{(a)} \sim 1) \tag{7.10.29}$$

where a_{32} is found from (7.4.21) with a coefficient $k_{32} \approx 100$. The dimensionless constant $k^{(a)}$ must be viewed as a constant of a model of critical regime of the mixture outflow, whose magnitude may be found from the condition of best agreement between analytical and experimental flow rates and pressure profiles along the channel.

A set of experimental data (H. Fauske, 1965; H. Ogasawara, 1969) on critical flow rates of steam-water mixtures in long tubes was processed based on the above-described technique for conditions when a dispersed-annular regime of flow is realized in the following range of vapor-concentration and pressure at the channel inlet: $x_{30} = 0.03$–1.0, $p_0 = 1$–5 MPa, and with the tube size: $D = 3$–10 mm, $L/D = 20$–385. Departure of the predicted flow-rate values in the range $m_*^0 = (3\text{–}12.5) \cdot 10^3$ kg/(m$^2 \cdot$ s) from experimental results does not exceed 10%, and, in general, within experimental error of their determination. In this case, the constant $k^{(a)}$ in formula (7.10.29) should be taken as $k^{(a)} \approx 4$.

It should be noted that even a slight change of flow rate by as low as

3–5% (which is within the experimental error of its measurement) leads to sufficiently large changes of critical lengths z_* of channels. Thus, uncertainty in prescribing the constants defining the interphase interactions, chosen on the premise of coincidence between critical lengths z_* and pressure profiles $p(z)$, will not lead to significant errors in determining the critical flow rates m_*.

With an increase of flow rate from its subcritical to near critical value, pressure at the inlet begins to drop so that at near-critical conditions its significant change is observed even at a very small variation of flow rate. Accordingly, a strong fluctuation of pressure at the tube outlet may be observed during experiment when the outflow regime approaches its critical state.

Analyses indicate an extremely small increase of specific critical flow rates m_* with growth of the channel diameter at fixed L/D and other parameters at the inlet. This flow rate m_g^0, in the above-indicated range, is proportional to $D^{\varkappa}$, $\varkappa = 0.02$–0.04.

Figure 7.10.7 shows the predicted results illustrating the possibility of reducing (choking) gas discharge by means of feeding liquid at the channel inlet. Such "choking" may be implemented during an emergency gas outflow. It can be seen that feeding liquid leads first to a rapid reduction of the critical flow rate m_{g*}. Then, with the growth of delivered flow rate of liquid, this reduction slows down. For a complete gas choking by liquid, it is necessary to ensure the liquid flow rate m_l higher than flow rate $m_{i\infty}$ at which hydraulic resistance equals the prescribed pressure drop $p_0 - p_\infty$ during a single-phase flow of liquid phase. However, even this flow rate of liquid may be insufficient for a complete choking (stopping) of gas. This is associated with the feasibility of realization, at small concentration of gas, of inversed dispersed-annular structure of a turbulent gas-liquid flow with gas film on the tube wall accounting for reduction of pressure losses caused by friction. Then, at small gas concentrations, the relationship m_{g*} (m_l) may become ambiguous (see Fig. 7.10.7).

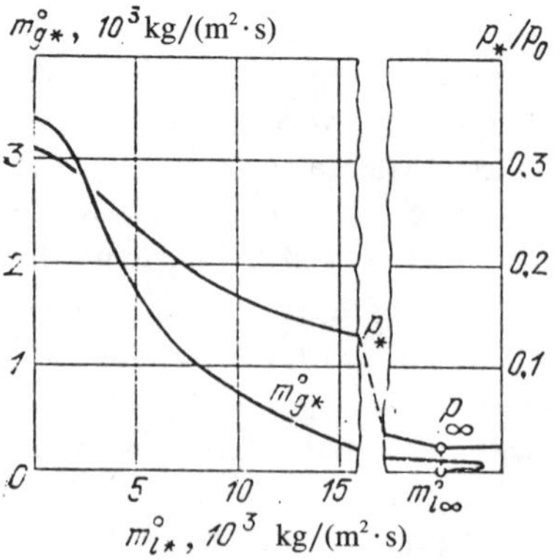

Figure 7.10.7 Effect of flow rate of liquid (water) $m_{l*}^0 = m_{2*}^0 + m_{3*}^0$ upon the flow rate of gas (air) and pressure p_* at the tube outlet ($D = 3.18$ mm, $L = 1225$ mm, $p_0 = 4.25$ MPa, $T_0 = 300$ K) at critical regime of outflow.

CHAPTER

EIGHT

THEORY OF INERTIA-FREE AND PERCOLATION FLOWS OF HETEROGENEOUS MEDIA

This chapter discusses flows of heterogeneous liquids when the inertial forces due to acceleration $d_i v_i/dt$ of material particles may be ignored. Generally, it takes place in processes of slow motions, i.e., at small Reynolds numbers, e.g., in the settling of small particles or drops, filtration of gases and liquids through porous media.

§8.1 ONE-DIMENSIONAL VERTICAL INERTIA-FREE FLOWS OF TWO-PHASE MEDIUM WITH INCOMPRESSIBLE PHASES. KINEMATIC WAVES

We will now examine flows of two-phase mixture when both inertial forces due to phase acceleration and phase compressibility can be neglected. Such flows are realized at small, compared to speed of sound in phases, velocities and in the absence of sharp changes of flow parameters; in particular, when the induced disturbances are sufficiently smooth or of shock-free nature, i.e., the following evaluation is satisfied

$$\rho_i\,(d_i v_i/dt) \sim \alpha_i \overset{\circ}{\rho}_i v_0 t_0^{-1} \ll \alpha_1 \alpha_2 K_\mu^* v_0,\ \ \overset{\circ}{\rho}_i \alpha_i g \qquad (8.1.1)$$

where t_0 is the characteristic time of the parameter variation. This evaluation states that there exists equilibrium of pressure forces, interphase forces, and gravity forces in the process of flow. In the interest of certainty, the axis z, parallel to phase velocity vectors, is directed upward, i.e., opposite to the direction of gravity forces. Then, equations of one-dimensional inertia-free motion are written in the form

$$\frac{\partial \alpha_1}{\partial t} + \frac{\partial \alpha_1 v_1}{\partial z} = 0, \quad \frac{\partial \alpha_2}{\partial t} + \frac{\partial \alpha_2 v_2}{\partial z} = 0 \qquad (\alpha_1 + \alpha_2 = 1)$$

$$-\alpha_1 \frac{\partial p}{\partial z} - K_\mu^* \alpha_1 \alpha_2 (v_1 - v_2) - \overset{\circ}{\rho}_1 \alpha_1 g = 0 \tag{8.1.2}$$

$$-\alpha_2 \frac{\partial p}{\partial z} + K_\mu \alpha_1 \alpha_2 (v_1 - v_2) - \overset{\circ}{\rho}_2 \alpha_2 g = 0 \qquad \left(K_\mu^* = \frac{9\mu_1}{2a^2} \psi_\alpha(\alpha_2)\right)$$

Here, the force on interphase interaction is predetermined as in a monodispersed mixture in a quasi-stationary approximation according to Stokes law, and with consideration given to the particle constraint defined by the coefficient $\psi_\alpha(\alpha_2) = (1 - \alpha_2)^{-m}$ ($m = 3$–5) (see Eq. (1.3.42)).

Drift model. Summation of the first two equations (8.1.2) yields a conservation equation for the volumetric flow rate of mixture, and summation of the third and fourth equations leads to an equation of the mixture equilibrium

$$\alpha_1 v_1 + \alpha_2 v_2 = W(t), \quad \partial p / \partial z = -\rho g \qquad (\rho = \overset{\circ}{\rho}_1 \alpha_1 + \overset{\circ}{\rho}_2 \alpha_2) \tag{8.1.3}$$

The expression for the phase slip velocity may be easily obtained from the equation for the phase motion

$$w_{21} = v_2 - v_1 = \frac{w_0}{\psi_\alpha(\alpha_2)}, \qquad w_0 = \frac{(\overset{\circ}{\rho}_1 - \overset{\circ}{\rho}_2) g a^2}{9/2\, \mu_1} \tag{8.1.4}$$

where w_0 is a drift velocity of a single particle. For gas with particles $w_0 < 0$, and for liquid with bubbles $w_0 > 0$.

Description of a nonstationary flow reduces to a quasi-linear equation of the first order, known as the drift equation

$$\frac{\partial \alpha_2}{\partial t} + \frac{\partial W_2}{\partial z} = 0, \qquad W_2(t, z) = \alpha_2 v_2 = \alpha_2 W(t) + w_0 \frac{\alpha_1 \alpha_2}{\psi_\alpha(\alpha_2)}$$

which may be represented in the form (N. Zuber et al., 1965 and 1966)

$$\frac{\partial \alpha_2}{\partial t} + W_2'(t, \alpha_2) \frac{\partial \alpha_2}{\partial z} = 0$$

$$W_2' = \frac{\partial W_2}{\partial \alpha_2} = W(t) + w_0 \frac{d}{d\alpha_2} J(\alpha_2), \qquad J(\alpha_2) = \frac{\alpha_1 \alpha_2}{\psi_\alpha(\alpha_2)} \tag{8.1.5}$$

Function $J(\alpha_2)$ is referred to as a drift-flux function (G. Wallis, 1969) and is assumed to be known. For predetermined $J(\alpha_2)$ and $W(t)$, the drift equation permits $\alpha_2(t, z)$ to be found. Then, functions $W_2(t, z)$ and $v_2(t, z)$ are determined, and from (8.1.4) the value w_{12} is found, by means of which the relationship $v_1(t, z)$ is determined. Before analyzing the equation of nonstationary flow, we shall consider a particular case when all parameters are time-independent.

Stationary flows, sedimentation, fluidization, and gas-lift. A flow in a stationary regime is defined by fixed (with respect to both coordinate and time) volumetric flow rates of phases and mixture, which sometimes are referred to as reduced velocities

$$\alpha_1 v_1 = W_{10} = \text{const}, \quad \alpha_2 v_2 = W_{20} = \text{const} \qquad (W_{10} + W_{20} = W_0) \tag{8.1.6}$$

Then, from (8.1.5), we obtain the equation governing the possible stationary regimes at prescribed W_{10} and W_{20}

$$\overline{W}_{20} - \alpha_2 W_0 = \pm J(\alpha_2) \tag{8.1.7}$$

$$\overline{W}_{i0} = W_{i0}/|w_0| \quad (i = 1, 2), \quad \overline{W}_0 = \overline{W}_{10} + \overline{W}_{20}.$$

Sign "+" before J identifies a case $\rho_1^0 > \rho_2^0$ ($w_0 > 0$), i.e., when dispersed phase "ascends" relative to liquid, and sign "−" before J identifies a case $\rho_1^0 < \rho_2^0$ ($w_0 < 0$), i.e., when dispersed phase descends relative to liquid.

Figure 8.1.1 represents graphical images of solutions of Eq. (8.1.7), which determines $\alpha_2(\bar{W}_{10}, \bar{W}_{20})$ in stationary, vertical flows of two-phase mixtures with either direct-flow ($W_{10}W_{20} > 0$) or counter-flow ($W_{10}W_{20} < 0$) motion of phases. The solution is identified by a point of intersection of curve $J(\alpha_2)$ or $-J(\alpha_2)$ with a straight line (secant) interconnecting point W_{20} on axis $\alpha_2 = 0$ with point $-W_{10}$ on axis $\alpha_2 = 1$. Abscissa of the point of intersection determines α_2 for measured W_{10} and W_{20}. Values J, W_{10}, and W_{20} on represented graphs are related to a maximum value J_{max} of the drift-flux function $J(\alpha_2)$. For $\psi_\alpha = (1 - \alpha_2)^{-m}$, we have

$$J(\alpha_2) = \alpha_2 (1 - \alpha_2)^{m+1}, \quad J_{max} = \frac{1}{m+2}\left(1 - \frac{1}{m+2}\right)^{m+1} \tag{8.1.8}$$

Figure 8.1.1 illustrates three characteristic regimes.

1. *Sedimentation*—regime at which the heavy particles or drops settle in light liquid or gas ($w_0 < 0$), and the volumetric flow rate of mixture is

$$W_0 = 0, \quad W_{10} = -W_{20}, \quad w_0 < 0 \tag{8.1.9}$$

The above-mentioned secants parallel to axis α_2 cut off two points (see S and S' in Fig. 8.1.1*b*). Thus, two stationary regimes with different α_2 correspond to each volumetric flow rate $W_{20} = -W_{10}$. The maximum possible flow rate in the process of settlement is defined by point M when $W_{20} = |w_0|J_{max}$.

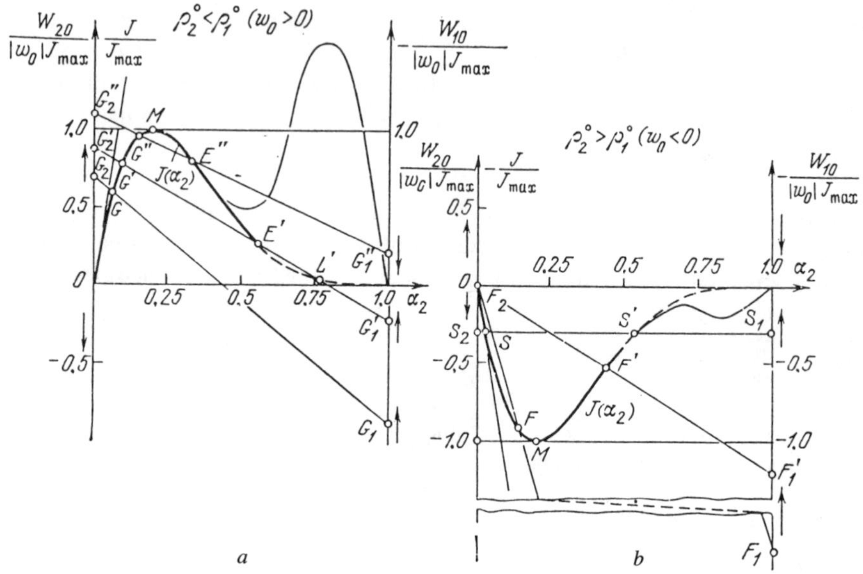

Figure 8.1.1 Schematic for determining the volumetric concentrations α_i of phases as functions of phase volumetric flow rates W_{10} and W_{20} related to the drift velocity w_0 in the process of stationary vertical flow in the field of gravity forces of a two-phase mixture of incompressible phases (see (8.1.7)) using an example $\psi_\alpha = (1 - \alpha_2)^{-3}$, when $J_{max} = 0.082$; *a*) for $\rho_2^0 < \rho_1^0$, *b*) for $\rho_2^0 > \rho_1^0$. Vertical arrows at axes W_{10} and W_{20} indicate whether the points on vertical semi-axes correspond to ascending or descending motion of the related phase. Thin lines qualitatively reflect the behavior of $J(\alpha_2)$ for a gas-liquid flow at $\alpha_2 > 0.5$.

2. *Fluidization*—regime at which the gas flow from below suspends a layer of particles; the latter are not displaced along vertical line

$$W_{20} = 0, \quad W_{10} = W_0 > 0, \quad w_0 < 0 \tag{8.1.10}$$

The related secants emanate from the origin (see F_2FF_1 and $F_2'F'F_1'$ in Fig. 8.1.1*b*) and cut-off points F and F' on the curve $J(\alpha_2)$. The value α_2 for fluidization is uniquely determined using W_{10}. The maximum possible velocity of gas in a regime of fluidization is determined by a secant which is tangent to line $J(\alpha_2)$ at the point of origin. Value $\alpha_2 = 0$ relates to this limiting regime. Having taken into consideration that $dJ/d\alpha_2 = 1$ at $\alpha_2 = 0$, we obtain $\bar{W}_{10} < 1$, i.e., the velocity of gas in a stationary regime of fluidization should not exceed the drift velocity: $v_1 < |w_0|$.

3. Stationary direct-flow uplift motion of liquid with light particles or bubbles (including the so-called gas-lift regime)

$$W_{10} > 0, \quad W_{20} > 0, \quad w_0 > 0 \tag{8.1.11}$$

Secants G_2G_1 and $G_2'G_1'$ which cut off points G, G', E, and L' (Fig. 8.1.1*a*) correspond to this regime. Here, the function $\alpha_2(W_{10}, W_{20})$ may be either a single-valued (see G_2G_1) or triple-valued ($G_2'G_1'$) relation.

Realization of stationary regimes under consideration depends on their stability. The stability investigation with regard to one-dimensional disturbances is based on equations of one-dimensional nonstationary flow of type (8.1.5). Some results of similar study are set forth in §4.1. The one-dimensional stationary regimes, stable to one-dimensional disturbances, may lose stability also caused by non-one-dimensional perturbations.

It should be remembered that with $\alpha_2 > 0.5$, the disperse structure of a flow with no contacts between particles and with the presence of force interaction, defined by original size a of particles and by viscosity μ_1 of the first phase, is disturbed. For indicated concentrations α_2 the function $J(\alpha_2)$ used in the form (8.1.8) is no longer in conformity with the physical implication of the process under discussion. Therefore, this segment of relation (8.1.8) is shown by dashed lines, and the part $J(\alpha_2)$, which relates to $\alpha_2 >$ 0.5 and reflects inversion of the structure of a gas-liquid flow (the second phase becomes continuous, and the first phase dispersed), is shown by thin lines.

Nonstationary flows with continuous waves and shocks. Equation (8.1.5) is equivalent to the condition of constancy along the characteristic direction

$$dz = (W(t) + C(\alpha_2))\,dt:\ d\alpha_2 = 0 \qquad (8.1.12)$$
$$(C(\alpha_2) = \partial W_2/\partial\alpha_2 - W(t) = w_0(dJ/d\alpha_2))$$

By means of transformation of coordinates

$$z' = z - \int_0^t W(t)\,dt, \qquad t' = t \qquad (8.1.13)$$

which is predetermined beforehand, since $W(t)$ is a prescribed function, Eq. (8.1.5) and its characteristic representation (8.1.12) assume a simpler form[9] that corresponds to

$$W = 0 \qquad (8.1.14)$$

Therefore, we may confine ourselves to analysis of this particular case of Eq. (8.1.5) only. Then, the characteristics are straight lines and the solution of differential equation (8.1.5) is

$$\alpha_2 = \alpha_2(\zeta), \qquad \zeta = z - C(\alpha_2)t \qquad (8.1.15)$$

Let us now consider the Cauchy problem with initial data

$$t = 0:\ \alpha_2 = \alpha_{20}(z) \qquad (8.1.16)$$

[9]Transition of (8.1.13) to an inertia-free system of coordinates for used equations is invariant, since in these two equations the inertial force is ignored.

We will single out two characteristics emanating at $t = 0$ from points with coordinates $z^{(1)}$ and $z^{(2)}$, at which the values α_2 at $t = 0$ equal $\alpha_{20}^{(1)}$ and $\alpha_{20}^{(2)}$, respectively

$$z - z^{(1)} = C\left(\alpha_{20}^{(1)}\right)t, \quad z - z^{(2)} = C\left(\alpha_{20}^{(2)}\right)t$$

We shall now find t_*, at which these characteristics intersect

$$t_* = -\frac{z^{(2)} - z^{(1)}}{C\left(\alpha_{20}^{(2)}\right) - C\left(\alpha_{20}^{(1)}\right)}$$

Intersection of characteristics results in the violation of single-valued nature of the continuous solution of type (8.1.15), since at point of intersection t_*, z_*, function $\alpha_2(t, z)$ assumes two values simultaneously, $\alpha_{20}^{(1)}$ and $\alpha_{20}^{(2)}$. This physically impossible solution is referred to as a solution of *overturned wave* (see also §6.4).

If points *1* and *2* are taken sufficiently close to each other, then

$$\alpha_{20}^{(2)} = \alpha_{20}^{(1)} + \frac{d\alpha_{20}}{dz}\left(z^{(2)} - z^{(1)}\right)$$

$$C\left(\alpha_{20}^{(2)}\right) = C\left(\alpha_{20}^{(1)}\right) + \frac{dC}{d\alpha_2}\left(\alpha_{20}^{(2)} - \alpha_{20}^{(1)}\right)$$

Now, we obtain a necessary and sufficient condition on $\alpha_{20}(z)$ that prevents intersection of characteristics at $t > 0$, to ensure a single-valued (unique) solution of (8.1.15)

$$\frac{d\alpha_2}{dz}\frac{dC}{d\alpha_2} \geqslant 0, \quad \text{or} \quad w_0 \frac{d\alpha_2}{dz}\frac{d^2 J}{d\alpha_2^2} \geqslant 0 \tag{8.1.17}$$

Otherwise, if for any z the following takes place

$$\frac{d\alpha_2}{dz}\frac{dC}{d\alpha_2} < 0, \quad \text{or} \quad w_0 \frac{d\alpha_2}{dz}\frac{d^2 J}{d\alpha_2^2} < 0 \tag{8.1.18}$$

then, to find a unique solution, as well as in gas dynamics, surfaces of discontinuity must be introduced on which the function $\alpha_2(t, z)$, together with functions $v_1(t, z)$ and $v_2(t, z)$, experiences a shock.

Equations on the surface of discontinuity in a flow model under discussion comprise the phase-mass conservation equations (see (1.1.62) and (1.4.17)) and the phase slip (drift) conditions before and after the shock. Let D be the shock velocity, and superscripts "$-$" and "$+$" relate to parameters, respectively, before and after the shock. Then

$$\begin{gathered} \alpha_1^-\left(v_1^- - D\right) = \alpha_1^+\left(v_1^+ - D\right) \\ \alpha_2^-\left(v_2^- - D\right) = \alpha_2^+\left(v_2^+ - D\right) \quad \left(\alpha_1^- + \alpha_2^- = 1, \quad \alpha_1^+ + \alpha_2^+ = 1\right) \\ v_2^- = v_1^+ + w_0/\psi_\alpha^-, \quad v_2^+ = v_1^+ + w_0/\psi_\alpha^+ \\ \left(\psi_\alpha^- = \psi_\alpha\left(\alpha_2^-\right), \quad \psi_\alpha^+ = \psi_\alpha\left(\alpha_2^+\right)\right) \end{gathered} \tag{8.1.19}$$

Following elementary transformations, we obtain expressions for the shock velocity in terms of conditions before and after the shock, and also for the phase-velocity jump

$$D = \frac{\alpha_i^+ v_i^+ - \alpha_i^- v_i^-}{\alpha_i^+ - \alpha_i^-} = \frac{W_i^+ - W_i^-}{\alpha_i^+ - \alpha_i^-}$$
$$v_i^+ - v_i^- = (-1)^i w_0 \left(\frac{1 - \alpha_i^+}{\psi_\alpha^+} - \frac{1 - \alpha_i^-}{\psi_\alpha^-} \right) \qquad (i = 1, 2) \tag{8.1.20}$$

As for shock waves in gas, we have $D \to C$ in a drift model for weak discontinuity ($\alpha_2^+ \to \alpha_2^-$). In this case, the drift function $W_2(\alpha_2) = w_0 J(\alpha_2)$ is an analog of adiabat $p(\rho^{-1})$ in a barotropic gas, namely, the shock velocity D is proportional to the tangent of the angle of inclination to axis α_2 of chord interconnecting points on the diagram $W_2(\alpha_2)$ corresponding to states before and behind the wave, and the velocity $C(\alpha_2)$ of the propagation of disturbances is proportional to the slope of the line tangent to the diagram $W_2(\alpha_2)$ at the corresponding point.

Break-up of an arbitrary discontinuity. Let the boundary $z = 0$ divide areas with different volumetric concentrations of phases at $t = 0$. It is necessary to determine a flow governed by Eq. (8.1.5). The boundary, on which there is a discontinuity, may be represented as a narrow transitional zone of thickness δ, across which α_2 undergoes a continuous and monotonic change, assuming values α_2 on different sides of discontinuity on the boundaries of this zone. Considering a limiting operation $\delta \to 0$, we obtain a boundary with a discontinuity in α_2. This representation of discontinuity is helpful for employment of criteria (8.1.17) and (8.1.18).

The formulated problem is considered using an example when on one side of $z = 0$ there is a single-phase liquid ($\alpha_2 = 0$) which is a continuous phase in a two-phase domain where at $t = 0$, we have $\alpha_2 = \alpha_{20}$, and when the dispersed phase is lighter than the continuous ($w_0 > 0$) phase, i.e., it rises. Other situations are solved in a like manner.

Figure 8.1.2*a* depicts the corresponding diagram $W_2(\alpha_2)$ and relation $C(\alpha_2) = dW_2/d\alpha_2$ ($W = 0$). Two specific points are shown on the diagram: point M, where $W_2 = W_{2,\max}$, $C = 0$, and the flex point K, where $d^2W_2/d\alpha_2^2 = 0$, $C = C_{\min}$.

Let us first consider a case when the two-phase domain with a light dispersed phase is situated below ($z < 0$). Then, in the zone of transition near $z = 0$, we have $d\alpha_{20}/dz < 0$. If $\alpha_{20} < \alpha_{2K}$, then, we have $dC/d\alpha_{20} < 0$ everywhere, and the condition (8.1.17) for realization of a continuous solution (8.1.15) with divergent characteristics (waves) is satisfied. In the case under consideration, all characteristics with $\alpha_2 < \alpha_{20}$ ($\alpha_{20} > 0$) originate from the transitional zone of thickness δ near $z = 0$. At $\delta \to 0$, character-

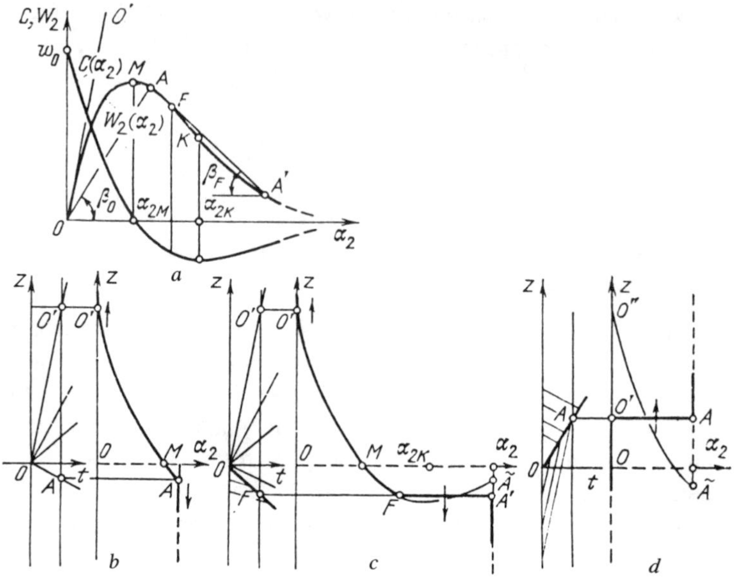

Figure 8.1.2 Schemes of both break-up of an arbitrary discontinuity of phase concentrations (which are predetermined by dashed lines defining $\alpha_{20}(z)$ at $t = 0$) and formation of kinematic (inertia-free) waves (centered waves and shocks) for the case of rising dispersed particles or bubbles ($\rho_2^0 < \rho_1^0$, $w_0 > 0$). For the case of depositing particles or drops ($w_0 < 0$), similar schemes are realized with the opposite motion of waves and phases.

istics originate at $t = 0$ from the point $z = 0$, which leads to a similarity solution of type of a centered wave

$$\frac{z}{t} = w_0 \frac{d}{d\alpha_2} J(\alpha_2) \tag{8.1.21}$$

which is similar to the rarefaction wave in gas. It can be solved in the form $\alpha_2 = \alpha_2(z/w_0 t)$. This solution is illustrated in Fig. 8.1.2*b*, in the form of trajectories of waves in coordinates t, z, and a diagram $\alpha_2(z)$ at fixed point in time $t > 0$. The boundary of a two-phase area is displaced upward with velocity w_0, because of the rise of particles. If $\alpha_{20} < \alpha_{2M}$, the disturbance does not penetrate downward, and the wave with $\alpha_2 = \alpha_{20}$ is displaced upward. If $\alpha_2 > \alpha_{2M}$, disturbance also propagates downward, and with $z = 0$, we have $\alpha_2 = \alpha_{2M}$. The zone in which α_2 changes from 0 to α_{20} (centered rarefaction wave for the dispersed phase and, simultaneously, compression wave for the continuous phase) expands proportional to t.

If $\alpha_{20} > \alpha_{2K}$, then for some of characteristics moving downward (for which $\alpha_2 > \alpha_{2K}$ ($\alpha_2 > \alpha_{20}$)), we have $dC/d\alpha_2 > 0$, and $d\alpha_{20}/dz < 0$, and the condition (8.1.8) is satisfied for the intersection of characteristics and the multiple-value continuous solution of type (8.1.15) and (8.1.21). The corresponding diagram $\alpha_2(z)$ is shown by line $MF\tilde{A}$ in Fig. 8.1.2*c*. Therefore, part of the difference α_2 from α_{20} to $\alpha_2(F)$ must be realized in a shock which overtakes all remaining characteristics directed downward and carrying $\alpha_2 < \alpha_2(F)$. Line $A'F$ corresponds to this fastest shock propagating downward; this line emanates from point A' on diagram $W_2(\alpha_2)$ (which relates to the original state $\alpha_2 = \alpha_{20}$) and tangent to the curve $W_2(\alpha_2)$ at point F (see Fig. 8.1.2*a*). The indicated plotting permits the determination of $\alpha_{2F} = \alpha_2(F)$. Inclination of the rectilinear trajectory of this shock (line OS) to axis t is (in conformity with (8.1.20)) proportional to angle and is determined by the shock velocity equal to $C(\alpha_{2F})$. The remaining characteristics directed downward and forming the centered wave FMO' (on which α_{20} reduces from α_{2F} to α_{2M}) move with lower velocity than that of shock. Note that the shock $A'F$ in Fig. 8.1.3*c* is a rarefaction shock for the dispersed phase and, simultaneously, compression shock for continuous phase.

We will now consider a case when a two-phase domain with rising dispersed phase is located above, i.e., at $t = 0$, the two-phase medium occupies domain $z > 0$, and single-phase medium occupies area $z < 0$. Then, in the zone of transition near $z = 0$, we have $d\alpha_{20}/dz > 0$, and for characteristics with $\alpha_2 < \alpha_{2K}$, $dC/d\alpha_2 < 0$, i.e., the condition (8.1.18) is satisfied for a multiple-value continuous solution of type (8.1.21) associated with the fact that strong disturbances of the initial two-phase state ($\alpha_2 = \alpha_{20}$) propagate upward faster than weaker perturbations. The corresponding multiple-value diagram $\alpha_2(z)$ is shown by line $OO'O''\tilde{A}A$ in Fig. 8.1.2*d*. A single-valued solution $\alpha_2(z, t)$ may be realized with a rarefaction shock for phase AO''; in this shock α_2 varies from α_2 to zero, and its velocity D_A is proportional to angle β_0 between axis α_2 and chord interconnecting point A of the initial state ($\alpha_2 = \alpha_{20}$) and point O of the final single-phase state ($\alpha_2 = 0$). In accordance with the obtained solution, the boundary of two-phase domain together with rising dispersed phase moves upward with velocity D_A.

Note that shocks $A'F$ (Fig. 8.1.2*c*) and AO' (Fig. 8.1.2*d*) are referred to as evolutionary[1] shocks.

For heavy depositing ($w_0 < 0$) dispersed particles, the same schemes are realized, but with opposite motion of waves and phases such that orientation of these motions is maintained the same relative to the direction of velocity w_0. Figure 8.1.3 illustrates the process of deposition of the heavy-particle cloud of final height, which accounts for the fact that flow as a whole is not of a self-similar nature. The following dimensionless independent variables are used

$$\bar{z} = z/L_0, \quad \bar{t} = t|w_0|/L_0 \tag{8.1.22}$$

where L_0 is the initial height of the cloud.

[1]The concept of evolutionary nature (see L. D. Landau and E. M. Lifshits, 1986) is discussed in §8.3 for more complicated kinematic waves.

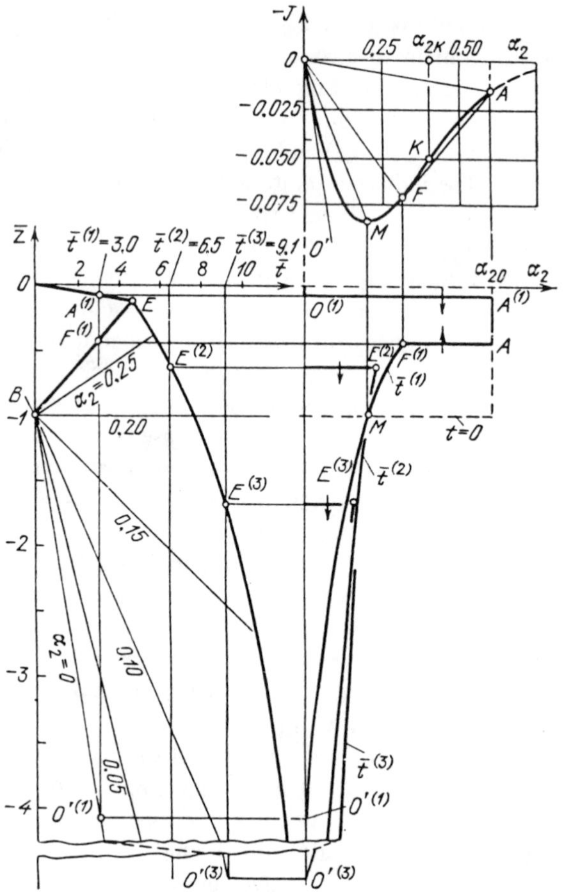

Figure 8.1.3 Schematic for analytical prediction of inertia-free deposition of a finite-height cloud of dispersed particles or drops in liquid or gas ($w_0 < 0$, $\rho_2^0 > \rho^0$). At the initial time $t = 0$, concentration of the dispersed phase in a cloud is $\alpha_{20} = 0.6$, and the cloud occupies a domain $-1 < \bar{z} < 0$ (that corresponds to dashed line $\alpha_{20}(\bar{z})$. The drift-flux function $J = \alpha_2(1 - \alpha_2)^{\max}$ ($m = 3$). The wave diagram is shown in dimensionless (see (8.1.22)) variables $\bar{t}$, $\bar{z}$, and diagram $\alpha_2(\bar{t}, \bar{z})$ at four points in time: $\bar{t} = 0$, $\bar{t}^{(1)} = 3.0$, $\bar{t}^{(2)} = 6.5$, and $\bar{t}^{(3)} = 9.1$. Thin lines on wave diagrams are characteristics (numerical labels on curves indicate values of α_2), and heavy lines correspond to shock trajectories.

The specific feature of the flow is characterized by the meeting of two rarefaction shocks OE and BE of dispersed phase, and also, formation of an accelerating shock $EE^{(2)}E^{(3)}$ directed downward, above which a single-phase medium ($\alpha_2 = 0$) is found. Thus, the motion of the upper boundary of a two-phase domain is described by line $OEE^{(2)}E^{(3)} \ldots$, and that of the lower boundary by line $BO'^{(1)}$. In this case, velocity of the upper boundary monotonically increases, while that of the lower boundary is fixed and equal to w_0.

§8.2 EQUATIONS OF PERCOLATION OF A MULTICOMPONENT MIXTURE OF TWO INCOMPRESSIBLE LIQUIDS

We shall now derive the principal equations of isothermal, nonequilibrium percolation of a multiphase mixture of several incompressible liquids with dissolved active admixtures (affecting rheologic characteristics of phases); by way of example, mixtures of two liquids are considered: hydrocarbons (petroleum, surface-active substances (SAS), salts, and other components dissolved in petroleum), and aqueous (water, SAS, polymers, salts, etc. dissolved in water). These are the most specific mixtures essential from the standpoint of development of oil fields.

Deformations of solid phase of porous systems, compressibility and temperature variation of liquids are very small, and the major effects defining motions of such systems are the nonequilibrium joint motion of several liquid phases, molecular and convective diffusion of components dissolved in phases, absorption of components by solid phase, mass exchange between phases, etc. It is these effects that are given the most consideration in this section in connection with percolation processes essential for practical applications.

Equations reflecting effects of deformation of a porous medium saturated with liquid or gas, phase inertia, and thermal effects are discussed in detail in §4.4 in R. I. Nigmatulin (1978).

Principal parameters of a saturated porous medium. The most important characteristic of a porous medium is its porosity m, which determines the volume of voids occupied by liquid. In the absence of deformation of the solid phase, which will be referred to as a zero-phase ($i = 0$) with a volumetric concentration $\alpha_0 = 1 - m$, porosity is time-independent and is a function only of spatial coordinates: $m = m(r)$. In the case of a homogeneous, nondeformable porous medium, porosity m is independent of coordinates, i.e., m = constant.

Parameters of organic and aqueous liquids are given subscripts p (petroleum) and w (water), respectively. In particular, S_p and S_w are volumetric concentrations of organic and aqueous liquids in voids (pores), or, respectively, saturation with oil and water. Accordingly, the volumetric concentrations of indicated liquids in the entire saturated porous medium equal, respectively, $\alpha_p = mS_p$ and $\alpha_w = mS_w$, such that the following holds

$$\alpha_p + \alpha_w + \alpha_0 = 1 \quad \text{or} \quad S_p + S_w = 1 \qquad (8.2.1)$$

The real porous media are generally micrononhomogeneous, i.e., besides pores of a characteristic, or average, cross-section area, pores of significantly smaller size are present. Extremely fine pores are often unavailable for motion of some phases or components. So, in hydrophilic (see below) porous media, part of the porous space consisting of the finest capillaries is unavailable for percolation of organic liquid because of the action of capillary forces. This part of porous space is occupied by immobile, or so-called bonded, water. An analogous effect has been indicated by H. Dawson and R. Lantz, (1972) during percolation of aqueous solutions of polymers. In connection with the above-indicated properties of porous medium, it is advisable to introduce an adjustment

$$m^{(d)} = m(1 - \Delta m), \quad S_p^{(d)} = \frac{S_p}{1 - \Delta m}, \quad S_w^{(d)} = \frac{S_w - \Delta m}{1 - \Delta m} \tag{8.2.2}$$

where Δm is the volume of voids (pores) occupied by bonded water, which is referred to as *inaccessible volume of voids;* the superscript d corresponds to so-called *dynamic values* of porosity and saturation with oil and water. Superscript d is later omitted, but m and S_i are understood as dynamic values.

The wetting angle θ between the liquid interface and solid surface of a porous medium (Fig. 8.2.1) is an important characteristic defining the phase interaction in a two-phase liquid between each other and porous medium; it, in particular, determines the minimum size of pores available for percolation of a particular phase in a mixture. In the interest of certainty, angle θ is measured from the surface wetted by water. If $\theta < 90°$, the porous medium is wetted by water, or hydrophilic, if $\theta > 90°$, the surface is hydrophobic (water-repellent).

From Fig. 8.2.1*a* and *b*, it can be seen that oil entrapping in hydrophilic media takes place due to surface tension between water and petroleum, which counteracts the hydrodynamic pressure drop. In hydrophobic media (Fig. 8.2.1*c*), petroleum is entrapped in the form of both drop entrapped by capillary forces and films. Thus, the two-phase nature of percolation liquid may lead to an increase of immobile mass compared to a single-phase percolation. In this case, the greater the surface tension between liquids, the stronger the entrapping.

Figure 8.2.2 schematically represents the distribution of velocities of particles of organic and aqueous liquids, which shows that some parts of liquids have zero velocity because of the presence of dead-end pores, ad-

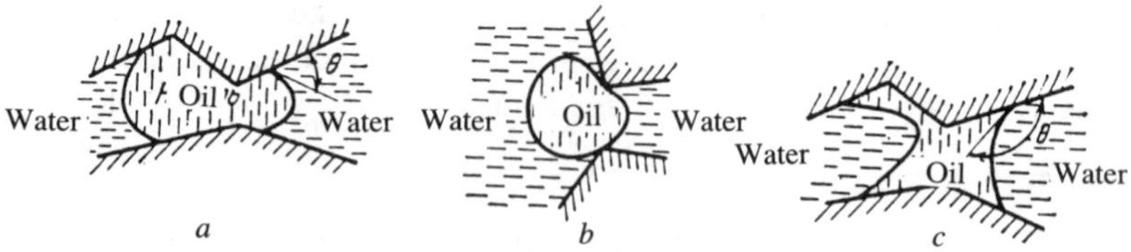

Figure 8.2.1 Wetting angles θ and qualitative pictures of displacement of petroleum out of porous volume and its entrapping due to capillary forces in hydrophilic (*a* and *b*) and hydrophobic (*c*) media.

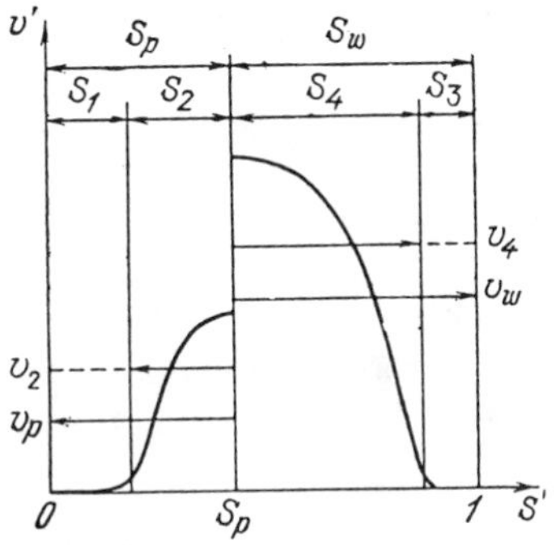

Figure 8.2.2 Distribution of microvelocities during percolation of a two-phase liquid.

herence to the solid skeleton, and effects of the above-mentioned capillary forces on interfaces between liquids.

If in each liquid, mobile and immobile masses (phases) can be singled out, then, in a porous medium saturated with a two-phase liquid, four liquid phases, in addition to immobile solid phase ($i = 0$), may be introduced.

1—organic immobile phase
2—organic mobile phase
3—aqueous immobile phase
4—aqueous mobile phase

where each of these phases has its saturation S_i or volumetric concentration $\alpha_i = mS_i$ and average velocity υ_i ($i = 0, 1, 2, 3, 4$), and $\upsilon_0 = \upsilon_1 = \upsilon_3 = 0$. Thus, at every point for each liquid, there are three kinematic parameters: for hydrocarbon liquid S_1, S_2, and υ_2, and for aqueous liquid S_3, S_4, and υ_4. In this case, both oil saturation and aqueous saturation are determined by formulas

$$S_1 + S_2 = S_p, \quad S_3 + S_4 = S_w, \quad \sum_{i=1}^{4} S_i = 1 \tag{8.2.3}$$

The method of describing percolation by separating liquid into mobile and immobile parts was first proposed by V. Ya. Bulygin (1965), and then developed by G. I. Barenblatt and V. M. Entov (1972). This separation renders an additional detailing of the process compared to the traditional approach when such separation is not envisioned, and there are only two kinematic parameters for each liquid: S_p and υ_p for hydrocarbon liquid, and S_w and υ_w for aqueous liquids.

It should be remembered that separation of moving multicomponent liquid into phases, or combination of components (species) into phases, may be accomplished by various methods depending on a particular issue or on the goals of investigation.

The above-indicated kinematic values determine the volumetric flow rates W_p and W_w of liquids, which can be readily measured

$$W_p = mS_p\upsilon_p = mS_2\upsilon_2, \quad W_w = mS_w\upsilon_w = mS_4\upsilon_4 \tag{8.2.4}$$

The reduced densities of liquid phases (phase mass per unit volume of the medium) can also be easily found

$$\rho_i = \overset{\circ}{\rho}_i \alpha_i, \quad \alpha_0 = 1 - m \quad (i = 0), \quad \alpha_i = mS_i \quad (i = 1, 2, 3, 4) \tag{8.2.5}$$

The above-mentioned phases may consist of several components (species).

Parameters of the k-th component in the i-th phase are given a subscript $i(k)$, and its content is determined by a reduced density $\rho_{i(k)}$, or mass concentration $c_{i(k)}$

$$c_{i(k)} = \frac{\rho_{i(k)}}{\rho_i}, \quad \rho_i = \sum_{k=1}^{N} \rho_{i(k)}, \quad \sum_{k=1}^{N} c_{i(k)} = 1 \tag{8.2.6}$$

where N is a total number of species in a mixture.

Conservation equations for phase mass and volume of mixture. The conservation equations for phase mass are written in a traditional form

$$\frac{\partial \rho_i}{\partial t} + \nabla^l \rho_i v_i^l = \sum_{j=0}^{4} J_{ji}, \qquad i = 0, 1, 2, 3, 4$$
$$J_{ji} = -J_{ij}, \quad v_0 = v_1 = v_3 = 0 \tag{8.2.7}$$

where J_{ji} is an intensity of the mass phase transition from the j-th to the i-th phase. In the absence of chemical reactions, transforming one component to another within a phase, the conservation equations for mass of components are

$$\frac{\partial \rho_{i(k)}}{\partial t} + \nabla^l \rho_{i(k)} \left(v_i^l + w_{i(k)}^l\right) = \sum_{j=0}^{4} J_{ji(k)}$$
$$i = 0, 1, 2, 3, 4; \quad k = 1, \ldots, N$$
$$J_{ji(k)} = -J_{ij(k)}, \quad \sum_{k=1}^{N} J_{ji(k)} = J_{ji}, \quad \sum_{k=1}^{N} \rho_{i(k)} w_{i(k)} = 0 \tag{8.2.8}$$

where w_{ik} is the velocity of the k-th component within the i-th phase.

Apparently, a simplifying assumption of coincidence between component concentrations in both aqueous and organic phases is admissible

$$c_{1(k)} = c_{2(k)} = c_{p(k)}, \quad c_{3(k)} = c_{4(k)} = c_{w(k)} \tag{8.2.9}$$

Then, introducing densities of components in both organic and aqueous liquids, and also, the total velocities of phase transitions

$$\rho_{p(k)} = \rho_{1(k)} + \rho_{2(k)}, \quad \rho_{w(k)} = \rho_{3(k)} + \rho_{4(k)}$$
$$\overset{\circ}{\rho}_1 = \overset{\circ}{\rho}_2 = \overset{\circ}{\rho}_p, \quad \overset{\circ}{\rho}_3 = \overset{\circ}{\rho}_4 = \overset{\circ}{\rho}_w$$
$$J_{0p(k)} = -J_{p0(k)} = J_{01(k)} + J_{02(k)}, \quad J_{0w(k)} = -J_{w0(k)} = J_{03(k)} + J_{04(k)} \tag{8.2.10}$$
$$J_{pw(k)} = -J_{wp(k)} = J_{13(k)} + J_{23(k)} + J_{14(k)} + J_{24(k)}$$

and adding in pairs Eqs. (8.2.8) in supposition that components move only in mobile phases ($w_{1(k)} = w_{3(k)} = 0$) (see the comment following Eq. (8.2.17) below), we obtain

$$\frac{\partial \rho_{p(k)}}{\partial t} + \nabla^l \rho_{2(k)} \left(v_2^l + w_{2(k)}^l\right) = J_{wp(k)} + J_{0p(k)}$$

$$\frac{\partial \rho_{w(k)}}{\partial t} + \nabla^l \rho_{4(k)} \left(v_4^l + w_{4(k)}^l\right) = J_{pw(k)} + J_{0w(k)} \tag{8.2.11}$$

The conservation equations for (8.2.7) for phase mass and components (8.2.11) may be rewritten in a different form in terms of volumetric concentrations α_i of phases, and mass concentrations $c_{p(k)}$ and $c_{w(k)}$ of components

$$\frac{\partial \left(mS_i\overset{\circ}{\rho}_i\right)}{\partial t} + \nabla^l \left(mS_i\overset{\circ}{\rho}_i v_i^l\right) = \sum_{j=0}^{4} J_{ji}, \qquad i = 0, 1, 2, 3, 4$$

$$m\overset{\circ}{\rho}_p S_p\left[\frac{\partial c_{p(k)}}{\partial t} + \left(v_2^l + w_{2(k)}^l\right)\nabla^l c_{p(k)}\right] = J_{0p(k)} + J_{wp(k)} - c_{p(k)}\nabla^l\left(mS_2\overset{\circ}{\rho}_p w_{p(k)}^l\right)$$

$$m\overset{\circ}{\rho}_w S_w\left[\frac{\partial c_{w(k)}}{\partial t} + \left(v_4^l + w_{4(k)}^l\right)\nabla^l c_{w(k)}\right] = J_{0w(k)} + J_{pw(k)} \tag{8.2.12}$$

$$- c_{wk}\nabla^l\left(mS_4\overset{\circ}{\rho}_w w_{w(k)}^l\right), \qquad k = 1, 2, \ldots, N$$

Note that even in the case of incompressible components, the true phase-density ρ_i^0 is generally not constant because of variation of phase composition (with regard to species) due to phase transitions. If component densities are similar, or the predominant part of a phase consists of a single component, and the remaining components are represented only by small admixtures (but, nevertheless, significantly affecting the percolation characteristics, e.g., affecting surface tension), true densities ρ_i^0 of phases may be considered constant

$$\overset{\circ}{\rho}_p = \text{const}, \quad \overset{\circ}{\rho}_w = \text{const} \tag{8.2.13}$$

We adopt this assumption. Then, having divided the first equations (8.2.12) by ρ_i^0, and summing them for liquid phases, we obtain a conservation equation for volume of liquid

$$\sum_{i=1}^{4} \nabla^l m S_i v_i^l = \sum_{i=1}^{4}\sum_{j=0}^{4} \frac{J_{ji}}{\overset{\circ}{\rho}_i} \tag{8.2.14}$$

The addends on the right-hand side of this equation determine the change of volume of liquid due to sorption and desorption on solid phase and due to phase transitions between liquid phases

$$J_{ji}\left(\frac{1}{\overset{\circ}{\rho}_i} - \frac{1}{\overset{\circ}{\rho}_j}\right) \qquad (i, j = 1, 2, 3, 4) \tag{8.2.15}$$

When evaluating these addends, it must be kept in mind that in some cases, phase transitions between solid phase and liquid ($p \rightleftarrows 0$, $w \rightleftarrows 0$), between organic and aqueous phases ($p \rightleftarrows w$), i.e., between phases with different densities ρ_p^0 and ρ_w^0 proceed only due to transitions of small admixtures (salts, surfactants, polymers, etc.) from one phase to another. The indicated transitions of small masses cannot significantly change the volume of liquid. The major portion of mass of percolating liquid in the form of oil and water undergo, in this case, transitions only between mobile and immobile phases of type $1 \rightleftarrows 2$ or $3 \rightleftarrows 4$. These transitions, though occurring in substantial volumes of mass, nevertheless cannot change the volume of liquid (which is in conformity with (8.2.15)), since they take place between phases with identical true densities of phases $\rho_1^0 = \rho_2^0 = \rho_p^0$, and $\rho_3^0 = \rho_4^0 = \rho_w^0$. Thus, the right-hand side of Eq. (8.2.14) may be ignored. Then, we have a conservation equation for volume of liquid in the form

$$\frac{\partial}{\partial x^l}\left(mS_2 v_2^l + mS_4 v_4^l\right) = 0 \tag{8.2.16}$$

Affected by some admixtures (e.g., surfactants soluble in petroleum), petroleum and water may produce a *micellar solution* in the form of mixture between petroleum and very small micelles (microdrops) of water; it is reasonable to relate this solution to an organic liquid (see §8.4 below). Such micellar solution may absorb a significant amount of water, which conforms with transition $w \to p$ between phases of various true densities $\rho_w^0 \neq \rho_p^0$. In this case, the terms of type (8.2.15) pertinent to transition $w \rightleftarrows p$ must be taken into consideration in the equation for the conservation of mass of liquid.

Velocity equations for phases and species (Darcy law of percolation and diffusion law); equation of piezoconductivity for pressure. The volumetric flow rate or velocity of inertia-free motion of liquid phases is governed by the Darcy law

$$W_i^l = mS_i v_i^l = -\frac{S_i k}{\mu_i}\nabla^l p, \text{ or } v_i^l = -\frac{k}{m\mu_i}\nabla^l p \quad (i = 2, 4)$$

$$W_i^l = v_i^l = 0 \qquad (i = 0, 1, 3) \tag{8.2.17}$$

where k is the absolute permeability of a porous medium, μ_i is the viscosity of the i-th phase, p is pressure. The relative permeability of the i-th mobile phase is assumed here to be equal to its volumetric concentration within pores (voids), or its saturation.

The motion of components within phases is described by the diffusion equation which, for the isotropic nature of diffusion, is written in the form

$$\rho_{i(k)} w_{i(k)}^l = -\rho_i \nu_i \nabla^l c_{i(k)}, \quad i = 0, 1, 2, 3, 4, \quad k = 1, 2, \ldots, N \tag{8.2.18}$$

Here, the diffusion coefficients ν_i consist of the molecular component $\nu_{(m)i}$, which depends only on the composition of the i-th phase, and a convective component $\nu_{(v)i}$, which depends on both micrononhomogeneities of the porous medium and velocity v_i of percolation (A. Sheidegger, 1957; V. N. Nikolaevskii et al., 1968; N. N. Verigin, 1969)

$$\nu_{(v)i} \sim L_0 v_i$$

where L_0 is the characteristic length of dispersion of a porous body. For percolation in oil-bearing beds, the molecular diffusion is small, and an evaluation $\nu_i \approx \nu_{(v)i} \approx 10^{-10}$–$10^{-7}\mathrm{m}^2/\mathrm{s}$ takes place.

Using the Darcy law for velocities of mobile phases, we obtain the equation for the pressure field

$$\frac{\partial}{\partial x^l}\left(\Lambda^{(p)}\frac{\partial p}{\partial x^l}\right) = 0, \qquad \Lambda^{(p)} = k\left(\frac{S_2}{\mu_2} + \frac{S_4}{\mu_4}\right) \tag{8.2.19}$$

where $\Lambda^{(p)}$ may be referred to as the coefficient of piezoconductivity of a porous medium saturated by two-phase liquid.

Kinetic equations for mass exchange in percolating liquid: equations of sorption and desorption for admixed species. In intensities J_{12} and J_{34} of mass transfer between mobile and immobile phases are described as linear kinetic equations defined by relaxation times $t_{12} = t_{21}$ and $t_{34} = t_{43}$, equilibrium volumetric concentrations, or saturations (of mobile phases) $S_2^{(e)}$ and $S_4^{(e)}$, that are referred to as *phase permeabilities*

$$J_{21} = \frac{S_2 - S_2^{(e)}}{t_{21}}, \quad J_{43} = \frac{S_4 - S_4^{(e)}}{t_{43}} \tag{8.2.20}$$

The phase permeabilities of petroleum $S_2^{(e)} = K_p$ and water $S_4^{(e)} = K_w$ are determined from stationary tests when the equilibrium saturations $S_2 = S_2^{(e)}$ and $S_4 = S_4^{(e)}$ of mobile phases are realized. In this case, for homogeneous, along the length, specimens, the following magnitudes are measured: volumetric flow rates W_p and W_w of liquids, pressure gradients, and saturation gradients for petroleum S_p and water S_w. Having determined the absolute permeability k for a specimen, and liquid viscosities μ_i, the appropriate equilibrium values $S_2^{(e)}$ and $S_4^{(e)}$ are readily found using the Darcy law (8.2.17). Experiments prove that phase permeabilities depend on oil-saturation $S_p = 1 - S_w$ and concentration c_{ik} of active admixtures in both liquids. In particular, for oil displacement by the micellar solution, the following takes place

$$S_4^{(e)} = K_p\left(S_p, c_{p(3)}, c_{w(4)}\right), \quad S_2^{(e)} = K_w\left(S_w\right)/R\left(S_{0(5)}\right) \tag{8.2.21}$$

where $c_{p(3)}$ is the concentration of petroleum-soluble surfactants in an organic liquid, $c_{w(4)}$ is the concentration of salt in water, and $S_{0(5)}$ is part of the volume of solid phase related to polymers adsorbed from water and acting as a thickening agent; this characteristic determines the resistance-factor R for water accounted for by the indicated polymers (G. G. Vakhitov et al., 1980). The phase permeability for water containing polymer is generally independent of the presence of surface-active admixtures (components $k = 3$ in petroleum).

We shall confine ourselves to a discussion on displacement of petroleum from porous medium (see Fig. 8.2.1*a* and *b*). The Berea sandstone is an example of a well-studied case of such a medium. Data on the phase permeability of sandstones of this type for both water and organic liquids at various surface tensions are reported by M. Gilliland, 1975; M. Talash, 1976; and R. Larson, 1978, and represented in Fig. 8.2.3. It is evident that in the case under consideration, in the process of displacement of petroleum by water ($\Sigma_{pw} \approx 30$ g/s^2), some oil is entrapped in rock; the amount of entrapped oil is determined by a residual dynamic oil-saturation $S_{pr}^{(d)} \approx 0.44$. As was indicated before, the superscript d will thereafter be omitted.

Experimental studies and their analysis showed (R. Healy and R. Reed, 1977; R. Larson, H. Davis, and S. Scriven, 1980) that phase permeabilities and, in particular, residual oil-saturation S_{pr}, depend on the dimensionless capillary number N_Σ, which equals the ratio between hydrodynamic and capillary forces

$$N_\Sigma = \frac{k \operatorname{grad} p}{\Sigma} \sim \frac{a \nabla p}{\Sigma / a} \tag{8.2.22}$$

where $a \sim \sqrt{k}$ is the characteristic radius of pores.

An example of this relationship for the Berea sandstone is depicted in Fig. 8.2.4.

When performing the ordinary flooding of oil-bearing beds, the capillary numbers are considerably smaller than unity ($N_\Sigma \sim 10^{-7}$). The effective entraining of oil entrapped by capillary forces is possible, as follows from Fig.

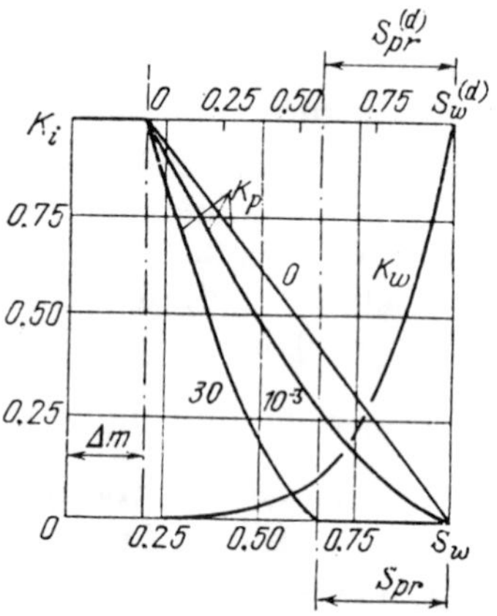

Figure 8.2.3 Phase permeabilities for petroleum ($K_p \equiv S_2^{(e)}$ and water ($K_w = S_4^{(e)}$) in a hydrophilic sandstone of Berea type (R. Larson, 1978) as a function of true S_w or dynamic $S_w^{(d)}$ (see (8.2.2)) water-saturations and concentration of oil-soluble surfactants which reduce the coefficient of surface tension between water and petroleum. Phase permeability K_w for water varies insignificantly (less than by 10–20%) with a change of Σ_{pw} ranging from 10^{-3}–30 g/s^2. The numerical labels on curves $K_p(S_w)$ indicate values of Σ_{pw} in g/s^2. The straight line for $\Sigma_{pw} = 0$ corresponds to perfect mixing and displacement.

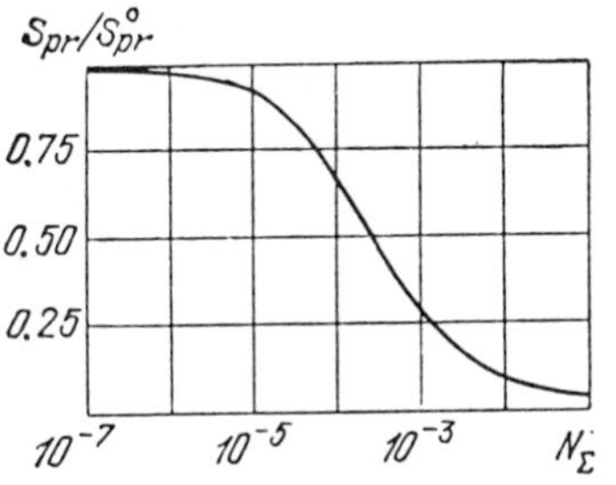

Figure 8.2.4 Dependence of residual oil-saturation of a hydrophilic porous medium (Berea sandstone) on the capillary number N_Σ. Here, S^0_{pr} is the residual oil-saturation at $N_\Sigma = 10^{-7}$.

8.2.4, only at $N_\Sigma > 10^{-4}$. To achieve such magnitude of N_Σ only by an increase of pressure gradient is feasible only on small-scale specimens in laboratory conditions. Displacement of petroleum out of large-scale layers (beds) with such N_Σ is technically realistic only through reducing the coefficient of surface tension on interfaces between injected liquid and liquid in the oil-bearing bed. The latter is achieved by pumping in micellar solutions.

In order to give consideration to additional (besides nonlinearity of relations between phase permeabilities and phase saturation) effects associated with capillary forces and wetting angle θ (see Fig. 8.2.1) which lead to the spread of shocks or zones of sharp variation of phase saturations (see §8.3 below), the inequality of pressures in phases due to surface tension Σ is taken into account in Darcy's equations for a two-phase liquid. Then, the generalization (8.2.17) assumes the form (see R. Collins, 1961; I. A. Charnyi, 1963; G. I. Barenblatt, V. M. Entov, and V. M. Ryzhik, 1984)

$$W_p^l = -\frac{S_2 k}{\mu_p}\nabla^l p_p, \quad W_w^l = -\frac{S_4 k}{\mu_w}\nabla^l p_w, \quad p_p - p_w = \frac{\Sigma\cos\theta}{\sqrt{k/m}} I \qquad (8.2.23)$$

where $I(S_p)$ is a monotonically decreasing Leverett function with the growth of saturation of the wetting phase. For processes being discussed below (see §§8.3 and 8.4), the account of pressure inequality in liquids leads to the spread of shocks in equations for motions; this spread is the more considerable, the larger the difference $p_p - p_w$.

To express the rates of transitions of admixed species between oil, water, and solid phase, we assume, similar to (8.2.20), linear kinetic equations defined by the relaxation times $t_{ij(k)}$ and equilibrium reduced densities $\rho^{(e)}_{i(k)}$ and $\rho^{(e)}_{i0(k)}$ of admixed species in percolating liquid and solid phase (due to adsorption), respectively

$$J_{ji(k)} = \frac{\rho^{(e)}_{i(k)} - \rho_{i(k)}}{t_{ji(k)}}, \quad J_{i0(k)} = \frac{\rho^{(e)}_{i0(k)} - \rho_{0(k)}}{t_{i0(k)}} \qquad (8.2.24)$$

$$\rho_{0(k)} = \overset{\circ}{\rho}_{0(k)}\alpha_{0(k)} = (1-m)\, S_{0(k)}\overset{\circ}{\rho}_{0(k)}, \quad \rho^{(e)}_{i0(k)} = (1-m)\, S^{(e)}_{i0(k)}\overset{\circ}{\rho}_{0(k)}$$

where $\rho^0_{0(k)}$ is the true density of the k-th component in solid phase, $S_{0(k)}$ is part of the volume of solid phase related to components adsorbed from liquid, $S^{(e)}_{i0(k)}$ is a correspondent equilibrium value. Recall that sorption of admixed species does not affect, because of its smallness, the balance of mass and volumes of percolating phases; however, change in small concentrations of active components may significantly affect the percolation characteristics (viscosity and fraction of mobile phases).

The equilibrium concentration of components $S^{(e)}_{i0(k)}$ adsorbed from a single-phase liquid ($S_i = 1$) is usually described by the Langmuir adsorption isotherm (Ya. I. Gerasimov, 1970; S. S. Voyutskii, 1976; A. A. Abramzon, 1981), which is used here also for two-phase liquid as an approximation

$$S^{(e)}_{i0(k)} = c_{i0(k)} \frac{b_{i0(k)} c_{i(k)}}{1 + b_{i0(k)} c_{i(k)}} \tag{8.2.25}$$

where $c_{i(k)}$ is the concentration of the k-th species in the i-th liquid, $c_{i0(k)}$ is the concentration of the adsorbed species in solid phase corresponding to saturation (at large $c_{i(k)} \gg 1/b_{i0(k)}$), $b_{i0(k)}$ is the adsorption activity of the k-th species on contact between the i-th liquid and solid phase. At small $c_{i(k)}$, we have $S^{(e)}_{i0(k)} = \Gamma_{i0(k)} c_{i(k)}$, where $\Gamma_{i0(k)} = c_{i0(k)} b_{i0(k)}$ is referred to as Henry's constant; this case is realized when water is thickened by polyacrylamide ($\Gamma_{w0(9)} = 0.1$–0.4) with small mass concentrations $c_{i0(k)} = 10^{-4}$–10^{-3}, which increases the viscosity of water by a factor of several tens. In micellar solutions (see §8.4 below), situations of sufficiently large concentrations $c_{p(k)} \gg 1/b_{p0(k)}$ of components such as surfactants may be realized.

Viscosities of solutions and microemulsions. Viscosities of percolating liquids may vary, affected by dissolved admixtures (thickening polymer, surfactants, etc.). The so-called micellar "solution" is a microemulsion, the surface tension in which between water and oil affected by surfactants may be reduced by thousand times, and the size of drops, or micelles (internal phase), of water or oil ($a = 10^{-3}$–10^{-2} micrometers) become much smaller than the characteristic size of pores a_0 (generally $a_0 \gtrsim 1$ μm). The micellar solution which is almost a homogeneous mixture of petroleum, water, and active components, may, at certain conditions, "dissolve" both petroleum and water (R. Healy, 1976; M. L. Surguchev, V. A. Shevtsov, and V. V. Surina, 1978). With an increase of concentration of water, the viscosity of micellar solution significantly grows, and with a sufficient concentration of water, it may become by many times higher than viscosity of petroleum and water. With an excess of some critical concentration of water, the growth and agglomeration of micelles cause the solution inversion when water becomes an external phase, and petroleum becomes an internal phase in the form of microdroplets. Further increase of water content reduces the viscosity of the microemulsion, which tends to the viscosity of pure water.

The thickening high-polymers of type of polyacrylamide form with water colloidal solutions with a larger size of microparticles ($a = 10^{-1}$–1 μm) than in micellar solutions.

The discussed rheological relations and equations of kinetics close the general system of equations of nonequilibrium percolation of two-phase liquid with active admixtures.

§8.3 EQUILIBRIUM PERCOLATION OF A TWO-PHASE MULTICOMPONENT MIXTURE

This section presents analytical solutions for some one-dimensional problems based on equations discussed in §8.2 and associated with analysis of the displacement of oil out of porous medium in an equilibrium approximation; under such conditions, times t_{ij} for establishing an equilibrium quasi-stationary distribution of the phase local velocities in pores, which define the phase permeability, are small compared to inherent time of the entire process.

In addition, one more simplification in employed, which is associated with percolation processes characterized by large Peclet numbers ($\mathbf{Pe} = v_0 L/D \gg 1$) when the role of diffusion transfer is small compared to convective transfer of the whole phase.

Percolation of a two-phase mixture of two single-component liquids. In this case, the system of equations (8.2.12) and (8.2.17) may be transformed to

$$m\frac{\partial S_w}{\partial t} + m\frac{\partial (S_w v_w)}{\partial x} = 0$$

$$mS_w v_w = -\frac{kK_w}{\mu_w}\frac{\partial p}{\partial x}, \quad mS_p v_p = -\frac{kK_p}{\mu_p}\frac{\partial p}{\partial x} \tag{8.3.1}$$

$$\frac{\partial}{\partial x} m(S_w v_w + S_p v_p) = 0, \qquad S_w + S_p = 1$$

We introduce a time-dependent magnitude W defining the volumetric flow rate of mixture, but independent of x; values F_w and F_p (known as Buckley-Leverett functions) are also introduced, which determine, respectively, volumes of water and petroleum in the total volume of flow of mixture

$$W(t) = mS_w v_w + mS_p v_p \qquad (\partial W/\partial x = 0)$$
$$F_w = \frac{mS_w v_w}{W}, \quad F_p = \frac{mS_p v_p}{W} \qquad (F_w + F_p = 1) \tag{8.3.2}$$

Then, the system of equations of equilibrium percolation of a two-phase mixture of two incompressible liquids may be represented in the form of two subsystems: equations for evolution of phase-saturation distributions S_i, and for their parts F_i ($i = w, p$) in the flow of mixture

$$m\frac{\partial S_w}{\partial t}+W(t)\frac{\partial F_w}{\partial x}=0$$

$$F_w=\frac{K_w/\mu_w}{K_w/\mu_w+K_p/\mu_p},\quad F_p=\frac{K_p/\mu_p}{K_w/\mu_w+K_p/\mu_p} \tag{8.3.3}$$

$$(K_i=K_i(S_i),\quad i=w,\ p)$$

and equation for the pressure distribution

$$\frac{\partial p}{\partial x}=\frac{W(t)}{k(K_w/\mu_w+K_p/\mu_p)} \tag{8.3.4}$$

Let us now introduce time θ, which has the dimension of length and is determined by volume of injected liquid

$$\vartheta=\frac{1}{m}\int_0^t W\,dt \tag{8.3.5}$$

The quasi-linear first-order partial differential equation (8.3.3) is transformed to a form of (8.1.5) which has been considered in connection with analysis of the inertia-free kinematic waves

$$\frac{\partial S_w}{\partial\vartheta}+F'_w(S_w)\frac{\partial S_w}{\partial x}=0,\qquad F'_w(S_w)=\frac{dF_w}{dS_w}\geqslant 0 \tag{8.3.6}$$

This equation was obtained by S. Buckley and M. Leverett (1942) for percolation of a two-phase liquid. In conformity with this equation, each value of water-saturation S_w is transferred along the characteristic with velocity $F'_w(S_w)$ proportional to the tangent of angle of inclination of a line tangent to curve $F_w(S_w)$, which is referred to as the Buckley-Leverett curve. It is this particular consideration that determines a kinematic wave (see §8.1). Figure 8.3.1*a* schematically depicts both F_w and F'_w. The expression for F'_w is a nonmonotonic function. With evolution of the initial distribution of saturations $S_{w0}(x)$, when the following takes place everywhere (cf. (8.1.17))

$$\frac{\partial^2 F_w}{\partial x^2}=F''_w\frac{dS_{w0}}{\partial x}<0\qquad\left(F''_w=\frac{d^2F_w}{d^2S_w}\right) \tag{8.3.7}$$

the solution is obtained by transferring the initial distribution $S_{w0}(x)$ along characteristics (in a kinematic wave); it is usually written in the form

$$x-x_0=\vartheta F'_w(S_{w0}(x)) \tag{8.3.8}$$

When there are zones where (cf. (8.1.18))

$$\frac{\partial^2 F}{\partial x^2}=F''_w\frac{dS_{w0}}{dx}>0 \tag{8.3.9}$$

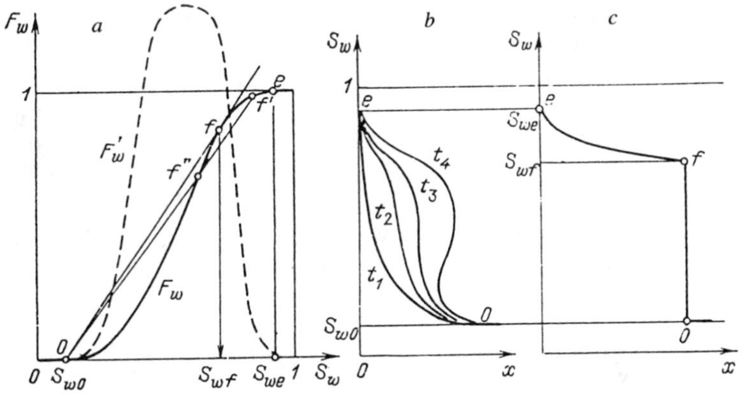

Figure 8.3.1 The Buckley-Leverett function F_w, and its derivative F'_w, defining the characteristic velocity in Eq. (8.3.7) for evolution of saturation S_w during a two-phase equilibrium percolation of incompressible liquids (*a*), and plotting the continuous (*b*) and discontinuous (*c*) solutions in a form of kinematic wave, in which transition from a state *o* to state *e* takes place and the water-saturation varies from S_{wo} to S_{we}.

the wave "overturning" takes place in analogy with the overturning of the Riemann compression wave in gas when large disturbances propagate with a higher velocity compared to weaker perturbations; this causes the ambiguity of $S_w(x)$. The absurdity of this solution leads to the necessity of introducing a surface, or wave (shock), of the parameter discontinuity.

Conditions of the phase mass conservation during transition through a shock permit the interrelation of the saturation values before and behind the shock

$$m\overset{\circ}{\rho}_i S_i^+ (D - v_i^+) = m\overset{\circ}{\rho}_i S_i^- (D - v_i^-) \qquad (i = w,\, p) \tag{8.3.10}$$

where D is the shock velocity, superscripts "$-$" and "$+$" correspond to the parameter values before and behind the shock ($\rho_i^{0-} = \rho_i^{0+}$). By introducing values F_i^- and F_i^+ (see Fig. 8.3.1*a*), we obtain that the shock velocity is proportional to the tangent of the angle of inclination of a line interconnecting points on the diagram $F_i(S_i)$; these points relate to states before and behind the shock

$$\frac{m}{W} D \equiv D^{(W)} = \frac{F_i^+ - F_i^-}{S_i^+ - S_i^-} \tag{8.3.11}$$

The latter equation interrelates three independent variables: S_i^-, S_i^+, and $D^{(w)}$, and it does not render unambiguous values for the shock parameters. To choose a physically realistic solution, it becomes imperative to employ the additional condition of the shock stability and its evolutional nature (B. L. Rozhdestvenskii and N. N. Yanenko, 1979). In accordance with the condition of evolutional nature, the number of characteristics arriving (carrying the initial values) in the section of a shock (the shock trajectory in coordinates x, t) must not be more than the number of required functions. In conformity with the stability condition, the Cauchy-problem solution with a "spread" original discontinuity in the interval of length δ must approach an "unspread" discontinuity at $\delta \to 0$. For a hyperbolic equation of the first order where the number of required functions equals unity, the stability criterion is proven (O. A. Oleinik, 1959): discontinuity is stable if its velocity is not larger than the velocity of any shock from a point S_w^+ behind discontinuity to any point S_w between points behind and before the discontinuity, i.e.

$$D^{(W)} = \frac{F_w(S_w^+) - F_w(S_w^-)}{S_w^+ - S_w^-} \leqslant \frac{F_w(S_w) - F_w(S_w^+)}{S_w - S_w^+}, \qquad S_w \in [S_w^-, S_w^+] \tag{8.3.12}$$

On a plane S_w, F_w, the stability criterion has the following geometrical interpretation: the slope of a chord interconnecting points behind and before discontinuity is not larger than the slope of any chord interconnecting a point behind discontinuity with any point lying on curve $F_w(S_w)$ between points behind and before discontinuity. It can be readily proven that the stability criterion is equivalent to the following two conditions: the slope of the chord is not larger than the slope of a line tangent to curve $F_w(S_w)$ at a point S_w^-, and not smaller than the slope of a line tangent to this curve at a point S_w^+; the chord does not intersect curve $F_w(S_w)$ within a segment $[S_w^-, S_w^+]$. The formulated stability criterion ensures both the existence and uniqueness of a solution of any Cauchy problem for a hyperbolic equation of the first order. It was proven by a method of "low viscosity" for Eq. (8.3.6) (I. M. Gel'fand, 1959) that the latter criterion is a condition for the existence of a discontinuity structure used when the capillary pressure difference between phases is introduced into model (8.3.6).

We shall now display the implication of the requirement of both stability and evolutionality using, by way of example, the Cauchy problem for Eq. (8.3.6) with homogeneous (with respect to coordinates) initial conditions and fixed boundary conditions

$$t = 0:\ S_w(0, x) = S_{w0}, \qquad x = 0:\ S_w(t, 0) = S_{we} \tag{8.3.13}$$

This problem concerned with a break-up of an arbitrary discontinuity fits the process of petroleum displacement out of a oil- and water-saturated bed. It allows a similarity solution $S_w = S_w(\eta)$, $\eta = x/\vartheta$, which is found from a solution of the boundary-value problem for an ordinary differential equation following from (8.3.6)

$$[\eta - F_w'(S_w)]\frac{dS_w}{d\eta} = 0$$

$$\eta = 0:\ S_w = S_{we}, \qquad \eta = \infty:\ S_w = S_{w0} \tag{8.3.14}$$

The obtained equations allows two types of continuous solutions: centered kinematic waves $\eta = F'_w(S_w)$, and segments of uniform saturation S_w = constant. In accordance with (8.3.8), the break-up velocity D is proportional to the value of η, at which discontinuity occurs. The solution describing a wave in which a transition is realized from state o with water-saturation S_{wo} to a state e with water-saturation S_{we} is generated in the following manner (see Fig. 8.3.1*c*). The wave front is a shock moving with a fixed velocity. The shock parameters are determined by a condition of type of Jouguet condition in detonation (see Chapters 3 and 5) specified by point (f) of tangency between a straight line drawn from point o and curve $F_w(S_w)$. The final increase of water-saturation from S_{wf} to S_{we} (transition $f \rightarrow e$) takes place in a continuous wave. The point characterizing a state behind a shock cannot be located to the right of point f, i.e., it cannot be a point of type f', since the shock of type of' is of a nonevolutional nature, or it is unstable because the characteristic disturbances before and behind the shock propagate slower than a shock which (in accordance with relationship $F(S_w)$) may "generate" a more "rapid" shock of lower intensity. The point characterizing a state behind a shock cannot be located to the left of point f either, i.e., it cannot be a point of type f'', since in this case, the second, the third, etc. shocks whose velocity is higher than that of shock of'' may be produced behind the shock of''. After these secondary shocks take over the shock of'', the solution assumes the form ofe with a shock of.

Thus, in conformity with a solution produced for a flow out of the oil-bearing bed, first comes the water-flooded oil ($F_w = 0$), then, flooding shock (jump) up to the magnitude $F_w(S_{wF})$, followed by flooding which monotonically grows to unity.

Percolation of a two-phase mixture of two multicomponent liquids using, by way of example, a mixture of water, oil, surfactants and a polymer. Let us discuss, using an equilibrium approximation, the displacement of oil by water containing active admixtures of type of surfactant affecting the phase-permeabilities of phases. In this case, the conservation equations for mass of water ($k = 1$), oil ($k = 2$), surfactant ($k = 3$) and percolation equations are modified to a form generalizing (8.3.6)

$$\frac{\partial S_w c_{w(1)}}{\partial \vartheta} + \frac{\partial F_w c_{w(1)}}{\partial x} = 0 \qquad \left(\vartheta = \frac{1}{m}\int_0^t W dt\right)$$

$$\frac{\partial \left(S_w c_{w(3)} + a_{(3)}\right)}{\partial \vartheta} + \frac{\partial F_w c_{w(3)}}{\partial x} = 0, \qquad c_{w(1)} + c_{w(3)} = 1 \tag{8.3.15}$$

$$F_w = F_w\left(S_w, c_{w(3)}\right), \quad m a_{(3)} = \rho_{0(3)} = (1 - m)\, S_{0(3)} \overset{\circ}{\rho}_{0(3)}$$

$$a_{(3)} = a_{(3)}\left(S_w, c_{w(3)}\right)$$

The conservation equations for mass of oil (phase p) and mass of the third component (surfactant) across the shock, which generalize (8.3.10), are written in the form

$$\begin{aligned} &\left(D - v_p^+\right) S_p^+ = \left(D - v_p^-\right) S_p^- \\ &\left(D - v_w^+\right) S_w^+ c_{w(3)}^+ + D a_{(3)}^+ = \left(D - v_w^-\right) S_w^- c_w^- + D a_{(3)}^- \end{aligned} \tag{8.3.16}$$

From these equations, it follows that two conditions must be simultaneously satisfied across the shock

$$\begin{aligned} &D^{(W)} = \frac{F_w^+ - F_w^-}{S_w^+ - S_w^-} \\ &D^{(W)} = \frac{F_w^+ c_w^+ - F_w^- c_w^-}{S_w^+ c_{w(3)}^+ - S_w^- c_{w(3)}^- + a_{(3)}^+ - a_{(3)}^-} \equiv \frac{F_w^\pm}{S_w^\pm + \left(a_3^+ - a_{(3)}^-\right)/\left(c_{w(3)}^+ - c_{w(3)}^-\right)} \end{aligned} \tag{8.3.17}$$

As in the case of a single equation (8.3.6), in order to select a unique discontinuity solution of the Cauchy problem for a system (8.3.15), the following stability-conditions are imposed upon discontinuity:

1) the total number of characteristics whose velocity is not higher than that of discontinuity in the zone before break-up, plus the number of characteristics whose velocity is not lower than that of discontinuity in the zone behind break-up equals three of four inequalities

$$\begin{aligned} &D^{(W)} \geqslant F'_w\left(S_w^-, c_{w(3)}^-\right), \quad D^{(W)} \geqslant \frac{F_w^-}{S_w^- + \left(\partial a_3/\partial c_{w(3)}\right)^-} \\ &D^{(W)} \leqslant F'_w\left(S_w^+, c_{w3}^+\right), \quad D^{(W)} \leqslant \frac{F_w^+}{S_w^+ + \left(\partial a_{(3)}/\partial c_{w(3)}\right)^+} \end{aligned} \tag{8.3.18}$$

three are simultaneously satisfied (see discussion of (8.3.12));

2) a segment of a straight line interconnecting points behind discontinuity (+) and before it (−) in plane ($c_{w(3)}$, $a_{(3)}$) does not intersect the sorption isotherm $a_{(3)} = a_{(3)}(c_{w(3)})$ at points other than (+) and (−).

Small perturbations propagate in hyperbolic systems along characteristics. Therefore, from the condition 1) it follows that with a small perturbation imposed upon discontinuity and with the linearization of a system (8.3.15), the characteristics on this discontinuity render as many conditions as required for an unambiguous determination of the motion of discontinuity. The magnitude of disturbance of the discontinuity trajectory in this case approaches zero, i.e., the discontinuity is stable with regard to small perturbations. The condition 1) is referred to as the condition of evolutionality. It provides an unambiguous solution of a linearized problem concerning the interaction between discontinuity and a small disturbance.

When considering a nonlinear problem on interaction between discontinuity and small disturbance, it turns out that with the condition 2) unsatisfied, the disturbance does not have sufficient time to reach the line of discontinuity: the perturbation wave-front overturns, generating new discontinuities whose intensities do not approach zero. Thus, the initial discontinuity is not stable.

Let us now analyze a similarity solution for a system of equations (8.3.15) which depends only on a single variable

$$S_w = S_w(\eta), \quad c_{w(k)} = c_{w(k)}(\eta), \qquad \eta = x/\vartheta \tag{8.3.19}$$

In this case, we have

$$\begin{aligned} &\left(\eta - \frac{\partial F_w}{\partial S_w}\right)\frac{dS_w}{d\eta} - \frac{\partial F_w}{\partial c_{w(3)}}\frac{dc_{w(3)}}{d\eta} = 0 \\ &\frac{\partial a_{(3)}}{\partial S_w}\eta\frac{dS_w}{d\eta} + \left[\eta\left(S_w + \frac{\partial a_{(3)}}{\partial c_{w(3)}}\right) - F_w\right]\frac{dc_{w(3)}}{d\eta} = 0 \end{aligned} \tag{8.3.20}$$

It may often be supposed that the concentration of adsorbed surfactant, determined by the sorption isotherm $a_{(3)}(S_w, c_{w(3)})$, is independent of water-saturation (that simplifies computations)

$$\frac{\partial a_{(3)}}{\partial S_w} = 0 \tag{8.3.21}$$

A linear homogeneous system (8.3.20) has a nontrivial solution if its determinant equals zero. This condition defines two families of characteristics

$$1)\ \eta = \frac{\partial F_w}{\partial S_w}, \quad 2)\ \eta = \frac{F_w}{S_w + \partial a_{(3)}/\partial c_{w(3)}} \tag{8.3.22}$$

Each value of η in coordinates ϑ, x corresponds to a rectilinear characteristic $x = \eta\vartheta$. The first characteristic is known as the S-characteristic, or S-wave, and along it we have $dS_w = 0$. The second characteristic is known as the c-characteristic, or c-wave, and along it we have $dc_{w(3)} = 0$.

Let us find a similarity wave configuration propagating through a medium with a homogeneous initial state determined by point o (S_{w0}, $c_{w(3)0} \approx 0$) when an aqueous solution of surfactant in state e ($S_{we} = 1$, $c_{w(3)e}$) is moving behind the wave (Fig. 8.3.2). According to the above-formulated condition for the shock stability, and to the condition of the solution similarity, the shocks must propagate in conformity with an analog of the Jouguet condition, namely, along one of characteristics (8.3.22) defined by values of parameters behind the shock

$$\frac{\partial F_w}{\partial S_w} = \frac{F_w^+ - F_w^-}{S_w^+ - S_w^-}, \qquad \frac{F_w^+}{S_w^+ + \left(\dfrac{\partial a_{(3)}}{\partial c_{w(3)}}\right)^+} = \frac{F_w^-}{S_w^+ + \dfrac{a_{(3)}^+ - a_{(3)}^-}{c_{w(3)}^+ - c_{w(3)}^-}} \tag{8.3.23}$$

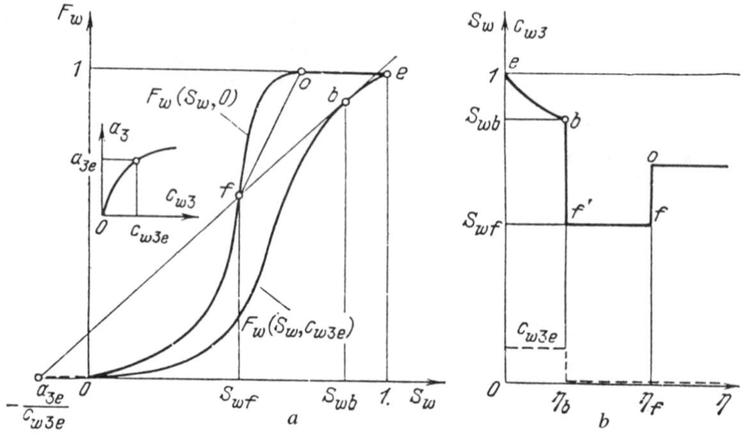

8.3.2 The similarity solution for an equilibrium model for displacement of residual oil (state *o*) by an aqueous solution of surfactant (state *e*). Here as well as in Figs. 8.3.3, 8.3.4, and 8.4.3, the parentheses in subscripts related to the component number are omitted ($a_3 = a_{(3)}$, $c_{w3} = c_{w(3)}$).

If the sorption isotherm for surfactant is convex ($\partial^2 a_{(3)}/\partial c_{w(3)}^2 < 0$), then from the latter expression it follows that the entire change of concentration of surfactant from 0 to $c_{w(3)e}$ is realized in a shock propagating along the fastest *c*-characteristic which is simultaneously an *S*-characteristic

$$\eta_b = \left(\frac{\partial F_w}{\partial S_w}\right)^+ = \frac{F_w^+ - F_w^-}{S_w^+ - S_w^-} = \frac{F_w^+}{S_w^+ + a_{(3)}^+/c_{w(3)}^+} \tag{8.3.24}$$

A straight line tangent to line $F_w(S_w, c_{w(3)})$ at $c_{w(3)} = c_{w(3)e}$ and passing through point $(-a_{(3)e}/c_{w(3)e}, 0)$ in the plane S_w, F_w (see Fig. 8.3.2*a*) corresponds to this shock. Under these conditions, the equalities (8.3.24) are satisfied. There is a domain $\eta < \eta_b$ behind the shock where the water-saturation monotonically grows as fast as we approach the location where the aqueous solution of surfactant is pumped in (line *be* in Fig. 8.3.2). The corresponding solution is described by the first equation (8.3.22); the concentration of surfactant in this domain being fixed and equal to $c_{w(3)}$, i.e., the concentration of surfactant in an injected aqueous solution. A solution for a domain of a two-phase, single-component ($c_{w(3)} = 0$) flow has already been examined above. The form of a solution at arbitrary point in time with two shocks *of* and *f'b* is shown in Fig. 8.3.2*b*.

A similarity solution of a problem concerned with oil displacement by a micellar solution with an external organic phase may be produced in an analogous manner; the injected reagent—micellar solution—consists of petroleum ($k = 1$) with dissolved surfactant ($k = 3$) and water ($k = 2$) in the form of micelles. The conservation equations for mass of surfactant and water in a flow of liquid are written in the form

$$\frac{\partial (S_p c_{p(3)} + a_{(3)})}{\partial \vartheta} + \frac{\partial (F_p c_{p(3)})}{\partial x} = 0$$

$$\frac{\partial (S_p c_{p(2)} + S_w c_{w(2)})}{\partial \vartheta} + \frac{\partial (F_p c_{p(2)} + F_w c_{w(2)})}{\partial x} = 0 \qquad (8.3.25)$$

$$S_w = 1 - S_p, \quad F_w = 1 - F_p, \quad c_{w(2)} = 1$$

It is supposed here that aqueous phase is a single-component liquid, and organic phase is a three-component system. In the case of an isothermal process and insignificant effects of pressure upon the component distribution over phases, the Gibbs phase rule shows that in equilibrium (this section is concerned with this particular situation), the number of degrees of freedom of a system equals unity, i.e., the concentration of only one species is independent. We assume

$$c_{p(2)} = B c_{p(3)}, \qquad B = \text{const} \sim 1$$

At small sorption of surfactant ($a_{(3)}B \ll S_p, S_w$), the system of equations (8.3.25) may be reduced to

$$\frac{\partial S_p}{\partial \vartheta} + \frac{\partial F_p}{\partial x} = 0, \quad \frac{\partial (S_p c_{p(3)} + a_{(3)})}{\partial \vartheta} + \frac{\partial F_p c_{p(3)}}{\partial x} = 0 \qquad (8.3.26)$$

This system of equations agrees exactly with a system (8.3.15); therefore, the solution is sought similar to what is shown in Fig. 8.3.2 but in the plane S_p, F_p. A solution for displacement of the residual oil (the homogeneous initial conditions at $S_{p0} < S_{pr}$, $c_{p3} = 0$) by a micellar solution is plotted in Fig. 8.3.3.

We will now also examine a similarity solution characteristic for an injection of a reagent (e.g., thickening reagent) affecting the phase viscosity. Changing the phase-viscosity ratio also affects the Buckley-Leverett function. The problem concerned with displacement of a micellar solution by a buffer liquid (which is an aqueous solution of a polymer) is formulated analogous to (8.3.15), but the conservation equation for the component mass is written, as distinct from (8.3.15), not for surfactant ($k = 3$) but for polymer ($k = 4$). The solution is depicted in Fig. 8.3.4 and has a form similar to that illustrated in Fig. 8.3.2.

A more detailed analysis of similarity solutions is reviewed in V. M. Entov (1981).

Stability of contact boundaries. Analysis of the behavior of two-dimensional weak disturbances on a flat contact boundary between two single-phase percolating liquids (P. Saffman and G. Taylor, 1958; R. Perrine, 1961) shows that these disturbances do not grow (i.e., the flat contact boundary is stable) if the liquid of higher viscosity is in the higher-pressure domain,

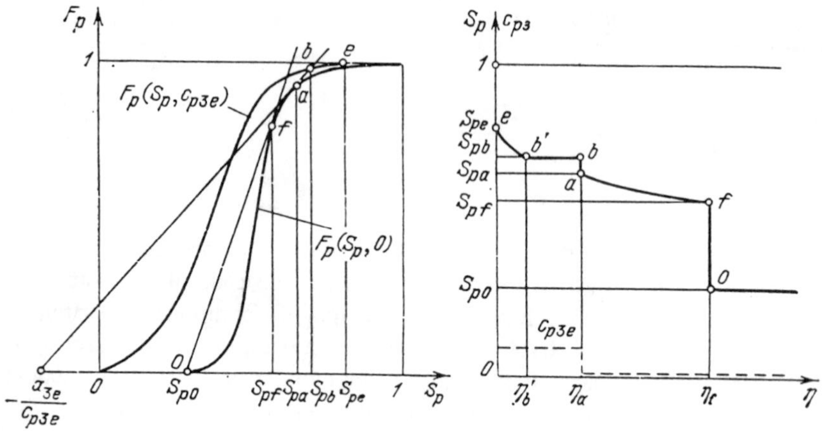

Figure 8.3.3 A similarity solution for an equilibrium model of displacement of residual oil (state *o*) by micellar solution (state *e*).

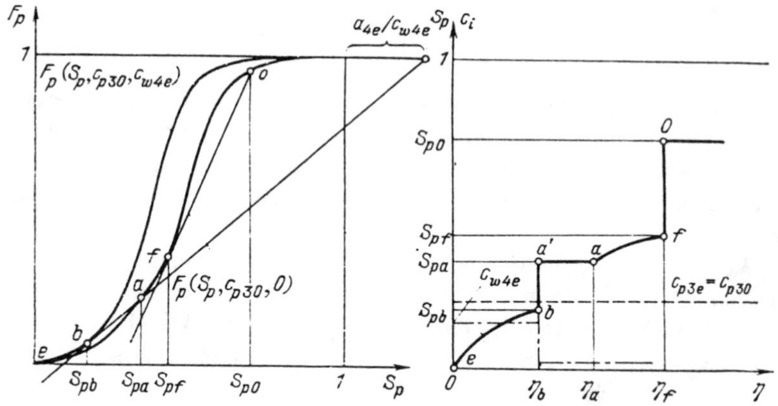

Figure 8.3.4 A similarity solution for an equilibrium model of the displacement of a micellar solution (state *o*) by a thickened aqueous solution of a polymer (state *e*). See also the caption for Fig. 8.3.2.

i.e., the higher-viscosity liquid displaces the lower-viscosity liquid. In cases when lower-viscosity liquid displaces the higher-viscosity liquid, the flat contact-boundary disturbances grow and can destroy it. The numerical investigation of the development of nonlinear two-dimensional disturbances of the contact boundary is presented in G. P. Tsibul'skii (1975), P. V. Indel'man et al., (1979), V. M. Entov and V. B. Taranchuk (1979).

§8.4 MATHEMATICAL MODELLING OF THE PROCESS OF FLOODING OF AN OIL-BEARING BED BY A MICELLAR-POLYMER MIXTURE

The method of a micellar-polymer displacement (recovery) of oil remaining after the oil has been displaced by regular water consists of a successive pumping into the bearing bed of fairly small volumes of a micellar solution (MS) and buffer liquid (BL) which is a solution of a thickening polymer in water (thickened water). "Propulsion" of these small volumes[2] (about 10% of the total volume of voids in oil-bearing bed) of MS and BL along the bed is usually accomplished by water (Fig. 8.4.1).

Giving consideration to the described process, we shall confine ourselves to a case when the micellar solution is used with external organic phase (see §8.2), and a low-concentration solution of high polymer (of type of polyacrylamide) in water is used as a buffer liquid. The role of a micellar solution consists of "dissolving" immobile petroleum on account of a significant reduction of surface tension, thereby making it mobile. The buffer liquid, because of its high viscosity, protects the advanced MS from being broken through by water (which has lower viscosity) which presses through the system of these solutions.

The organic liquid is divided into mobile and immobile phases, each consisting of three components: petroleum, water, and surface-active admixtures (surfactant, alcohol, etc.) which produce the micellar solution from the hydrocarbon liquid. The aqueous phases (both mobile and immobile) contain three components, too: water, salt, and thickening polymer (see Table 8.4.1).

In the interest of simplifying the calculations, we assume that true densities of both ($i = p, w$) liquid phases are identical, and that the diffusion motion of components may be ignored

$$\overset{\circ}{\rho}_w = \overset{\circ}{\rho}_p = \text{const}; \qquad w_{i(k)} \equiv 0 \tag{8.4.1}$$

Then, the system of equations represented in §8.1 may be written in the form

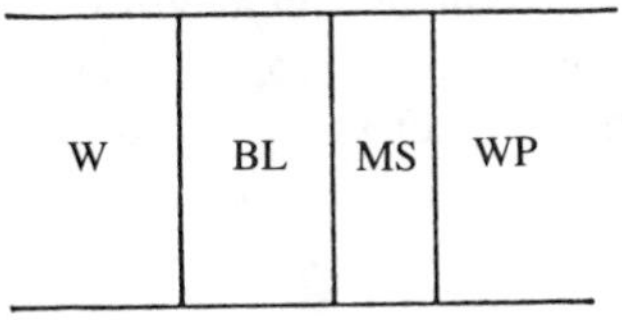

Figure 8.4.1 One-dimensional schematic for oil displacement by means of micellar-polymeric solutions; WP is the original water-flooded petroleum; MS is the micellar solution; BL is the buffer liquid (water + thickening agent); W is the water used for "propulsion" of MS and BL.

[2]Small volumes of these ingredients is accounted for by the high cost of both MS and BL.

Table 8.4.1.

Phases and their saturation S_i ($i = p, w, 1, 2, 3, 4$)	Components and their number $k = 1, 2, 3, 4, 5$	Mass concentration of components $c_{i(k)}$ ($i = p, w; k = 1, 2, 3, 4, 5$)
Organic phase ($i = p$; $S_p = S_1 + S_2$)	Petroleum ($k = 1$)	$c_{p(1)}$
immobile ($i = 1$); S_1	Water ($k = 2$)	$c_{p(2)}$
mobile ($i = 2$); S_2	Surface-active admixtures ($k = 3$)	$c_{p(3)} \ll 1$
Aqueous phase ($i = w$; $S_w = S_3 + S_4$)	Water ($k = 2$)	$c_{w(2)}$
immobile ($i = 3$); S_3	Thickening polymer ($k = 4$)	$c_{w(4)} \ll 1$
mobile ($i = 4$); S_4	Salt ($k = 5$)	$c_{w(5)}$

$$\frac{\partial S_p}{\partial \bar{t}} + \frac{\partial F_p}{\partial \bar{x}} = -\bar{J}_{pw(2)}, \quad \frac{\partial S_w}{\partial \bar{t}} + \frac{\partial F_w}{\partial \bar{x}} = \bar{J}_{pw(2)}, \quad \frac{\partial S_p c_{p(1)}}{\partial \bar{t}} + \frac{\partial F_p c_{p(1)}}{\partial \bar{x}} = 0$$

$$\frac{\partial S_p c_{p(2)}}{\partial \bar{t}} + \frac{\partial F_p c_{p(2)}}{\partial \bar{x}} = -\bar{J}_{pw(2)}, \quad \frac{\partial S_p c_{p(3)}}{\partial \bar{t}} + \frac{\partial F_p c_{p(3)}}{\partial \bar{x}} = -\bar{J}_{p0(3)}$$

$$\frac{\partial S_w c_{w(2)}}{\partial \bar{t}} + \frac{\partial F_w c_{w(2)}}{\partial \bar{x}} = \bar{J}_{pw(2)}, \quad \frac{\partial S_w c_{w(4)}}{\partial \bar{t}} + \frac{\partial S_w c_{w(4)}}{\partial \bar{x}} = -\bar{J}_{w0(4)} \tag{8.4.2}$$

$$\frac{\partial S_w c_{w(5)}}{\partial \bar{t}} + \frac{\partial F_w c_{w(5)}}{\partial \bar{x}} = \bar{J}_{0w(5)}$$

$$\frac{\partial S_{0(3)}}{\partial \bar{t}} = \bar{J}_{p0(3)}, \quad \frac{\partial S_{0(5)}}{\partial \bar{t}} = -\bar{J}_{0w(5)}, \quad \bar{J}_{pw(2)} \gg \bar{J}_{p0(3)}, \ \bar{J}_{w0(k)}$$

$$F_p = \frac{S_2/\mu_p}{S_2/\mu_p + S_4/\mu_w}, \quad F_w = \frac{S_4/\mu_w}{S_2/\mu_p + S_4/\mu_w} \qquad (F_p + F_w = 1)$$

Here, there are employed: dimensionless spatial coordinate $\bar{x}$ related to the characteristic oil-bearing bed length, dimensionless time $\bar{t}$ proportional to the volume of injected liquid, dimensionless intensities of phase transitions $J_{ij(k)}$ related to volumetric flow rate of liquid

$$\bar{x} = \frac{x}{L}, \quad \bar{t} = \frac{V(t)}{V_m} \equiv \frac{\vartheta}{L}, \quad \bar{J}_{ij(k)} = \frac{J_{ij(k)} L}{\overset{\circ}{\rho}_k W(t)}$$

$$\left(W(t) = m S_w v_w + m S_p v_p, \ V(t) = \int_0^t W(t')\, dt', \ V_m = Lm \right) \tag{8.4.3}$$

The system of equations (8.4.2) takes into consideration: transition of water between organic and aqueous phases ($J_{wp(2)}$), adsorption and desorption of active admixtures soluble in oil ($J_{p0(3)}$) and those of thickener ($J_{w0(4)}$), and dissolving of salt contained in oil-bearing bed by water ($J_{0w(5)}$); all these are prescribed in accordance with (8.2.24). In this case, separation of water from the micellar solution takes place when the organic liquid is diluted by active admixtures below some critical concentration $c^*_{(3)}$

$$J_{pw(2)} = \begin{cases} 0, & c_{p(3)} > c^*_{(3)}, \\ c_{p(2)}/t_{pw}, & c_{p(3)} \leqslant c^*_{(3)} \end{cases} \tag{8.4.4}$$

Dependence of phase permeability K_p of organic liquid on oil-saturation S_p and concentration $c_{p(3)}$ of active admixture dissolved in petroleum, and also that of phase permeability K_w of water on water-saturation S_w, depicted in Fig. 8.2.3, may be approximated in the form

$$K_p(S_p, 0) = \begin{cases} \left(\dfrac{S_p - S_{pr}}{1 - S_{pr}}\right)^{\gamma}, & S_p \geqslant S_{pr} \\ 0, & S_p \leqslant S_{pr} \end{cases}$$

$$K_p(S_p, c_{p(3)}) = S_p^{\varkappa}, \quad c_{p(3)} \geqslant c^*_{(3)}, \quad K_w = S_w^{\beta} \tag{8.4.5}$$

where S_p is the dynamic oil-saturation $S_p^{(d)}$ (see (8.2.2)); S_{pr}, γ, $\varkappa$, $c^*_{(3)}$, and β are parameters of the porous medium ($\varkappa$ and c^*_3 depend also on properties of active admixtures). In the intermediate range of concentrations of active component ($k = 3$), i.e., when $0 < c_{p(3)} < c^*_{(3)}$, a linear interpolation with respect to $c_{p(3)}$ may be used between $K_p(S_p, 0)$ and $K_p(S_{p(3)}, c^*_{(3)})$.

As in the case of relationship between K_p and $c_{p(3)}$, we shall use linear approximations for viscosities of organic (as a function of water-content) and aqueous (as a function of concentration of the thickening polymer) liquids

$$\mu_p = \mu_{p(1)} + \left(\overset{\circ}{\mu}_p - \mu_{p(1)}\right)\frac{c_{p(2)}}{\overset{\circ}{c}_{p(2)}}, \quad \mu_w = \mu_{w(2)} + \left(\overset{\circ}{\mu}_w - \mu_{w(2)}\right)\frac{c_{w(5)}}{\overset{\circ}{c}_{w(5)}} \tag{8.4.6}$$

where $\mu_{p(1)}$ and $\mu_{w(2)}$ are viscosities of pure oil ($c_{p(2)} = 0$) and water ($c_{w(5)} = 0$), respectively; μ_w is the viscosity of MS with concentration of water in it equal to $c^0_{p(2)}$; μ^0_w is the viscosity of "thickened" water with concentration of the thickener in it equal $c^0_{w(5)}$.

Parameters characterizing phase permeability and viscosities of phases in variants analyzed below were assumed as follows

$$\gamma = 1.5, \quad \varkappa = 1.0, \quad \beta = 3.75, \quad S_{pr} = 0.44, \quad \mu_{p1} = 7.0 \text{ g/(cm)(s)}(10^2)$$

$$\overset{\circ}{\mu}_p = 25.2 \text{ g/(cm)(s)}(10^2), \quad \mu_{w2} = 1.0 \text{ g/(cm)(s)}(10^2), \quad \overset{\circ}{\mu}_w = 30 \text{ g/(cm)(s)}$$

$$(10^2)\ (1 \text{ g/(cm)(s)}(10^2) = 10^{-3} \text{ Pa}\cdot\text{s}) \tag{8.4.7}$$

Results only for an equilbrium approximation in describing the variation of phase permeabilities, or separation of percolating liquids to mobile and immobile phase ($t_{ij} = 0$) are outlined. This approximation holds good if $\bar{t}_{ij} \ll 1$. Henry's law constants for adsorption of surface-active admixtures ($k = 3$) and thickening polymer ($k = 4$) were varied in the range of actual values (W. Gogarty, 1968; A. M. Polishchuk and E. M. Surkova, 1979).

The oil-bearing bed in its original state contained water and oil remaining after flooding ($S_p(0, x) = S_{pr} = 0.44$). Both solutions MS and BL have been successively fed at the inlet

$$0<\bar{t}<\bar{t}_*:\ S_p(\bar{t}, 0) = 1, \quad c_{p(k)} = \overset{\circ}{c}_{p(k)} \quad (k = 1, 2, 3)$$
$$c_{w(k)} = 0 \quad (k = 2, 4, 5)$$
$$\bar{t}_*<\bar{t}<\bar{t}_{**}:\ S_w(\bar{t}, 0) = 1, \quad c_{w(k)} = \overset{\circ}{c}_{w(k)} \quad (k = 2, 4, 5) \tag{8.4.8}$$
$$c_{p(k)} = 0 \quad (k = 1, 2, 3)$$

The problem was solved using a numerical approach based on an explicit model of the first order of accuracy with orientation with respect to characteristics. The number of cells was chosen so that not less than 20 difference cells were used for the indicated solutions.

Let us first examine (based on numerical computations) the wave picture of displacement of residual oil by small amounts of solutions. The process may be divided into four stages schematically illustrated in Fig. 8.4.2. Pumping

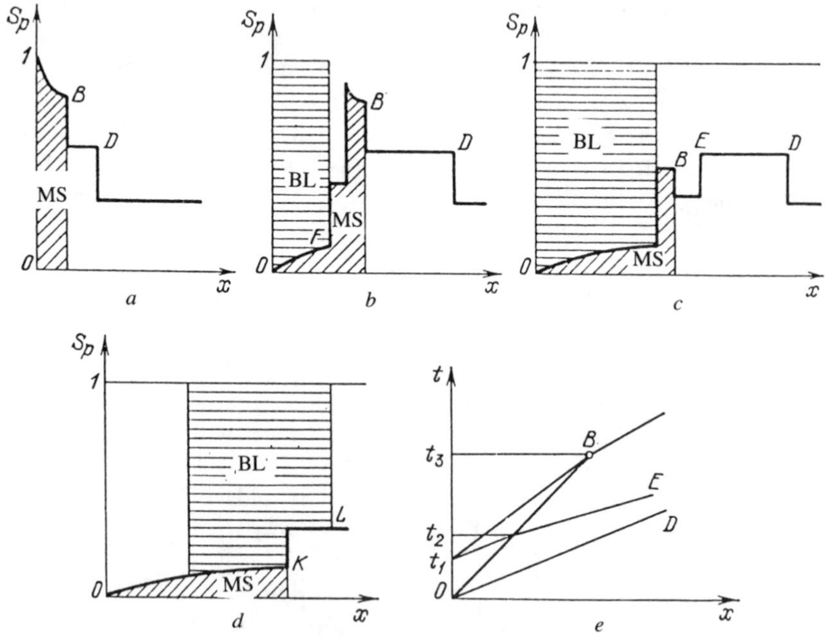

Figure 8.4.2 Evolution of kinematic waves or that of oil-saturation distribution (solid lines), zones of micellar liquid (oblique hatching) and thickener (horizontal hatching) in the oil-displacement process in accordance with Fig. 8.4.1.

MS into the oil-bearing bed is the first stage of the process (Fig. 8.4.2*a*). Distribution of oil-saturation and active admixtures along the oil-bearing bed on this stage is described by the similarity solution depicted in Fig. 8.3.3. Pumping-in the BL is the second stage of the process; BL is being fed until the water wave formed before the BL solution propagates in the MS-solution zone (8.4.2*b*). Distribution of oil-saturation in the zone where BL arrived is described by the similarity solution shown in Fig. 8.3.4. As follows from this solution, part of MS is retained in the BL-zone, which leads to the subsequent breaking of the MS-solution layer (edging).

The third stage is characterized by intrusion of water from the MS zone into the petroleum zone (Fig. 8.4.2*c*), thereby reducing the amount of oil saturation. As a consequence, a domain of combined motion of water and petroleum is formed, which is referred to as the water-oil wave.

On the fourth stage, the BL-layer (edging) front leaves behind the MS layer, virtually breaking it, and displaces the water-oil wave, thereby reducing the oil-saturation in front of the MS edging up to the residual oil saturation S_{pr} (Fig. 8.4.2*d*). This stage takes place only in the case of sufficiently thin layers. In this case, part of the petroleum and MS is retained in a porous medium, which reduces the effectiveness of the process.

Figure 8.4.2*e* shows trajectories of nonlinear waves on *xt* diagram illustrating development of the above-described process. Time $\bar{t} = 0$ corresponds to the beginning of injection of the MS solution, $\bar{t} = \bar{t}_1$ is the beginning of injection of BL. After the volume of liquid equal to $\bar{t}_2 V_m$ is pumped-in (V_m is the volume of voids in the oil-bearing bed), formation of the water-petroleum wave takes place. Time $\bar{t}_3$ corresponds to breaking the MS edging.

Figure 8.4.3 represents results of numerical computations[3] illustrating the above-described wave process in the form of diagrams of oil-saturation and concentrations of components at four points in time. It should be remembered that because of "numerical diffusion," which is characteristic for difference methods of computations, the saturation shocks and component concentrations experience spread as a function of time as $\bar{t}^{1/2}$. In order to avoid such spreading, the discontinuity surfaces must be specified using the shock equations. From Fig. 8.4.3, it is evident that for prescribed conditions, the MS edging of size equal to 4% of the volume of voids breaks during its motion through the oil-bearing bed, causing rupture of buffer-liquid BL and its penetration into the zone of water-oil wave.

Variation of phase flow rates for water, petroleum, and components at the oil-bearing bed outlet as functions of time are generally measured in experiments on a core sample. It is for these particular values that the numerical and experimental results are compared in Figs. 8.4.4 and 8.4.5.

Figure 8.4.4 relates to a thick MS layer ($V_{MS} = 0.8V_m$) when the process

[3]Some analytical solutions for the displacement of liquids by edgings can be found in P. G. Bedrikovetskii et al., (1982).

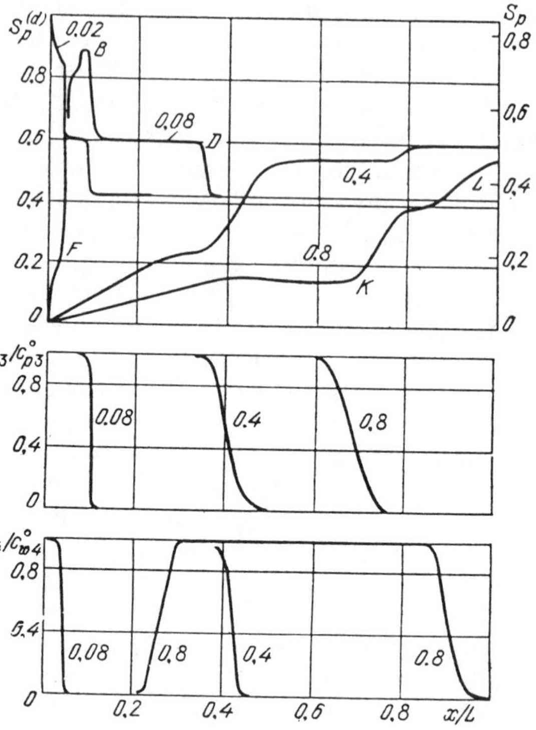

Figure 8.4.3 Evolution of kinematic waves of oil-saturation and concentrations of active component ($c_{p3} \equiv c_{p(3)}$), and polymer ($c_{w(y)} \equiv c_{w(y)}$) during the micellar-polymer flooding (V_{MS} = 0.04 V_m), V_{BL} = 0.4V_m; for the remaining data see (8.4.7). The wave notation is the same as in Fig. 8.4.2. The numerical labels on curves indicate time $\bar{t}$.

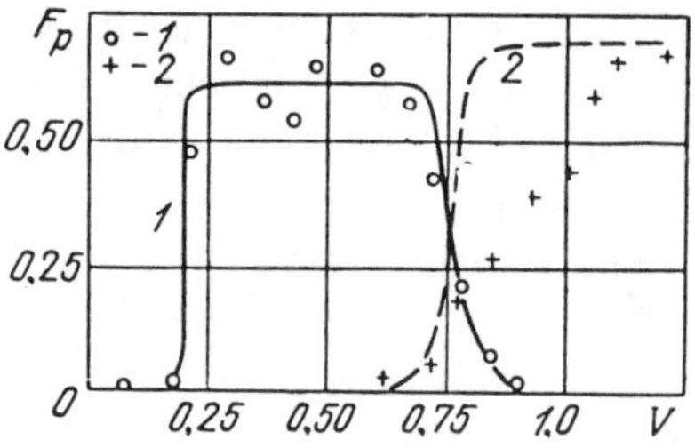

Figure 8.4.4 Variation of the volumetric share of both oil (*1*) and micellar solution (*2*) at the outlet of the oil-bearing bed with time (or as a function of growth of relative volume V/V_m of pumped liquid). Volume of MS layer: V_{MS} = 0.8V_m. For the remaining data, see (8.4.7). Lines *1* and *2* correspond to predicted analytical data; dots *1* and *2* correspond to experimental data (S. Davis and C. Jones, 1968).

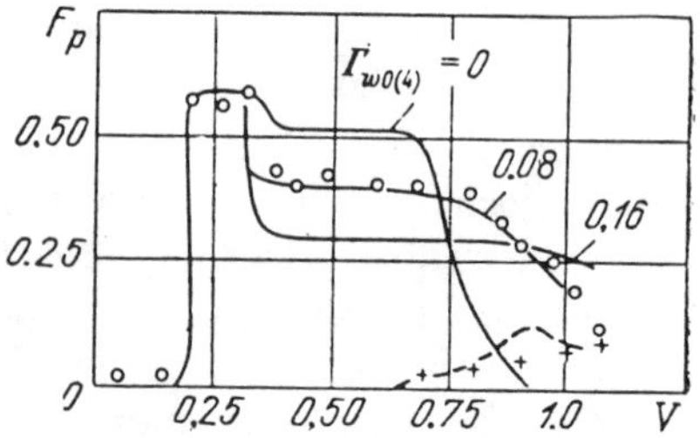

Figure 8.4.5 The same as in Fig. 8.4.4, but for the volume of the MS solution equal to V_{MS} = 0.1V_m, and with variation of Henry's coefficient $\Gamma_{w0(4)}$ in the polymer (thickener) adsorption law used in analytical calculations (labels on curves indicate values of Henry's law coefficients).

is described by a similarity solution concerned with oil displacement by a micellar solution (see Fig. 8.4.3). In conformity with this solution, oil is displaced out of the bed in the form of a fixed-saturation wave, which breaks through near the bed outlet after a volume of liquid equal to 0.2 of the volume of voids V_m ($\bar{t} = 0.2$) is pumped through the bed. Up to a point in time $\bar{t} = 0.8$ (which corresponds to a volume of liquid of $0.8V_m$ pumped through the oil-bearing bed), the oil recovery (in terms of its part of the volume in the total flow of liquid) becomes constant. Figure 8.4.5 relates to a real regime with a thin layer of edging MS ($V_{\mathrm{MS}} = 0.1V_m$), when formation of the petroleum wave of variable oil-saturation takes place; namely, the share of oil in the flow of discharged liquid grows to $F_p \approx 0.6$ immediately after arrival of the petroleum wave at time $\bar{t} \approx 0.2$, followed by a sharp decrease at time $\bar{t} \approx 0.3$, as distinct from the regime depicted in Fig. 8.4.4. From Fig. 8.4.5, it is seen that intensity of adsorption of thickening polymer ($k = 4$) determined by Henry's law constant $\Gamma_{w0(4)}$ significantly affects the dynamics of the process. Note that the value $\Gamma_{w0(4)} = 0.4$ relates to polyacrylamides.

Evaluation of the impact of various effects contributing to rupture of the MS edging (which was described in the discussion of Figs. 8.4.2*d*, *e*, 8.4.3, and 8.4.5) permitted the identification, by means of numerical analysis, of the principal mechanism of the process: retention of MS in the zone of BL, which causes the rupture of the buffer liquid through the MS edging into the zone of the petroleum wave (Fig. 8.4.2*e*). Considering only this mechanism of rupture of the MS-solution edging and ignoring other possible mechanisms (for instance, sorption of surface-active admixtures dissolved in petroleum, which transforms petroleum to a micellar solution) permits results which are in satisfactory agreement with experimental data (illustrated in Fig. 8.4.6.) to be obtained. It is evident that in the case under consideration, the MS-solution edging of a size of about 5% of the void volume recovers almost the entire volume of petroleum from a homogeneous porous medium. Layers of smaller size are retained in the oil-bearing bed, thereby reducing the coefficient of oil displacement.

Thus, the mathematical model of a micellar-polymer flooding based on

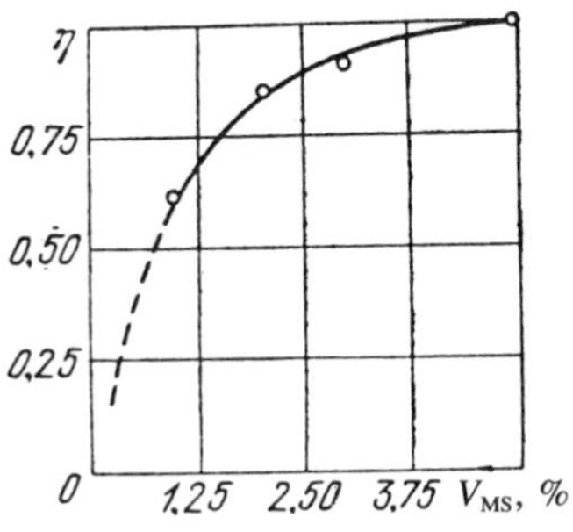

Figure 8.4.6 Effectiveness of oil recovery as a function of the MS-solution volume determined by the percentage of displaced oil (η) relative to the total volume of petroleum in a homogeneous bore-core sampled in the Berea sandstone. The size of the BL layer is $V_{\mathrm{BL}} = 0.6V_m$. Solid lines represent prediction (ignoring sorption of admixtures ($k = 3$) soluble in petroleum), dots represent experiment (S. Davis and C. Jones, 1968).

the implementation of equations of mechanics of a multiphase, multicomponent filtration in porous medium permits the satisfactory description of the principal features of the process. The discussed model in its one-dimensional variant may be used for analysis of laboratory experiments, and, then, in two- and three-dimensional variants, for analysis of the process and its optimization directly in industry. It should be remembered that it is in laboratory experiments on relatively small lengths ($L \sim 0.1$–1.0 m), when duration of the process is short ($t \sim 0.1$–1.0 hour), that the nonequilibrium effects of the transition from immobile to mobile phases may play a certain role.

APPENDIXES

Table A.1. Thermophysical parameters of selected gases and liquids (N. B. Vargaftik, 1972; M. P. Vukalovich et al., 1969; I. K. Kikoin, 1976; S. L. Rivkin et al., 1973)

Name of substance	Conditions		Density, ρ kg/m³	Speed of sound C, m/s	Adiabatic index, γ	Specific heat capacity c_p, m²/(s²·K)	Viscosity μ, 10^{-5} kg/(m·s)	Heat conductivity λ, 10^{-2} kg·m/(s³·K)
	p,MPa	T, K						
Nitrogen (gas, N_2, μ_m = 28.0)	0.1	77 *	4.61	175	1,44	1 173	0,53	0,76
	0.1	293	1.15	346	1.40	1 041	1,77	2,56
	0,1	1773	0.188	830	1,31	1 269	5.56	~10,5
Nitrogen (liquid)	0,1	77 *	807	867	—	1 955	14,6	2,4
Water (gas, H_2O, μ_m = 18.0, i_0 = −13.45 × 10^6 m²/s²	0.1	373 *	0.592	470	1.30	2 034	1,21	2.48
	1.0	453 *	5.15	500	1,29	2 615	1.49	3.41
	5.0	537 *	25.4	500	1,27	4 360	1.80	5.43
	7.0	559 *	36.4	490	1.25	5 285	1,84	6.38
	10.0	584 *	55.5	470	1.22	7 070	2,07	8.00
	15.0	615 *	96.6	440	1,25	12 720	2.42	11.6
	0.1	1773	0.118	~1000	1.20	2 750	6,07	26,1
Water (liquid)	0,1—1.0	293	998	1500	—	4 180	100,1	60.2
	0.1	373 *	958	1540	—	4 220	27.9	68.0
	1.0	453 *	887	1405	—	4 410	14.9	67.6
	5.0	537 *	780	1080	—	5 040	10.2	59.8
	7.0	559 *	746	990	—	5 410	9.44	56,4
	10.0	584 *	688	860	—	6 170	8.62	52.0
	15.0	615 *	604	680	—	8 630	7.38	45.0

Table A.1. Cont'd

Name of substance	Conditions		Density, ρ kg/m^3	Speed of sound C, m/s	Adiabatic index, γ	Specific heat capacity c_p, m^2/(s$^2\cdot$K)	Viscosity μ, 10^{-5} kg/(m$\cdot$s)	Heat conductivity λ, 10^{-2} kg$\cdot$m/(s$^3\cdot$K)
	p,MPa	T, K						
Hydrogen (H_2, μ_m = 2.0)	0.1	1773	0.0148	2650	1,33	16 580	3,1	64
Air (μ_m = 29.0)	0.1	293	1.19	343	1,40	1 007	1.81	2,58
Helium (He, μ_m = 4.0)	0.1	293	0.164	1005	1.67	5 190	1,94	14.9
Oxygen (0_2, μ_m = 32, i_0 = 0)	0.1	293	1,31	326	1.41	915	2.02	2.47
Methane (liquid, CH_4, μ_m = 16)	2.0 3,0	161 * 178 *	337 288	800 540	—	~4 000 ~4 000	4.4 3,3	9,1 9.1
Sodium (liquid, Na, μ_m = 23)	0.1—1.0 1.0	373 1156 *	928 740	2530	— —	1 383 1 284	68.6 15.7	8600 5085
Nitrogen oxide (NO, μ_m = 30.0)	0.1	1773	0.19	830	1.30	1 285	6.8	~10
Carbon monoxide (CO, μ_m = 28.0)	0.1	1773	0,19	820	1.30	1 280	6.0	10.7
Oleinic acid	0.1—1.0	293	895	—	—	1 890	—	23.1

Table A.1. Cont'd

Name of substance	Conditions		Density, ρ kg/m³	Speed of sound C, m/s	Adiabatic index, γ	Specific heat capacity c_p, m²/(s²·K)	Viscosity μ, 10^{-5} kg/(m·s)	Heat conductivity λ, 10^{-2} kg·m/(s³·K)
	p, MPa	T, K						
Powder (model)	0.1—1.0	293	1550	—	—	1 466	—	68
Propane (liquid, C_3H_8, μ_m = 44.0)	2.0 3.0	330 * 351 *	436 385		— —	3 270 4 340	~7.0 ~7.0	7.0 7.0
Mercury (liquid, μ_m = 201)	0.1—1.0 1.0	293 630 *	13600 12740	1450	— —	139 136	154 89	804 1220
Tetraline (liquid, $C_{10}H_{12}$, μ_m = 132, i_0 = $3.13 \cdot 10^6$ m²/s²)	0.1—1.0	293	969	—	—	1 640	—	—
Carbon dioxide (CO_2, μ_m = 44.0, i_0 = $-8.95 \cdot 10^6$ m²/s²)	0.1 0.1	293 1773	1.815 0.30	266 620	1.29 1.16	844 1 359	1.46 6.2	1.615 7.3
Carbon (solid)	0.1—1.0	293	2200	—	—	714	—	~600
Ethanol (liquid, C_2H_5OH, μ_m = 46)	0.1—1.0 0.1	293 351 *	807 757	1170	— —	2 410 3 010	120 43.0	17.0 15.5

*Corresponds to saturation temperature $T_S(p)$.

Table A.2. Interphase thermophysical characteristics of selected gas-liquid and vapor-liquid systems (N. B. Vargaftik, 1972; S. L. Rivkin et al., 1973; I. K. Kikoin, 1976)

Name of substance	Conditions		Coefficient of surface tension Σ, 10^{-3} kg/s	Heat of evaporation l, 10^6 m²/s²
	p, MPa	T, K		
Nitrogen	0.1	77 *	8.85	0.198
Water	0.1 0.1 1.0 5.0 7.0 10.0 15.0	293 * 373 * 453 * 537 * 559 * 584 * 615 *	73.0 58.9 42.2 22.8 17.6 11.8 5.41	— 2,26 2,01 1,64 1,50 1,32 1,01
Methane	2,0 3.0	161 * 178 *	<5 <5	0.371 0.265
Sodium	0,1—1,0 0.1	373 1156 *	206 118	— 3.88
Propane	2,0 3,0	330 * 351 *	<5 <5	0.263 0,199
Mercury	0,1—1,0 0.1	293 630 *	465 393	— 0.295
Tetraline	0.1	477 *		0.406
Ethanol (ethyl alcohol, C_2H_5OH)	0,1—1.0 0.1	293 351 *	22,8 17.1	— 0.963

*Corresponds to saturation temperature $T_S(p)$.

Table A.3. Thermophysical and kinetic parameters for one of the raw materials used in production of petroleum coke in an installation for a retarded coking (IRC) (see §7.9)

Notation	Units of measure-ments	Magnitude	Notation	Units of measure-ments	Magnitude
$R_{1(0)}$	$m^2/(s^2 \cdot K)$	461.9	$T^{\circ}_{S(2)}$	K	723
$R_{1(1)}$	same	83.1	$l_{(1)}$	m^2/s^2	$1.26 \cdot 10^5$
$R_{1(2)}$	same	20.8	$l_{(2)}$	m^2/s^2	$2.09 \cdot 10^5$
$c_{1(0)}$	same	2200	K°	s^{-1}	$0.5 \cdot 10^{10}$
$c_{1(1)}$	same	2600	$T_{(21)}$	K	24 839
$c_{1(2)}$	same	2800	$T_{(23)}$	K	26 338
c_2	same	3470	$T_{(31)}$	K	25 332
$\overset{\circ}{\rho}_2$	kg/m^3	700	$T_{(41)}$	K	25 448
$\mu_1/\overset{\circ}{\rho}_1$	m^2/s	$1.0 \cdot 10^{-6}$	$T_{(34)}$	K	24 753
$\mu_2/\overset{\circ}{\rho}_2$	m^2/s	$3.3 \cdot 10^{-6}$	$T_{(45)}$	K	25 448
Σ	kg/s^2	$2.5 \cdot 10^{-3}$			

REFERENCES

Abramson, A. A. Surface-active substances. Properties and applications. Leningrad, Khimiya, 1981 (in Russian).

Avdeev, A. A., Maidanik, V. N., and Shanin, V. K., Methods of analysis of flashing adiabatic flows. Teploenergetika, 1977, No. 8, pp. 67–69 (in Russian).

Aidagulov, R. R., Khabeev, N. S., and Shagapov, V. Sh. Structure of a shock wave in a liquid with gas bubbles in the presence of a nonstationary interphase heat exchange. PMTF, 1977, No. 3, (in Russian).

Andreevskii, A. A. Wave flow of thin layers of viscous fluids. Thermal regime and hydraulics of steam generators. Leningrad, Nauka, 1978, pp. 181–230 (in Russian).

Apshtein, E. Z., Grigoryan S. S., and Yakimov, Yu. L. On stability of a cluster of air bubbles in oscillating liquid. Izv. A.N. S.S.S.R., MZhG, 1969, No. 3, pp. 100–104 (in Russian).

Armand, A. A. Resistance in the process of flow of a two-phase system through horizontal tubes. Izv. Vsesoyuznogo Teplotekhn. Inst., 1950, No. 2 (in Russian).

Barenblatt, G. I. and Vinichenko, A. P. Nonequilibrium filtration of immiscible liquids. Uspekhi mekhaniki, 1980, Vol. 3, No. 3, pp. 35–50 (in Russian).

Barenblatt, G. I., Entov, V. M., and Ryzhik, V. M. Flow of liquids and gases in natural seams. Moscow, Nedra, 1984, 211 pp. (in Russian).

Blinkov, V. N., Petukhov, I. I., and Bespyatov, M. A. Experimental study of flow of flashing water in the Laval nozzle. Gas thermodynamics of multiphase flows in power-generating installations. Kharkov, 1981, Vol. 4, pp. 71–78 (in Russian).

Blinkov, V. N. and Frolov, S. D. Model of flashing liquid flow in nozzles. IFZh, 1982, Vol. 42, No. 5, pp. 741–746 (in Russian).

Borisov, A. A., Gel'fand, B. E., Nigmatulin, R. I., et al. Amplification of shock waves in a liquid with vapor bubbles. Nonlinear wave processes in two-phase media. Ed. by S. S. Kutateladze. ITF, Novosibirsk, 1977 (in Russian).

Borisov, A. A., Gel'fand, B. E., Nigmatulin, R. I., et al. Amplification of shock waves in liquids with vapor bubbles and dissolved gas. DAN SSSR, 1982, Vol. 263, No. 3, pp. 592–598 (in Russian).

Bulygin, V. Ya. Simultaneous percolation of two liquids in an oil-bearing bed. Uch. zapiski, Kazan. Univ. Problems in hydrodynamics of ground water flow. 1965, Vol. 125, Book 8, (in Russian).

Burdukov, A. P., Koz'menko, B. K. and Nakoryakov, V. E. Distribution of velocity profiles for liquid phase in a gas-liquid flow at low gas concentrations. PMTF, 1975, No. 6 (in Russian).

Bykov, V. I. and Lavrent'ev, M. E. Formation of the drop size spectra in a gas-liquid flow. IFZh, 1976, Vol. 31, No. 5 (in Russian).

Valyavin, G. G., Gimaev, R. N., et al. Development of a kinetic model for the process of destruction of petroleum residua. Trudy BashNIINP, Ufa, 1975, Vol. 13, pp. 44–51 (in Russian).

Vargaftik, N. B. Manual of thermophysical properties of gases and liquids. Moscow, Nauka, 1972 (in Russian).

Vakhitov, G. G., Ogadzhanyants, V. G., and Polishchuk, A. M. Experimental investigation of the effect of polymer admixtures to water upon relative permeability of porous media. Izv. AN SSSR, MZhG, 1980, No. 4, pp. 163–166 (in Russian).

Verigin, N. N. Diffusion and mass exchange during percolation of liquids in porous medium. Development of studies in the theory of percolation in the USSR. Moscow, Nauka, 1969 (in Russian).

Voyutskii, S. S. Colloidal chemistry (text book). Moscow, Khimiya, 1976, 512 pp. (in Russian).

Vukalovich, M. P., Rivkin, S. L., and Alexandrov, A. A. Tables of thermophysical properties of water and steam. Moscow, Izd. Standartov, 1969 (in Russian).

Gazizov, R. K. and Gubaidulin, A. A. Intensification of shock waves in a bubbly liquid with a gas-concentration gradient. Izv. AN SSSR, MZhG, 1988, No. 1 (in Russian).

Gazizov, R. K. and Gubaidulin, A. A. On enhancement of shock-wave disturbance effects upon a barrier in gas during its flow through a bubbly screen of variable gas concentration. TVT, 1989, Vol. 27, No. 5 (in Russian).

Galiev, Sh. U. Dynamics of interaction between structural elements and pressure wave in liquid. Kiev, Naukova Dumka, 1977, 170 pp. (in Russian).

Ganiev, R. F. and Kobasko, N. I., et al. Oscillation phenomena in multiphase media and their applications in technology. Kiev, Tekhnika, 1980, 143 pp. (in Russian).

Ganiev, R. F. and Lapchinskii, V. F. Problems of mechanics in space technology. Moscow, Mashinostroenie, 1978, 120 pp. (in Russian).

Gel'fand, B. E., Gubin, S. A., Kogarko, B. S., and Kogarko, S. M. Investigation of compression waves in mixtures of liquids with gas bubbles. DAN SSSR, 1973, Vol. 213, No. 5 (in Russian).

Gel'fand, B. E., Gubin, S. A., et al. Investigation of gas bubble breaking in a liquid by shock waves. Izv. AN SSSR, MZhG, 1975, No. 4 (in Russian).

Gel'fand, B. E., Gubin, S. A., Nigmatulin, R. I., and Timofeev, E. I. Effects of gas density on bubble fragmentation by shock waves. DAN SSSR, 1977, Vol. 235, No. 2, pp. 292–294 (in Russian).

Gel'fand, B. E., Stepanov, V. V., Timofeev, E. I., and Tsyganov, S. A. Amplification of shock waves in a nonequilibrium system liquid-bubbles of soluble gas. DAN SSSR 1978, Vol. 239, No. 1, pp. 71–73 (in Russian).

Gel'fand, B. E., Timofeev, E. I., and Stepanov, V. V. On a structure of weak shock waves in a system gas bubbles-liquid. TVT, 1978, Vol. 16, No. 3, pp. 569–575 (in Russian).

Gerasimov, Ya. I. Physical chemistry (text book). Moscow, Khimiya, 1970, 592 pp. (in Russian).

Gimatudinov, Sh. K. and Shirkovskii, A. I. Physics of petroleum and gas strata. Moscow, Nedra, 1982 (in Russian).

Gofman, G. V., Kroshilin, A. E., and Nigmatulin, B. I. Nonstationary wave flow of flashing liquid out of vessels. TVT, 1981, Vol. 19, No. 6, pp. 1240–1250 (in Russian).

Grigoryan, S. S., Yakimov, Yu. L., and Apshtein, E. Z. Behavior of air bubbles in a liquid under vibration. Fluid dynamics transactions. Warsaw, Vol. 2, 1965.

Gubaidulin, A. A., Ivandaev, A. I., and Nigmatulin, R. I. Nonstationary waves in a liquid with gas bubbles. DAN SSSR, 1976, Vol. 226, No. 6 (in Russian).

Gubaidulin, A. A., Ivandaev, A. I., and Nigmatulin, R. I. Investigation of nonstationary shock waves in gas-liquid mixtures of bubbly structure. PMTF, 1978, No. 2, pp. 78–86 (in Russian).

Gubaidulin, A. A., Ivandaev, A. I., Nigmatulin, R. I., and Khabeev, N. S. Waves in bubbly liquids. Itogi nauki i tekhniki, (Mechanics of liquids and gas), Moscow, VINITI, 1982, Vol. 17 (in Russian).

Deich, M. E. and Filippov, G. A. Gas dynamics of two-phase media. Moscow, Energoizdat, 1981, 472 pp. (in Russian).

Deksnis, B. K. Propagation of moderately strong shock waves in a two-phase medium. Izv. AN Latv. SSR, Ser. fiz. i tekhn. nauk, 1967, No. 1, pp. 75–81.

Doroshchuk, V. E. Heat exchange crises during boiling of water in tubes. Moscow, Energoatomizdat, 1983, 120 pp. (in Russian).

Doroshchuk, V. E., Levitan, L. L., and Lantsman, F. P. Recommendations for analysis of heat exchange crises in a circular tube during uniform heat release. Teploenergetika, 1975, No. 12 (in Russian).

Entov, V. M. and Taranchuk, V. B. Numerical modelling of the process of unstable displacement of petroleum by water. Izv. AN SSSR, MZhG, 1979, No. 5, pp. 58–63, (in Russian).

Zhivaikin, L. Ya. On the thickness of a liquid film in film-type devices. Khimicheskoe mashinostroyeniye, 1961, No. 4, p. 47 (in Russian).

Zabrudskii, V. T. and Kholpanov, L. P. Investigation of wave parameters of film flow along a channel in a regime of upward flow at various physical properties of liquid phase. TOKhT, 1979, Vol. 13, No. 2 (in Russian).

Zel'dovich, Ya. B. Chemical physics and hydrodynamics. Moscow, Nauka, 1984, 374 pp. (in Russian).

Zuong Ngok Hai. Investigation of wave processes in vapor-liquid media of bubbly structure. Dissertation, Candidate of math. science, Moscow, MGU (Moscow University), 1982, 127 pp. (in Russian).

Zuong Ngok Hai and Musaev N. D. Results of numerical investigation of shock-wave intensification in bubbly liquids. Doclady AN AzSSR, 1988, Vol. 10, No. 7 (in Russian).

Zuong Ngok Hai, Nigmatulin, R. I., and Khabeev, N. S. Nonstationary waves in liquids with vapor bubbles. Izv. AN SSSR, MZhG, 1984, No. 5, pp. 109–118 (in Russian).

Zuong Ngok Hai and Khabeev, N. S. On a unique approach to a problem concerning vapor-liquid medium of bubbly structure. TVT, 1983, Vol. 21, No. 1, pp. 137–145 (in Russian).

Zuong Ngok Hai and Khabeev, N. S. Evolution of shock waves and finite-length impulses in a liquid with bubbles of vapor. Nonstationary outflows of multiphase systems with chemical-physical transformations. Ed. by R. I. Nigmatulin and A. I. Ivandaev. Moscow, MGU, 1983, pp. 43–50 (in Russian).

Zuong Ngoc Hai. Propagation of shock waves in gas- or vapor-liquid mixtures. J. Mech., 1987, Vol. 9, No. 4, pp. 3–8.

Zyong Ngok Khai, Nigmatulin, R. I., and Khabeev, N. S. Structure of shock waves in a liquid with vapor bubbles. Izv. AN SSSR, MZhG, 1982, No. 2, pp. 109–118 (in Russian).

Zysin, V. A., Baranov, G. A., Barilovich, V. A., and Parfenova, T. N. Flashing adiabatic flows. Moscow, Atomizdat, 1976, 152 pp. (in Russian).

Ivandaev, A. I. and Gubaidulin, A. A. Investigation of a nonstationary outflow of flashing liquid in a thermodynamic equilibrium approximation. TVT, 1978, Vol. 16, No. 3 (in Russian).

Ivandaev, A. I. and Nigmatulin, B. I. Application of a model of a dispersed-annular flow for analysis of two-phase critical flows. TVT, 1977, Vol. 15, No. 3, pp. 573–580 (in Russian).

Ivandaev, A. I. and Nigmatulin, B. I. Propagation of weak disturbances in vapor-liquid dispersed-annular flows. TVT, 1980, Vol. 18, No. 2, pp. 359–366 (in Russian).

Ivandaev, A. I. and Nigmatulin, R. I. An elementary theory of critical maximum flow rates of two-phase mixtures. TVT, 1972, Vol. 10, No. 5 (in Russian).

Ivandaev, S. I. Determining the laws of interaction between components of gas-liquid dispersed-annular flow. Nonlinear wave processes in two-phase media. Ed. by S. S. Kutateladze. Novosibirsk, ITF, 1977, pp. 244–255 (in Russian).

Ivandaev, S. I. Prediction of the heat-transfer crisis in uniformly heated tubes. National (USSR) conf. on thermophysics and hydrogasdynamics of boiling and condensation. Vol. 1. Riga, 1982 (in Russian).

Isaev, O. A. and Pavlov, P. A. Flashing of liquid in a large volume during rapid pressure drop. TVT, 1980, Vol. 18, No. 4 (in Russian).

Kaznovskii, S. P., Pomet'ko, R. S., and Pashichev, V. V. Heat-transfer crisis and distribution of liquids in a dispersed-annular regime of flow. TVT, 1978, Vol. 16, No. 1, pp. 94–100 (in Russian).

Kapitsa, P. L. Wave flow of thin layers of liquid. ZhETF, 1948, Vol. 18, No. 1 (in Russian).

Karpman, V. I. Nonlinear waves in dispersion media. Moscow, Nauka, 1973, 176 pp. (in Russian).

Kafarov, V. V. and Dorokhov, I. N. System analysis of processes in chemical technology. Moscow, Nauka, 1976, 500 pp. (in Russian).

Kedrinskii, V. K. Propagation of perturbations in liquids with gas bubbles. PMTF, 1968, No. 4, pp. 29–35 (in Russian).

Kedrinskii, V. K. Dynamics of the cavitation zone during explosion near the free surface. PMTF, 1975, No. 5 (in Russian).

Kedrinskii, V. K. Shock waves in a liquid with gas bubbles. FGV, 1980, No. 5, pp. 14–25 (in Russian).

Kikoin, I. K., Tables of physical values. Moscow, Atomizdat, 1976 (in Russian).

Kirillin, V. A., Sychev, V. V., and Sheindlin, A. E. Technical thermodynamics. Moscow, Energiya, 1974, 447 pp. (in Russian).

Kirillov, P. L., Komarov, N. M., Subbotin, V. I., et al. Measurement of some characteristics of a vapor-liquid flow in a circular tube under pressure of 68.6 bar. FEI (preprint)-431, Obninsk, 1973, 104 pp. (in Russian).

Kirillov, P. L., Yur'ev, Yu. S., and Bobkov, V. P. Handbook on thermohydraulic analysis and design (nuclear reactors, heat-exchangers, steam generators). Moscow, Energoatomizdat, 1984, 296 pp. (in Russian).

Kochin, N. E., Kibel', I. A., and Rose, N. V. Theoretical hydrodynamics. Vols. I and II. Moscow, Fizmatgiz, 1963 (in Russian).

Koshkin, V. K., Kalinin, E. K., Dreitser, G. A., and Yarkho, S. A. Nonstationary heat exchange. Moscow, Mashinostroyeniye, 1973 (in Russian).

Kreimer, M. L., Ilembitova, R. N., et al. Algorithm for analysis of phase-equilibrium constants and pressure of saturated vapor of hydrocarbons and petroleum fractions. Algorithmization for analysis of processes and equipment for processing and transportation of petroleum and gas. Kiev, Naukova dumka, 1974, No. 6, pp. 12–36 (in Russian).

Kroshilin, A. E., Kukharenko, V. N., and Nigmatulin, B. I. Deposition of particles on the channel wall in a gradient turbulent disperse flow. Izv. AN SSSR, MZhG, 1985, No. 4, pp. 51–63 (in Russian).

Kuznetsov, V. V., Nakoryakov, V. E., and Pokusaev, B. G. Interaction between a solution and gas bubbles in a liquid. Letters to ZhETF, 1978, Vol. 28, No. 8, pp. 520–523 (in Russian).

Kuznetsov, V. V. and Pokusaev, B. G. Pressure-wave evolution in a liquid with gas bubbles.

Transition of a laminar boundary layer to a turbulent. Two-phase flows. Ed. by S. S. Kutateladze. Novosibirsk, ITF, 1978, pp. 61–67 (in Russian).

Kutateladze, S. S. Fundamentals of the heat exchange theory. Moscow, Atomizdat, 1979, 265 pp. (in Russian).

Kutateladze, S. S. and Leont'ev, A. I. Heat and mass exchange, and friction in a turbulent boundary layer. Moscow, Energiya, 1972, 342 pp. (in Russian).

Kutateladze, S. S., Mironov, B. P., Nakoryakov, V. E., and Khabakhpasheva, E. M. Experimental study of near-wall turbulent flows. Novosibirsk, Nauka, 1975, 166 pp. (in Russian).

Kutateladze, S. S. and Nakoryakov, V. E. Heat and mass exchange, and waves in gas-liquid systems. Novosibirsk, Nauka, 1984, 302 pp. (in Russian).

Kutateladze, S. S. and Styrikovich, M. A. Hydrodynamics of gas-liquid systems. Moscow, Energiya, 1976, 296 pp. (in Russian).

Kutepov, A. M., Sterman, L. S., and Styushin, N. G. Hydrodynamics and heat exchange in the process of vapor formation. Moscow, Vysshaya shkola, 1977, 352 pp. (in Russian).

Labuntsov, D. A. Heat transfer in the process of film condensation of pure vapor on a vertical surface and in horizontal tubes. Teploenergetika, 1957, No. 7, pp. 72–79 (in Russian).

Lavrent'ev, M. A. and Shabat, B. V. Methods of the theory of complex variables. Moscow, Nauka, 1973 (in Russian).

Leonchik, B. I. and Mayakin, V. P. Measurements in disperse flows. Moscow, Energetika, 1971, 248 pp. (in Russian).

Loitsyanskii, L. G. Mechanics of liquids and gas. Moscow, Nauka, 1973, 848 pp. (in Russian).

Mamaev, V. A., Odishariya, G. E., Klapchuk, O. V., Tochigin, A. A., and Semenov, N. I. Flow of gas-liquid mixtures in tubes. Moscow, Nedra, 1978, 271 pp. (in Russian).

Marinov, M. I. and Kabanov, L. P. Investigations of heat exchange deterioration under lowered pressure and fairly low bulk velocities. Teploenergetika, 1977, No. 7, pp. 81–83 (in Russian).

Mednikov, E. P. Turbulent transfer and deposition of aerosols. Moscow, Nauka, 1981, 176 pp. (in Russian).

Methods of analysis of thermophysical properties of gases and liquids. Moscow, Khimiya, 1974, 248 pp. (in Russian).

Mirzadzhanzade, A. Kh., et al, Theory and practice of application of nonequilibrium systems in petroleum production industry. Baku, 1985 (in Russian).

Mironov, Yu. V. Predicting the critical flow rate of steam-water mixture. TVT, 1975, Vol. 13, No. 1 (in Russian).

Miropol'skii, Z. L., Shneerova, R. I., and Karamysheva, A. I. Concentration of steam during a pressure flow of a steam-water mixture with heat influx under adiabatic conditions. Teploenergetika, 1971, No. 5, pp. 60–63 (in Russian).

Mozharov, N. A. Investigation of critical velocity of the film break-away off a steam pipe. Teploenergetika, 1959, No. 2, pp. 50–63 (in Russian).

Mukhachev, G. A., Pavlov, B. M., and Tonkonog, V. G. Flow of vaporizing liquid in nozzles. Trudy KAI, 1973, No. 158, pp. 50–54 (in Russian).

Nakorchevskii, A. I. Heterogeneous turbulent jets. Kiev, Naukova dumka, 1980, 142 pp. (in Russian).

Nakoryakov, V. E., Burdukov, A. P., et al. Investigation of turbulent flows of two-phase media. Ed. by S. S. Kutateladze. Novosibirsk, Nauka, 1973, 314 pp. (in Russian).

Nakoryakov, V. E., Pokusaev, B. G., and Shreiber, I. R. Propagation of waves in a gas-liquid and vapor-liquid media. Novosibirsk, ITF, 1983, 238 pp. (in Russian).

Nakoryakov, V. E., Sobolev, V. V., and Shreiber, I. R. Long-wave disturbances in a gas-liquid mixture. Izv. AN SSSR, MZhG, 1972, No. 5 (in Russian).

Nevstrueva, E. I. Heat and mass exchange in atomic-power installations with water-cooled reactors. Itogi nauki i tekhniki. (Heat and mass exchange). Moscow, VINITI, 1978, Vol. 1, 112 pp. (in Russian).

Netunaev, S. V., Nigmatulin, B. I., and Goryunova, M. Z. Investigation of intensity of drop deposition onto a liquid film in a vertical air-water flow. Teploenergetika, 1982, No. 3, pp. 61, 62 (in Russian).

Nigmatulin, B. I. Hydrodynamics of a two-phase flow in a dispersed-annular regime of flow. PMTF, 1971, No. 6, pp. 141–153 (in Russian).

Nigmatulin, B. I. Investigation of characteristics of two-phase dispersed-annular flows in heated tubes. PMTF, 1973, pp. 78–88 (in Russian).

Nigmatulin, B. I. The heat-transfer crisis and flow rate of liquid in a film in the process of a dispersed-annular flow. TVT, 1979, Vol. 17, No. 6, pp. 1254–1258 (in Russian).

Nigmatulin, B. I., Vinogradov, A. A., Vinogradov, V. A., and Kurbanov, Sh. E. Methods of measurements of thickness and wave characteristics of a liquid-film surface in a steam-water dispersed-annular flow. TVT, 1982, Vol. 20, No. 6 (in Russian).

Nigmatulin, B. I., Goryunova, M. Z., and Vasil'ev, Yu. V. Generalization of experimental data on heat transfer during flow of liquid films along solid surfaces. TVT, 1981, Vol. 19, No. 5, pp. 991–1001 (in Russian).

Nigmatulin, B. I., Goryunova, M. Z., Guguchkin, V. V., and Markovich, E. E. Effects of gas injection upon motion of drops along the wall of a horizontal channel in a gas-liquid disperse flow. TVT, 1982, Vol. 20, No. 2 (in Russian).

Nigmatulin, B. I., Dolinin, I. V., Rachkov, V. I., and Semenov, V. P. Investigation of drop deposition on a liquid film in a vertical steam-water flow. Teploenergetika, 1978, No. 6, pp. 82–84 (in Russian).

Application of the salt-tracer method for determining the intensity of moisture exchange and distribution of liquid between core and film in a dispersed-annular steam-water flow. TVT, 1978, Vol. 18, No. 4, pp. 832–839 (in Russian).

Nigmatulin, B. I. and Ivandaev, A. I. Investigation of the crisis phenomenon in a hydrodynamic two-phase flow. TVT, 1977, Vol. 15, No. 1, pp. 129–136 (in Russian).

Nigmatulin, B. I. and Soplenkov, K. I. Elementary theory of critical (maximum) flow rate of a two-phase mixture in channels of variable cross section. TVT, 1978, Vol. 16, No. 2, pp. 370–376 (in Russian).

Nigmatulin, B. I. and Soplenkov, K. I. Investigation of a nonstationary flow of flashing liquid out of channels in a thermodynamic nonequilibrium approximation. TVT, 1980, Vol. 18, No. 1, pp. 118–131 (in Russian).

Nigmatulin, R. I. Fundamentals of mechanics of heterogeneous media. Moscow, Nauka, 1978, 336 pp. (in Russian).

Nigmatulin, R. I. Effects of wave propagation in bubbly media, and their mathematical description. Selected problems of advanced mechanics (dedicated to 50th anniversary of S. S. Grigoryan). Ed. by G.G. Chernyi. Moscow, NII mekhaniki MGU (Moscow University), 1981, pp. 64–89 (in Russian).

Nigmatulin, R. I., Surguchev, M. L., Fedorov, K. M., Khabeev, N. S., and Shevtsov, V. A. Mathematical modelling of the micellar-polymer flooding process. DAN SSSR, 1980, Vol. 255, No. 1, pp. 52–56 (in Russian).

Nigmatulin, R. I. and Shagapov, V. Sh. Structure of shock waves in a liquid with gas bubbles. IZv. AN SSSR, MZhG, 1974, No. 6, pp. 30, 31 (in Russian).

Nigmatulin, R. I., Shagiev, R. G., et al. Mathematical modelling in the hydraulic approximation of gas-liquid flows with chemical reactions, and analysis of heating of petroleum materials in tube furnaces. DAN SSSR, 1977, Vol. 237, No. 6, pp. 1311–1314 (in Russian).

Nigmatulin, Rs. I. Experimental study of the heat-release nonuniformity effects along a channel upon the heat-transfer crisis in two-phase flows. Voprosy gasotermodinamiki energoustanovok, Kharkov, KhAI, 1975, pp. 117–121 (in Russian).

Nigmatulin, Rs. I. Investigation of the heat-exchange crisis during a nonuniform heat release.

Nonlinear wave processes in two-phase media. Ed. by S. S. Kutateladze. Novosibirsk, ITF, 1977, pp. 300–312 (in Russian).

Nikolaevskii, V. N. Flow of organic mixtures in porous medium. Moscow, Nedra, 1968 (in Russian).

Petukhov, B. S., Genich, L. G., and Kovalev, S. A. Heat exchange in nuclear power installations. Moscow, Atomizdat, 1974, 403 pp. (in Russian).

Petukhov, B. S., Zhukov, V. M., and Shil'dkret, V. M. Experimental investigation of hydraulic resistance in the process of a two-phase flow of helium in a vertical tube. TVT, 1980, Vol. 18, No. 5, pp. 1040–1045 (in Russian).

Pokusaev, B. G. Pressure waves in bubbly gas-liquid and vapor-liquid media. Hydrodynamics and heat exchange in single-phase and two-phase media. Ed. by V. E. Naloryakov. Novosibirsk, IT SO AN SSSR, 1979, pp. 26–36 (in Russian).

Popov, V. P. and Pokryvailo, N. A. Experimental investigation of a nonstationary mass exchange by an electrochemical method. Teplomassoobmen, Minsk, 1966 (in Russian).

Pudovkin, M. A., Salamatin, A. N., and Chugunov, V. A. Thermal processes in operating wells. Kazan University, 1977 (in Russian).

Rassokhin, N. G., Kuzevanov, V. S., et al. Critical conditions during a nonstationary outflow of a two-phase medium caused by failure of a pipe line. TVT, 1977, Vol. 15, No. 3 (in Russian).

Selivanov, V. G. Acceleration of liquid by gas in nozzles. Voprosy gasotermodinamiki energoustanovok, Kharkov, 1976 (in Russian).

Recommendations for prediction of the heat-transfer crisis during boiling in circular tubes. Preprint 1–57, Moscow, High-temperature Institute (Institut vysokikh temperatur), AN SSSR, 1980, 67 pp. (in Russian).

Remizov, O. V., Sergeev, V. V., and Yurkov, Yu. I. Experimental study of the heat-transfer deterioration during both upward and downward flow of water in a tube. Teploenergetika, 1983, No. 9, pp. 64–66 (in Russian).

Rivkind, S. L. Thermodynamic properties of gases. Moscow, Energiya, 1973 (in Russian).

Rozhdestvenskii, B. L. and Yanenko, N. N. Systems of quasi-linear equations. Moscow, Nauka, 1979 (in Russian).

Rozenberg, M. D. and Kundin, S. A. Multiphase, multicomponent percolation in production of petroleum and gas. Moscow, Nedra, 1976 (in Russian).

Sagdeev, R. Z. Collective processes and shock waves in dilute plasma. Problems in theory of plasma. Moscow, Atomizdat, 1964, Vol. 20, No. 4 (in Russian).

Semenov, N. I. and Kosterin, S. I. Investigation of speed of sound in flowing gas-liquid mixtures. Teploenergetika, 1964, No. 6 (in Russian).

Smirnov, O. K., Pashkov, L. T., and Zaitsev, V. N. Analysis of parameters of a medium in a heated channel during strong perturbations of flow rate. Trudy MEI, 1973, Vol. 157 (in Russian).

Smolin, V. N. et al. Experimental data and methods of prediction of the heat-transfer crisis during boiling of water circulating in tubes with both uniform and nonuniform heat release. Problems in atomic science and engineering. Issue: Physics and engineering of atomic reactors. Moscow, Central Scientific-and-Research Institute Atominform, 1970, Vol. 5(9) (in Russian).

Stekol'shchikov, E. V. and Fedorov, A. S. Experimental study of propagation parameters of acoustic waves in boiling water. Teploenergetika, 1974, No. 9, pp. 76, 77 (in Russian).

Styrikovich, M. A., Polonskii, V. S., and Tsiklauri, G. V. Heat and mass exchange and hydrodynamics in two-phase flows of atomic power stations. Moscow, Nauka, 1982, 270 pp. (in Russian).

Subbotin, V. I., Remizov, O. V., and Vorob'ev, V. A. Thermal regimes and heat transfer in a zone of deteriorated heat exchange. TVT, 1973, Vol. 11, No. 6, pp. 1220–1226 (in Russian).

Surguchev, M. L., Shevtsov, V. A., and Surina, V. V. Application of micellar solutions for augmenting oil recovery from oil-bearing beds. Moscow, Nedra, 1977 (in Russian).

Syunyaev, Z. I. Petroleum carbon. Moscow, Khimiya, 1980, 272 pp. (in Russian).

Tarasova, N. V. and Leont'ev, A. I. Hydraulic resistance during flow of steam-water mixture in a heated vertical tube. TVT, 1965, Vol. 3, No. 1 (in Russian).

Tikhnenko, L. K., Kevorkov, L. R., and Lutovinov, S. Z. Critical discharge of hot water during flow out of a tube. Teploenergetika, 1979, No. 5, pp. 32–36 (in Russian).

Tolubinskii, V. I. Heat exchange in the process of boiling. Kiev, Naukova dumka, 1980, 316 pp. (in Russian).

Fisenko, V. B. Critical two-phase flows. Moscow, Atomizdat, 1978, 159 pp. (in Russian).

Khlestkin, D. A., Korshunov, A. S., and Kanishchev, V. P. Determination of discharge of high-parameter water during its outflow to atmosphere through cylindrical channels. Izv. AN SSSR, Energetika i transport, 1978, No. 5, pp. 126–135 (in Russian).

Tsiklauri, G. V., Danilin, V. S., and Seleznev, L. I. Adiabatic two-phase flows. Moscow, Atomizdat, 1973, 447 pp. (in Russian).

Tsiklauri, G. V., Dzhishkariani, G. S., and Kipshidze, M. E. Heat transfer in a post-crisis region under low pressure and small mass velocities. IFW, 1982, Vol. 43, No. 5, pp. 709–715 (in Russian).

Tsibul'skii, G. P. Plane problem on a two-phase percolation of immiscible liquids with no consideration given to capillary forces. Izv. AN SSSR, MZhG, 1975, No. 1 (in Russian).

Chan, V. Ch. and Shkadov, V. Ya. Instability of a layer of a viscous liquid affected by a boundary gas flow. Izv. AN SSSR, MZhG, 1979, No. 2, pp. 28–36 (in Russian).

Charnyi, I. A. Underground hydro- and gasdynamics. Moscow, Gostoptekhizdat, 1963 (in Russian).

Shagapov, V. Sh. Structure of shock waves in a polydisperse mixture of liquid and gas bubbles. Izv. AN SSSR, MZhG, 1976, No. 6 (in Russian).

Shagapov, V. Sh. Propagation of small perturbations in bubbly liquids. PMTF, 1977, No. 1, pp. 90–101 (in Russian).

Yakimov, Yu. D., Eroshin, V. A., and Romanenkov, N. I. Simulating motion of a body in water with consideration given to compressibility. Selected problems in continuum mechanics (didicated to 70th birthday of Acad. L. I. Sedov). Ed. by S. S. Grogoryan. Moscow, NII Mekhaniki MGU, 1978 (in Russian).

INDEX